ALLE · ZEIT · WACH · 1842

Turbulent Shear Flows 8

Selected Papers from the Eighth International Symposium on Turbulent Shear Flows, Munich, Germany, September 9 - 11, 1991

Editors:

F. Durst R. Friedrich B. E. Launder
F. W. Schmidt U. Schumann J. H. Whitelaw

With 269 Figures

Springer-Verlag
Berlin Heidelberg NewYork London Paris
Tokyo HongKong Barcelona Budapest

Franz Durst
Lehrstuhl für Strömungsmechanik, Friedrich-Alexander-Universität,
Cauerstraße. 4, W–8520 Erlangen, Germany

Rainer Friedrich
Lehrstuhl für Fluidmechanik, Technische Universität München,
Arcisstraße 21, W–8000 München 2, Germany

Brian E. Launder
Department of Mechanical Engineering, University of Manchester,
Institute of Science and Technology, P.O. Box 88,
Manchester M60 1QD, UK

Frank W. Schmidt
Department of Mechanical Engineering, The Pennsylvania State University,
University Park, PA 16802, USA

Ulrich Schumann
Deutsche Forschungsanstalt für Luft- und Raumfahrt (DLR),
Institut für Physik der Atmosphäre,
W–8031 Oberpfaffenhofen, Germany

James H. Whitelaw
Department of Mechanical Engineering, Imperial College of Science and Technology,
Exhibition Road, London SW7 2BX, UK

ISBN-13:978-3-642-77676-2 e-ISBN-13:978-3-642-77674-8
DOI: 10.1007/978-3-642-77674-8

Library of Congress Cataloging-in-Publication Data.
International Symposium on Turbulent Shear Flows (8th : 1991 : Munich, Germany) Turbulent shear flows 8 : selected papers from the eight International Symposium on Turbulent Shear Flows, Munich, FRG, September 9-11, 1991 / editors, F. Durst ... [et al.]. Includes index.
ISBN-13:978-3-642-77676-2 (New York)
1. Turbulence–Congresses. 2. Shear flow–Congresses. 3. Turbulent boundary layer–Congresses.
I. Durst, F. II. Title. III Title: Turbulent shear flows eight.

Softcover reprint of the hardcover 1st edition 1993

Typesetting with TEX: Danny Lee Lewis, Berlin

27/3020 - 5 4 3 2 1 0 – Printed on acid-free paper

Preface

This volume contains a selection of the papers presented at the Eighth Symposium on Turbulent Shear Flows held at the Technical University of Munich, 9-11 September 1991. The first of these biennial international symposia was held at the Pennsylvania State University, USA, in 1977; subsequent symposia have been held at Imperial College, London, England; the University of California, Davis, USA; the University of Karlsruhe, Germany; Cornell University, Ithaca, USA; the Paul Sabatier University, Toulouse, France; and Stanford University, California, USA. The purpose of this series of symposia is to provide a forum for the presentation and discussion of new developments in the field of turbulence, especially as related to shear flows of importance in engineering and geophysics.

From the 330 extended abstracts submitted for this symposium, 145 papers were presented orally and 60 as posters. Out of these, we have selected twenty-four papers for inclusion in this volume, each of which has been revised and extended in accordance with the editors' recommendations. The following four theme areas were selected after consideration of the quality of the contributions, the importance of the area, and the selection made in earlier volumes:

- wall flows,
- separated flows,
- compressibility effects,
- buoyancy, rotation, and curvature effects.

As in the past, each section corresponding to the above areas begins with an introduction by an authority in the field that places the individual contributions in context with one another and with related research.

This symposium was organized in cooperation with the American Meteorological Society, the American Society of Mechanical Engineers, and the Deutsche Meteorologische Gesellschaft.

Financial support for the Eighth Symposium was generously provided by BMW AG, Convex Computer GmbH, Cray Research GmbH, Deutsche Aerospace AG, Deutsche Lufthansa AG, Deutsche Forschungsanstalt für Luft- und Raumfahrt, Deutsche Meteorologische Gesellschaft, the European Office of Aerospace Research and Development of the U.S. Air Force, the European Research Office of the U.S. Army, IBM Deutschland GmbH, Linde AG, and Mercedes Benz AG.

Local arrangements were made by staff of the Technical University and the Deutsche Forschungsanstalt für Luft- und Raumfahrt in Oberpfaffenhofen. Their considerable efforts ensured a well-organized and pleasant meeting.

The members of the Advisory Committee, who provided reviews of the extended abstracts, comprised:

R.J. Adrian, USA
H. Alfredsson, Sweden
J.-C. André, France
R.A. Antonia, Australia
G. Bergeles, Greece
R. Blackwelder, USA
P. Bradshaw, USA
M. Coantic, France
S. Correa, USA
J. Cousteix, France
H. Fernholz, Germany
I. Gartshore, Canada
W.K. George, USA
V.W. Goldschmidt, USA
H. Ha-Minh, France
K. Hanjalic, Yugoslavia
T.J. Hanratty, USA
J.R. Herring, USA
J.A.C. Humphrey, USA
F. Hussain, USA
N. Kasagi, Japan
J.F. Keffer, Canada
L. Kleiser, Germany
W. Kollmann, USA
E. Krause, Germany
B. Laschka, Germany
A. Leonard, USA
M. Lesieur, France
M.A. Leschziner, UK
J.L Lumley, USA
P.J. Mason, UK
J.J. McGuirk, UK
G.E.A. Meier, Germany
P. Moin, USA
Y. Nagano, Japan
I. Nakamura, Japan
M. Peric, Germany
S.B. Pope, USA
W.C. Reynolds, USA
W. Rodi, Germany
A.J. Smits, USA
M. Sokolov, Isreal
M. Sommerfeld, Germany
R.L. Street, USA
K. Suzuki, Japan
A. Taylor, UK
H. Wengle, Germany
M. Wolfshtein, Israel

Many of these colleagues later served as session chairmen at the symposium and have throughout been valuable sources of advice and helpful criticism.

For the first time, this volume has been produced based on manuscripts provided in electronic form and using LaTeX for text editing. Some files were converted to this format by Springer-Verlag. This helped to reduce the price of the volume and the time required for publication. We thank the authors for their cooperation.

The editors wish to thank the staff of Springer-Verlag for their advice and considerable help in producing a book in consonance with our aims.

Oberpfaffenhofen, May 1992 The Editors

Contents

Part I Wall Flows

Part II Separated Flows

Part I

Wall Flows

Introductory Remarks

R. A. Antonia

Department of Mechanical Engineering, University of Newcastle, N.S.W., 2308, Australia

Turbulence research on wall flows has in the last few years benefited significantly from direct numerical simulations (DNS) [e.g. Moin and Spalart, 1989; Robinson, 1991]. The data from the simulations have been helpful in guiding the development of turbulence models (e.g. Mansour, 1991), especially in the near-wall region. There is ample evidence of this in many of the presentations at this conference (e.g. Session 26 in the Soft Cover Proceedings) and in one of the papers (Launder and Tselepidakis) included in this section. The simulations can also provide detailed data in situations, for example when the molecular Prandtl number, Pr, is small (the subject of the Kasagi et al. paper), which are practically inaccessible to measurement. The other papers in this chapter deal with a diversity of issues (*inter alia* the effect of a pressure gradient on a turbulent boundary layer, the origin of streaks, the vorticity field associated with a horseshoe-type vortex), all of which are certainly of current interest.

The paper by Nagano et al. concentrates on the effect of an adverse pressure gradient on a turbulent boundary layer, the emphasis being mainly on the wall region. The Clauser pressure gradient parameter β was in the range 0.76 to 4.66, while R_θ varied between 1290 and 3350. The method used to obtain U_τ assumes that the normalised mean velocity distribution, as measured by a hot wire (in the region $1.1 \leq y^+ \leq 5$) is independent of the magnitude of U_τ and of the pressure gradient. In the log region, the Kármán constant is unaffected but the additive constant decreases as β increases. When normalised by wall variables, the Reynolds stresses (especially $\overline{v^2}$ and $\overline{w^2}$) increase with β in the wall region. Increasing β results in a significant increase in the wall-limiting value of $\overline{u^2}^{\frac{1}{2}}/\overline{U}$, implying, *inter alia*, an increase in the average turbulent energy dissipation close to the wall. The third order moments $\overline{u^2v}$ and $\overline{uv^2}$ also change with β, a result which the authors attribute to the drastic change in turbulence structure. As noted above, the increase in β is accompanied by an increase in R_θ and, since low R_θ effects are not negligible for this range of R_θ, it is possible that the observed increase in the Reynolds stresses are, in part, caused by Reynolds number effects. This effect may need to be taken into account if the data are to be used for modelling.

Makita and Sassa present conditional velocity and vorticity measurements following the introduction of an artificially induced vortex in a self-preserving turbulent boundary layer. Spanwise vorticity contours in the $x-y$ plane of symmetry show that the head of the vortex ($\omega_z < 0$) lies above a positive ω_z region, resulting from the flow induced by the main vortex and the no-slip condition at the wall. The circulations associated with the $+\omega_z$ and $-\omega_z$ regions have significantly different streamwise developments; the former decays continuously while the latter increases quickly to a self-preserving state before decaying. At a sufficient distance downstream of its introduction, the vortex assumes a horseshoe-like shape. The authors underline that the vorticity is stronger in the legs than in the head of the horseshoe. The vorticity data should provide an interesting basis of comparison when analysing the relatively more complicated vorticity data which have been obtained in a natural turbulent boundary layer.

Waleffe and Kim address the interesting question of what imposes the spanwise scale of about 100 wall units for low- and high-speed streaks in turbulent shear flows. Numerical experiments on the linear streak generation mechanism show that it is not scale-selective. The direct resonance theory (Jang et al., 1986) assumes that the generation of streamwise vorticity is caused by non-linear interactions of certain oblique vorticity components but numerical experiments indicate that non-linear velocity interactions dominate; the theory does not therefore explain scale selection. While the paper rules out two possibilities for scale selection, it proposes that the scale of about 100 wall units corresponds to a critical Reynolds number below which turbulent flows cannot be maintained. The proposal does not explain however why streaks cannot exist below a scale of 100 wall units.

Hwang and Brereton study the response of fully developed pipe flow to forced oscillations at frequencies in the range $0.3f_B$ to about $4f_B$, where f_B is the average bursting frequency. No evidence of a resonant behaviour was found (this is not perhaps surprising in view of the complicated spatio-temporal relationships associated with bursting). There is, however, evidence of weak organisation due to the forcing. The authors report some stabilization of the near-wall flow due to a reduction in the spanwise meandering of low-speed streaks. The possibility of further exploiting this stabilisation through spatio-temporal flow-control techniques should be interesting.

In Sandham's paper, a relatively simple model equation is developed expressing a balance between viscosity, the pressure gradient and "momentum redistributions" due to ejections, sweeps and outer-layer motions. The sweep term (opposite in sign to the ejection, except near the channel centreline) has its maximum contribution at the first computational point ($y^+ = 9$, a linear profile is assumed below this point). The ejection term is zero at $y^+ = 9$, increases to a maximum near $y^+ = 30$ but remains important in the central region of the channel. The author notes the *ad hoc* nature of some of the terms in the formulation and several incorrect features (for example, a discontinuity in the velocity derivative at $y^+ = 9$) of the model. A comparison between the modelled ejection and sweep terms with those inferred from available channel flow data of contributions from the second and fourth quadrants to $\overline{uv}$ would help in assessing the assumed models.

Launder and Tselepidakis report on improvements to a second-moment closure model in order to reproduce the characteristics of near-wall turbulence correctly. The improvements include a reduced emphasis on the turbulence Reynolds number, a significant reduction in the importance of wall-reflection effects on the fluctuating pressure field and the inclusion of a model for ϵ_{ij}, the Reynolds stress dissipation, which has a correct limiting wall behaviour. The calculated Reynolds stresses are in good agreement with the fully developed channel flow DNS data of Kim et al. (1987) [So et al.'s 1991 review shows that the near-wall anisotropy of the normal stresses is reproduced by only a few models]. However, the increase with Reynolds number of all the Reynolds stresses (Wei and Willmarth, 1989; Antonia et al., 1992) is not reliably reflected in the calculations. As the authors note, more work is needed on modelling ϵ_{ij} (the suggestion that anisotropy invariants of ϵ_{ij} should be included could be worth pursuing).

Kasagi et al. have carried out a direct numerical simulation of the fully developed thermal field in a two-dimensional turbulent channel flow with $Pr = 0.025$. This simulation extends in a significant way the DNS data bases (same flow but marginally higher Reynolds number) of Kim and Moin (1989) which were done for $Pr = 0.1$, 0.71 and 2. Relative to $Pr = 0.71$, Kasagi et al. observe significant differences, in the mean temperature, rms temperature, turbulent heat fluxes and the turbulent Prandtl number. Near-wall thermal streaks are shorter and thicker at $Pr = 0.025$. This suggests that the nearly perfect coincidence between the thermal streaks and the low-speed streaks for $Pr \simeq 1$, no longer applies at small Pr. Equivalently, the nearly perfect correlation coefficient between u and θ which is observed for $Pr \simeq 1$ (Kim and Moin, 1989; Antonia et al., 1988) should not hold at small Pr. The significant increase in the turbulent Prandtl number appears to be in qualitative agreement with Jischa and Rieke's (1979) empirical formula.

References

Antonia, R. A., Krishnamoorthy, L. V. and Fulachier, L. (1988): Correlation Between the Longitudinal Velocity Fluctuation and Temperature Fluctuation in the Near-Wall Region of a Turbulent Boundary Layer, *Int. J. Heat Mass Transfer*, **31**, 723-730.

Antonia, R. A., Teitel, M., Kim, J. and Browne, L. W. B. (1992): Low Reynolds Number Effects in a Fully Developed Turbulent Duct Flow, *J. Fluid Mech.* , **236**, 579-605.

Jang, P. S., Benney, D. J. and Gran, R. I. (1986): On the Origin of Streamwise Vortices in a Turbulent Boundary Layer, *J. Fluid Mech.*, **169**, 109-123.

Jischa, M. and Rieke, H. B. (1979): About the Prediction of Turbulent Prandtl and Schmidt Numbers from Modeled Transport Equations, *Int. J. Heat Mass Transfer*, **22**, 1547-1555.

Kim, J. and Moin, P. (1989): Transport of Passive Scalars in a Turbulent Channel Flow, *Turbulent Shear Flows 6*, Berlin, Springer, 85-96.

Kim, J., Moin, P. and Moser, R. (1987): Turbulence Statistics in Fully Developed Channel Flow at Low Reynolds Number, *J. Fluid Mech.*, **177**, 133-166.

Mansour, N. N. (1991): The Use of Direct Numerical Simulation in Turbulence Modelling, *Paper AIAA-91-0221*, presented at 29th Aerospace Sciences Meeting, Nevada.

Moin, P. and Spalart, P. R. (1989): Contributions of Numerical Simulation Data Bases to the Physics, Modeling and Measurement of Turbulence, in W. K. George and R. Arndt (eds.) *Advances in Turbulence*, New York, Hemisphere, 11-38.

Robinson, S. K. (1991): Coherent Motions in the Turbulent Boundary Layer, *Ann. Rev. Fluid Mech.*, **23**, 601-639.

So, R. M. C., Lai, Y. G., Zhang, H. S. and Hwang, B. C. (1991): Second-order Near-Wall Turbulence Closures : A Review, *AIAA Jnl.*, **29**, 1819-1835.

Wei, T. and Willmarth, W. W. (1989): Reynolds-Number Effects on the Structure of a Turbulent Channel Flow, *J. Fluid Mech.*, **204**, 57-95.

Effects of Adverse Pressure Gradients on Mean Flows and Turbulence Statistics in a Boundary Layer

Y. Nagano, M. Tagawa and T. Tsuji

Department of Mechanical Engineering, Nagoya Institute of Technology, Gokiso-cho, Showa-ku, Nagoya 466, Japan

Abstract

Measurements in boundary layers with 'moderate' to 'strong' adverse pressure gradients are presented and discussed. With increasing adverse pressure gradients, the velocity profile in $\overline{U}^+ \sim y^+$ coordinates lies below the standard log law, thus indicating a reduction in the thickness of the sublayer. Correspondingly, the turbulence energy components as well as the Reynolds shear stress peak in the outer region of the boundary layer. Higher-order moments of velocity fluctuations are also seriously affected by the adverse pressure gradient. In strong adverse-pressure-gradient flows, the triple products of velocity have completely opposite signs to those in zero-pressure-gradient flows over most of the boundary layer.

Nomenclature

C_p	wall static pressure coefficient, $= (\overline{P} - \overline{P}_{ref})/(\rho \overline{U}^2_{ref}/2)$
H	shape factor, $= \delta^*/\theta$
$\overline{P}$	mean pressure
P^+	dimensionless pressure gradient parameter, $= \nu(d\overline{P}/dx)/\rho u_\tau^3$
R_θ	Reynolds number based on momentum thickness, $= \overline{U}_e\theta/\nu$
$S(\chi)$	skewness factor of χ, $= \overline{\chi^3}/(\overline{\chi^2})^{3/2}$
$\overline{U}$	mean velocity in x direction
$\overline{U}_e$	free stream velocity
u_τ	friction velocity, $= \sqrt{\tau_w/\rho}$
$u,\ v,\ w$	fluctuating velocity components in x, y and z directions
X	distance from tripping plate
$x,\ y,\ z$	streamwise, wall-normal and lateral coordinates
y^+	dimensionless distance from wall, $= u_\tau y/\nu$
β	Clauser pressure gradient parameter, $= (\delta^*/\tau_w)d\overline{P}/dx$
δ_{99}	boundary layer thickness
$\delta^*,\ \theta$	displacement and momentum thicknesses
ν	kinematic viscosity
ρ	density
τ_w	wall shear stress
$\overline{(\ )}$	time mean value

Introduction

Since adverse-pressure-gradient flows often occur in various kinds of fluid machinery, it is of both fundamental and practical importance to investigate the effects of adverse pressure gradients on the characteristics of turbulent boundary layers.

In this paper, we present the first report of our study undertaken to quantitatively investigate the effects of an adverse pressure gradient (APG) on the statistics and structures of turbulent boundary layers and to obtain the fundamental data which provide a guide for the development and improvement of turbulence theory and modeling.

The main objectives of the present investigation are to answer the following unresolved questions: (i) Does the standard log-law in zero-pressure gradient (ZPG) flows:

$$\overline{U}^+ = 2.44 \ln y^+ + 5.0 \tag{1}$$

still hold in APG flows?; (ii) How do the near-wall turbulence statistics behave in APG flows?; and (iii) How do the higher-order statistics respond to the APG?

There have been several papers investigating the basic characteristics of APG flows. For example, Samuel and Joubert (1974) and Cutler and Johnston (1989) investigated the mean flow and fluctuating quantities. On the other hand, Kline et al. (1967) and Lian (1990) studied turbulence structures by using flow visualization techniques. However, the above-mentioned problems are still unsettled. In particular, the effects of the APG on near-wall behavior of turbulence are totally unclear, which is mainly due to the difficulties in measurement. Furthermore, although we need detailed knowledge of the relationship between turbulence structures and turbulence statistics to develop appropriate turbulence models (Robinson, Kline and Spalart 1990; Nagano and Tagawa 1988; 1990a; 1991), there have been few studies on the APG flows which dealt with this subject.

Experimental Apparatus

The test section is shown in Fig. 1. Working fluid (air) flows successively through the settling chamber, two-dimensional contraction, test section, plenum chamber and blower. At the entrance to the settling chamber, a superhigh performance air filter (Japan Muki, Model ATMC51PA) is equipped to avoid a temporal change in hot-wire responses caused by adhesion of dust. This filter eliminates, for example, almost 99.97 percent of 0.3 μm-dia. fine particles contaminated in an entering air flow. To reduce free-stream turbulence and to establish a desirable two-dimensional flow field, special care is taken in arranging honeycomb and screens in the light of previous works (e.g., for honeycomb, see Lumley (1964) and Loehrke and Nagib (1976); for screen, Schubauer et al. (1950) and Bradshaw (1965)). Thus, in the settling chamber, we installed one honeycomb (cell size: 4.7 mm; length 50 mm) and five screens (mesh: 32; s.w.g.: ♯38; open area ratio: 65.4 %). The shape of wind-tunnel contraction is designed to conform to the requirements cataloged by Uberoi (1956) and Morel (1977). The contraction ratio is 9 : 1 and the cross-sectional area of its exit is 50.7 mm × 700 mm. Then, the aspect ratio at the inlet to the test section is 13.8 : 1. As a result, nearly ideal, two-dimensional uniform inflow is obtained, whose free-stream turbulence level is below 0.1 % and velocity non-uniformities in the y- (normal to the wall) and z- (lateral) directions are within 0.17 % (2.2 mm $\leq y \leq$ 48.5 mm) and 0.5 % (-200 mm $\leq z \leq$ 200 mm), respectively, under the present

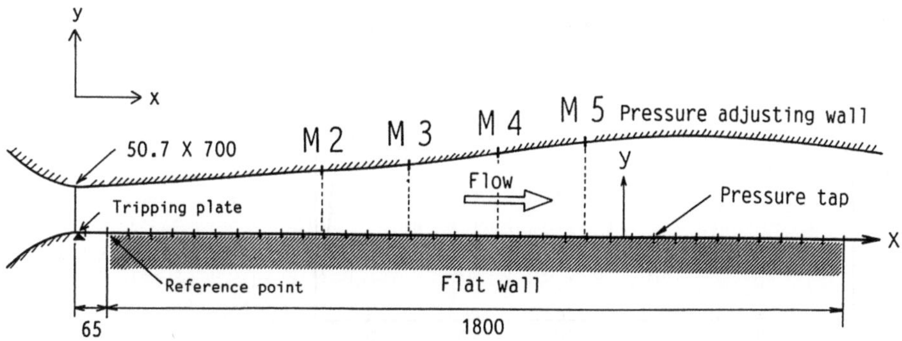

Fig. 1. Test section (flow is from left to right)

measurement conditions. To generate a stable and fully-developed turbulent boundary layer, a row of equilateral triangle plates (length of one side: 10 mm; thickness: 1 mm) is located at the inlet to the test section. The test section is composed of a flat-plate on which a turbulent boundary layer develops, and a roof-plate to adjust pressure gradients. The material of both plates is acrylic acid resin and its thickness is 10 mm.

Wall static pressures were measured with a Göttingen-type manometer equipped with a microscope (measurement accuracy: ±0.01 mm). The pressure taps with 0.5 mm diameter are located at 20 mm apart from the centerline of the flat-plate and at 50 mm intervals from the inlet.

Velocity measurement was done with hot-wire probes, i.e., a handmade subminiature (Ligrani and Bradshaw 1987) normal hot-wire (dia.: 3.1 μm; length: 0.6 mm), and two types of specially devised X-probes for measurement of u and v, and for u and w (dia.: 3.1 μm; length: $\ell = 0.6$ mm $\simeq 7.5\,\nu/u_\tau$; wire spacing: $\Delta z = 0.23$ mm $\simeq 2.9\,\nu/u_\tau$). All these hot-wires were operated using constant-temperature mode hot-wire anemometers (Hayakawa Model HC-30W). As a traversing mechanism, we used a finely adjustable positioning device equipped with a micrometer having the positioning accuracy of ± 0.005 mm. The absolute distance from the wall is determined by measuring, with a telescope, the distance between the real wire and its reflected image on the flat-plate.

All data were recorded in analog form with an FM data recorder (TEAC SR-31) and then reproduced for conversion to digital form. The data digitized with 16 bit analog-to-digital converter (Canopus ADF-16) were stored in a 40 Mbyte hard-disk unit (Logitec LHD-34V). The real-time sampling frequency was 34.08 kHz, and the number of data points per measurement was 262,144. The statistical analysis of the data was performed on an NEC PC-9801 personal computer.

Results and Discussion

Zero Pressure Gradient Flows

Since the object of the present measurement is concerned with a boundary layer itself which develops on a flat-plate, this boundary layer should be unaffected by the presence

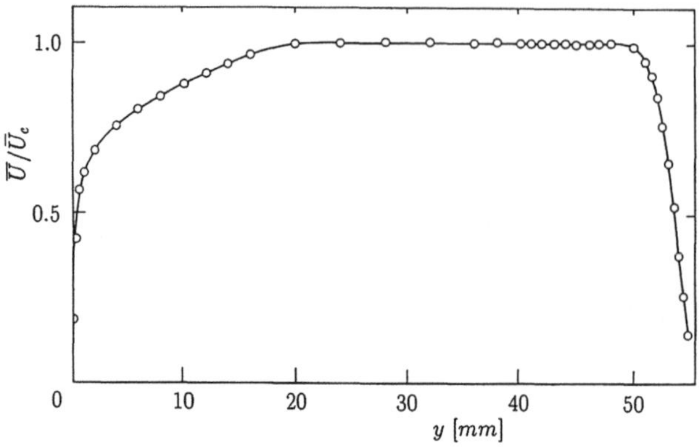

Fig. 2. Velocity profile at the end of test section

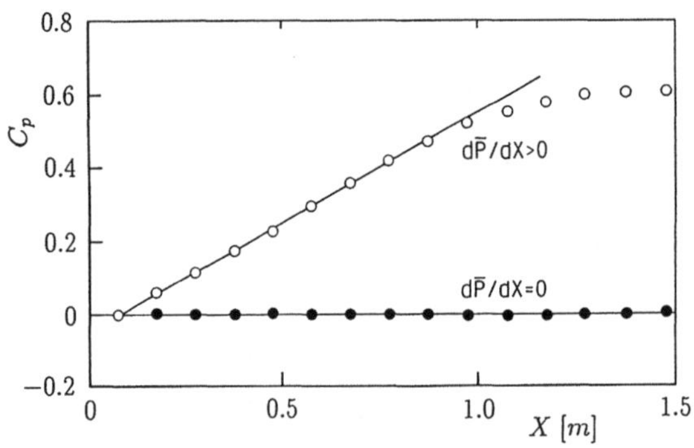

Fig. 3. Development of the pressure coefficient

of an opposite pressure-adjusting roof-plate. Thus, we first investigated a boundary layer formed on the roof-plate under a zero-pressure gradient condition. The result at the measurement location farthest downstream is shown in Fig. 2. As clearly seen from this figure, a highly uniform free stream is formed over a wide portion of the channel. Furthermore, it is confirmed that the velocity profile in the boundary layer on the roof-plate agrees quite well with the Blasius' laminar-flow solution. Thus, this boundary layer (boundary-layer thickness: $\delta_{99} = 5.3$ mm) is completely laminar even at the end of the test section and separated by the uniform free stream, so that we have concluded that there exists no influence of the pressure-adjusting plate on the objective turbulent boundary layer. Also, as seen in Fig. 3, ideal zero-pressure-gradient (ZPG) flows are formed in the test section. Here, the pressure coefficient C_p is defined with respect to reference conditions at the position 65 mm downstream of the tripping plate (see Fig. 1).

It is well known that the standard log-law velocity profile given by (1) holds in ZPG turbulent boundary layers. Before starting the discussion on APG flows, we re-examine this fact experimentally. We estimate the friction velocity, u_τ, using a method similar to that of McEligot (1984). Thus, with the Van Driest damping function for the mixing

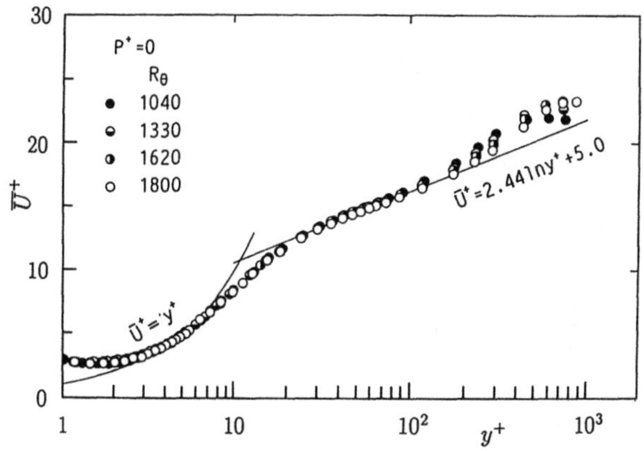

Fig. 4. Mean velocity profiles in zero-pressure-gradient flows

length, we may write the velocity profile in the wall region as:

$$\overline{U}^+ = 2\int_0^{y^+} \frac{dy^+}{1+(1+4\ell^{+2})^{1/2}} \tag{2}$$

where

$$\ell^+ = \kappa y^+[1-\exp(-y^+/A^+)] \tag{3}$$

In (2) and (3), $\overline{U}^+ = \overline{U}/u_\tau$ and $y^+ = u_\tau y/\nu$ are the dimensionless velocity and distance, where u_τ is unknown and to be determined. On the other hand, in the log-law region, we have

$$y(d\overline{U}/dy) = u_\tau/\kappa = \gamma \tag{4}$$

Here γ can be determined from measurement. From (2) – (4), together with the measurements of mean velocity, we can estimate values of κ, A^+ and u_τ so that (2) gives a best fit to measurements. The optimum values are: $\kappa \simeq 0.41$ and $A^+ \simeq 26.5$. The universal velocity profiles normalized with the friction velocity u_τ thus obtained are shown in Fig. 4. Obviously, Eq. (1) is valid in the log-law region. Therefore, it is adequate to estimate the friction velocity by using the established Clauser's method (Clauser 1954), i.e., the law-of-the-wall fit, in the cases of ZPG flows.

We applied the above method to nine ZPG flows different in free stream velocities, and then obtained nine friction velocities, u_τ. Figure 5 shows the near-wall distributions of the apparent velocities $\overline{U}_m^+$ $(= \overline{U}_m/u_\tau)$ versus y^+ normalized by the friction velocities obtained with the law-of-the-wall fit. As clearly seen in Fig. 5, in the vicinity of the wall ($y^+ < 3.6$), $\overline{U}_m^+$ deviates systematically from the 'law of the wall,' $\overline{U}^+ = y^+$, in the viscous sublayer as the wall is approached. In general, this can be ascribed to the wall proximity effect of hot-wire outputs. Previous extensive studies investigating this wall proximity effect (for example, Oka and Kostic 1972; Hebbar and Melnik 1978; Bhatia,

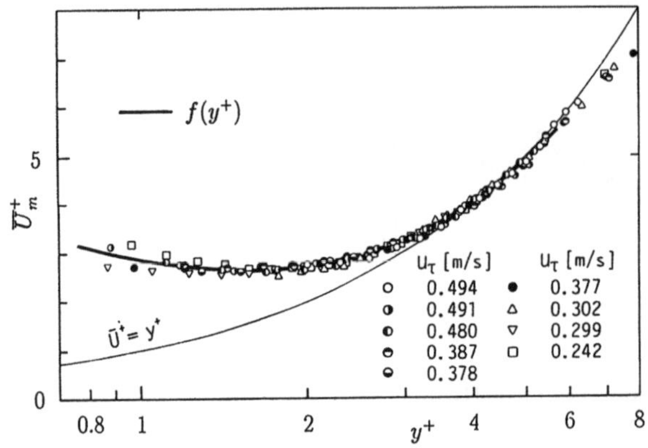

Fig. 5. Near-wall distributions of apparent velocities $\overline{U_m^+}$ in various zero-pressure-gradient flows

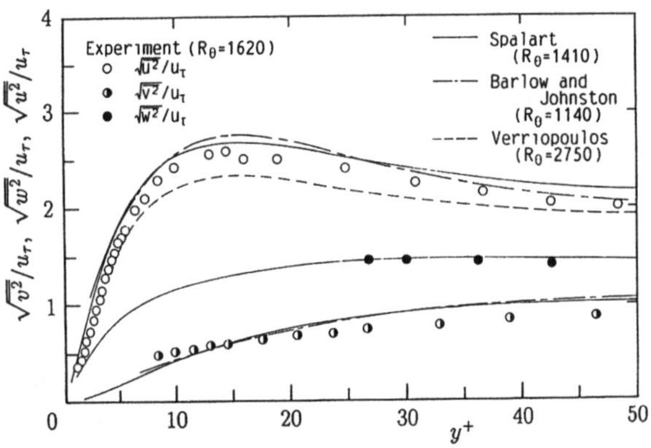

Fig. 6. R.m.s. velocity fluctuations in zero-pressure-gradient flow

Durst and Jovanovic 1982; Janke 1987) have reported that: once material of a wall, and geometrical factors of a hot-wire and a probe are given, the amount of this deviation can be determined universally, independent of values of wall shear stresses, so that $\overline{U_m^+} = f(y^+)$. As seen in Fig. 5, the apparent velocities $\overline{U_m^+}$ in the present experiments collapse very well onto a unique curve $f(y^+)$, being consistent with the previous findings. Therefore, in the following, we use this curve $f(y^+)$ regardless of pressure gradients or wall shear stresses, and determine u_τ using the least-squares-fitting method to the data in the range $1.1 \leq y^+ \leq 5$. Consequently, the following measurements were done with the same probe used in this calibration.

The profiles of r.m.s. values of streamwise, wall-normal and spanwise velocity fluctuations ($\sqrt{\overline{u^2}}/u_\tau$, $\sqrt{\overline{v^2}}/u_\tau$ and $\sqrt{\overline{w^2}}/u_\tau$) are shown in Fig. 6. It can be seen that all of the present results in the ZPG boundary layer are in good agreement with the existing experiments (Verriopoulos 1983; Barlow and Johnston 1988) and the DNS results (Spalart 1988), thus indicating that the present turbulent boundary layers are canonical.

Adverse Pressure Gradient Flows

Next, we shall present the results of experiments on adverse pressure gradient (APG) turbulent boundary layers. The distributions of C_p and $d\,C_p/dx$ are shown in Fig. 3, and the mean flow parameters are summarized in Table 1. In the present APG flow, the pressure gradient $d\,C_p/dx$ keeps a nearly constant value of 0.6 over 65 mm $\leq x \leq$ 700 mm and then decreases slowly. On the other hand, the pressure gradient parameter normalized by inner variables (Huffman and Bradshaw 1972), P^+ $(= \nu(d\overline{P}/dx)/\rho u_\tau^3)$, and the Clauser pressure gradient parameter, β $(= (\delta^*/\tau_w)d\overline{P}/dx)$, increase monotonously, thus giving 'moderate' to 'strong' adverse pressure gradients.

Table 1. Mean flow parameters ($\overline{U}_{ref}$ = 10.8 m/s)

X mm	$\overline{U}_e$ m/s	δ_{99} mm	u_τ m/s	R_θ	H	P^+	β
523	9.08	16.2	0.390	1290	1.52	8.98×10^{-3}	0.76
723	8.18	24.6	0.307	1880	1.61	1.81×10^{-2}	2.05
925	7.54	34.2	0.251	2660	1.74	2.30×10^{-2}	3.54
1121	6.68	46.1	0.197	3350	1.88	2.51×10^{-2}	4.66

Figure 7 shows the mean velocity distributions $\overline{U}^+$ in the APG boundary layers. These quantities are normalized by the friction velocities obtained with the above-mentioned method of least-squares curve fitting, in which corrections (Bhatia, Durst and Jovanovic 1982) were made for the mean velocity $\overline{U}^+$ near the wall ($y^+ < 3.6$) by subtracting the quantity $\Delta\overline{U}_m^+ = f(y^+) - \overline{U}^+$ shown in Fig. 5 from the apparent measurement $\overline{U}_m^+$ in APG flows. This figure presents the important findings that: (i) in the viscous sublayer, the typical linear relationship $\overline{U}^+ = y^+$ holds; but (ii) in the fully turbulent wall region, the velocity profiles $\overline{U}^+$ deviate downwards from the standard log-law profile given by (1). The amount of deviation increases with an increase of P^+. Using (2) – (4), the procedure described previously was applied so as to cross-check the present mean velocity profiles. As shown in Fig. 8, the Von Kármán constant was unchanged ($\kappa \simeq 0.4$), but the Van Driest damping constant A^+ decreased with increasing P^+, which agreed well with the semi-empirical formula obtained by Kays (1971) as:

$$A^+ = 26/(1 + 30.18P^+) \tag{5}$$

The velocity profiles in $\overline{U}^+$ v.s. y^+ coordinates were almost identical to those presented in Fig. 7.

These results in the APG flows ($P^+ > 0$) exhibit a striking contrast to those in favorable pressure gradient flows ($P^+ < 0$) in which the velocity profile deviates upwards from (1) (Blackwelder and Kovasznay 1972; Spalart 1986). Kline et al. (1967), on the other hand, reported that the $\overline{U}^+$ distributions lay above and below the standard log-law

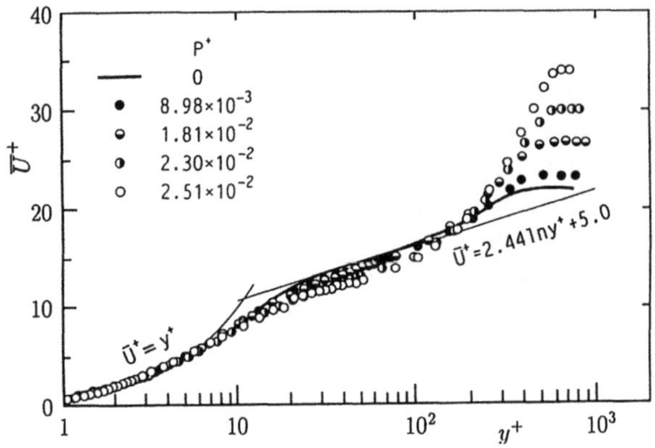

Fig. 7. Mean velocity profiles in wall coordinates in adverse-pressure gradient flows

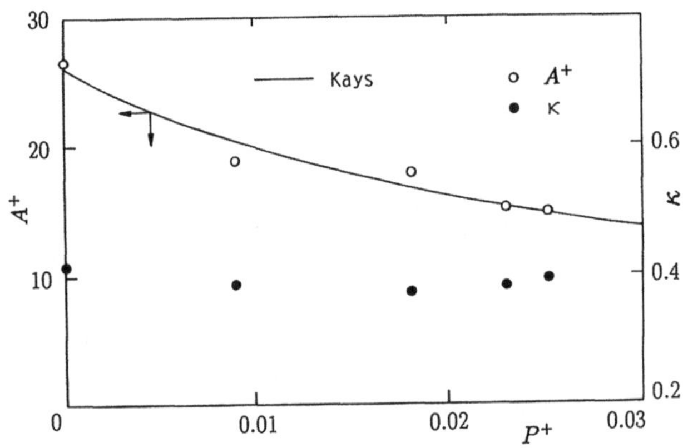

Fig. 8. Von Kármán and Van Driest constants in adverse-pressure-gradient flows

according as $P^+ > 0$ and $P^+ < 0$. The marked difference between their results and the present ones can be attributed mainly to their omission of wall proximity corrections to hot-film outputs and to their over-fitting of measurements to the linear relation $\overline{U}^+ = y^+$ up to $y^+ \simeq 10$. Such an over-fitting may cause the estimation of smaller friction velocity and result in the upward deviation of $\overline{U}^+$ from the standard log-law, (1), even in ZPG turbulent boundary layers (Blackwelder and Haritonidis 1983). In the light of the foregoing, we have concluded that the standard log-law, Eq. (1), does not hold in APG turbulent boundary layers. This accords exactly with the conjecture made by Launder (1986). Consequently, this result also warns against the use of the wall functions on which many turbulence models rely (for detailed discussion, see, for example, Nagano and Tagawa 1990b).

As seen in Fig. 9 a–c, the adverse pressure gradient also exerts a great influence on the distributions of turbulence quantities. As P^+ increases, all of the velocity fluctuation

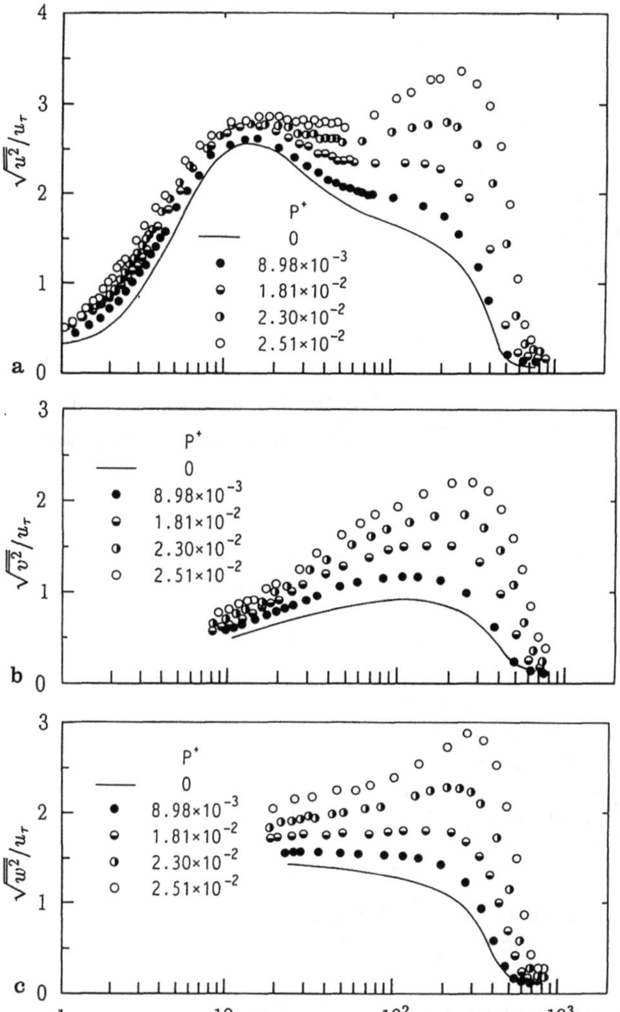

Fig. 9 a–c. Turbulence intensities in adverse-pressure-gradient flows. **a:** Velocity fluctuations in the streamwise direction; **b:** Velocity fluctuations in the wall-normal direction; **c:** Velocity fluctuations in the lateral direction

components ($\sqrt{\overline{u^2}}/u_\tau$, $\sqrt{\overline{v^2}}/u_\tau$ and $\sqrt{\overline{w^2}}/u_\tau$) become remarkably large in the outer region of a boundary layer. The increase in $\sqrt{\overline{u^2}}/u_\tau$ penetrates into the inner region systematically with increasing P^+, and the universal profile of $\sqrt{\overline{u^2}}/u_\tau$ observed in the ZPG flows disappears in the cases of APG flows. As the wall is approached, the distributions of $\sqrt{\overline{u^2}}/u_\tau$ approach each straight line which coincides with the origin (Fig. 10). This is proper because the wall-limiting behavior of u is $u \propto y$ for $y \to 0$. However, the gradients of the linear lines become steeper as P^+ increases. Since the wall law, $\overline{U}^+ = y^+$, holds even in the APG flows very near the wall, this leads to a larger

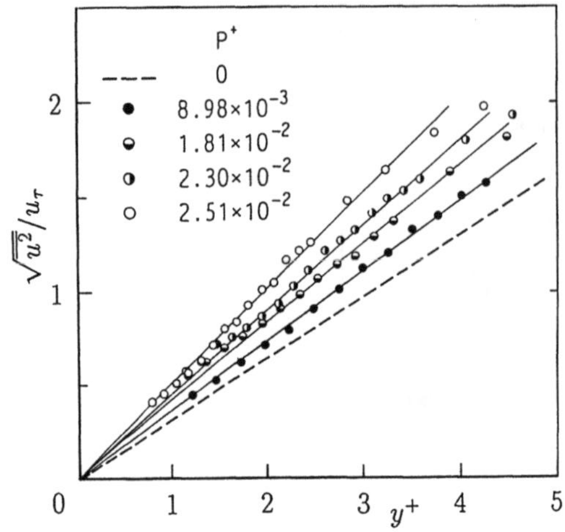

Fig. 10. Wall-limiting behavior of $\sqrt{\overline{u^2}}/u_\tau$ in adverse-pressure-gradient flows

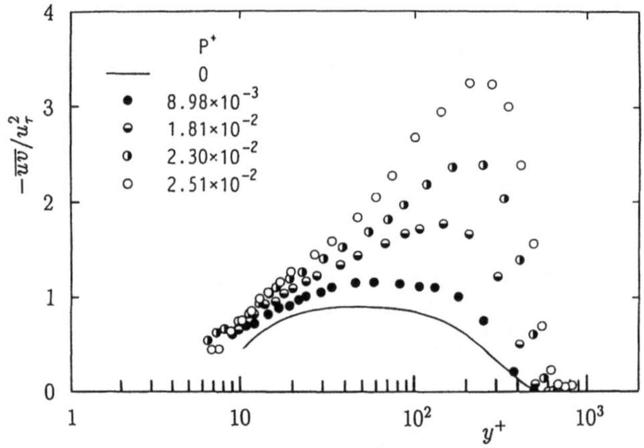

Fig. 11. Reynolds shear stress profiles in adverse-pressure-gradient flows

wall-limiting value of the relative turbulence intensity $\sqrt{\overline{u^2}}/\overline{U}$ with increasing P^+, thus indicating that an increase in the dimensionless pressure-gradient parameter P^+ makes the viscous region more turbulent and energetic.

The profiles of Reynolds shear stress, $-\overline{uv}/u_\tau^2$, are also seriously affected by the adverse pressure gradient. With increasing P^+, $-\overline{uv}/u_\tau^2$ increases in the outer region as shown in Fig. 11. Thus, it should be noted that the constant-stress-layer relationship $-\overline{uv}/u_\tau^2 \simeq 1$ observed in ZPG flows is no longer valid and the turbulent shear stress $-\rho\overline{uv}$ becomes considerably larger than the wall shear stress τ_w, i.e., $-\overline{uv} > u_\tau^2$, as P^+ increases. This too may account for the nonexistence of the universal law of the wall in APG boundary layers.

From the profiles of $\sqrt{\overline{u^2}}$, $\sqrt{\overline{v^2}}$ and $\sqrt{\overline{w^2}}$ in wall coordinates shown in Fig. 9, one may infer that the rate of turbulence production in APG flows will become much

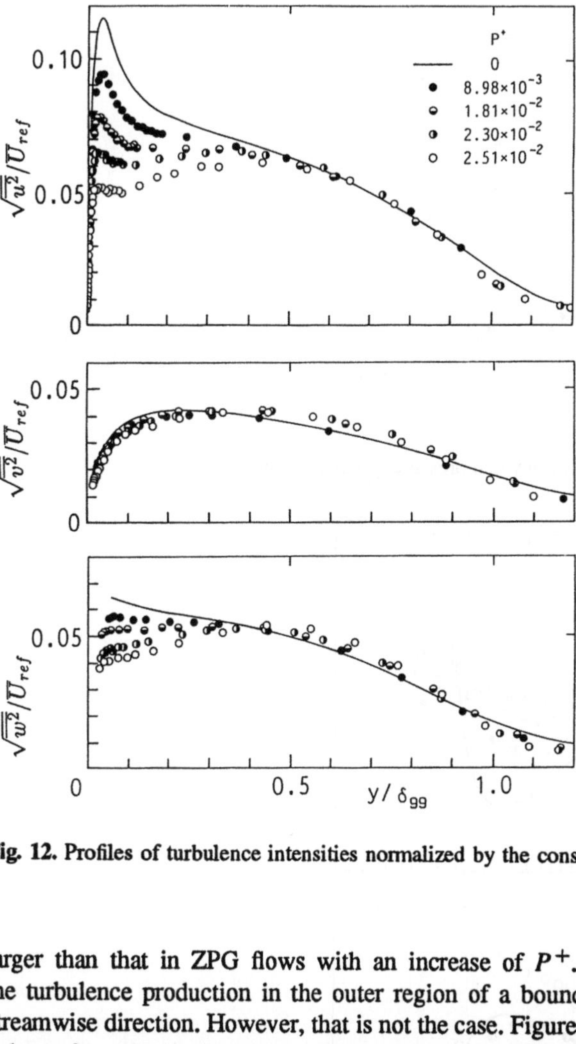

Fig. 12. Profiles of turbulence intensities normalized by the constant reference velocity $\overline{U}_{ref}$

larger than that in ZPG flows with an increase of P^+. In particular, it appears that the turbulence production in the outer region of a boundary layer is increasing in the streamwise direction. However, that is not the case. Figure 12 shows the profiles of r.m.s. values of streamwise, wall-normal and spanwise velocity fluctuations normalized by the 'constant' reference velocity $\overline{U}_{ref}$ at the inlet (see Fig. 1 and Table 1). Obviously, all the profiles in the outer region are kept unchanged and almost identical with those in ZPG flows. The turbulence-intensity reduction is found to occur first in the wall region. The reduction becomes greater with increasing P^+, which results in a gradual increase of the APG affected region away from the wall. Also, it should be noted that the existence of the APG changes the r.m.s. values of velocity fluctuations in the order, u, w, v. This means a kind of memory effects of turbulence can be seen in APG flows. The pronounced decrease of $\sqrt{\overline{u^2}}$ in the wall region is the consequence of the deterioration in the turbulence production rate when an adverse pressure gradient is imposed or a flow becomes retarded, as seen from Fig. 13.

Now, there is little knowledge of what effects the adverse pressure gradient produces on the higher-order turbulence statistics. Figures 14(a) and 14(b) show the turbulent transport, i.e., third-order moments, of turbulence energy component u^2 and Reynolds

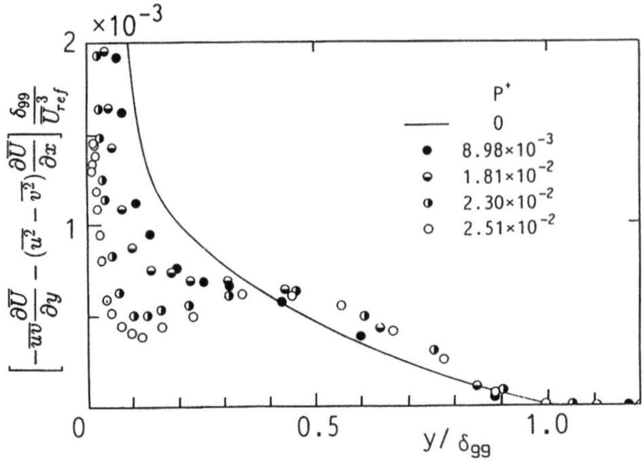

Fig. 13. Turbulence production rate in adverse pressure gradient flows

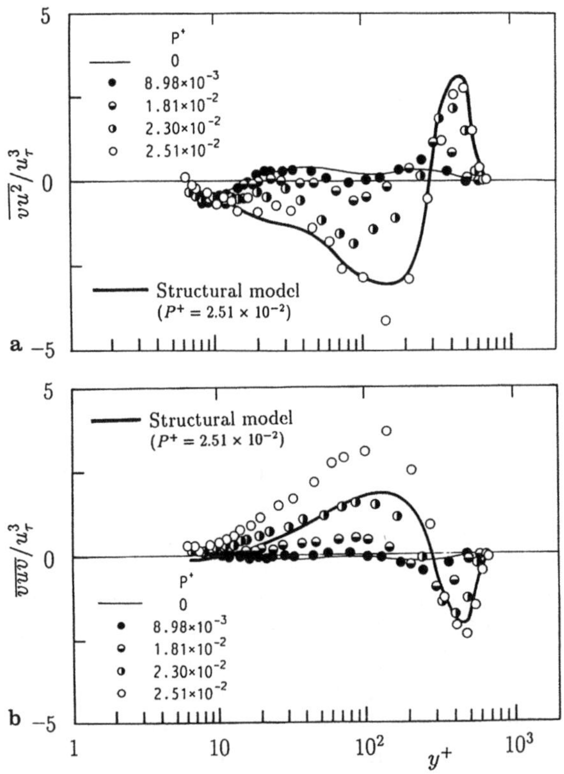

Fig. 14a, b. Distributions of turbulent transport (third-order moments) in adverse-pressure-gradient flows. **a:** $\overline{vu^2}$; **b:** $\overline{vuv}$

shear stress uv, respectively. The marked effects of the APG are seen on both third-order moments. From Fig. 14(a), the positive region of $\overline{vu^2}$ observed in the ZPG flows for $y^+ > 15$ disappears partly as P^+ increases. Since third-order moments (i.e., triple products) are very sensitive to the change of coherent turbulence structures such as ejections and sweeps (Nagano and Tagawa 1988; 1990a), this result suggests that internal structural changes occur in the APG boundary layers. Also, negative values of $\overline{vu^2}$ in the near-wall to outer regions indicate the existence of turbulence energy transport toward the wall from the regions away from the wall. This is consistent with the results of Bradshaw (1967) and Cutler and Johnston (1989). From Fig. 14(b), it is found that a similar inward transfer takes place in the turbulent diffusion of the Reynolds shear stress. In particular, as P^+ increases, turbulent diffusion in the APG boundary layer occurs in the direction completely opposite to that in the ZPG case.

The structural change in the higher-order turbulence statistics in the APG flows may be predicted by a 'structural' turbulence model of Nagano and Tagawa (1990a; 1991):

$$\overline{\hat{v}\hat{u}^2} = C\left[\sigma_{\overline{uv}}\,\frac{1}{2}\,\pi S(u) + S(v)\right] \tag{6}$$

$$\overline{\hat{v}\hat{u}\hat{v}} = C\left[S(u) + \sigma_{\overline{uv}}\,\frac{1}{2}\,\pi S(v)\right] \tag{7}$$

with

$$C = 1/\{3[(\pi/2)^2 - 1]\} \tag{8}$$

where a circumflex denotes the normalization by the respective r.m.s. value, $S(\chi)$ is the skewness factor of a stochastic variable χ, and $\sigma_{\overline{uv}}$ represents a sign function, i.e., $\sigma_{\overline{uv}} = 1$ for $\overline{uv} \geq 0$ and $\sigma_{\overline{uv}} = -1$ for $\overline{uv} < 0$. As can be seen from Figs. 14(a) and (b), Eqs. (6) and (7) well represent the structural change in the APG boundary layer, although the conventional gradient-type diffusion model has failed to do so.

Conclusions

The present experiment reveals that the mean flow and turbulence statistics are greatly affected by increasing adverse pressure gradients. The well-known 'law-of-the-wall' established for flows with a zero pressure gradient does not hold in the adverse-pressure-gradient flows. This finding is very important since most existing turbulence models in applications are constructed on the basis of this law. The existence of the APG changes the profiles of the r.m.s. intensities of velocity fluctuations near the wall in the order of streamwise, lateral and normal components. A much more drastic change in the structure of turbulence is found in the statistics of higher-order moments of velocity. Large changes in triple products cannot be reproduced by the conventional gradient-type diffusion model. A structural turbulence model developed by Nagano and Tagawa (1990a; 1991), on the other hand, can predict these changes.

References

Barlow, R. S., Johnston, J. P. (1988): Structure of a turbulent boundary layer on a concave surface. J. Fluid Mech. **191**, 137–176

Bhatia, J. C., Durst, F., Jovanovic, J. (1982): Corrections of hot-wire anemometer measurements near walls. J. Fluid Mech. **122**, 411–431

Blackwelder, R. F., Kovasznay, L. S. G. (1972): Large-scale motion of a turbulent boundary layer during relaminarization. J. Fluid Mech. **53**, 61–83

Blackwelder, R. F., Haritonidis, J. H. (1983): Scaling of the bursting frequency in turbulent boundary layers. J. Fluid Mech. **132**, 87–103

Bradshaw, P. (1965): The effect of wind-tunnel screens on nominally two-dimensional boundary layers. J. Fluid Mech. **22**, 679–687

Bradshaw, P. (1967): The turbulence structure of equilibrium boundary layers. J. Fluid Mech. **29**, 625–645

Clauser, F. H. (1954): Turbulent boundary layers in adverse pressure gradients. J. Aeronautical Sciences **21**, 91–108

Cutler, A. D., Johnston, J. P. (1989): The relaxation of a turbulent boundary layer in an adverse pressure gradient. J. Fluid Mech. **200**, 367–387

Hebbar, K. S., Melnik, W. L. (1978): Wall region of a relaxing three-dimensional incompressible turbulent boundary layer. J. Fluid Mech. **85**, 33–56

Huffman, G. D., Bradshaw, P. (1972): A note on von Kármán's constant in low Reynolds number turbulent flows. J. Fluid Mech. **53**, 45–60

Janke, G. (1987): "Hot wire in wall proximity," in *Advances in Turbulence* (G. Comte-Bellot, J. Mathieu, eds) (Springer, Berlin, Heidelberg, New York), pp. 488–498

Kays, W. M. (1971): Heat transfer to the transpired boundary layer. *ASME Paper No. 71-HT-44*

Kline, S. J., Reynolds, W. C., Schraub, F. A., Runstadler, P. W. (1967): The structure of turbulent boundary layers. J. Fluid Mech. **30**, 741–773

Launder, B. E. (1986): Low-Reynolds-number turbulence near walls. *UMIST Dept. of Mech. Eng., TFD/86/4*

Lian, Q. X. (1990): A visual study of the coherent structure of the turbulent boundary layer in flow with adverse pressure gradient. J. Fluid Mech. **215**, 101–124

Ligrani, P. M., Bradshaw, P. (1987): Subminiature hot-wire sensors: development and use. J. Physics E – Scientific Instruments **20**, 323–332

Loehrke, R. I., Nagib, H. M. (1976): Control of free-stream turbulence by means of honeycombs: a balance between suppression and generation. Trans. ASME, J. Fluids Engng. **98**, 342–353

Lumley, J. L. (1964): Passage of a turbulent stream through honeycomb of large length-to-diameter ratio. Trans. ASME, J. Basic Engng. **86**, 218–220

McEligot, D. M. (1984): Measurement of wall shear stress in accelerating turbulent flows. *Max-Planck-Institut für Strömungsforschung (Göttingen) Bericht* 109

Morel, T. (1977): Design of two-dimensional wind tunnel contractions. Trans. ASME, J. Fluids Engng. **99**, 371–378

Nagano, Y., Tagawa, M. (1988): Statistical characteristics of wall turbulence with a passive scalar. J. Fluid Mech. **196**, 157–185

Nagano, Y., Tagawa, M. (1990a): A structural turbulence model for triple products of velocity and scalar. J. Fluid Mech. **215**, 639–657

Nagano, Y., Tagawa, M. (1990b): An improved $k-\varepsilon$ model for boundary layer flows. Trans. ASME, J. Fluids Engng. **112**, 33–39

Nagano, Y., Tagawa, M. (1991): "Turbulence model for triple velocity and scalar correlations," in *Turbulent Shear Flows 7* (F. Durst, B. E. Launder, W. C. Reynolds, F. W. Schmidt, J. H. Whitelaw, eds) (Springer, Berlin, Heidelberg, New York), pp. 47–62

Oka, S., Kostic, Z. (1972): Influence of wall proximity on hot-wire velocity measurements. *DISA Information*, No. 13, pp. 29–33

Robinson, S. K., Kline, S. J., Spalart, P. R. (1990): "Quasi-coherent structures in the turbulent boundary layer: part II: verification and new information from a numerically simulated flat-plate layer," in *Near-Wall Turbulence* (S. J. Kline, N. H. Afgan, eds) (Hemisphere), pp. 218–247

Samuel, A. E., Joubert, P. N. (1974): A boundary layer developing in an increasingly adverse pressure gradient. J. Fluid Mech. **66**, 481–505

Schubauer, G. B., Spangenberg, W. G., Klebanoff, P. S. (1950): Aerodynamic characteristics of damping screens. *NACA Technical Note* 2001

Spalart, P. S. (1986): Numerical study of sink-flow boundary layers. J. Fluid Mech. **172**, 307–328

Spalart, P. S. (1988): Direct simulation of a turbulent boundary layer up to $R_\theta = 1410$. J. Fluid Mech. **187**, pp. 61–98

Uberoi, M. S. (1956): Effect of wind-tunnel contraction on free-stream turbulence. J. Aeronautical Sciences **23**, 754–764

Verriopoulos, C. A. (1983): Effects of convex surface curvature on heat transfer in turbulent flow. *Ph.D. Thesis*, Imperial College.

Three-Dimensional Vortical Structure of a Large-Scale Coherent Vortex in a Turbulent Boundary Layer

H. Makita and K. Sassa

Dept. of Energy Engineering, Toyohashi University of Technology, 1-1 Tenpaku-cho, Toyohashi 441, Japan

Abstract

The streamwise evolution of three-dimensional structure of an artificially induced vortex was pursued by measuring velocity and vorticity fields through a simplified conditional sampling method. The artificial vortex developed downstream through the growth, the self-preserving and the decay stages. In the growth stage, the vortex grew up to have a horseshoe sculpture and the vorticity confined in it was increased downstream. The grown-up large-scale horseshoe vortex reached about 1.2δ in height, 1.0δ in width and 5.0δ in streamwise length. The vortex became more inclined forward and lost the ability of maintaining its self-similar construction in the decay stage. The convection velocity of the vortex's head was $0.56\,U_\infty$ in the foregoing two stages and was raised to $0.85\,U_\infty$ in the decay stage.

1. Introduction

Large-scale coherent structures are thought to play important roles in momentum and energy transfer across a turbulent boundary layer. As Kline & Robinson (1989) summarized, the large-scale u-discontinuity spontaneously observed in the turbulent boundary layer reported by Brown & Thomas (1977), Eckelmann et al. (1977) and Praturi & Brodkey (1978) and the turbulent bulge measured by Kovasznay et al. (1970) can be related to the large-scale coherent structures. These observations, however, have caught a large-scale coherent structure only by some characteristic features and also suggest the existence of a large-scale vortical motion in the boundary layer. Actually, many vortical structures, such as an arch, a horseshoe, or a hairpin, have been proposed based on the results of flow visualizations and hot-wire measurements, as reviewed by Robinson (1989).

The coherent vortex has a complex three-dimensional configuration and appears randomly in time and space in a turbulent boundary layer. Moreover, its vortical structure is ambiguous buried in the background turbulence. It is difficult for the hot-wire measurement to chatch an arbitrary portion of the vortical structure reaching some stage of the vortex's streamwise growth during its lifetime. As for the authors' knowledge, several researchers, such as Fukunishi & Sato (1987) and Antonia & Bisset (1991) have tried to clarify the three-dimensional structure of the coherent vortex. They gave the coherent vortex similar configurations like a horseshoe, of which the detailed features were, however, still dependent upon the sampling conditions employed in each measurement. The hot-wire measurements have given only obscure images for the large-scale coherent vortex in a turbulent boundary layer.

It is easy to get a two-dimensional image of the coherent vortex by flow visualization. Usually, photographs on visualized planes can not give so much quantitative information, which sometimes causes the lack in objectivity in the interpretation of the observed results. With such experimental difficulties, it seems hopeless to pursue, the coherent vortex changing its three-dimensional configuration from its birth to death. Antonia & Bisset (1991) suggested that there was no way except for numerical simulations to solve these difficulties. Kim (1987) and Robinson et al. (1989) showed the vortical structure by using the DNS numerical method. Their results, however, seem to have a few doubts because of the assumptions in their calculations, such as the periodic boundary condition.

The present experiment aims to describe the aspect of streamwise evolution of the large-scale coherent vortex by determining its three-dimensional vortical structure. In order to make the detection by a hot-wire easy, we tried to induce an artificial coherent vortex having the structure same as a natural one in a fully developed turbulent boundary layer, as shown by Makita et al. (1988, 1989). The other types of the artificial generation of coherent structures were examined by Viets et al. (1981) and Gad-el-Hak et al. (1986, 1987). But for Gad-el-Hak & Hussain (1986), these works seem more interested in the problem of boundary layer control than the detection of the characteristic quantities of the coherent structures, In the present experiment, a seed of the coherent vortex was formed in the lower part of the fully developed turbulent boundary layer. Concentrated spanwise vorticity, ω_z, was induced there through the injection of a pair of fine swirling jets from the flat plate beneath. The existence of the spanwise vorticity in a natural large-scale coherent motion was reported by Praturi & Brodkey (1978). Fukunishi et al. (1987) showed the possibility that a small perturbation of the spanwise vorticity placed in a shear layer grew up to be a large-scale horseshoe vortex, using the vortex filament method. In the present experiment, the vorticity seed was observed to grow op to be a large-scale horseshoe vortex as it drifted downstream. The velocity vectors and the vorticity components were measured about the artificially induced vortex to pursue the streamwise development of its three-dimensional configuration.

2. Experimental Procedure

As shown in Fig. 1, a turbulent boundary layer without a pressure gradient was developed on a smooth flat plate made of glass settled in a wind tunnel test section of $0.42 \times 0.42 \times 5.7\,\mathrm{m}^3$. A No.40 sand roughness was pasted near the leading edge for 100 mm in streamwise length. The free-stream velocity, U_∞, was 5 m/sec and the turbulence level of the free-stream, u'/U_∞, was less than 0.1%. The Reynolds number based on the momentum thickness, Re_θ, was about 1820 at 2 m downstream from the leading edge, where a pair of injection slots was placed on the flat plate. The slots were 2 mm in diameter and placed 20 mm apart from each other. They were inwardly facing at the injection angle of ±45 deg. Swirlers were inserted in the conduits to the slots. As a loud

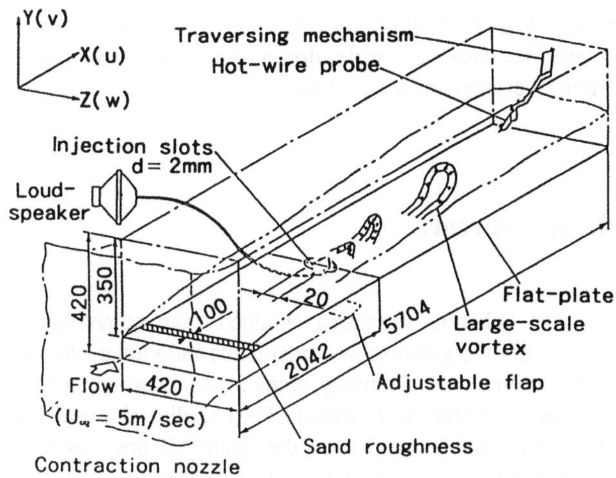

Fig. 1. Experimental setup (unit: mm)

speaker pump was driven, small puffs of swirling jets were ejected from the injection slots. The puffs were connected at about $Y = 0.26\delta$ ($\delta = 40$ mm) above the center of the slots and immediately formed a seed of a coherent vortex, having a top spin rotation. The induced spanwise vorticity, ω_Z, had the same sign as that of the mean velocity gradient there. The volume of puffs per each injection was about 0.16 cc, and confirmed not to affect the macroscopic structure of the boundary layer if the jets were not swirled. The coordinates system was set as shown in Fig. 1 and its origin was at the center of the slots.

Measurements were made by using an X-wire or a triple (double-X) wire probe and constant temperature hot-wire anemometers. The probes were made of tungsten wire of 2.5 μm in diameter and their measuring volumes were $0.7 \times 0.7 \times 0.2\,\text{mm}^3$ (X-wire) and $0.7 \times 0.7 \times 0.3\,\text{mm}^3$ (triple wire). The triple wire was used to eliminate the effect of shear normal to the X-Z plane in the simulataneous measurement of u- and w-components. The anemometers had a good S/N ratio of more than 72 dB and a frequency response flat from DC to about 20 kHz. The data sampling was conditioned by the raising of electric pulses to drive a speaker pump and a proper delay time was taken to compensate the travel time of the coherent vortex from the origin to the measuring point. By the present experimental scheme, any arbitrary portion of the coherent vortex can be detected when it reaches the previously imagined stage of its streamwise growth only by traversing the hot-wire probe. The sampled data were ensemble averaged for 1024 times by a 12-bit digital signal analyzer. The temporal resolution of the analyzer was 0.4 msec. The ensemble-averaged velocity fluctuations, $\langle u \rangle$, $\langle v \rangle$ and $\langle w \rangle$, were obtained at several streamwise positions between $X/\delta = 0.13 - 45.0$. At each streamwise position, the measurements were made at 14×13 points locating in every 0.1δ steps in both normal and spanwise directions. In this way, the timewise charts of the ensemble-averaged velocity fields were obtained for the coherent structure; see Makita et al. (1989). The ensemble-averaged values of the vorticity components, $\langle \omega_X \rangle$, $\langle \omega_Y \rangle$ and $\langle \omega_Z \rangle$, were numerically calculated from these averaged velocities and their distributions were mapped. The streamwise derivative, $\partial/\partial X$, was obtained from the time derivative, $\partial/\partial t$, assuming Taylor's frozen pattern hypothesis. The error by applying the Taylor's hypothesis on the shear layer was almost negligible in the present case, because the contribution of the term, $\partial/\partial t$, was at most 10% of each coherent vorticity component, $\langle \omega_Y \rangle$ or $\langle \omega_Z \rangle$.

3. Symmetry of the Vortical Structure

Blackwelder (1988) pointed out that coherent structures obtained from conventional ensemble-averaged data were apt to become symmetrical even if the individual realizations of the coherent structure were asymmetrical. Actually, some of the recent numerical (Robinson et al. 1989) and experimental (Antonia & Bisset 1991) studies showed that most of the individual vortices had asymmetric structure and the symmetric one was rare. In the present study, the magnitude of the initially given artificial disturbance was in the same order as the background turbulence level. Therefore, the individual artificial vortex was also asymmetrical strongly deformed by the background turbulence.

Adding further sampling conditions so as to detect the asymmetry would spoil the expediency of the present method and we can show only averaged symmetric images of the coherent structure. We prefer the averaged symmetric structure as an idealized macroscopic model of the large-scale coherent vortex to the individual asymmetric one. Different from the case of small-scale coherent vortices in the inner layer, the large-scale vortices involve smaller turbulent eddies or random velocity fluctuations in itself. We assume that the asymmetric configuration of the individual vortex results from the superposition of the smaller turbulent eddies on the large-scale symmetric coherent structure. It is more useful to grasp the averaged structure in order to clarify the dynamics of the large-scale vortex than to determine the minute structure of the individual vortices deformed by such fine-scale structures.

4. Results and Discussion

4.1 Streamwise Evolution of the Large-Scale Vortex

Figures 2a-e show the distributions of velocity vectors at several streamwise locations to explain the aspect of the streamwise evolution of the large-scale artificial vortex. The velocity field around the artificial vortex on the T-Y plane at $Z/\delta = 0$ is shown in each figure. The abscissa of the figures represents the time, t, passing from right to left. The right-hand side also correspond to the downstream side of X-direction in each figure. Each arrow in the figures denotes the relative velocity induced by the artificial vortex from which the local mean velocity of the turbulent boundary layer is subtracted; $(\langle u \rangle, \langle v \rangle)$. Kovaszanay et al. (1970) and Brown & Thomas (1977) showed the velocity vectors as ensemble-averaged velocities subtracted the convection velocity of the coherent structure, $(\langle U+u \rangle - U_c, \langle v \rangle)$. Their representation may show the vortical motion more properly, if

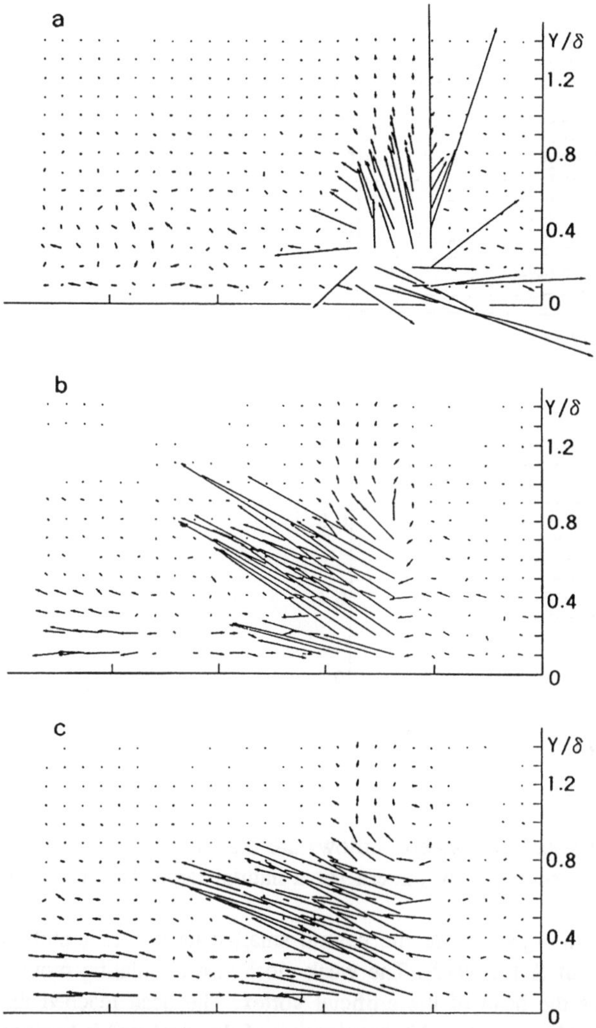

Fig. 2a-c. Induced velocity vectors around the artificial vortex on the symmetrical T-Y plane ($Z/\delta = 0$) at several streamwise locations. a: $X/\delta = 0.5$, b: $X/\delta = 2.5$, c: $X/\delta = 5$, d: $X/\delta = 14$, e: $X/\delta = 35$. The magnitude of arrow shown bottom of the Fig. e denotes $0.05\,U_\infty$.

the convection velocity of the large-scale vortex is uniform across the turbulent boundary layer. As described later, the convection velocity varies with the portion of the large-scale vortex. The present method of representation gives at least the fluid motion induced by the coherent vortex correctly, though it can not give the completely accurate image of the vortical motion about the coherent structure.

As shown in Fig. 2a, a flow field overlapped by upward and circulating fluid motions is observed at $X/\delta = 0.5$ just downstream of the injection slots. Such fluid motions, directly induced by the injection of the swirling jets, have not yet formed a typical pattern of the induced flow field around the coherent vortex. In the further downstream

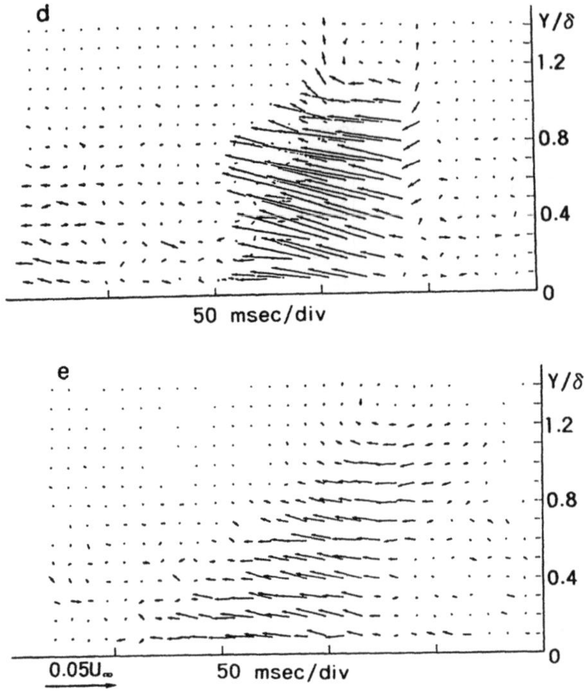

Fig. 2d-e. For caption see last page.

region of $X/\delta = 2.5 - 5$, the strong low-speed outflows are observed for $Y/\delta \leq 0.8$. Fukunishi & Sato (1987) got a similar flow pattern for a natural large-scale coherent vortex.

The outflow resembles to the ejection for the natural small-scale coherent motion in the inner layer (e.g Willmarth & Lu 1972). The weak clockwise circulating motion existing for $Y/\delta > 0.8$ shows the head of the artificial vortex. The head reached the outer edge of the superlayer at about $X/\delta \approx 12$ downstream of the slots and it became a turbulent bulge (Makita et al. 1988). The momentum in the free-stream is entrained by such fluid motion. The outward motion of the upstream border (back) shows so-called the 'large-scale (δ-scale) u-discontinuity' as observed by Brown & Thomas (1977). In Figs. 2b-e, the magnitude and upward angles of the arrows showing low-speed outflows decrease downstream for about $X/\delta > 2.5$ It shows that the induced velocity decays downstream and the $\langle v \rangle$-component decays more rapidly. In more downstream region of $X/\delta > 12$, the artificial vortex becomes gradually fragile and more inclined forward as also reported by Antonia & Bisset (1991), keeping its size almost unaltered; see Figs. 2d-e. The convection velocity is known to vary with the portion of the vortex. Judging from the velocity fields, the artificial vortex kept its coherency until about $X/\delta = 45$.

Figures 3a-e present the contour maps of the spanwise vorticity component, $\langle \omega_z \rangle$, on the symmetry plane of the artificial vortex ($Z/\delta = 0$). If the artificial vortex has a horseshoe configuration, it is shown by the distribution of clockwise (negative) vorticity on its symmetry plane. At $X/\delta = 0.5$, the vorticity distributes in a quite complicated

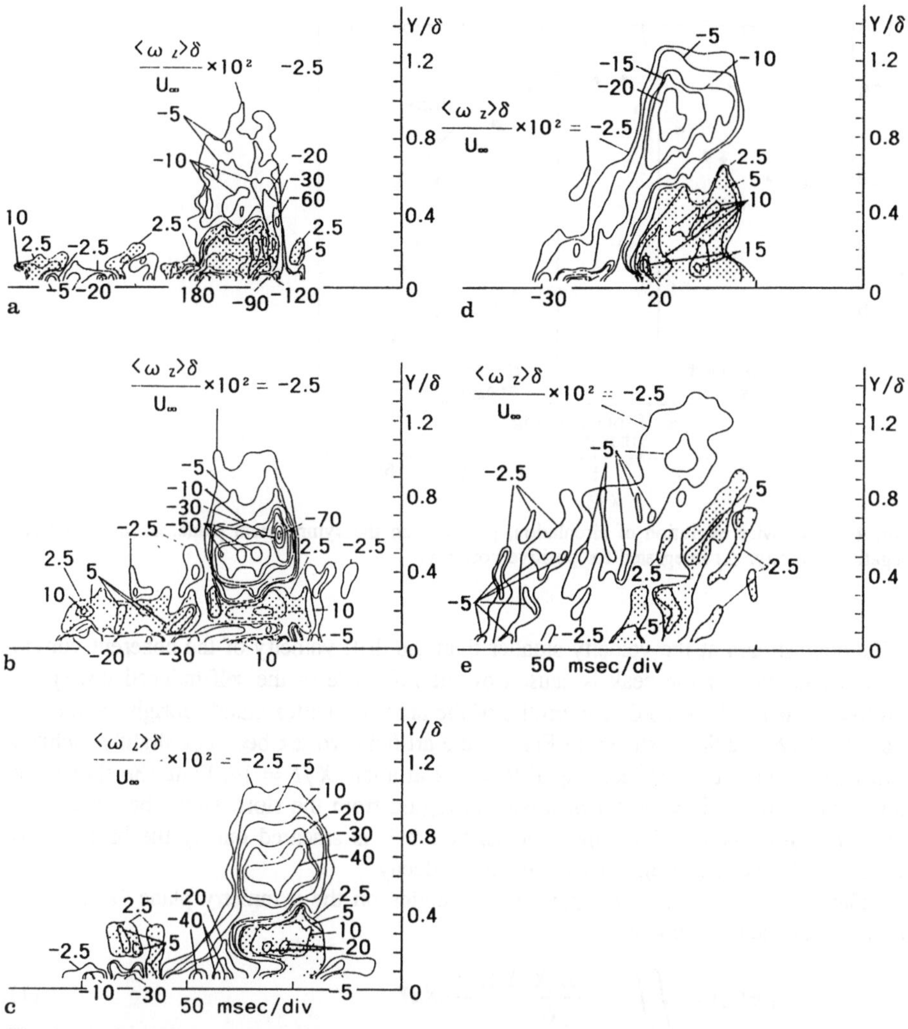

Fig. 3a-e. Contour maps of spanwise vorticity, ω_z, on symmetrical T-Y plane ($Z/\delta = 0$) at several streamwise locations. a: $X/\delta = 0.5$, b: $X/\delta = 2.5$, c: $X/\delta = 5$, d: $X/\delta = 14$, e: $X/\delta = 35$. Hatched area denotes $+\omega_z$ distribution.

manner maintaining the effect of the injection of the swirling jet. Roughly speaking, $-\langle\omega_z\rangle$ distributes upper side of the structure of about $Y/\delta > 0.4$ and $+\langle\omega_z\rangle$ is observed for about $Y/\delta < 0.4$. Such features of the distributions tell us that the induced vorticity has not yet attained the self-similar configuration of the horseshoe vortex there. The upper area of $-\langle\omega_z\rangle$ grows to be the head of a horseshoe vortex as it flows downstream. The lower side $+\langle\omega_z\rangle$ may be a secondary vortex induced between the head of the vortex and the flat plate as suggested by Jiménez et al. (1988). The peak point of $-\langle\omega_z\rangle$ shifts upstream in the artificial vortex with the outward growth of the vortex downstream; see Figs. 3a-d. This fact says that the vorticity peak in the large-scale coherent vortex

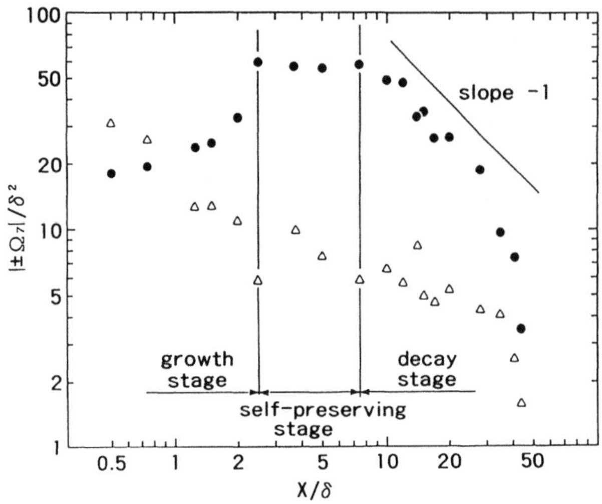

Fig. 4. Streamwise evolution of circulation, $|\pm\omega_Z|$, on the symmetrical plane. •: integrated for negative vorticity, △: integrated for positive vorticity.

shifts downstream at the velocity smaller than the drift velocity of the coherent vortex. This deceleration of the peak is caused by the influence of the self-induced low-speed outflow shown in Figs. 2b-d. The profile of the artificial vortex stands upright on the flat plate for $X/\delta \leq 2.5$. As shown in Fig. 3d, the artificial vortex becomes slightly inclined downstream with its head reaching $Y/\delta \approx 1.2$ at about $X/\delta = 14$. Figure 3e shows the contour map of $X/\delta = 35$; far downstream region from the slots. Here, the large-scale coherent vortex is split into small blocks before it is smeared out by the background turbulence in the final stage of its streamwise decay.

The positive and negative spanwise vorticities on the symmetry plane ($Z/\delta = 0$) were independently integrated as

$$|\pm\Omega_Z| = \iint \frac{|\pm\langle\omega_z(X,Y)\rangle|\,\delta}{U_\infty} dXdY \tag{1}$$

where Ω_Z denotes circulation on the symmetry plane. The streamwise integration was converted from the timewise one. The streamwise change in $|\pm\Omega_Z|$ is plotted in Fig. 4. It must be noticed that $|-\Omega_Z|$ increases in almost proportion to X/δ in the initial growth stage of about $X/\delta \leq 2.5$. The seed of artificial vortex seems to be fostered to be a grown-up coherent vortex having naturally receptible structure in the turbulent boundary layer by the velocity gradient, as it travels downstream in the region. It is inferred that the artificial vortex becomes able to act as a natural large-scale vortex before $X/\delta = 2.5$. In the self-preserving stage of about $X/\delta = 2.5 - 7.5$, the height of the coherent vortex gradually increases downstream but $|-\Omega_Z|$ is kept almost unaltered. In the final decay stage for $X/\delta \geq 7.5$, $|-\Omega_Z|$ decays monotonously downstream at the slope of about -1.

The convection velocity, U_C, of the large-scale vortex was estimated from the streamwise shift of the vorticity peak in the head. It was about $0.56\,U_\infty$ in the growth and the self-preserving stages. In the decay stage, U_C was raised to $0.85\,U_\infty$ which was almost equal to the local mean velocity, U, at the peak position.

Good coherency in vortical structure of the large-scale vortex is maintained at most until the beginning of the decay stage. There, the large-scale vortex is known to play some active roles in the momentum and energy transport across the turbulent boundary layer. In the decaying stage, the large-scale vortex seems to become merely conveyed downstream on the local mean velocity as a weak structure having only passive functions for the transport process. The decay stage is the longest in the total lifetime of the artificial vortex. Not the smaller part of the natural turbulent bulges ever observed in boundary layers are, we guess, such passive structures. The large-scale coherent structure is weakly organic, except for younger one, and does not play important rolls in dynamics of the transport mechanism in a turbulent boundary layer.

At the beginning of the initial growth stage, $|+\Omega_Z|$ apparently exceeds $|-\Omega_Z|$. This may be because some excessive positive circulation is induced by some extra initial disturbances, such as the mass effect of the injected puffs and the backward motion of the speaker pump in addition to the direct secondary induction by the vortex. $|+\Omega_Z|$ decreases monotonously downstream to be smaller than the negative for $X/\delta \geq 1.25$. The fact implies that it is difficult for the excessive positive vorticity given to the counter clockwise vortex to persist so long in the turbulent boundary layer.

4.2 Three-Dimensional Structure of the Large-Scale Vortex

As mentioned before, the artificial vortex has only passive structure in the decay stage. Most of the natural large-scale vortices ever observed are, however, those in the decay stage. To show the three-dimensional structure of the artificial vortex, the distributions of vorticity components at $X/\delta = 14$ are illustrated with respect to $\langle \omega_Z \rangle$, $\langle \omega_X \rangle$ on the T-Y planes in Figs. 5, 6 and to $\langle \omega_Y \rangle$ on the T-Z planes in Fig. 7. Though it is already at the beginning of the decay stage, the coherent vortex still holds the typical self-consistent structure of a horseshoe vortex so well. The distribution of $-\langle \omega_Z \rangle$ in Fig. 5 shows the head of the horseshoe vortex on the symmetry plane. Outside of the artificial vortex at $Z/\delta = 0.6$, the island of $+\langle \omega_Z \rangle$ seems to be a dominant structure, which shows a neighbouring secondary vortex induced by the original artificial one.

Different from the case of $\langle \omega_Z \rangle$, the distributions of $+\langle \omega_X \rangle$ and $+\langle \omega_Y \rangle$ locate apart from the symmetry plane as shown in Figs. 6 and 7, respectively. These distributions show the existence of a pair of the vortex tube inclined downstream, constituting the legs of the horseshoe vortex. The rate of contribution of three vorticity components to the total vorticity contained in the coherent vortex is nearly the same among them. Antonia & Bisset (1991) obtained the arrangement of the head and the legs similar to that presented here for the natural large-scale vortex.

Three-dimensional distributions of vorticity vectors and ensemble averaged enstrophy, $|\omega|^2 = \langle \omega_X \rangle^2 + \langle \omega_Y \rangle^2 + \langle \omega_Z \rangle^2$, obtained from three vorticity components are shown in Figs. 8 and 9, respectively. The vorticity vectors show a half view of the horseshoe configuration of the large-scale vortex cut by the symmetry plane. Kim (1987) obtained a similar configuration for a large-scale vortex in a parallel pipe flow by the numerical simulation. The axes of the vortex tubes are located at $Z/\delta = \pm 0.2$ and inclined forward at the angle of about 50 deg. to the flat plate. The inclination angle is close to the value given by Head & Bandyopadhyay (1981). But, the coherent vortex becomes more inclined in more downstream region, as described before. The distribution of $|\omega|^2$ shows that the

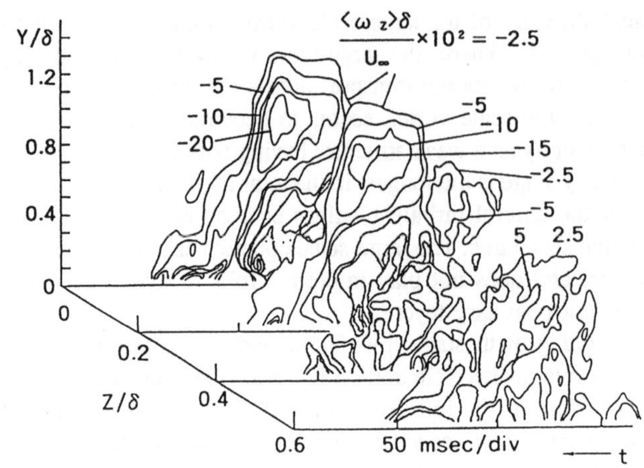

Fig. 5. Three-dimensional distribution of spanwise vorticity, ω_z, at $X/\delta = 14$. Hatched area denotes $+\omega_z$ distribution.

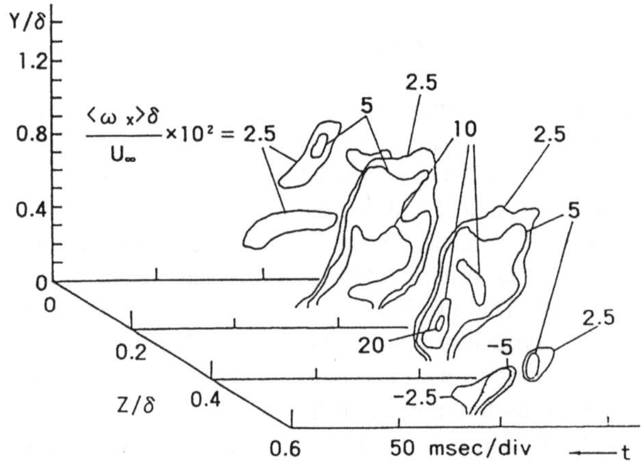

Fig. 6. Three-dimensional distribution of streamwise vorticity, ω_x, at $X/\delta = 14$. Hatched area denotes $-\omega_x$ distribution.

strength of the vortical motion is highest at the leg. The large-scale coherent structure was apt to be grasped as a spanwise vortex, as observed by Praturi & Brodkey (1978). The present result shows that the dominant part of the vortical motion exists not at the head shown by the spanwise vortex but at the legs of the horseshoe sculpture.

4.3 Relation Between the Vortex and the Other Coherent Structures

We propose a model of the large-scale coherent vortex as illustrated in Fig. 10. It has a horseshoe sculpture. The scales of the horseshoe vortex determined from the vorticity

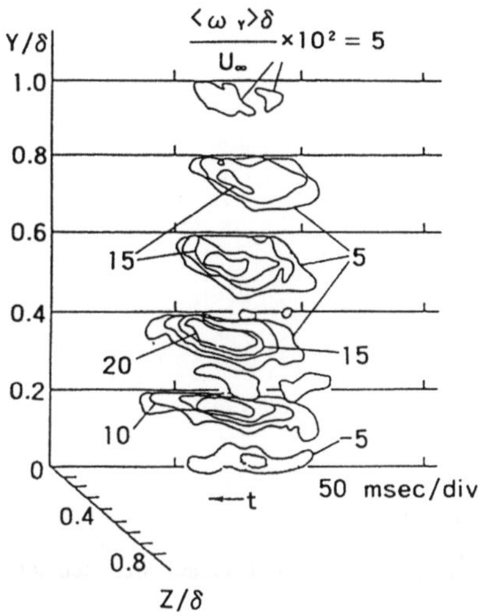

Fig. 7. Three-dimensional distribution of normal vorticity, ω_Y, at $X/\delta = 14$. Hatched area denotes $-\omega_Y$ distribution.

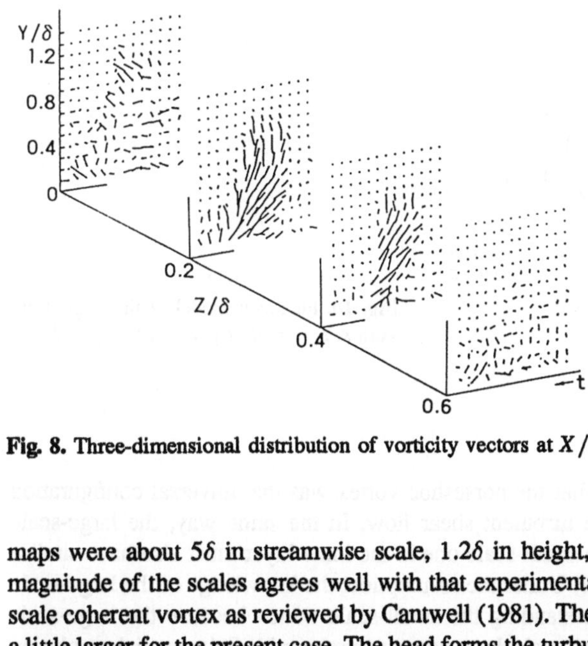

Fig. 8. Three-dimensional distribution of vorticity vectors at $X/\delta = 14$.

maps were about 5δ in streamwise scale, 1.2δ in height, and δ in horizontal span. The magnitude of the scales agrees well with that experimentally given for the natural large-scale coherent vortex as reviewed by Cantwell (1981). The longitudinal scale is, however, a little larger for the present case. The head forms the turbulent bulge in the superlayer and the fluid motion on the upstream side (the back) causes the u-discontinuity, as described in Section 1. The large-scale vortex is known to induce the coherent motions, such as the outflow around the symmetry plane as shown in Fig. 2d and the 'inrush' in the outside part of the legs for $Z/\delta > 0.4$ as shown in the velocity fields at $Z/\delta = 0.6$ in Fig. 11. The inrush representing the inward motion of high-speed fluid occurs in the layer lower than the outflow. The Reynolds stress is produced around the outflow and the inrush (Makita et al. 1989).

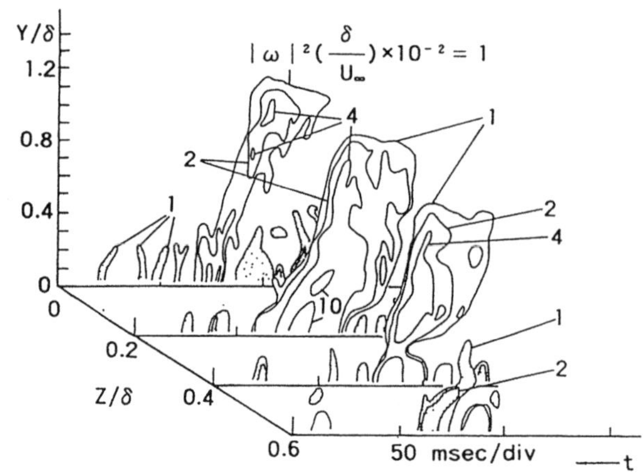

Fig. 9. Three-dimensional distribution of enstrophy, $|\omega|^2$, at $X/\delta = 14$. Hatched area denotes secondary induced vorticity.

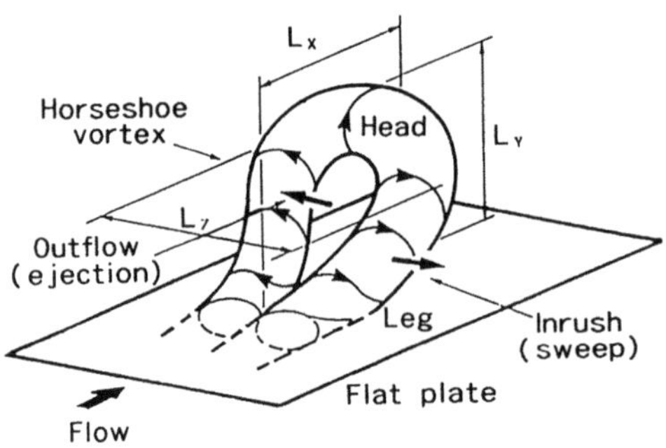

Fig. 10. Idealized model of the large-scale vortex. $L_X = 5\delta$, $L_Y = 1.2\delta$, $L_Z = \delta$.

Moin (1989) demonstrated that the horseshoe vortex was the universal configuration of the coherent structure in the turbulent shear flow. In the same way, the large-scale vortex in the outer layer and the small-scale one in the inner layer must also have similar vortical structure with each other. Under the assumption, the knowledge of the large-scale vortex may be helpful for understanding the small-scale vortex because the large-scale vortex can be measured more minutely by a hot-wire probe with finite spatial resolution than the small-scale one. From the present experimental results, the ejection and the sweep around the small-scale coherent vortex is analogous to the low-speed outflow and the high-speed inrush of the large-scale horseshoe vortex as follows. The ejection occurs around symmetry plane between the two legs, and the sweep does on the outside of the legs. Such arrangement of the ejection and the sweep around the coherent vortex was also obtained numerically by Robinson et al. (1989). The facts give good similarity between the coherent motions in the outer and the inner layers.

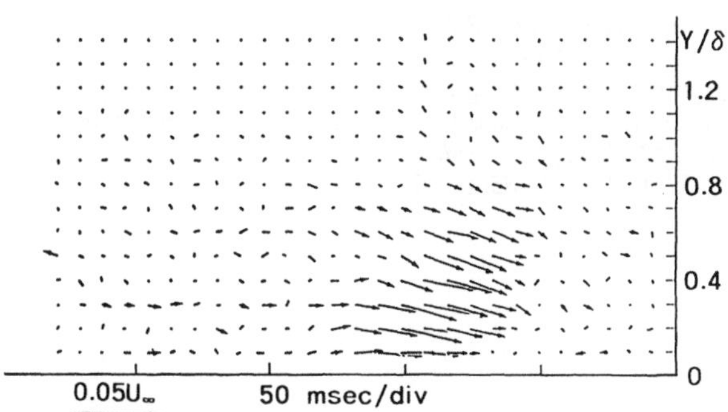

Fig. 11. Induced velocity vectors around the artificial vortex on the T-Y plane at $Z/\delta = 0.6$ at $X/\delta = 14$. The magnitude of arrows are same as those of Figs. 2.

5. Conclusion

The three-dimensional vortical structure of the large-scale coherent vortex in a fully developed turbulent boundary layer was determined. The coherent vortex had a horseshoe-like sculpture. The turbulent bulge in the superlayer corresponds to its head and the large-scale u-discontinuity shows the fluid motion on its back.

Though the generation mechanisms of the large-scale vortex could not be made clear, the aspects of the streamwise development of the vortical structure were understood throughout its growth, self-preserving, and decay stages in detail. The large-scale horseshoe vortex kept its coherency until the end of the self-preserving stage drifting at the convection velocity of $0.56\,U_\infty$, but it came to flow downstream on the local mean velocity of $0.85\,U_\infty$ and then dispersed into small blocks in the decay stage.

References

Antonia, R.A. & Bisset, D.K. (1991): Turbulence and Coherent structures, 141, Kluwer Academic.
Blackwelder, R.F. (1988): Transport Phenomena in Turbulent Flows, 89, Hemisphere Pub.
Brown, G.L. & Thomas, A.S.W. (1977): Phys. Fluids **20**, s243.
Cantwell, B.J. (1981): Ann. Rev. Fluid Mech. **13**, 457.
Eckelmann, H., Nychas, S.G. Brodkey, R.S., & Wallace, J.M. (1977): Phys. Fluids **20(10-2)**, 5225.
Fukunishi, Y. & Sato, H. (1987): Fluid Dyn. Res. **2(2)**, 113.
Fukunishi, Y, Sato, H., & Inoue, O. (1987): AIAA Pap. **87-1253**.
Gad-el-Hak, M. & Blackwelder, R.F. (1987): AIAA J. **25**, 1207.
Gad-el-Hak, M. & Hussain, A.K.M.F. (1986): Phys. Fluids **29**, 2124.
Head, M.R. & Bandyopadhyay, P. (1981): J. Fluid Mech. **107**, 297.
Jiménez, J., Moin, P., Moser, R., & Keefe, L. (1988): Phys. Fluids **31**, 1311.
Kim, J., (1987): Turbulent Shear Flows **5**, 221, Springer-Verlag.
Kline, S.J. & Robinson, S.K. (1989): Near-Wall Turbulence, 200, Hemisphere Pub.
Kovasznay, L.S.G., Kibens, V. & Blackwelder, R.F. (1970): J. Fluids Mech. **41**, 283.
Makita, H., Sassa, K. & Abe, M. (1988): Frontiers of Fluid Mechanics, 104. Pergamon Press.
Makita, H., Sassa, K., Abe, M. & Itabashi, A. (1989): AIAA J. **27(2)**, 155.

Moin, P. (1989): Near-Wall Turbulence, 2, Hemisphere Pub.
Praturi, A.K. & Brodkey, R.S: (1978): J. Fluid Mech. **89**, 251.
Robinson, S.K. (1990): Structure of Turbulence and Drag Reduction, 23, Springer-Verlag.
Robinson, S.K., Kline, S.J., & Spalart, P.R. (1989): Near-Wall Turbulence, 218, Hemisphere Pub.
Viets, S., Piatt, M., & Ball, M. (1981): AIAA Pap. **81-0256**.
Willmarth, W.W. & Lu, S.S. (1972): J. Fluid Mech. **55(1)**, 65.

On the Origin of Streaks in Turbulent Shear Flows

Fabian Waleffe, John Kim and James M. Hamilton

Center for Turbulence Research, NASA Ames Research Center, Moffett Field, CA 94035, USA

Abstract

It is shown that the ideas of *selective amplification* and *direct resonance*, based on linear theory, do not provide a selection mechanism for the well-defined streak spacing of about 100 wall units (referred to as 100^+ hereafter) observed in wall-bounded turbulent shear flows. For the direct resonance theory (Benney & Gustavsson, 1981; Jang et al., 1986), it is shown that the streaks are created by the nonlinear self-interaction of the vertical velocity rather than of the *directly forced* vertical vorticity. It is then proposed that the selection mechanism must be inherently nonlinear and correspond to a *self-sustaining* process. The streak formation is only one stage of the complete mechanism and cannot be isolated from the rest of the process. The 100^+ value should be considered as a critical Reynolds number for that self-sustaining mechanism. For the case of plane Poiseuille flow the 100^+ criterion corresponds to a critical Reynolds number of 1250, based on the centerline velocity and the channel half-width, which is close to the usually quoted value of about 1000. In plane Couette flow, it corresponds to a critical Reynolds number of 625, based on the half velocity difference and the half-width.

1. Introduction

An intriguing feature of wall-bounded turbulent flows is the existence of bands of low- and high-speed fluid, elongated in the streamwise direction, and with a very consistent spanwise spacing of about one hundred wall units, i.e. $100\nu/u_*$, where $u_* = (\nu dU/dy|_w)^{1/2}$ is the friction velocity. The streaks are known to initiate a localized instability that leads to a "bursting process," during which most of the turbulence production takes place (Kline *et al.*, 1967; Kim *et al.*, 1971). Large momentum is exchanged between the wall and the outer fluid during the bursting process, thus sustaining the turbulent flow.

It is now fairly well accepted that the mechanism for streak generation is a rapid one, meaning that the linear distortion of fluctuations by the mean shear strongly dominates the nonlinear effects (Lee *et al.*, 1990). The physical process is a redistribution of downstream momentum by cross-stream motions (downstream rolls) which are decoupled from the mean flow. The important question remaining is to determine what imposes the characteristic streak spacing. A natural suggestion is that the linear mechanism itself leads to a favored scale (selective amplification). The inner-layer of a typical turbulent boundary layer being about 50 wall units thick, the scale of the largest downstream roll would be also around 50^+ and these would induce streaks with a spanwise wavelength around 100^+. The flaw in that reasoning is that it rests on an intuitive, yet false, assumption that rolls are necessarily circular while in fact they can be elliptical. Our analysis indicates that while the linear mechanism provides a scale selection, it is much too weak to be significant and does not correspond to the value of 100^+. The weak scale selection shows up in the vertical vorticity spectra but not in the downstream velocity spectra. In a turbulent boundary layer a peak appears in both spectra.

In the direct resonance theory (Gustavsson 1981; Benney and Gustavsson 1981), oblique vertical vorticity modes are forced by the vertical velocity. The physical process is entirely similar to that for the streak generation mentioned above and consists of a redistribution of momentum in oblique planes by oblique rolls. Because of the obliqueness, the two types of fluctuations (oblique rolls and oblique streaks, or in mathematical terms vertical velocity and vorticity) are, in general, influenced by the mean flow in two different ways. The small vertical velocity fluctuations are governed by the Orr-Sommerfeld equation, while the vertical vorticity fluctuations obey an advection-diffusion equation. The argument is that the forcing of the vertical vorticity by the velocity should be most effective when the eigenvalue of the velocity mode is identical to an eigenvalue of the homogeneous vorticity equation (a so-called direct resonance). The resonance condition is only satisfied for well-defined streamwise (x) and spanwise (z) scales, thus providing a scale selection. If the "resonance" condition is met, the vorticity is expected to reach "large" values. The nonlinear self-interaction of the "large" vertical vorticity then gives rise to downstream rolls (downstream vertical velocity) as shown by Jang, Benney and Gran (1986), and these rolls lead to streaks as explained in the previous paragraph. Unfortunately, our analysis does not support this approach either. The direct resonance criterion does not necessarily select the most amplified vorticity fluctuations. This is because the "resonance" always occurs for damped modes. So the maximum vorticity amplitude that can be obtained is a function of the damping rate, and there are non-resonant modes with a lower damping rate which can reach larger amplitudes. In fact, the largest vorticity amplitudes are obtained for downstream modes, as reported recently by other authors also (Gustavsson 1991; Henningsson 1990), but, as discussed in the previous paragraph, the downstream fluctuations do not yield a significant scale selection. So this approach can not explain the streak spacing either. In any case, our analysis also indicates that the generation of downstream rolls is a result of the nonlinear interaction of the oblique vertical velocities and not of the vorticities, thus bypassing completely the "resonant" amplification of the oblique streaks.

The final part of this paper presents an introduction to ongoing research aimed at providing an understanding of the scale selection and the mechanisms taking place in the near-wall region of turbulent boundary layers. We propose that the streak spacing should be considered as a critical Reynolds number for a self-sustaining nonlinear process, of which streak formation is one of the elements. The nonlinear process would not *persist* for scales lower than about 100^+. This is the case, of course, if the largest scale allowed in the domain is smaller than 100^+. Therefore a link is established between the 100^+ characteristic streak spacing and the critical Reynolds numbers, above which turbulence can be maintained. The 100^+ criterion has the potential to be a more universal value for shear flows than other measures of the critical Reynolds number. It translates into a value of 1250 based on the channel half-width and the centerline velocity in plane Poiseuille flow, and 625 based on the half width and half velocity difference in plane Couette flow (or 1000 and 400 respectively, if the minimum streak spacing is 80^+). The self-sustaining nonlinear process in question would consist of the following elements. First, streaks are created by downstream rolls, the streaks then break down due to an instability of inflexional type initiated by the spanwise inflexions (the vertical shear might significantly influence the nature of that instability). The streak instability leads to the formation of vortices which reinforce the original downstream rolls, and the process repeats itself provided the scale is larger than 100^+; otherwise it loses intensity and eventually decays. The complete process is expected to be itself unstable, or at least "broadband", so that the flow appears disordered if the largest admissible scale is sufficiently bigger than 100^+. In order to capture the process in its simplest form it is necessary to experiment in a domain whose largest scale is close to 100^+.

The idea of selective amplication by the linear nechanism is examined in section 2. The direct resonance theory of Jang et al. (1986) is reexamined in section 3. The self-sustaining mechanism for marginal turbulent flows is presented in section 4, followed by a short summary in section 5.

In this paper, x, y, and z denote streamwise, normal (to the wall), and spanwise directions, respectively, while u, v, and w denote corresponding velocities, respectively.

2. Linear Analysis and Selective Amplification

A number of papers (e.g. Lee *et al.*, 1990) show that the mechanism for streak generation is linear. The argument is that in the near wall-region the time scale for the mean $(dU/dy)^{-1}$ is much shorter than the time scale for the nonlinear effects, measured by q^2/ϵ, where q is a turbulent velocity scale and ϵ is the dissipation rate. The evolution is then dictated by linear equations and streaks are created from the redistribution of the downstream momentum by vertical and spanwise motions. The mechanism is a simple advection and is most efficient when the fluctuating fields are elongated downstream. The question here is to examine whether the linear mechanism favors spanwise scales of about 100^+. The mathematical description of the mechanism is briefly stated in the next few paragraphs.

The governing equations for the fluctuating field, obtained by eliminating pressure and the continuity constraint, are:

$$(\frac{\partial}{\partial t} + U\frac{\partial}{\partial x} - \frac{1}{R}\nabla^2)\nabla^2 v - \frac{d^2U}{dy^2}\frac{\partial}{\partial x}v = NL_v \tag{1}$$

$$(\frac{\partial}{\partial t} + U\frac{\partial}{\partial x} - \frac{1}{R}\nabla^2)\eta + \frac{dU}{dy}\frac{\partial}{\partial z}v = NL_\eta \tag{2}$$

where v and η denote respectively the y-component of velocity and vorticity, and U is the mean velocity. The right-hand sides, NL_v, NL_η, represent nonlinear terms. Flows in a channel will be considered in this paper (plane Poiseuille or plane Couette flow) with the boundary conditions $v = \partial v/\partial y = \eta = 0$ at the walls, located at $y = \pm 1$.

In the linear case ($NL_v = 0$), the equation for v is homogeneous and admits eigen-solutions of the form:

$$v = \hat{v}(y)e^{i(\alpha x+\beta z-\omega t)} \tag{3}$$

where $\hat{v}(y)$ satisfies the *Orr-Sommerfeld* equation. In general, for a turbulent mean profile, all of these eigensolutions are decaying. The η-equation on the other hand, is non-homogeneous for v fluctuations with a spanwise variation. When forced by an eigenmode of the v equation, the linear response of the vertical vorticity has the form:

$$\eta = \hat{\eta}(y,t)e^{i(\alpha x+\beta z)} \tag{4}$$

with $\hat{\eta}(y,t)$ given by:

$$\hat{\eta}(y,t) = \beta \sum_n \lambda_n \frac{e^{-i\omega t} - e^{-i\mu_n t}}{\omega - \mu_n} \eta_n(y) \tag{5}$$

Where

$$\lambda_n = \frac{\int U' \hat{v}\, \eta_n\, dy}{\int \eta_n \eta_n dy} \tag{6}$$

and $\eta_n(y)$, μ_n represent the eigenmodes and eigenvalues of the homogeneous η-equation. Note that the vertical vorticity response corresponds to streaks, as opposed to vortices. The physical mechanism is a redistribution of momentum or vorticity *tilting*, not *stretching*.

When the OS eigenvalue ω is close to the Squire eigenvalue μ_n, the n-th coefficient will behave as $t\exp(-i\mu_n t)$. This corresponds to an algebraic growth followed by exponential decay, as μ_n corresponds to a viscously decaying mode. For a significant algebraic growth to occur the real parts of ω and μ must be sufficiently close. Otherwise the modes will decorrelate. Furthermore, the viscous damping must be small. Thus one expects and verifies numerically that the largest responses occur for downstream modes ($\alpha = 0$) for which the real parts of the eigenvalues vanish (downstream modes do not feel the mean flow) and the decay rate is inversely proportional to the Reynolds number.

The possibility of a scale selection by the linear mechanism was investigated numerically, by introducing a downstream OS mode, normalized such that the maximum vertical velocity was unity, onto a turbulent mean profile with $R = 180$, based on the friction velocity and the channel half-width. The mean velocity profile was chosen as a Reynolds-Tiederman profile, defined by

$$\frac{dU}{dy} = -\frac{Ry}{(1+\nu_t)}$$

$$\nu_t = \frac{1}{2}\left\{1 + \left[\frac{1}{3}KR\left(1-y^2\right)\left(1+2y^2\right)\right.\right.$$

$$\left.\left.\left(1-\exp\left(-\left(1-|y|\right)R/A\right)\right)\right]^2\right\}^{1/2} - \frac{1}{2}$$

where K and A were chosen respectively as 0.525 and 37.

The forced responses of the vertical vorticity are shown in Fig.1. It can be seen that there is a peak around $\lambda_z^+ = 35$, but it is too weak to represent a significant selection. The

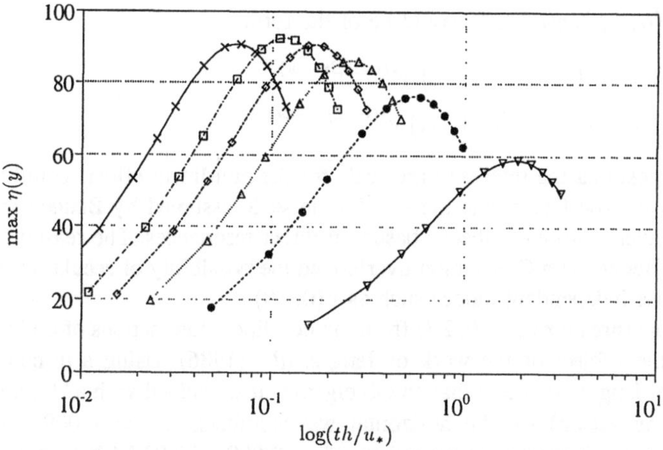

Fig. 1. Maximum pointwise vertical vorticity response to a single downstream even OS mode. ∇, $\lambda_z^+ = 188$; $\bullet$, $\lambda_z^+ = 94$; $\triangle$, $\lambda_z^+ = 63$; $\diamond$, $\lambda_z^+ = 47$; $\square$, $\lambda_z^+ = 38$; $\times$, $\lambda_z^+ = 27$.

streamwise *velocity* responses is obtained by multiplying the vorticity by the wavelength and the largest response corresponds to the largest wavelength. This does not match the experimental observations, which show a scale selection both in the velocity and vorticity spectra. In any case, the "peak" does not correspond to the typical value for the streak spacing, which is between 80^+ and 100^+. We must conclude that the linear mechanism does not provide a scale selection.

3. Direct Resonance Theory

In the direct resonance scenario, streaks originate from a three-step process (Benney and Gustavsson, 1981; Jang *et al.*, 1986). The first step is linear and consists of the resonant forcing of the vertical vorticity by the velocity, exactly as in the previous section, but focuses on oblique disturbances, for which the nonlinear effects can be less trivial. The second step is the nonlinear interaction of the vorticity with its mirror image across a vertical downstream plane. This would create downstream vortices which, finally, give rise to the streaks. That sequence of interactions is illustrated by the following diagram,

$$\begin{aligned} \epsilon\, v(\alpha, \pm\beta) &\longrightarrow \epsilon\, t\, \eta(\alpha, \pm\beta) \\ \epsilon^2 t^2\, \eta\eta^* &\longrightarrow \epsilon^2 t^3\, v(0, 2\beta) \\ \epsilon^2 t^3\, v(0, 2\beta) &\longrightarrow \epsilon^2 t^4\, \eta(0, 2\beta) \end{aligned}$$

One can imagine that this would then be completed by the non-linear interaction of the downstream streaks ($\eta(0, 2\beta)$) with the oblique ones ($\eta(\alpha, \pm\beta)$) to generate $v(\alpha, \pm\beta)$ which then would recreate the oblique streaks and so on,

$$\epsilon^3 t^5\, \eta(\alpha, \pm\beta)\eta(0, 2\beta) \longrightarrow \epsilon^3 t^6 v(\alpha, \pm\beta) \longrightarrow \epsilon^3 t^7 \eta(\alpha, \pm\beta)$$

The expansions for $v(\alpha, \pm\beta), \eta(\alpha, \pm\beta)$ would be of the form

$$\begin{aligned} v(\alpha, \pm\beta) &= \epsilon \ \left(1 + \epsilon^2 t^6 + \ldots\right) \ e^{i(\alpha x \pm \beta z - \omega t)} \\ \eta(\alpha, \pm\beta) &= \epsilon t \ \left(1 + \epsilon^2 t^6 + \ldots\right) \ e^{i(\alpha x \pm \beta z - \omega t)} \end{aligned}$$

These expansions suggest that the relevant time scale for the non-linear effects could be as fast as $\epsilon^{-1/3}$. This is even faster than the $\epsilon^{-1/2}$ time scale assumed by Benney and Gustavsson (1981) in their non-linear theory based on direct resonances. The reason for this difference is that Benney and Gustavsson overlooked the possibility of secular terms arising from streamwise independent modes such as $v(0, 2\beta)$.

The creation of downstream rolls $v(0, 2\beta)$ from the non-linear interactions of oblique streaks $\eta(\alpha, \pm\beta)$ is the subject of the work of Jang *et al.* (1986). Using a turbulent boundary layer profile, Jang *et al.* found that an OS eigenvalue coincided with a "Squire" eigenvalue (a *direct resonance*) for the horizontal wavenumbers, $\alpha^+ = 0.0093$ and $\beta^+ = 0.035$. The common eigenvalue was equal to $\omega^+ = 0.090 - i\,0.037$. They showed that the interaction of that mode with its spanwise reflexion induced streamwise vortices with a spanwise wavelength around $\lambda^+ = 2\pi/\beta^+ \simeq 90^+$.

One problem with the direct resonance concept is that it must assume not only that the eigenvalues of the two linear homogeneous operators are close, but also that the damping rates are small. Otherwise, as discussed above, the algebraic growth will be quickly shut off by the exponential viscous decay. Another problem comes from multiple resonances or near-resonances. There is no a priori criterion for which one should prevail.

The possibility of the direct resonance route to streaks has been explored in the turbulent channel flow (Kim *et al.*, 1987). The Reynolds number is about 180, based on the channel half-width and the friction velocity. A direct resonance is found for

$$\alpha^+ \simeq 0.005 \qquad \beta^+ \simeq 0.039$$

with the corresponding eigenvalues

$$\begin{aligned} \omega_{14}^+ &= 0.07871485 - i\,0.02168581 \\ \mu_{15}^+ &= 0.07871461 - i\,0.02168558 \end{aligned}$$

These values are different but close to the values reported above for the turbulent boundary layer. The nonlinear interaction of the pair of modes $(\alpha, \pm\beta)$ leads to streaks with a spanwise spacing of about 80^+. The vertical vorticity responses are displayed in figure 2. The initial conditions were such that the maximum v amplitude was 0.1 with no vertical vorticity. As indicated by the subscripts, this direct resonance occurs at the 14th OS mode and the 15th Squire mode, where the modes are ordered according to their decay rate. Although this result looks encouraging, the picture is not as clear when one analyzes other modes.

For instance the linear response obtained from the 13th OS mode for slightly different wavenumbers is very similar to the direct resonance mode, but the nonlinear response is even larger (Fig.2). The response to the 19th OS mode for $(\alpha, \beta) = (1.6, 12)$ revealed rather confusing results (Fig.3). The linear response is almost four times larger than the direct resonance one, but the nonlinear response is much smaller. One would expect that if the linear response was four times larger than for the direct resonance modes, the nonlinear response should be sixteen times larger than in the direct resonance case.

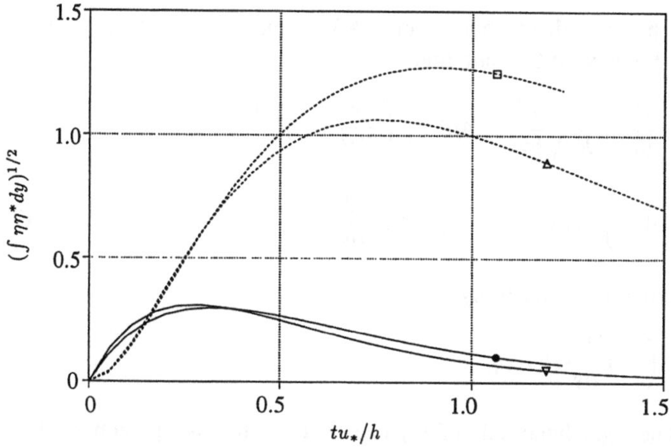

Fig. 2. Linear (solid curves) and non-linear (dashed curves) vertical vorticity response for direct resonance: ∇, $(\alpha, \beta) = (0.9, 7.07)$ (forced by 14th OS mode); △, $(0, 14.14)$; •, $(0.8, 6)$ (forced by 13th OS mode); □, $(0, 12)$.

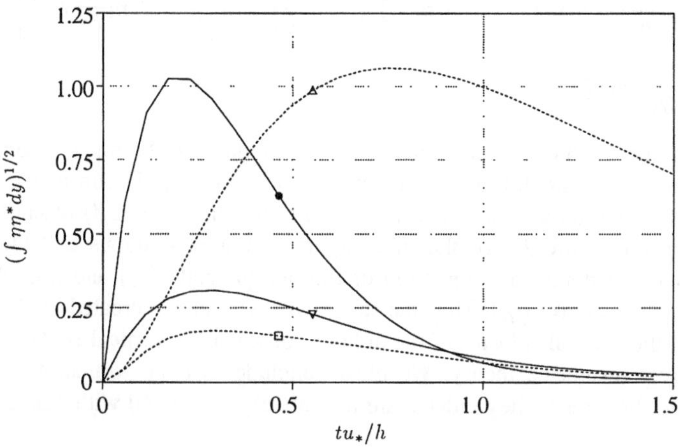

Fig. 3. Linear (solid curves) and nonlinear (dashed curves) vertical vorticity response for direct resonance: ∇, $(\alpha, \beta) = (0.9, 7.07)$ (forced by 14th OS mode); △, $(0, 14.14)$; •, $(1.6, 12)$ (forced by 19th OS mode); □, $(0, 24)$.

To understand what goes on it is necessary to analyze the nonlinear interactions. The computations and the theory are started with a pair of oblique rolls such as:

$$v = \cos \beta z \left(v_1(y,t)e^{i\alpha x} + v_1^*(y,t)e^{-i\alpha x} \right)$$

The linear processes then introduce a pair of oblique streaks:

$$\eta = \sin \beta z \left(\eta_1(y,t)e^{i\alpha x} + \eta_1^*(y,t)e^{-i\alpha x} \right)$$

The nonlinear effect of the primary interest is the generation of streamwise vortices $V(y,t) \cos 2\beta z$. The complete equation for $V(y,t)$ is obtained from (1) with the nonlinear forcing provided by the pair of oblique rolls. After some manipulations (see e.g. Benney, 1961; Lin and Benney, 1962), one finds:

$$\left[\frac{\partial}{\partial t}-\frac{1}{R}\left(\frac{\partial^2}{\partial y^2}-4\beta^2\right)\right]\left(\frac{\partial^2}{\partial y^2}-4\beta^2\right)V=$$

$$\beta\left(\frac{\partial^2}{\partial y^2}+4\beta^2\right)(v_1w_1^*+v_1^*w_1)+4\beta^2\frac{\partial}{\partial y}(v_1v_1^*+w_1w_1^*) \tag{7}$$

where w_1 is given in terms of v_1 and η_1 by

$$w_1=k^{-2}\left(-\beta\frac{\partial v_1}{\partial y}+i\alpha\eta_1\right)$$

with $k^2=\alpha^2+\beta^2$. The right hand side of (7) can be rewritten as the sum of three forcing terms $F_{vv}+F_{v\eta}+F_{\eta\eta}$, where

$$F_{vv} = 4\frac{\alpha^2\beta^2}{k^2}\frac{\partial}{\partial y}(v_1v_1^*)+4\frac{\beta^4}{k^4}\frac{\partial}{\partial y}\left(\frac{\partial v_1}{\partial y}\frac{\partial v_1^*}{\partial y}\right)-\frac{\beta^2}{k^2}\frac{\partial^3}{\partial y^3}(v_1v_1^*)$$

$$F_{v\eta} = \frac{i\alpha\beta}{k^2}\left[\left(\frac{\partial^2}{\partial y^2}+4\beta^2\right)(\eta_1v_1^*-\eta_1^*v_1)+\frac{4\beta^2}{k^2}\frac{\partial}{\partial y}\left(\eta_1^*\frac{\partial}{\partial y}v_1-\eta_1\frac{\partial}{\partial y}v_1^*\right)\right]$$

$$F_{\eta\eta} = \frac{4\alpha^2\beta^2}{k^4}\frac{\partial}{\partial y}(\eta_1\eta_1^*)$$

In the direct resonance theory, the oblique streaks (η_1) are supposed to be much larger than the rolls so that only their nonlinear interaction, i.e. the term $F_{\eta\eta}$, is considered. The three forcing terms for the direct resonance case at the time when $\eta(\alpha,\pm\beta)$ reaches its maximum amplitude (cf. figure 2) are shown in figure 4. Figure 4 shows that the dominant forcing term is the nonlinear interaction of the oblique rolls (v_1) and not of the "directly forced" oblique streaks (η_1). Thus the streamwise vorticity is created by the nonlinear interaction of the vertical *velocity* of the oblique rolls, not the vertical *vorticity* as proposed in the direct resonance theory. We must conclude that the formation of streamwise vortices, and the streaks they induce, are not directly associated with "direct resonances."

4. Marginal Flow

In view of the failure of the linear mechanism to provide a scale selection, we hypothesize that the selection must come from the original disturbance. In a turbulent flow the disturbance seems to arise from the breakdown of the streaks themselves (the bursting process). Thus the 100^+ selection must come from the complete *self-sustaining* mechanism. Our conjecture is that disturbances with a spacing smaller than about 100^+ cannot be maintained. The 100^+ should then be considered as a *critical Reynolds number*. This conjecture is consistent with the work of Jimenez and Moin (1991). They show that for a channel flow at three different Reynolds numbers, $U_ch/\nu=5000$, 3000 and 2000, (U_c

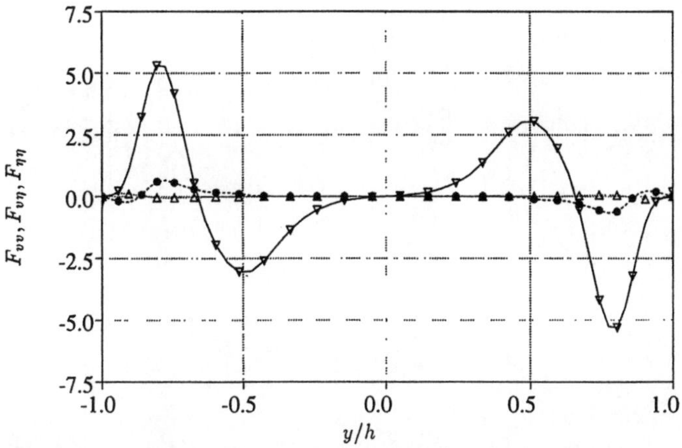

Fig. 4. Forcing terms for the direct resonance mode at time when η is maximum: ∇, F_{vv}; $\bullet$, $F_{v\eta}$; and $\triangle$, $F_{\eta\eta}$.

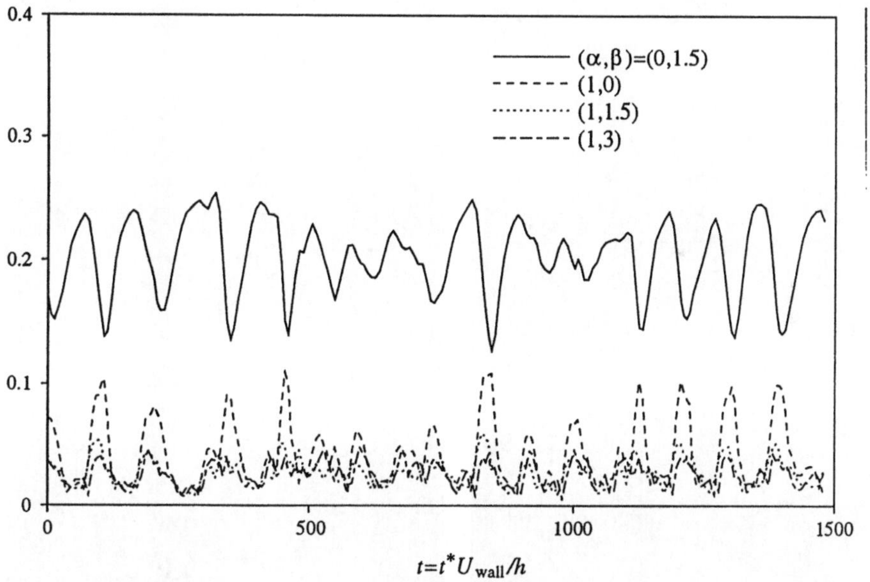

Fig. 5. Time history of the y-averaged energies of various Fourier components in a marginal Couette flow at Re=400.

is the centerline velocity of the equivalent laminar flow, h is the half-height), the flow returns to a laminar state when the spanwise width of the periodic box is reduced below about 100^+.

We observe that for channel width of about 100^+ the channel is very narrow. The characteristic length for the scaling of the disturbance should then be taken as the half width as opposed to the half height. In more general terms, the characteristic length should be taken as the smallest of half the channel height and width. Doing so it turns out that turbulence can not be maintained if $u_* \lambda_z / \nu < 100$, which corresponds to $U_c \lambda_z / (2\nu) <$

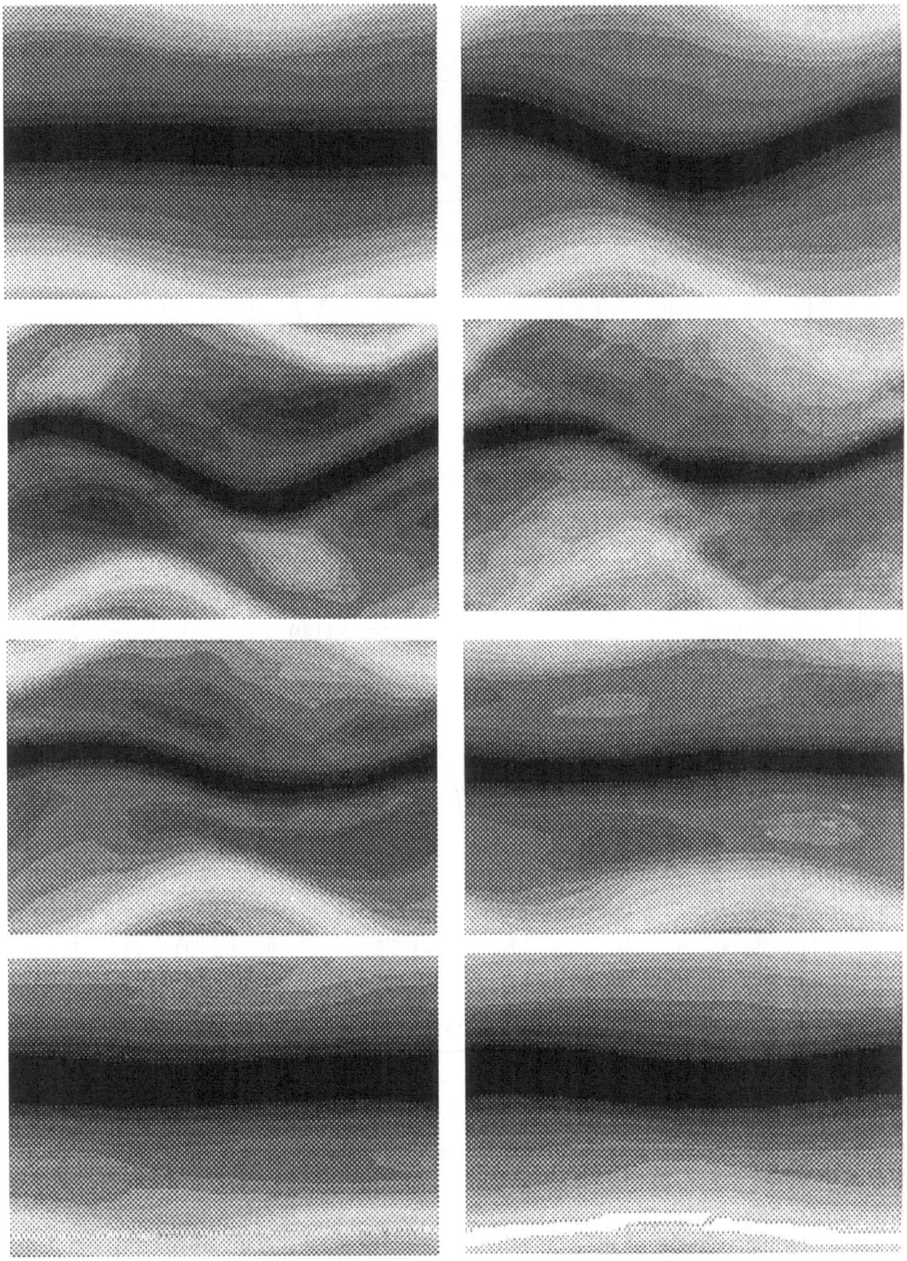

Fig. 6. Contours of streamwise velocity at eight successive times (time is from left to right and from top to bottom) illustrate the near-periodic formation and breakdown of streaks in a marginal plane Couette flow. Contours shown are drawn in the (x, z)-plane in the *middle* of the channel. Top wall moves to the right, bottom one to the left.

1250 ($\lambda_z/2$ is the half-width), irrespective of the value of $U_c h/\nu$. But 1250 is close to the usually quoted value for the critical Reynolds number in channel flow. If one had reduced the height h as opposed to the width λ_z, "turbulence" would have also disappeared when $U_c h/\nu < 1250$ or $u_* h/\nu < 100$.

In order to capture the self-sustaining process in its simplest form, the next step is to reduce both dimensions ($2h$ and λ_z) to their minimum value, so as to eliminate all unnecessary scales. One should be reminded at this point that the streaks are expected to be an essential element of the whole mechanism. The streaks are created from the redistribution of downstream momentum. From the distribution of the mean shear in a channel flow, the simplest self-sustaining non-laminar flow should then consist of a pair of opposite streaks in the spanwise direction and in the direction perpendicular to the walls as well, i.e. one pair of streaks in either half of the channel. In Couette flow, where the mean shear has only one sign, the simplest solution should correspond to only one pair of streaks in the middle of the channel. Thus, the simplest marginal channel flow should have dimensions $u_*\lambda_z/\nu \simeq 100$ and $u_* 2h/\nu \simeq 100$ (corresponding to Re=1250, based on the centerline velocity), while the simplest marginal Couette flow should have $u_*\lambda_z/\nu \simeq 100$ and $u_* 2h/\nu \simeq 50$ (corresponding to Re=625, based on the half velocity difference). According to Smith and Metzler (1983), the smallest value for the streak spacing is rather about 80+, a value which then corresponds to a critical Reynolds number of 400 in Couette flow.

A number of simulations of both flows were performed, and the results support the above reasoning. Turbulent Couette flow, for instance, could not be maintained at Reynolds numbers of 330 and below (based on the half height and half velocity difference) but was maintained for over 2000 convective time units ($2h/2U_w$) at a Reynolds number $U_w h/\nu = 400$. Thus the non-laminar flow was maintained for over 5 viscous units (h^2/ν), a time scale over which the slowest decaying scales, i.e. $u(y,z) = \cos(\pi y/2h)\cos(2\pi z/\lambda_z)z)$ with $\lambda_z = 4h$, would decay by a factor $\exp(-2.5\pi^2) \simeq 1.9\times 10^{-11}$ if they were not sustained. The computed flow fields indicate strong similarity to those obtained in the near-wall region at a higher Reynolds number (e.g. Lee & Kim 1991). The main mechanism appears to be the breakdown of the streaks caused by a spanwise inflectional instability. This is illustrated in Fig.5 which displays a partial time history of the y-averaged energy of the main Fourier components of the flow. The most significant aspect of this plot is the near-periodic behavior on a time-scale of the order $100h/U_w$, which is of the same order as the intermittency cycle observed by Jimenez and Moin (1991). The streaks ($\alpha = 0$) and the other modes ($\alpha \neq 0$) have nearly opposite phases in support of the contention that it is the breakdown of the streaks which energizes these x-dependent modes. The energy of the $\alpha = 0$ modes is entirely dominated by the streamwise component, note then that the ratio of the energies in the streaks to the other modes is of the same order as the ratio of the energies in the streamwise fluctuating velocity to the other components in the near-wall region of a turbulent flow (e.g. Kim *et al.* (1987)). Figure 6 shows contours of streamwise velocity at eight successive times between t=162 and t=262, corresponding to the third cycle on Fig.5.

5. Summary

The streaks, well-defined bands of low- and high-speed fluid present in the near-wall region of turbulent shear flows, are an integral part of the bursting process, by which momentum is exchanged between the wall and the outer fluid. The formation of the streaks is well described by linear theory and corresponds to the rotation of the vertical mean shear by downstream rolls to create a spanwise shear. The linear theory, however, does not provide an explanation for the well-defined scale selection (100^+). The amplitudes of the streaks obtained from the linear equations for a range of scales around the observed 100^+ do not show any significant peak. The direct resonance theory (Jang et al., 1986) is based on the same linear mechanism but for oblique disturbances, whereby oblique rolls create large oblique streaks. A pair of these large oblique streaks then interact nonlinearly to yield downstream rolls, which themselves create downstream streaks. In the direct resonance theory, the scale selection comes from the fact that the creation of oblique streaks by oblique rolls would be more efficient for some particular scales. Unfortunately, this is not found to be the case. In fact the largest responses occur for downstream modes which do not show any significant scale selection as stated above. Furthermore the analysis of the numerical experiments indicates that the downstream rolls are created by the self-interaction of the oblique rolls rather than that of the oblique streaks.

Our conclusion is that the scale selection comes from the complete self-sustaining nonlinear mechanism which consists of the creation, destruction and regeneration of the streaks. The observed characteristic scale of about 100 wall units corresponds to a critical Reynolds number for that process, below which it would not be self-sustaining. This establishes a link between the 100^+ streak spacing and critical Reynolds numbers below which turbulent flows can not be maintained. Numerical investigations support this view. Our goal is to determine the structure of the flow in order to deduce a simple dynamical theory of the whole non-linear process.

References

Benney, D.J. (1961): "A non-linear theory for oscillations in a parallel flow", J. Fluid Mech. Vol. 10, pp. 209-236.

Benney, D.J. & Chow, K. (1989): "A mean flow first harmonic theory for hydrodynamic instabilities", Studies in Applied Math. Vol 80, pp. 37-74.

Benney, D.J. & Gustavsson, L.H. (1981): " A New Mechanism for Linear and Nonlinear Hydrodynamic Stability", Studies in Appl. Math. Vol. 64, pp. 185-209.

Gustavsson, L. H. (1981): "Resonant growth of three dimensional disturbances in plane Poiseuille flow", J. Fluid Mech. Vol. 112, pp. 253-264.

Gustavsson, L. H. (1991): "Energy growth of three dimensional disturbances in plane Poiseuille flow", J. Fluid Mech. Vol. 224, pp. 241-260.

Henningsson, D. S. (1990): "An eigenfunction expansion of localized disturbances", Poster presented at 3rd European Conference, Stockholm.

Jang, P.S., Benney, D.J. & Gran, R.L. (1986): "On the origin of streamwise vortices in a turbulent boundary layer", J. Fluid Mech. Vol. 169, pp. 109-123.

Jimenez, J. & Moin, P. (1991): "The minimum flow unit in near-wall turbulence", J. Fluid Mech. Vol. 225, pp. 213-240.

Kim, J., Moin, P. & Moser, R.D. (1987): "Turbulence statistics in fully developed channel flow at low Reynolds number", J. Fluid Mech. Vol. 162, pp. 339-363.

Kim, H.T., Kline, S.J. & Reynolds, W.C. (1971): "The production of turbulence near a smooth wall in a turbulent boundary layer", J. Fluid Mech. Vol. 50, pp. 133-160.

Kline, S.J., Reynolds, W.C., Schraub, F.A. & Runstadler, P. (1967): "The structure of turbulent boundary layers", J. Fluid Mech. Vol. 30, pp. 741-773.
Lee, M.J., Kim, J. & Moin, P. (1990): "Structure of turbulence at high shear rate", J. Fluid Mech. Vol. 216, pp. 561-583.
Lee, M.J. & Kim, J. (1991): "The structure of turbulence in a simulated plane Couette flow", Proceedings of Eighth Symposium on Turbulent Shear Flows, Munich.
Lin, C.C. & Benney, D.J. (1962): "On the instability of shear flows", AMS, Proc. Symp. Appl. Math. Vol. 13, pp. 1-30.
Smith, C.R. & Metzler, S.P. (1983): "The characteristics of low-speed streaks in the near wall region of a turbulent boundary layer", J. Fluid Mech. Vol. 129, pp. 27-54.

Turbulence in High-Frequency Periodic Fully-Developed Pipe Flow

J.-L. Hwang and G. J. Brereton

Department of Mechanical Engineering and Applied Mechanics, The University of Michigan, Ann Arbor, Michigan 48109, USA.

Abstract

The effects of high frequency organized unsteadiness on wall-bounded turbulence have been studied experimentally at higher frequencies than have been achieved in previous investigations. A detailed examination was made of the contention that non-linear resonant interactions might be found if oscillation were induced at a frequency characteristic of turbulence in the parent boundary layer. It was found that the response of turbulence to forced unsteadiness around and above the burst frequency constituted a monotonic approach to a "frozen state." No resonant behavior was detected. However, when the period of forced oscillation reached a timescale characteristic of the lifetime of a low-speed streak in the sublayer, the streaks underwent some spatial organization with a commensurate reduction in their spanwise meandering motion. This effect is a minor one and does not appear to be a precursor to any resonant behavior.

1. Introduction

While unsteady turbulent flows are of prime importance in large numbers of engineering systems, in which unsteady forcing covers a broad range of frequencies, studies of their response to organized unsteadiness have focused predominantly on low-frequency effects, where information is accessible with relative ease. The purpose of this study is to carry out a detailed investigation of the response of wall-bounded turbulence to forced oscillation at high frequencies, superposed upon a mean flow, in order to address the issue of whether forced oscillation around the burst frequency in the parent boundary layer can cause interaction with the turbulence producing motions of the mean flow.

The motivation for this study is that it has long been speculated that non-linear resonant interactions might be found if oscillation were induced at a frequency characteristic of turbulence in the parent boundary layer. While a number of investigators have reported measurements in the neighbourhood of the burst frequency in unsteady flow experiments, they have, in almost all cases, been restricted to a single data set at a frequency thought to be comparable to the burst frequency – usually the highest attainable in the particular apparatus. A survey of the relevant literature reveals differing opinions on whether such any interaction takes place. The pipe-flow experiments of Ramaprian & Tu (1983) revealed an orderly decrease in amplitude of phase-dependent turbulence intensity with increasing frequency, towards a "frozen" state, as did the channel flow experiments of Binder & Kueny (1981). In the adverse pressure gradient boundary-layer experiments of Brereton *et al.* (1990, 1991), when the frequency of forced oscillation was of the order of the burst frequency, phase-conditioned turbulent motions tended towards a "frozen" state everywhere except in the region of peak production of the parent boundary layer ($y^+ \simeq 11$, at $Re_\theta = 3200$), where a small first-harmonic oscillation in $\langle u'u' \rangle$ could still be detected (where $\langle \ \rangle$ represents the phase average and u' the turbulent component of streamwise velocity). It was not clear if this peak would disappear at higher frequencies of oscillation. No modification of profiles of mean velocity or of time-averaged components of the Reynolds-stress tensor was observed. The conclusions of this study were that all periodic production of turbulence was primarily a modulation of the robust turbulence producing motions of the parent (mean) boundary layer, and that any resonant interaction might only come about if both spatial and temporal control of the turbulence producing motions could be achieved. The conclusion that there is no noticeable interaction over other ranges of frequency has been supported by a number of other studies in other unsteady turbulent boundary-layer, channel, and pipe flows when the frequency of forced oscillation approaches the estimated burst frequency (Shemer *et al.* (1985), (Cousteix & Houdeville 1983)). At frequencies of forced unsteadiness very much higher than characteristic turbulence frequencies, the observation of a smooth approach to a "frozen" state has been supported by the experiments of Brown *et al.* (1969) – in these experiments phase-conditioned friction factor measurements exhibited a quasi-laminar dependence on frequency, and turbulence was reported as "frozen."

Amongst the numerous observations of comparability between steady and mean unsteady effects, several contradictory findings have been reported at high frequencies of forced unsteadiness. Amongst the first to look at the detailed structure of unsteady turbulent pipe flow were Mizushina *et al.* (1973, 1975), who inferred that significant changes in turbulence structure occur when the frequency of oscillation reached a critical frequency which scaled with estimates of the burst frequency. While the trustworthiness of their data has been the subject of some criticism (Tardu *et al.*, 1987), their findings have still to be contradicted by a comparable study. Ramaprian & Tu (1983) studied oscillatory fully-developed pipe flow at different frequencies of forced unsteadiness and reported slight difference in profiles of mean quantities for oscillation close to the burst frequency, attributing these effects to non-linear interactions caused by the imposed oscillatory flow. More recently, Mao & Hanratty (1986) measured the wall shear in oscillatory pipe flow, at frequencies which appeared to be as high as the mean-flow burst frequency. They noted that the phase-conditioned turbulence intensity in wall shear constituted a distinctly non-linear response to a harmonic phase-conditioned wall shear stress, which they attributed to an unexplained interaction between organized flow oscillation and turbulence close to the wall. Abrams & Hanratty (1985) also noted similarities between flows with forced spatial variation and those subjected to forced temporal variation. In their studies of flow over a wavy wall, they observed relaxation effects in the phase of the response of turbulence to a wavy surface of short wavelength, consistent with non-linear behavior observed in their high-frequency smooth-wall pipe-flow results (Mao & Hanratty, 1986). In the channel-flow study of Acharya & Reynolds (1975), the response of a fully-developed turbulent channel flow to forced oscillation was compared at a low frequency and at a frequency of the order of the burst frequency. Although their correlation technique allowed measurement only of the first-harmonic response, some of

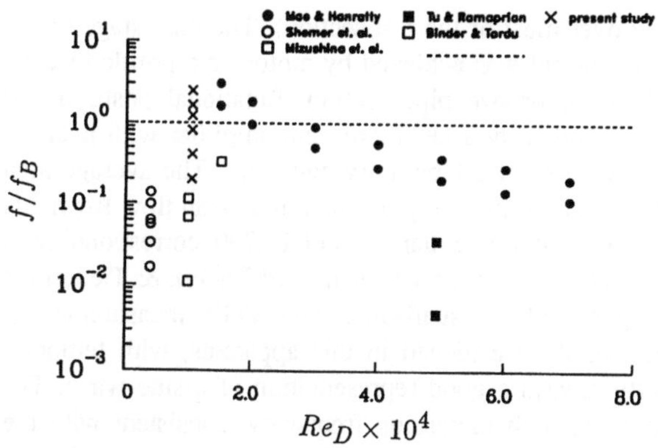

Fig. 1. Oscillation frequency vs. Reynolds number parameter space for unsteady turbulent pipe flow experiments.

the discrepancies between data at the two frequencies were attributed to interaction with the turbulence-producing motions. More recently, Tardu & Binder (1987, 1989) have considered aspects of turbulence structure in these flows. Using conditional sampling techniques to monitor bursting activity, they observed a strong modulation of bursting frequency during each oscillation and identified differences in educed bursting characteristics when the oscillation frequency approached the burst frequency. Their results showed that one effect of imposed oscillation at a frequency of the order of the burst frequency was to decrease the average ejection frequency, while maintaining the same burst frequency. Moreover, bursts which result in multiple ejections, and bursts which produce single ejections appeared to react differently to imposed oscillation, leading to the notion that, in unsteady flow, each might be governed by somewhat different mechanisms.

In summary, there are a number of different experiments, conducted in different flows, from which contradictory conclusions have been drawn concerning the interaction between turbulence and forced oscillatory motion in the neighborhood of the burst frequency of the mean flow. The objective of this study was to conduct a systematic study of effects of forced oscillation on wall-bounded turbulence, over a wide range of frequencies which would exceed by an order of magnitude the mean-flow burst frequency. A turbulent pipe flow is a good candidate for a study of this kind because the compact geometry of this flow translates to lower mechanical power requirements for forced oscillation at high frequencies than its channel-flow or boundary-layer counterparts. The parameter space of the study (Reynolds number vs. mean flow burst frequency (as estimated by the correlation of Luchik & Tiederman (1987)) is shown in Fig. 1, together with the parameter space covered by other pertinent pipe-flow studies.

2. Experimental Apparatus

The test facility comprised a standard recirculating water loop with a constant-head tank, sump, pump, and a test section of translucent pipe (57 mm in diameter, 160 diameters long, assuring fully-developed flow over most of the test section). The final stage of the test section was of bronze, and flow control was achieved by motoring a profiled sleeve around a longitudinal slot milled in the bronze pipe section. Rotational position and speed of the profiled sleeve were controlled by a DC motor and amplifier with position feedback, which operated under the control of a laboratory computer. The average area presented for outflow from the test section to the sump resulted in a mean flow Reynolds number (referenced to centerline velocity and pipe diameter) of 11,700, corresponding to a burst frequency of about 1.7 Hz (according to the correlation of Luchik & Tiederman (1987), and consistent with hydrogen-bubble visualization and VITA measurements). Forced oscillation at up to 10 Hz could be achieved in this apparatus, with temporal variation of phase-averaged velocity always a good representation of a sine wave. The amplitude of flow oscillation decreased with increasing frequency, consistent with the impedance characterization of the pipe by Shemer *et al.* (1985). In these experiments the amplitude of oscillation at the pipe centerline varied between 19% of the mean centerline velocity at 0.25 Hz and 12% at 3.0 Hz.

Phase-conditioned and time series measurements of streamwise velocity were made using a laser-Doppler anemometer with frequency shifting and a counter, which was interfaced with a laboratory computer to allow phase-resolved measurements of the instantaneous velocity of the flow. Since highly repeatable periodic motion could be imposed by the flow-control apparatus, a phase-averaging procedure was adopted for decomposition of flow variables into mean, oscillatory and turbulent components. All measurements were averaged over at least 1000 ensembles. Statistical convergence in data was assumed to have been reached when the fractional tolerance (a measure of differences in $\langle u'u' \rangle$ over the first and second halves of the data set, normalized by the rms level in $\langle u'u' \rangle$) reached 0.1%. For oscillation at 3.0 Hz, the typical number of cycles required to satisfy this tolerance was around 5000.

The relatively low Reynolds number of the flow allowed the use of hydrogen-bubble visualization to study timelines and streaklines of the unsteady flow, in the manner outlined by Schraub *et al.* (1965). In order to view timelines initiated a constant distance from the wall, a very thin flat strip was inserted at the surface, at the observation station in the manner described by Achia & Thompson (1977). Visual information was recorded using either phase-conditioned photography or a high speed video camera (Kodak Ektapro Motion Analyzer).

3. Experimental Results

3.1 Results from Velocity Measurements

Measurements of the time-averaged values of mean velocity (U) and mean square of turbulence $\overline{u'u'}$ across a radius of the pipe were in good agreement with reference steady-flow data and showed no variation with frequency, even when oscillation was forced at six times the estimated burst frequency of the mean flow. These results are well established and consistent with those of nearly every unsteady flow experiment and are

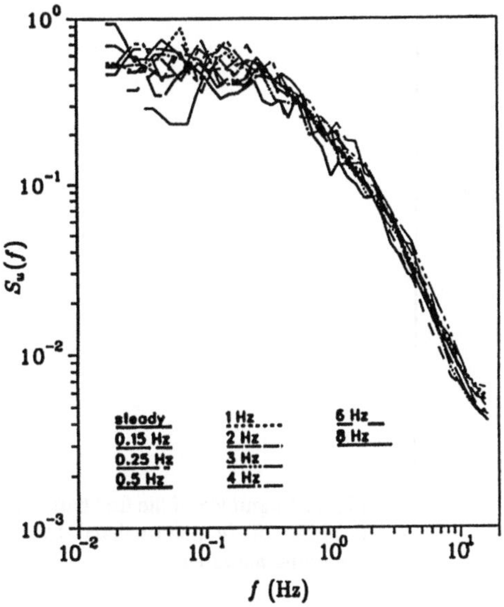

Fig. 2. Power spectra of u' at $y^+ = 15$ for steady and decomposed unsteady turbulent flow. The burst frequency of the parent boundary layer is about 1.7 Hz.

reported in Hwang (1992). Of more interest were power spectra of u', as deduced from time series measurements from which the phase-averaged velocity had been subtracted. These power spectra are shown in Fig. 2, for forced oscillation at frequencies as high as 8 Hz, measured at $y^+ \simeq 15$ – close to the position of peak production of turbulence in the parent boundary layer. While there is considerable low-frequency noise in these data, caused by the differing lengths of time series processed and the variance of the FFT as a power-spectral estimator, the general trend is one of qualitative agreement. There is no convincing evidence of any re-arrangement of the spectral content of the flow, reinforcing the notion of invariance of all scales of motion of the parent boundary layer to forced sinusoidal oscillation.

The high-frequency response of the organized component of streamwise velocity is illustrated in Fig. 3, in which the amplitude of $\langle u \rangle$ at its first harmonic is plotted in the near-wall region. For frequencies above 1 Hz, these profiles are in good agreement with the asymptotic high-frequency (Stokes) solution to the streamwise momentum equation in the oscillatory field of flow. They also illustrate clearly that, at these frequencies, all oscillatory shear is confined to a very thin region within the viscous sublayer and so is unlikely to provide direct interaction with the turbulence producing motions of the parent boundary layer.

The first harmonic response of the streamwise component of turbulence is shown in Fig. 4, normalized by u_τ, the friction velocity of the mean flow. In this figure, the amplitude of the square root of the oscillatory component of $\langle u'u' \rangle$ is plotted – the square root represents a more robust statistical measure which is less sensitive to false measurements than the phase average of the square of u' (though it is more indirect in relevance to decomposed forms of the Navier-Stokes equations). It is clear from this figure that the trend is a monotonic one towards "frozen" turbulence with increasing frequency, with the peak in the turbulence response coincident with the peak of turbulence production

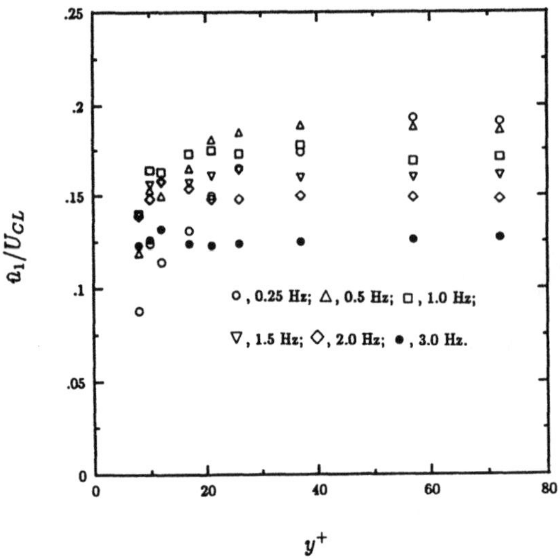

Fig. 3. Amplitude of the first harmonic response of $\langle u \rangle$, normalized by the centerline amplitude.

in the parent boundary layer ($y^+ \simeq 11$). It is worth pointing out that the amplitudes of the phase-conditioned turbulence response are extremely small. Therefore they are very sensitive to minor cycle-to-cycle variations in the organized forcing, which are interpreted as turbulence by the phase-average decomposition. We were initially misled by such organized variation (in the form of a noisy signal from a position encoder, which produced the effect of an occasional phase shift in $\langle u \rangle$) which manifested itself as a second-harmonic response when it exceeded in amplitude the true turbulence response. However, once this problem was resolved, second harmonic motion was always an order of magnitude smaller than that at the fundamental frequency.

The phase dependence of the deviation of $\langle u'u' \rangle$ from its mean is shown in Fig. 5 for three selected frequencies, measured at $y^+ = 12$. The attenuation of phase-dependent variation in turbulent intensity with increasing frequency may be seen plainly, together with the obvious variance from a sinusoidal response. For the phase average of $u'u'$ measured at 2.0 Hz, a Fourier decomposition still reveals a dominant first-harmonic response. The similarity between these data and the equivalent measures of turbulence in wall shear stress (Mao & Hanratty (1986)) prompted a more thorough examination of whether the variance of these data from a sinusoidal response was due to an inability to distinguish between harmonic components and background noise, or whether it did indeed represent the growing importance of higher harmonics with increasing frequencies.

This issue was examined by estimating the power spectrum of the square of u' (and thus organization in magnitude of u') from a time series of $u(t)$ from which the phase-averaged velocity $\langle u \rangle$ had been subtracted. Spectra of the square of u' are shown in Fig. 6 for time series measured at $y^+ = 12$, with oscillation at 0.5 Hz and 2.0 Hz. From the high-frequency content of the spectra, it appears that there is no obvious broadening due to instrumentation noise for the case of forced oscillation at 2 Hz, even though the signal level (shown in Fig. 5) is significantly lower. The spectra reveal enhanced organization in magnitude of u' at the fundamental and second-harmonic and hint at the possibility of weak peaks appearing at higher harmonics. The second harmonic arises through the

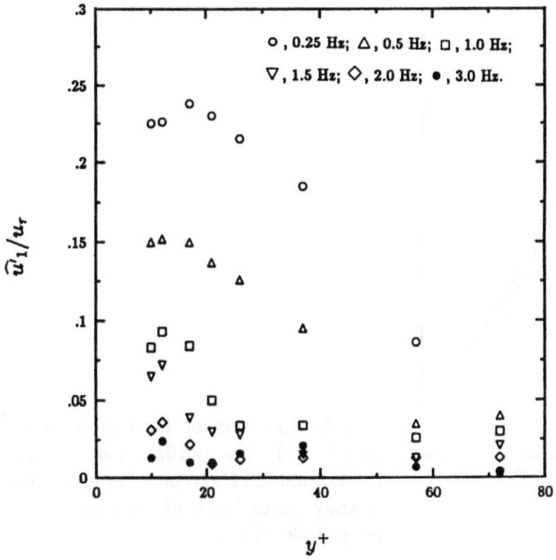

Fig. 4. Amplitude of the first harmonic response of $\langle u' \rangle$, normalized by u_τ.

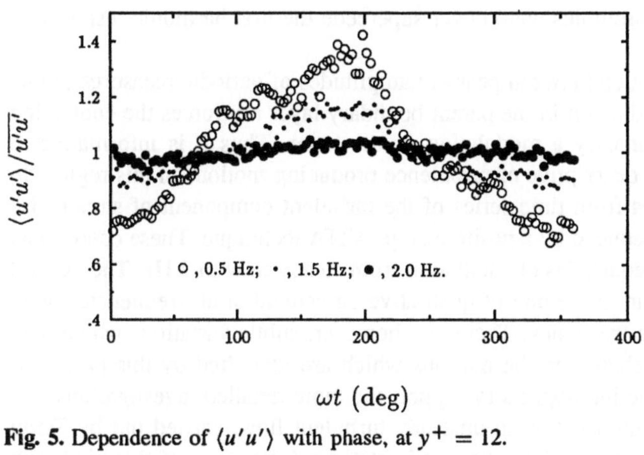

Fig. 5. Dependence of $\langle u'u' \rangle$ with phase, at $y^+ = 12$.

non-linear character of the periodic Reynolds-stress equations in coupling first harmonic shear ($\langle \partial u / \partial y \rangle$) and first harmonic Reynolds stress ($\langle u'v' \rangle$) to yield a second-harmonic production term for $\langle u'u' \rangle$. Through the same coupling, multiple higher harmonics in $\langle u'u' \rangle$ may be generated in this kind of flow. However, it appears that the energy in $\langle u'u' \rangle$ is only transferred from the fundamental to higher harmonics when the period of oscillation approaches a time scale characteristic of the parent flow (burst frequency, large-eddy turn-over time, viscous time scale, *etc.*), and the allowed relaxation time is no longer sufficient for a purely fundamental response.

The indication from these results is that variance about the sinusoidal response shown in Fig. 5 is turbulence at higher harmonics and not noise. Thus the proportional role of higher harmonics does appear to grow as turbulence approaches a "frozen" state, though

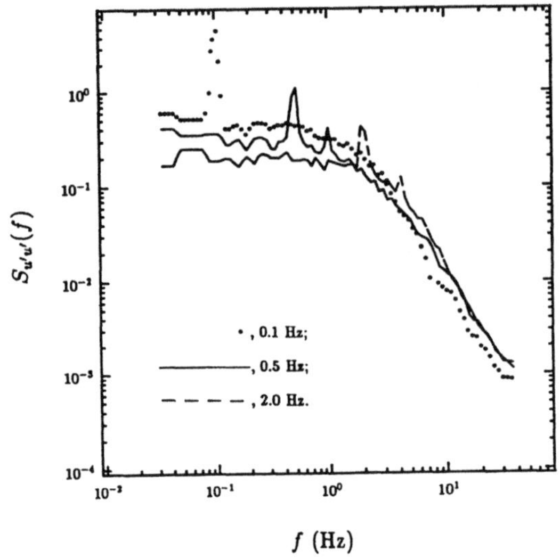

Fig. 6. Spectrum of the square of u' at $y^+ = 12$ for oscillation at 0.1 Hz, 0.5 Hz and 2.0 Hz, representing the frequency dependence of organization in magnitude of u'.

it is not clear that higher harmonics would ever supercede the first harmonic response in importance.

The coincidence in position between peaks in amplitudes of periodic measures of turbulence and turbulence production in the parent boundary layer reinforces the contention that the former is predominantly a modulation of the latter. Thus it is informative to examine statistical features descriptive of turbulence producing motions in this region. In Fig. 7, conditional averages from time series of the turbulent component of streamwise velocity are presented, as deduced by a positive-slope VITA technique. These conditional averages span a range of frequencies of oscillation from steady flow to 2 Hz. The general trend in these results appears to be one of qualitative agreement at all frequencies, with no dramatic dependence on frequency. There are, however, subtle variations which may be indicative of structural change in the motions which are identified by this particular statistical technique. This notion appears to support the more detailed investigations into identification of structural information in unsteady turbulent flow, carried out by Tardu *et al.* (1989) which characterized subtle changes in statistical measures of this kind with frequency variation.

3.2 Results from Visualization Studies

Further information on possible modifications of the structure of the near-surface turbulence field was sought using hydrogen bubble visualization. A high-speed video camera was focused on the flowfield downstream of a platinum wire positioned at $y^+ \simeq 15$ and the evolution of timelines during unsteady oscillation was recorded. Video footage demonstrated clearly that streak initiation (commensurate with peak levels in $u'u'$) became organized in time by the superposed oscillation, favoring the beginning of the deceleration phase when the flow is most susceptible to the as-yet-unidentified destabilization effect which drives the bursting phenomenon. This result was consistent with

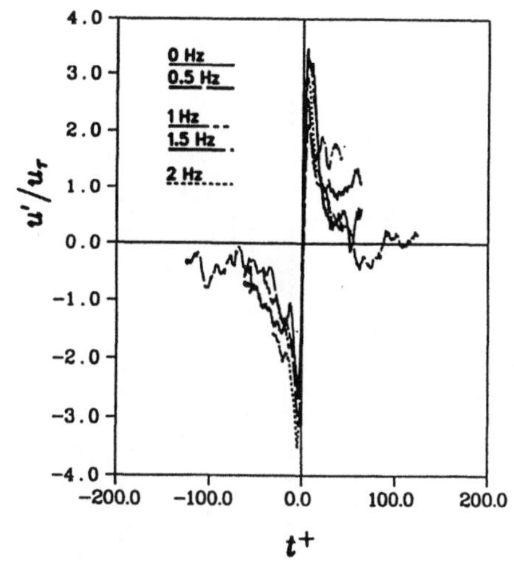

Fig. 7. VITA positive-slope conditional averages for forced oscillation over a range of frequencies, at $y^+ \simeq 15$ – the abscissa is in viscous time units.

measurements of the phase of $\langle u'u' \rangle$ relative to that of the oscillatory forcing, and in agreement with similar measurements made by Tardu *et al.* (1987). As the period of forced oscillation was decreased, the streak signature (indicative of features of the robust turbulence producing motions of the mean flow) occasionally survived whole periods of oscillation. Moreover, the dying signature of a streak generated during the deceleration phase of the previous unsteady cycle occasionally marked the position upstream of which new streaks were most likely to appear during the next deceleration phase of the unsteady cycle.

Selected frames from these recordings are shown in Figs. 8, 9 and 10 for the cases of steady flow, low-frequency unsteady flow (at one seventh of the mean-flow burst frequency) and high frequency unsteady flow (at a frequency approximately equal to three times the mean flow burst frequency). In the steady flow sequence, attention is drawn to the apparently random appearance of a new streak (at the left of the frame) below the existing one (near the right of the frame), as it twists and loses its coherence. For the sequence at low-frequency unsteadiness, the streak (towards the right of the frame) forms during the deceleration phase and maintains its characteristic "V" signature as it meanders before being swept downstream during the acceleration phase of the unsteady cycle (during which the final frame was taken).

In the high-frequency case, the sequence of images spans a complete period, beginning with a well-established region of low-speed flow in the center of the frame. As this region is swept downstream during the acceleration phase, the characteristic "V" signature of a newly initiated streak begins to form in the wake of the previous one. The "V" signature continues to develop during the deceleration phase, establishing a new broad region of low-speed fluid. The subsequent advection downstream during the acceleration phase then precedes the initiation of a new streak in its wake until, eventually, spanwise meandering breaks this pattern. This pattern of events was observed with some regularity in the case of "high-frequency" forced oscillation, while scarcely observed at all for steady flow and "low-frequency" forcing. These visualizations exhibited some repeatability at frequencies at and above the burst frequency, when viewed as a continuous video

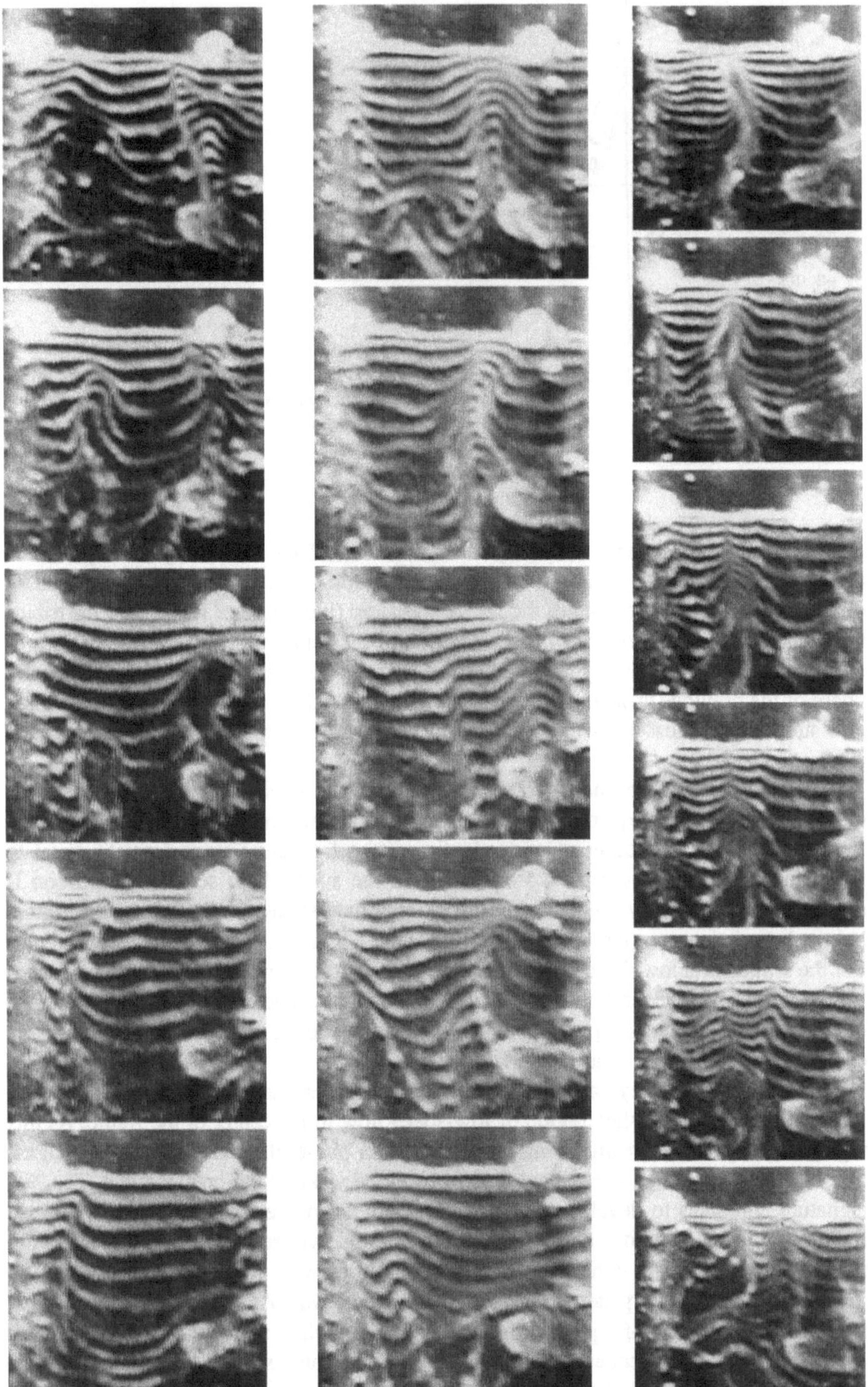

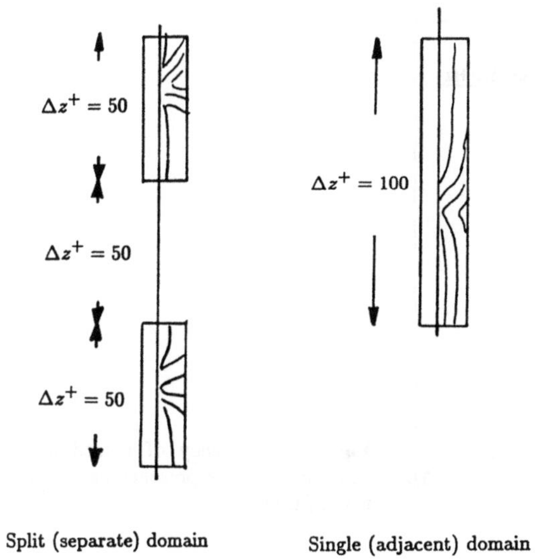

Fig. 11. The domains used for interrogation of the frequency of continuous streak observation over prescribed time periods.

sequence. A companion observation from these recordings was that spanwise meandering of streak signatures was somewhat reduced when organized oscillation was imposed.

In order to quantify the significance of this observed organizational effect, the frequency with which one or more streaks were observed continuously within a chosen domain, over prescribed time periods, was measured. This procedure involved making (by hand) a map of the momentary streak traces present in each frame of the video sequence and then analyzing statistics of streak presence from that map. The domains of observation were chosen as a single line at $y^+ \simeq 15$ of spanwise dimension $\Delta z^+ = 100$, and two regions of a line at $y^+ \simeq 15$, both of length $\Delta z^+ = 50$, separated by a void region of $\Delta z^+ = 50$, shown in Fig. 11. The total spanwise extend of each domain was then $\Delta z^+ = 100$ which corresponded to the expected average distance between streaks. Thus, on average, a single streak would be detected for each observation, regardless of the choice of domain. This result was confirmed in these experiments and reported in (Hwang, 1992). The robustness of the mean streak spacing of $\Delta z^+ = 100$ in both steady and unsteady flow, over a range of oscillation frequencies, concurred with insights from both experiments and full simulations that this length scale is of fundamental significance to the understanding of mechanisms of generation of turbulence.

The frequency of continuous observation of streak presence within the domain of span $\Delta z^+ = 100$ is shown in Fig. 12. The abscissa represents the period (in time) for

Fig. 8. The decay of one streak and the disorganized initiation of another, in steady flow – the time sequence is from top to bottom.

Fig. 9. The deceleration-phase growth and acceleration-phase downstream advection of a streak in low-frequency unsteady flow.

Fig. 10. The downstream advection of one streak and the subsequent initiation of a new one in its wake in high-frequency unsteady flow.

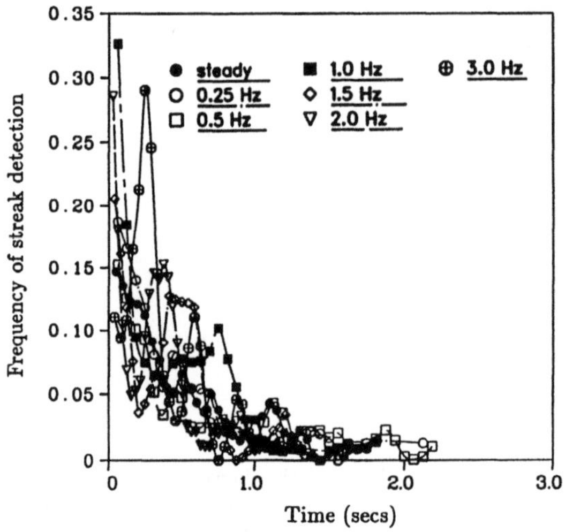

Fig. 12. Frequency of detection of continuous streak presence as a function of time.

which continuous streak presence must be observed (to within the nearest 1/8th of a cycle) for qualification as a valid measurement. It is clear that, regardless of the forcing frequency, there is scarcely any continuous presence of streaks for more than 1.5 seconds (150 viscous time units). At the highest frequencies of forced oscillation, the likelihood of continuous observation of streak presence over multiple cycles is greatest, as evidenced by the data Fig. 13 (with its abscissa recast in terms of cycles of forced unsteadiness). The improved collapse of these data indicates the effect of cyclic modulation on streak duration by the oscillatory flow. Thus, while there clearly is a modulation effect owing to the forced unsteadiness, the expected longevity of streak presence scales on the mean character of the parent boundary layer, consistent with the notion of turbulence being generated by robust motions which can withstand organized modulation without detriment to their role in producing new turbulence.

When statistics of observation of streak presence were examined, comparisons between frequency of continuous occurrence over prescribed times measured for each of the two domains provided an indication of the spatial localization of low-speed fluid marking streaks, by forced unsteadiness. If the observation frequency for the single domain exceeded that for the split domain, it would imply that the more favorable location for the appearance of new streaks is in the region *between* old streaks. On the other hand, if the observation frequency for the single domain were equal to that for the split domain, it would seem that new streaks were most likely to be detected *in the wake of*, or close to, old streaks. The third possibility (observation frequency for the split domain exceeding that for the single domain) would indicate that a favorable location for new streaks to be detected would be of the order of $\Delta z^+ = 100$ distant from existing streaks, though not necessarily in the wake of a decaying streak.

In Fig. 14, the frequency of continuous observation of streak presence is shown for oscillation at 2.0 Hz, for the single and split domains. It is clear from this figure that continuous streak detection over more than one cycle is significantly more likely in the split domain than in the single domain. Comparable results were found for other "high" frequencies for which continuous streak detection was found over more than

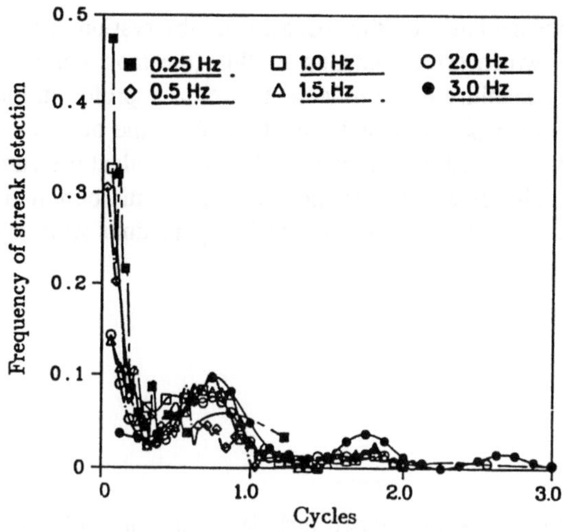

Fig. 13. Frequency of detection of continuous streak presence as a function of phase.

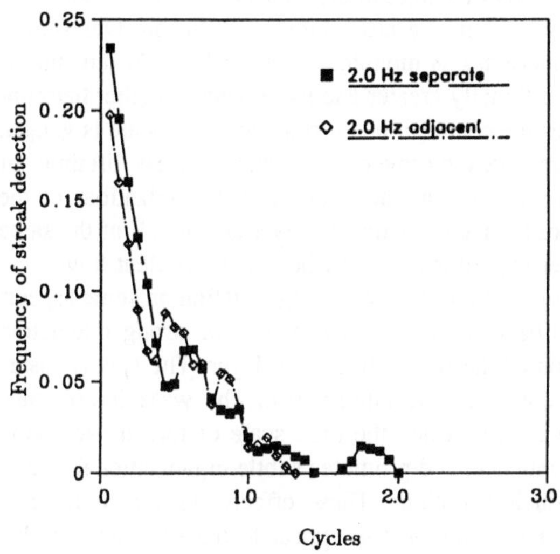

Fig. 14. Comparison of frequency of continuous streak detection for the single and split interrogation domains, for a forcing frequency of 2.0 Hz.

one cycle (Hwang, 1992). These results were not duplicated at low frequencies since streaks in low-frequency unsteady flows were far more susceptible to advection during the acceleration phase of the unsteady cycle. These findings imply that high-frequency forced unsteadiness appears to play the role of providing occasional stabilization of the location at which the turbulence producing motions (as marked by accompanying streak patterns) are generated close to the wall. In particular, when this occasional stabilization effect results in the continuous presence of streaks over more than one cycle of oscillatory motion, one is more likely to observe the appearance of a new streak either behind or around $\Delta z^+ = 100$ distant from an existing streak, than between existing streaks. One may infer that the turbulence producing motions, which are marked by the presence of streaks of low-speed fluid, undergo similar occasional organization.

It would seem that plausible explanations for this occasional observation should be linked with the elusive instability which drives bursting motion (the most common suggestion for which is a secondary instability of an inflectional velocity profile in the updraught between counter-rotating vortex pairs), or with feedback from the outer flow of the organized oscillatory pressure field, which would overwhelm the turbulent pressure fluctuations which act in steady flow. However, a better understanding of these motions is required at a fundamental level before one could propose such explanations with any sincerity.

4. Conclusions

The effect of high-frequency oscillation on wall-bounded turbulence appears to be predominantly one of reducing the response time afforded to the turbulence-producing motions of the mean flow to adapt to momentary changes in the bulk flow. Phase-conditioned production is thus weaker and loses its coherence after outward diffusion over shorter distances, as frequency of oscillation increases. A monotonic approach to "frozen" turbulence follows, characterized by a proportionally greater energy content in higher harmonics. It does not seem to matter whether the characteristic turbulence time scale is gauged as a reciprocal burst frequency, a large-eddy turnover time scale, a streak lifetime, or a turbulent kinetic energy/dissipation rate ratio, as these are all characterizations of the robust turbulence producing motions of the parent boundary layer and are all of the same order in a given low Reynolds number experiment in wall-bounded turbulent flow.

While it appears that temporal forcing can impose weak organization on some aspects of turbulent motion, no evidence has been found of resonance or of strong interaction with the turbulence-producing motions of the parent boundary layer. The spatial disorganization of these motions prohibits such simple interference. The weak interference which is achieved by temporal forcing arises when the time scale of forced oscillation approaches that of the response of turbulence, and promotes subtle organizational effects in the localization of turbulence-producing motions. These effects are essentially cosmetic changes to the signature of an active motion and appear to have no measureable effect on $u'u'$.

References

Abrams, J. & Hanratty, T. J. Relaxation effects observed for turbulent flow over a wavy surface. *J. Fluid Mech.* **151** (1985), 443-455.

Acharya, M. & Reynolds, W. C. Measurements and predictions of a fully developed turbulent channel flow with imposed controlled oscillations. *Report TF-8*, Department of Mechanical Engineering, Stanford University, Stanford, California, (1975).

Achia, B. U. & Thompson, D. W. Structure of the turbulent boundary layer in drag-reducing pipe flow. *J. Fluid Mech.* **81** (1977), 439-464.

Binder, G. & Kueny, J. L. Measurements of the periodic velocity oscillations near the wall in unsteady turbulent channel flow. In *Turbulent Shear Flows 3*, Springer-Verlag, New York, 1982.

Brereton, G. J. Reynolds, W. C. & Jayaraman, R. Response of a turbulent boundary layer to sinusoidal free-stream unsteadiness. *J. Fluid Mech.*, 221 (1990), 131-159.

Brereton, G. J. & Reynolds, W. C. Dynamic response of boundary-layer turbulence to oscillatory shear. *Phys. Fluids A* **3** (1) (1991), 178-187.

Brown, F. T., Margolis, D. L. & Shah, R. P. Small-amplitude frequency behavior of fluid lines with turbulent flow. *Trans. ASME* D: *J. Basic Engng.*, **91** (1969), 678-693.

Cousteix, J. & Houdeville, R. Couches limites instationnaires. *Rapport Technique 53/2259 AND*, Departement d'Etudes et de Recherches en Aerothermodynamique, Centre d'Etudes et de Recherches de Toulouse, 1983.

Hwang, J.-L. Ph. D. Thesis, Department of Mechanical Engineering and Applied Mechanics, The University of Michigan, 1992.

Luchik, T. S. & Tiederman, W. G. Timescale and structure of ejections and bursts in turbulent channel flows. *J. Fluid Mech.* **174** (1987), 529-552.

Mao, Z.-X. & Hanratty, T. J. Studies of the wall shear stress in a turbulent pulsating pipe flow. *J. Fluid Mech.* **170** (1986), 545-564.

Mizushina, T., Maruyama, T. & Hirasawa, H. Structure of the turbulence in pulsating pipe flows. *J. Chem. Engng Japan* **8** (1975), 210-216.

Mizushina, T., Maruyama, T. & Shiozaki, Y. Pulsating turbulent flow in a tube. *J. Chem. Engng Japan* **6** (1973), 487-494.

Ramaprian, B. R. & Tu, S. W. Fully developed periodic turbulent pipe flow. Part 2. The detailed structure of the flow. *J. Fluid Mech.* **137** (1983), 59-81.

Schraub, F. A., Kline, S. J., Henry, J., Runstadler, P. W. & Littell, A. Use of hydrogen bubbles for quantitative determination of time-dependent velocity fields in low-speed water flows. *Trans. ASME* D: *J. Basic Engng.* **87** (1965), 429-444.

Shemer, L., Wygnanski, I. & Kit, E. Pulsating flow in a pipe. *J. Fluid Mech.* **153** (1985), 313-337.

Shemer, L., Kit, E. & Wygnanski, I. On the impedance of the pipe in laminar and turbulent pulsating flows. *Expt. Fluids* **3** (1985), 185-189.

Tardu, S., Binder, G. & Blackwelder, R. Modulation of bursting by periodic oscillations imposed on channel flow. *Sixth Symposium on Turbulent Shear Flows*, Université Paul Sabatier, Toulouse, 1987.

Tardu, S. & Binder, G. Ejections and bursts in pulsatile turbulent wall flow measurements and visualizations. *Seventh Symposium on Turbulent Shear Flows*, Stanford University, Stanford, 1989.

Tardu, S., Binder, G. & Blackwelder, R. Response of turbulence to large amplitude oscillations in channel flow. In *Advances in Turbulence* (eds. G. Compte-Bellot & J. Mathieu), Springer-Verlag, New York, 1987.

A Model Equation for Transitional and Turbulent Plane Channel Flow

N. D. Sandham[1]

DLR, Institute for Theoretical Fluid Mechanics, Göttingen, Germany

Abstract

A predictive model for the mean velocity profile in wall-bounded turbulent flow is developed. The model is conceptually based on the dynamical processes known to occur in near-wall turbulence. Ejections, sweeps and outer-layer motions are modelled directly as momentum transfer processes. These combine with viscous and pressure gradient effects to keep the mean velocity profile in a state of balance. A quantitative model is developed for the case of plane channel flow and is shown to correctly give an inner law, a logarithmic region, and a wake region. Predictions of skin friction, channel centreline velocity and shape factor agree well with experiments over a wide range of Reynolds numbers, including the transitional and low-Reynolds-number regimes. The procedure adopted represents a new and fundamentally non-local approach to turbulence prediction.

1. Introduction

The problem of turbulent flow near a wall has been a very active research topic for many years. Significant progress has been made, resulting from studies using many different approaches. Although a complete understanding has not yet been achieved, there is substantial agreement on the main features of the flow. As a background to the current paper, a brief review of our current understanding will be presented.

Central to the current view of the turbulent boundary layer are the flow visualisations from Kline et al. (1967) which showed elongated streaks in the region very close to the wall, indicating a degree of organisation within the flow that was previously unsuspected. Streaks containing lower momentum fluid ('low-speed streaks') were observed to rise slowly, and break up in a high-activity process termed 'bursting'. Work since then has focussed on using increasingly sophisticated methods to try to understand this process. Motions have been identified based on the quadrant of $u'v'$ in which they occur (in the usual notation u' and v' are the velocity fluctuations relative to the mean in the streamwise and wall-normal directions respectively), as shown on Fig. 1. For example if u' is negative and v' is positive the event falls in quadrant 2, and is termed an ejection. It is interpreted as the movement of low momentum fluid away from the wall. An event

[1] Current address: Department of Aeronautical Engineering, Queen Mary and Westfield College, Mile End Road, London E1 4NS

with u' positive and v' negative falls in quadrant 4 and is termed a sweep. It represents the motion of high momentum fluid towards the wall. Ejections and sweeps are common in the buffer region between the linear sublayer and the logarithmic region of the mean velocity profile.

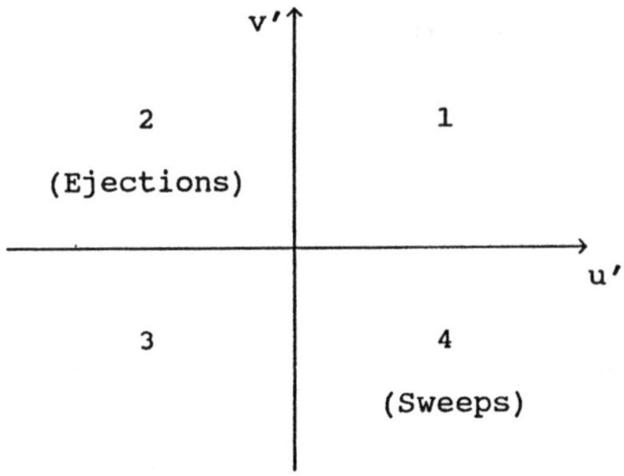

Fig. 1. Quadrant plot of velocity fluctuations. Ejections and sweeps are events which occur in quadrants 2 and 4 respectively

Another approach has been to construct conceptual models of turbulent flow based on the vortical structures in the flow. A recent review has been given by Robinson (1991). In the near-wall region a popular vortical structure has been the hairpin vortex. In the region between the legs of the hairpin fluid is moved away from the wall, which could correspond to an ejection process. As well as these near-wall vortices, there are also large-scale motions in the outer portion of the boundary layer. An example of the kind of motion proposed for this part of the flow can be found in Brown and Thomas (1977). The relation between the vortices in the inner and outer regions is relatively poorly understood.

The most recent input to studies of the turbulent boundary layer has come from direct numerical simulations of the Navier-Stokes equations for flows in simple geometries. The resulting databases provide a wealth of information, including quantities that cannot be measured in experiments. Statistics from a computation of turbulent channel flow have been presented by Kim, Moin and Moser (1987). One important result from the simulations has been the observation that near-wall vortical motions do not contain the symmetries that had been previously supposed. Instead of hairpin vortices, Robinson (1991) found asymmetric structures, with single quasi-streamwise vortices common in the buffer region. It is increasingly apparent (Kline, 1990) that there are many mechanisms at work and the processes and vortical structures may be non-unique.

What has been largely absent from all the studies, of both the inner and outer regions, has been the application of the new facts and concepts to generating new *prediction* schemes for turbulent flows. The intention of this paper is to present one possible approach to this problem. One of the simplest turbulent flows, incompressible two-dimensional channel flow, is considered. In Sect. 2 the procedure is introduced, and a quantitative model developed. Section 3 presents a comparison with experimental data. Further discussion is then made in Sect. 4. Standard notation is employed throughout. The mean velocity profile is $u(y)$, where y is the distance from the lower wall. Wall-coordinates are $y^+ = yu_\tau/\nu$ and $u^+ = u/u_\tau$, where $u_\tau^2 = \nu(du/dy)_w$, ν is the kinematic viscosity and the subscript w refers to the wall.

2. Model

2.1 Concepts

Turbulent flow in a channel can be considered as a state held in balance by the competing actions of eddies and viscosity. The eddying motions bring high-momentum fluid towards the wall, while viscosity tries to damp these motions, especially near the wall. The limiting cases resulting from these effects are sketched on Fig. 2. If the viscous effects act alone we would return to laminar flow with a parabolic velocity profile. If the eddying motions are allowed to act alone, unchecked by viscosity or walls, we would have a perfect redistribution of momentum, leaving a uniform velocity profile. The latter state can never practically be obtained, since there is always viscosity present, but it is conceptually useful. The state of turbulence can therefore be considered at a simple level as a balance between the two effects.

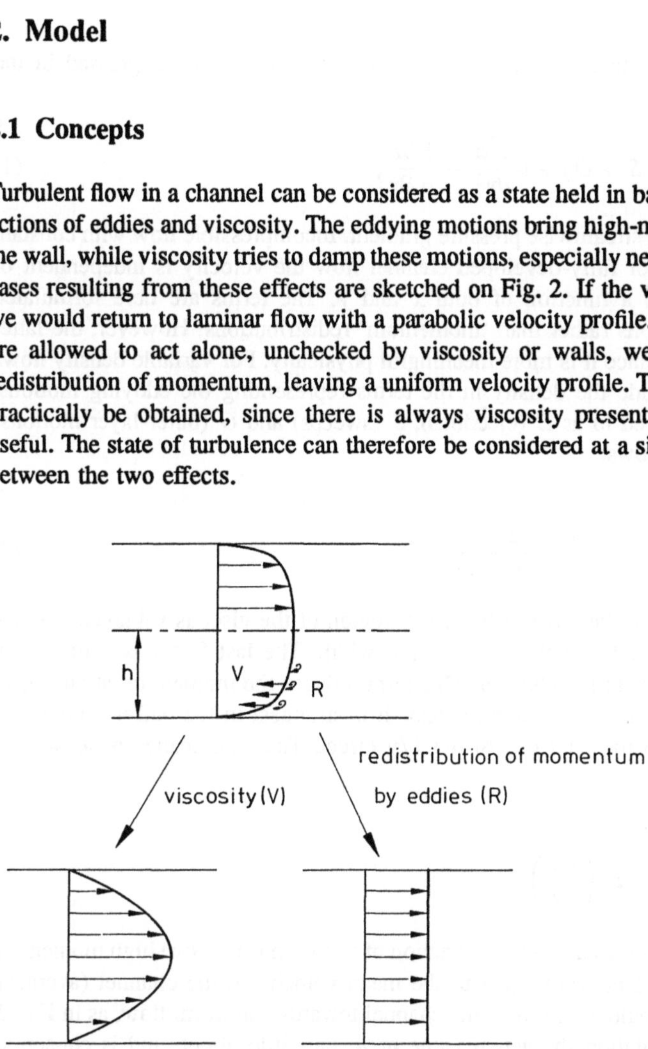

Fig. 2. Schematic of the mean turbulent flow in a channel, showing the balance between eddy motions and viscosity. Viscosity tries to return the mean flow to the parabolic laminar profile, while the eddying motions try to redistribute momentum uniformly across the channel

This conceptual picture suggests the approach of predicting the mean velocity profile in a turbulent flow by modelling individually the momentum redistributions caused by certain classes of eddy motions. The three types of eddying motions which will be considered are ejections, sweeps and outer-layer motions. The model terms, describing the momentum redistribution caused by each of these motions, were developed by experimentation.

2.2 Formulation

The balance between momentum redistribution and viscosity can be expressed in the following form

$$0 = T(E + S + O) + \nu\frac{d^2u}{dy^2} - \frac{1}{\rho}\frac{\partial p}{\partial x}, \tag{1}$$

where the last term is the streamwise pressure gradient. Incompressible flow with constant density ρ is assumed. For fully-developed channel flow the velocity is independent of x, while the pressure is a function of both x and y. The terms are here formulated as 'velocity' redistributions rather than 'momentum' redistributions. However, the latter terminology is retained since it is more meaningful physically. For variable density flows one would have to include the density in the terms representing the eddying motions. These motions are referred to as E (ejections), S (sweeps) and O (outer-layer motions) and are modelled as follows.
Ejections (E):

$$E = -c_1(u - u_e)^2\frac{2h}{y(2h - y)}, \tag{2}$$

where u_e is the velocity at the edge of the linear region of the viscous sublayer, y is the wall-normal coordinate and h is the channel half width. The last factor is simply $1/y$, but applied for both walls in the channel. This term reduces the momentum at any point in the channel, and models the ejection process. It is strongest in the buffer region, and small further away from the wall due to the $1/y$ effect. The significance of u_e will be discussed later.
Sweeps (S):

$$S = c_2(u_s - u)\left(\frac{du}{dy}\right)_w, \tag{3}$$

where the subscript w denotes the wall. The action of this term is to move high momentum fluid towards the wall. If u_s were equal to the mean velocity in the channel (averaged over y) then this term would act to force the channel towards a uniform flow, as in Fig. 2. In the current implementation the actual value of u_s has little effect, and is chosen so that global mass is conserved during the process of momentum redistribution.

The two terms above give a reasonable model for near-wall turbulence. The mean velocity profile is realistic up to $y^+ = 100$, but results in too strong a wake in the outer portion of the channel. To get the correct behaviour in this region an additional term is required.
Outer-layer motions (O):

$$O = \frac{c_3(du/dy)_w^2}{2h/[y(2h - y)] - c_4/h + c_5 O^*}, \tag{4a}$$

with

$$O^* = \frac{\nu}{u\overline{u}^2}\left(\frac{du}{dy}\right)^2_{\mathrm{w}}, \tag{4b}$$

where $\overline{u}$ is the mean velocity in the channel (averaged over y). Near the wall the first term in the denominator of (4a) is dominant and the whole term then acts like $y(du/dy)^2_{\mathrm{w}}$ and combines with E and S to generate the logarithmic law. The remaining terms in the denominator are the most ad hoc part of the model. They serve to give the correct Reynolds number trend of the wake size.

Transitional effects are handled in a simple manner, by reducing the size of the momentum redistribution term in accordance with the size of the ratio y_e/h, where y_e is the value of y at the edge of the linear region of the sublayer. Hence:

$$T = 1 - c_t\frac{y_e}{h}, \tag{5}$$

where c_t is a (single) transition constant. This term reduces the momentum redistribution as if it were cramped at the lower Reynolds numbers, where y_e becomes significant compared to the channel half-width. When T drops to zero laminar flow becomes the solution to (1).

2.3 Numerics

To focus attention on the model equation the numerical method for solution has been kept as simple as possible. The solution is obtained by two nested iteration processes. For the inner iteration (1) is reformulated into two steps:

$$\begin{aligned} u^{(1)} &= u^n + \Delta t\left\{T(E+S+O)+\nu\frac{\delta^2 u}{\delta y^2}\right\} \\ u^{n+1} &= u^{(1)} - \Delta t\frac{1}{\rho}\frac{\partial p}{\partial x}, \end{aligned} \tag{6}$$

where Δt is the time step and δ^2 represents a second-order central difference. The pressure gradient is independent of y, with the magnitude determined as a correction to the first step, to ensure that a constant channel throughflow is maintained. Equation (6) is iterated until a steady state is obtained.

The outer iteration consists of adjusting y_e until the solution to (6) gives a prescribed value of u_e^+, the velocity of the edge of the sublayer in wall coordinates. The velocity profile for $y < y_e$ is assumed to be linear, so $u^+ = y^+$ up to y_e. This procedure ensures that the point y_e is at the same position in wall coordinates for all cases. This fixes the behaviour in wall-coordinates and is essential in producing an inner law (u^+ a function of y^+ alone). The secant method was used for the iterations.

The basic solution can be made on a uniformly spaced mesh, with the first inner point at y_e. However a little more flexibility and speed is obtained if a stretched mesh is used. Results will be presented for a grid with the first mesh point at the wall, the second at y_e, the third at $2y_e$ and with succeeding points following:

$$y_{j+1} = y_j + \alpha(y_j - y_{j-1}). \tag{7}$$

In the computations $\alpha \approx 1.1$ was used, adjusted slightly so that the last point fell at the channel centreline. The results are not sensitive to α, which can be made as large as 1.4 with only a one percent error in skin friction. Tests showed that a complete resolution of the buffer region required adjustments to the constants. This did not lead to different results, but did lead to very small time steps. An implicit treatment of the viscous term would probably cure this problem. For the current explicit scheme a *CFL* number can be defined as

$$CFL = \frac{\nu \Delta t}{(\Delta y)^2_{\min}}. \tag{8}$$

The scheme was found by trial to be stable for $CFL < 0.38$.

As a start point for the iterations the laminar solution was used. Symmetry was exploited so that only one channel half needed to be computed. The number of points required for $\alpha = 1.1$ varied from 7 at $Re = 1700$ to about 60 at $Re = 10^6$. The whole iteration process required less than 5 seconds on a 386 personal computer, even at the highest Reynolds numbers considered.

2.4 Constants

Dimensionless constants used in the model are: $u_e^+ = 9.0$, $c_1 = 0.074$, $c_2 = 0.0051$, $c_3 = 4.3 \times 10^{-6}$, $c_4 = 2.0$, $c_5 = 2.6$ and $c_t = 5.4$. It is possible that c_4 doesn't need to be counted as a constant – it makes the combination of the first two terms in the denominator of (4*a*) equal to zero at the channel centreline. The constants u_e^+, c_1, c_2 and c_3 were determined so that (a) a logarithic region was obtained, and (b) the constants for this region matched the law of the wall constants. An explicit rule connecting the constants from the current model with the law of the wall constants has not so far been obtained. The remaining constants were determined in order to give a reasonble behaviour in the outer region of the channel and in the transition region.

3. Results

Throughout this section the Reynolds number is defined as:

$$Re = \frac{2h\overline{u}}{\nu}, \tag{9}$$

where h is the channel half-width, $\overline{u}$ is the average velocity and ν is the kinematic viscosity. For comparison purposes the standard logarithmic law is used

$$u^+ = \frac{1}{\kappa} \ln y^+ + b, \tag{10}$$

with $\kappa = 0.41$ and $b = 5.0$.

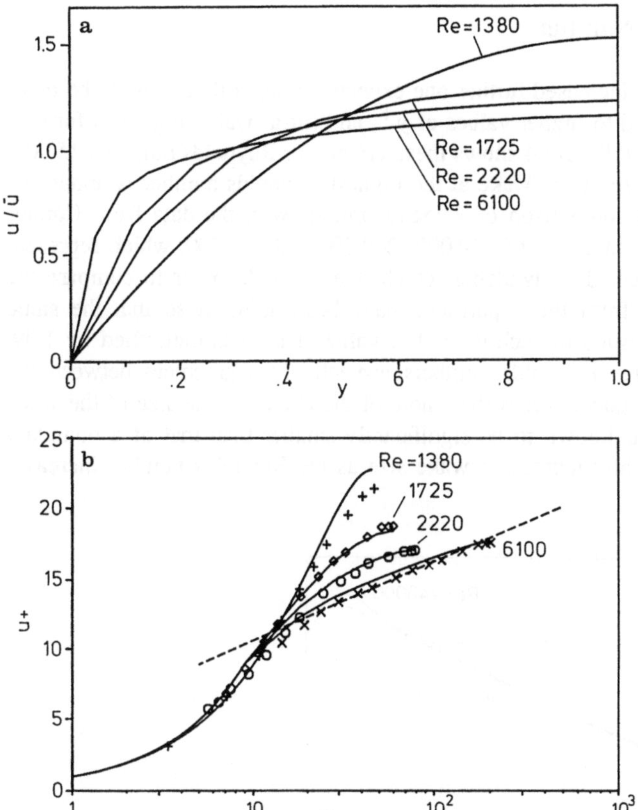

Fig. 3. Model results in the low-Reynolds-number regime: (a) in physical coordinates, (b) in wall units, compared with experimental data from Patel and Head (1969). The solid line is from the model at Reynolds numbers 1380 (laminar flow), 1725, 2220 and 6100. The symbols are the experimental data. The dashed line is the logarithmic law (10)

3.1 Transitional Flow

Solutions to (6) exist for Reynolds numbers greater than 1700. Below this the flow is taken to be laminar. The low Reynolds number region was investigated experimentally by Patel and Head (1969), who found that the trend from a laminar profile to a turbulent profile (approximately satisfying (10)) took place in the Reynolds number range 1380 to 6100. The change in the mean velocity profile from the model over this range is shown on Fig 3(a) in physical coordinates for Reynolds numbers 1380 (laminar), 1725, 2220 and 6100. The same profiles are shown in wall units on Fig. 3(b) and compared with the results from Patel and Head. Agreement with the experimental results is good, both for the trends in the mean flow towards the logarithmic law, and in the shapes of the profiles. Skin friction and channel centreline velocities also agree well with the experimental values.

3.2 Effect of Reynolds Number

As the Reynolds number is increased further one expects the logarithmic law to be more closely approched, to extend to higher values of y^+, and for a 'wake' region to form in the outer part of the channel. Figure 4 shows these effects for Reynolds numbers 15 000, 60 000 and 240 000. The size of the wake at the highest Reynolds number is about one u_τ unit. Figure 5 shows a comparison of velocity profiles with the data from Comte-Bellot (1965) for Reynolds numbers of 114 000, 240 000 and 460 000, which represent the highest Reynolds number data available for channel flow. In order to compare the profiles, the values of u_τ from the experiment have been adjusted so that the same logarithmic law (10) is satisfied in each case. The value of u_τ was multiplied by 1.09, 1.03 and 1.085 for the three Reynolds numbers respectively. Agreement between the model and the data is very good, across the whole of the channel. The size of the wake in turbulent channel flow is known to be significantly smaller than that of a boundary layer. A trend towards slightly increasing wake size as the Reynolds number increases is obtained.

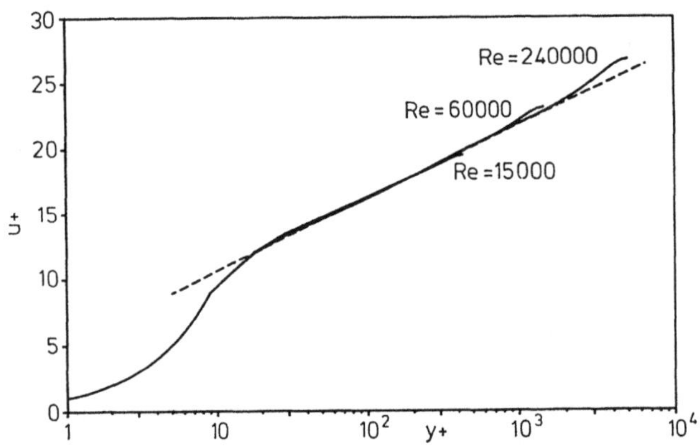

Fig. 4. Evolution of the mean velocity profile in wall coordinates, showing the development of a logarithmic region and a small wake

A large quantity of experimental data was collated by Dean (1978), who produced empirical laws for the variation of skin friction and centreline velocity with Reynolds number. On Fig. 6 the model predictions and Dean's correlations are compared with experimental data from Patel and Head (1969), Hussain and Reynolds (1975) and Comte-Bellot (1965), as well as direct numerical simulation results from Kim, Moin and Moser (1987) and Gilbert (1988). Figure 6(a) shows the skin friction $c_f = 2\nu/\overline{u}^2(du/dy)_w$, Fig. 6(b) the channel centreline velocity $u_c/\overline{u}$ and Fig. 6(c) the shape factor H. The shape factors for the experimental data were computed from the original velocity profiles using trapezoid integration. The model quantitatively predicts all the quantities correctly. Especially notable is the ability of the model to give the correct behaviour of skin friction at low Reynolds numbers. Below the last point of the curves the model predicts laminar flow, so the curves could be joined to the laminar solution by a vertical line. This is not quite correct, as evidenced by the deviation from the experimental data for a Reynolds number of 1380, shown on Fig. 3(b).

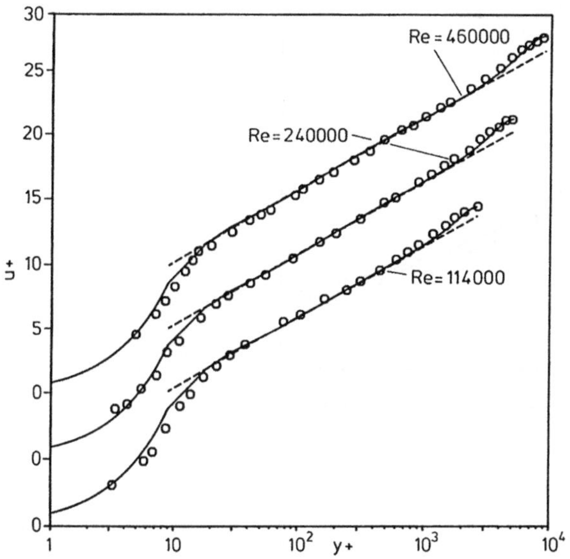

Fig. 5. Comparison of the model velocity profile (solid line) with data from Comte-Bellot (1965) (symbols). The experimental value of u_τ was adjusted so that the data matched the logarithmic law (10)

3.3 Model Terms

It is of interest to examine the behaviour of the individual terms in the model. The variations with y of the ejection term (E), the sweep term (S), the outer-layer term (O), the viscous term (V) and the pressure gradient term (P) at Reynolds number 114 000 are shown on Fig. 7(a) in the near-wall region and on Fig. 7(b) across half the channel width. At this Reynolds number the premultiplying transition factor T is equal to 0.996, and can be ignored. The ejection term is strongest in the buffer region $20 < y^+ < 40$ as expected. In fact the shape of this term is similar to the conditionally sampled streamwise velocity during a burst (see figure 6 in Blackwelder and Kaplan, 1976). In the model this term reduces to zero at $y^+ = u_e^+ = 9.0$. At y_e there is a balance between the sweep term S and the effect of viscosity V. The viscous term uses the physical viscosity ν, and should not be confused with a 'turbulent' viscosity. In the logarithmic region the viscous term has much less influence, and the important terms are the ejections, sweeps and outer-layer motions. In the central part of the channel the ejection term, the sweep term (here producing a decrease in momentum) and the outer-layer term all play a role. The pressure gradient acts uniformly across the channel, but is of negligible importance compared with the other terms.

4. Discussion

The idea for the current model came during a study of the late stages of transition to turbulence (Sandham and Kleiser, 1991). High wall shear was observed to develop,

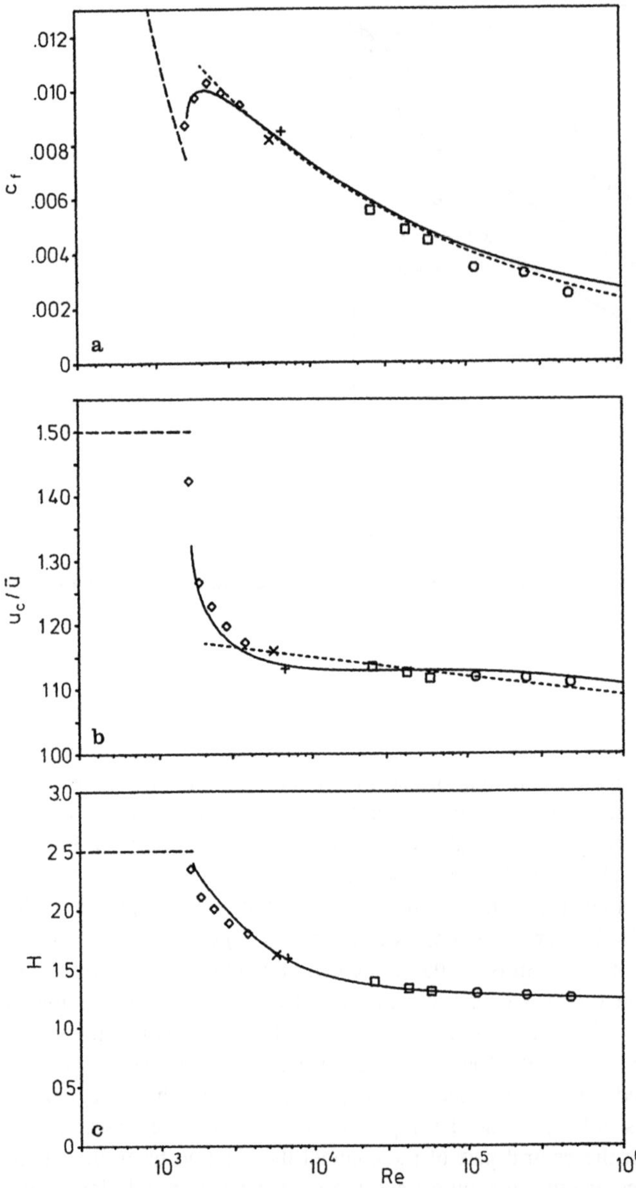

Fig. 6. Variation with Reynolds number of (a) skin friction $c_f = 2\nu/\bar{u}^2(du/dy)_w$, (b) centerline velocity $u_c/\bar{u}$ and (c) shape factor $H = \delta_1/\delta_2$, where δ_1 is the displacement thickness and δ_2 is the momentum thickness. ———, model; - - - -, Dean correlations ($c_f = 0.073Re^{-1/4}$, $u_c/\bar{u} = 1.28Re^{-0.0116}$); — —, laminar flow; ◇, Patel and Head (1969); □, Hussain and Reynolds (1975); ○, Comte-Bellot (1965); ×, Kim, Moin and Moser (1987); +, Gilbert (1988)

caused by vortices in the flow. The question was then posed: why does the shear at the wall not increase indefinitely? The obvious answer of viscosity led directly to the picture

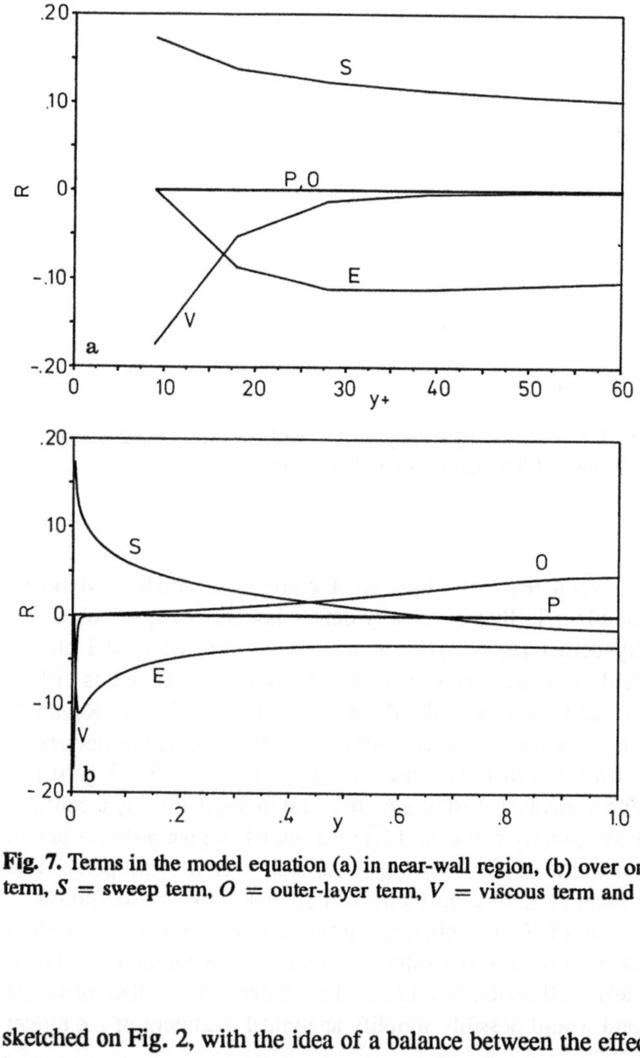

Fig. 7. Terms in the model equation (a) in near-wall region, (b) over one channel-half. E = ejection term, S = sweep term, O = outer-layer term, V = viscous term and P = pressure gradient term

sketched on Fig. 2, with the idea of a balance between the effects of eddies and viscosity in the near wall region. Further development of this picture was pursued in the hope that by considering the direct effects of certain eddy motions in the near-wall region some of the basic physics would be modelled, and the final model would be of more general applicability than the usual empirical approaches via the Reynolds-averaged equations. Whether this new approach can be turned into an engineering tool remains to be seen, but the experience from channel flow is encouraging.

Several informal rules were followed during this work. One was to keep all constants independent of Reynolds number, and another was to avoid the use of any functions (logarithms, exponentials etc.) in the model terms. Thus the logarithmic region of the velocity profile develops naturally rather than being imposed. This leads to the hope that turbulent flows in which there is no logarithmic region (for example 'relaxing' flows) may also be correctly treated. In this respect the low Reynolds number effects in the channel flow can be cited – the model, in agreement with the experiments, does not give a logarithmic region.

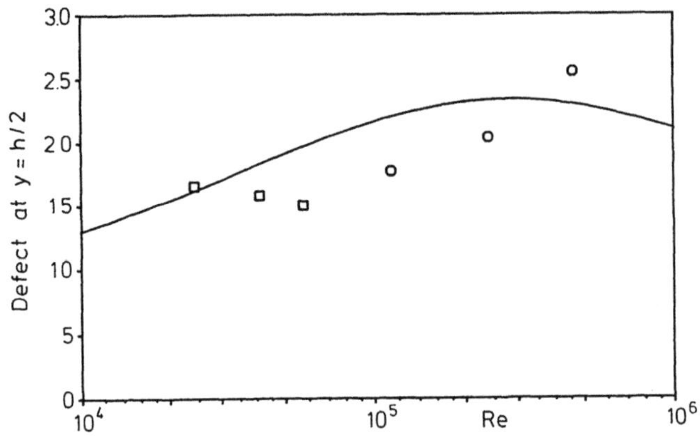

Fig. 8. Variation of the velocity defect $(u_C - u)/u_\tau$ at $y = h/2$ with Reynolds number, showing how a defect law is not quite obtained at the higher Reynolds numbers

The current model has several features which are incorrect, or which it would be preferable to avoid. One problem is that a velocity-defect law is not quite obtained. Figure 8 shows the velocity defect $(u_c - u)/u_\tau$ at $y = h/2$ from the model plotted against Reynolds number and compared with data from Hussain and Reynolds (1975) and Comte-Bellot (1965). It can be seen that the defect is not independent of Reynolds number, suggesting that the treatment of the outer layer is not quite correct. The deviations are generally too small to show up against the experimental velocity profiles (Fig. 5) and it should be noted that the experiments themselves do not convincingly display a constant velocity defect. However, conceptually one would like the model to give a defect law in the limit of infinite Reynolds number. Also, the very near-wall region could perhaps be improved. The current formulation applies only down to y_e, below which the profile is assumed to be linear. The y_e point is fundamental to the success of the model, but it does lead to a discontinuity in the derivative of the velocity profile. A continuous formulation would be preferable in the near-wall region, would avoid the numerical problem of where to put the first few points, and would possibly simplify analytical treatment of the model system.

The change from laminar to turbulent flow is an interesting part of the model. When the Reynolds number is too low there is no solution to the model system and the flow is assumed to be laminar. What the model is really saying is whether turbulent flow can be sustained. Thus the transition processes are really those of 'bypass' transition (where large initial disturbances lead directly to breakdown, bypassing primary and secondary instabilities) or of relaminarisation. Thus a simplified form of the model may perhaps be useful, not for predicting transition, but for prediction of locations in a flow where wall-bounded turbulence would self-sustain.

To classify the model it is helpful to make a comparison with the turbulent boundary-layer equations. The momentum equation reduces to

$$0 = -\frac{d\overline{u'v'}}{dy} + \nu\frac{d^2\overline{u}}{dy^2} - \frac{1}{\rho}\frac{\partial\overline{p}}{\partial x}, \tag{11}$$

where the mean velocity $\overline{u}$ and the Reynolds stress $-\rho\overline{u'v'}$ are functions of y only. The pressure $\overline{p}$ varies in both x and y but the derivative $\partial\overline{p}/\partial x$ is independent of y. Comparing this equation with the model (1) we see that the viscous and pressure gradient terms are identical and the Reynolds stress is related to the other model terms by

$$-\overline{u'v'} = \int T(E + S + O)dy. \tag{12}$$

Thus the Reynolds stress appears as an integral. Effectively this means that the model is forming the Reynolds stress by using information about the entire flow between the wall and the point of interest. Hence the model can be classified as non-local.

The term non-local has been used elsewhere in different contexts. The idea of non-local transport of mass appears in the meteorology literature, termed transilient turbulence by Stull (see for example Ebert et al., 1989). However, the applications have been to model the vertical motion of matter in the atmosphere, rather than the redistribution of momentum near a wall. Perhaps a step in this direction was taken by Rüdiger (1982), who proposed to model the Reynolds stress with a linear integral equation (the current model leads to a nonlinear integral for the Reynolds stress). However, to the best of the author's knowledge nothing like the current model has appeared in the literature.

5. Conclusions

A simple model equation has been developed, which contains as its solution many of the observed features of the mean velocity profile in transitional and turbulent channel flow. Solutions to the model equation exist above a certain Reynolds number. As the Reynolds number is increased the mean profile develops a logarithmic region, and at higher Reynolds numbers a small wake forms. Good quantitative agreement with experiments is obtained for the mean velocity profile, together with the associated measures of skin friction, centreline velocity and shape factor.

Individual terms in the model equation can be considered as modelling directly the transfer of momentum in a channel due to certain types of eddy motions. One term models ejections and transfers low momentum fluid away from the wall. Another models sweeps and carries high momentum fluid towards the wall. A third term is interpreted as modelling motions in the outer region. Each transfer of momentum is a non-local process, depending on conditions near the wall.

Acknowledgement

The author would like to thank L. Kleiser for constructive comments on a draft version of this paper.

References

Blackwelder, R.F., Kaplan, R.E. (1976): "On the wall structure of the turbulent boundary layer", J. Fluid Mech. **76**, 89-112

Brown, G.L., Thomas, A.S.W. (1977): "Large structure in a turbulent boundary layer", Phys. Fluids **20**, S243-S252

Comte-Bellot, G. (1965): "Écoulement turbulent entre deux parois parallèles", Publications Scientifiques et Techniques du Ministère de l'Air, No. 419

Dean, R.B. (1978): "Reynolds number dependence of skin friction and other bulk flow variables in two-dimensional rectangular duct flow", J. Fluids Eng. **100**, 215-223

Ebert, E.E., Schumann, U., Stull, R.B. (1989): "Nonlocal turbulent mixing in the convective boundary layer evaluated from large-eddy simulation", J. Atm. Sci. **46**, 2178-2207

Gilbert, N. (1988): "Numerische Simulation der Transition von der laminaren in die turbulente Kanalströmung", DFVLR FB 88-55

Hussain, A.K.M.F., Reynolds, W.C. (1975) "Measurements in fully developed turbulent channel flow", J. Fluids Eng. **97**, 568-580

Kim, J., Moin, P., Moser, R. (1987): "Turbulence statistics in fully developed channel flow at low Reynolds number", J. Fluid Mech. **177**, 133-166

Kline, S.J., Reynolds, W.C., Schraub, F.A., Runstadler, P.W. (1967): "The structure of turbulent boundary layers", J. Fluid Mech. **30**, 741-773

Kline, S.J. (1990): Comment in Whither Turbulence? Turbulence at the Crossroads. J.L. Lumley, ed., Lecture Notes in Physics **357**, Springer-Verlag, 383-388

Patel, V.C., Head, M.R. (1969): "Some observations on skin friction and velocity profiles in fully developed pipe and channel flows", J. Fluid Mech. **38**, 181-201

Robinson, S.K. (1991): "Coherent motions in the turbulent boundary layer", Ann. Rev. Fluid Mech. **23**, 601-639

Rüdiger, G. (1982): "A heuristic approach to a non-local theory of turbulent channel flow", ZAMM **62**, 95-101

Sandham, N.D., Kleiser, L. (1991): "Vortex formation in the late stages of transition to turbulence", Royal Aeronautical Society, Conference on Boundary-Layer Transition and Control, April 8-11, Cambridge, UK

Contribution to the Modelling of Near-Wall Turbulence

B.E. Launder and D.P. Tselepidakis

Dept. of Mech. Eng., UMIST, PO Box 88, Manchester M60 1QD, UK

Abstract

The paper considers the modelling, within the framework of second-moment closure, of the turbulent stress field in the wall-adjacent sublayer. Particular attention is given to handling wall-proximity effects on the pressure-strain (ϕ_{ij}) and stress-dissipation (ε_{ij}) processes. Comparisons of the resultant model with the direct numerical simulations of plane channel flow show close agreement of the Reynolds stress profiles at two Reynolds numbers.

1. Introduction

As research on wall turbulence shifts away from flows where the near-wall region is a strictly constant-stress layer, the limited validity of the 'universal' logarithmic laws is becoming increasingly recognized. This recognition gives impetus to efforts at devising a generally applicable model for the turbulent stresses within this viscosity-affected region. To achieve the desired width of applicability, it is widely acknowledged that full second-moment closure needs to be adopted. While models of this type have been in use for more than 20 years, only recently have computer resources become sufficiently abundant for one to contemplate applying models of this type to a wide range of engineering flows.

There is no shortage of *proposals* for second-moment closures that purport to apply right up to the wall itself. From Donaldson's [1] early scheme for predicting boundary layer transition to Shih & Mansour's [2] very recent contribution, around a dozen versions have appeared. Most of the proposals emerged before the direct numerical simulations (DNS) of turbulent shear flows became available, notably those of Kim et al [3] of channel flow and Spalart [4] of boundary layers. These simulations, provide unprecedented detail of the near-wall region, including the budgets of the Reynolds stresses. Although the very recent proposals of [2] *have* used the DNS channel flow data as the basis for model calibration, their proposals contain flow-specific features (such as the dependence of sublayer parameters on the overall boundary layer thickness) that make the model unsuitable for wider application. The subject of the present paper is the adaptation of the closure for free shear flows, presented at TSF 7 in Stanford [5,6], so as to be applicable in the vicinity of the wall. The particular focus is the problem of accounting for the wall's effect on the pressure-strain (ϕ_{ij}) and dissipation (ε_{ij}) processes.

Traditionally the turbulent Reynolds number, R_t, has been the principal (and often the only) parameter adopted for modifying these processes. There are, however, other routes for achieving the desired changes in ϕ_{ij} and ε_{ij} as the wall is approached and these may prove superior. While the use of R_t has not been entirely eliminated, it has been adopted here as a *last* rather than a first resort. Moreover, its influence is confined to regions where turbulent transport is no more than one and a half orders of magnitude greater than molecular transport, a limit that appears desirable on physical grounds.

The present contribution assembles into a single publication the modelling proposals and results reported in two recent conference proceedings [7,8]. It is thus the definitive publication from our study though, particularly with regard to ε_{ij}, further work is evidently needed.

2. Closure Proposals

2.1 Preliminary Considerations

The exact transport equations for the kinematic Reynolds stress, $\overline{u_i u_j}$, for a uniform density fluid medium unaffected by force fields may be obtained by taking a velocity-weighted moment of the Navier Stokes equations. The result may be expressed:

$$\begin{aligned}\frac{D\overline{u_i u_j}}{Dt} = & \underbrace{-\left(\overline{u_i u_k}\frac{\partial U_j}{\partial x_k} + \overline{u_j u_k}\frac{\partial U_i}{\partial x_k}\right)}_{P_{ij}} \\ & \underbrace{+\overline{\frac{p}{\rho}\left(\frac{\partial u_i}{\partial x_j} + \frac{\partial u_j}{\partial x_i}\right)}}_{\phi_{ij}} \underbrace{-2\nu\overline{\frac{\partial u_i}{\partial x_k}\frac{\partial u_j}{\partial x_k}}}_{\varepsilon_{ij}} \\ & \underbrace{+\frac{\partial}{\partial x_k}\left(\nu\frac{\partial\overline{u_i u_j}}{\partial x_k}\right)}_{d_{ij}^{\nu}} \underbrace{-\frac{\partial}{\partial x_k}\left(\overline{u_i u_j u_k} + \frac{\overline{pu_i}}{\rho}\delta_{jk} + \frac{\overline{pu_j}}{\rho}\delta_{ik}\right)}_{d_{ij}} ,\end{aligned} \tag{1}$$

where upper and lower case U's denote the mean and turbulent parts of the velocity vector and the remaining notation is standard. The closure problem consists of devising models for ϕ_{ij}, ε_{ij} and d_{ij} (since P_{ij} and d_{ij}^{ν} require no approximation). Of these the first two are generally regarded as the most crucial and it is on these that the present paper focuses.

The aim of adopting second-moment closure is logically to achieve greater width of applicability and a closer proximity to the real processes than is possible with eddy viscosity models. This means that non-local parameters such as the wall friction velocity, the distance from the wall or, worse, the boundary-layer thickness or pipe diameter ought not to appear as elements of these closures. Even the unit vector n_i marking the direction normal to the wall, that has been extensively used over the past decade, ought to be discarded, at least if one is serious about predicting strongly non-planar surfaces. In place of these quantities, one endeavours to mimic experimental behaviour by including invariant parameters of the turbulence field into the closure proposals.

A local parameter that *is* admissible is the turbulent Reynolds number, $R_t \equiv k^2/\nu\varepsilon$. However, this parameter, the valium of the turbulence modelling industry, has been freely (and sometimes irresponsibly) prescribed to fix all modelling problems from the decay of grid turbulence to the spreading rate of the round jet. Side effects are rarely considered. While, in a parallel flow, it may be used to hasten the fall off of the turbulent shear stress as the wall is approached in accord with observations, the variation of the normal-stress components remains indifferently captured. Yet R_t is just one parameter, from several, that may be used in the treatment of near-wall turbulence; a satisfactory 'cure' – if one is to be found – demands the use of a more extensive arsenal.

While it may be desirable in the future to consider invariants of the stress *dissipation tensor* [9], for the moment, in addition to R_t we include only the two invariants of the stress tensor itself: $A_2 \equiv a_{ij}a_{ij}$ and $A_3 \equiv a_{ij}a_{jk}a_{ki}$, where $a_{ij} \equiv (\overline{u_i u_j} - \tfrac{1}{3}\delta_{ij}\overline{u_k u_k})/k$. In fact, in place of A_3, it is convenient to adopt Lumley's [10] "flatness" parameter A:

$$A \equiv 1 - \frac{9}{8}(A_2 - A_3)$$

which ranges in value from unity to zero as the turbulent fluctuations are "flattened" from being isotropic to lying entirely in a plane (as they do in the limit as the wall is approached). The variation of these invariants and R_t across the near-wall region of plane channel flow appears in Fig. 1. Evidently, all the parameters undergo rapid change across the sublayer and within that region they assume values outside of the range encountered in free shear flows. These features are helpful to the aim of producing a model for near-wall turbulence that is also applicable in free flows. We note that in this particular flow the turbulent Reynolds number reaches its maximum value towards the edge of the buffer layer and decreases thereafter. This characteristic is due to the low bulk-flow Reynolds number of the simulation. At high Reynolds number (bulk Reynolds number greater than about 2×10^4) the peak value of R_t would occur in the core region.

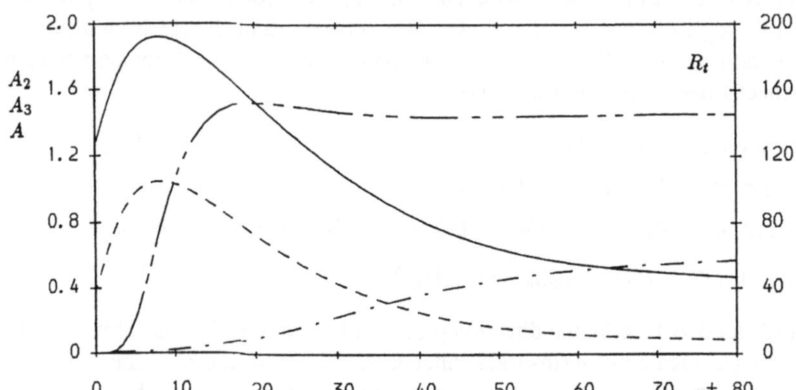

Fig. 1. Distribution of invariant parameters across the sublayer region at $Re = 5600$ (from DNS data, [3]). A_2: ———— ; A_3: — — — —; A: . —— . —— . ; R_t: - - ——— - -

In common with leading US centres of turbulence modelling research, the modelling effort at UMIST has for several years striven to adopt approximations for the different processes that are consistent with the *two-component limit* in which, in one direction, velocity fluctuations entirely vanish. Although this may, at first, seem a foolishly extreme state with which to constrain a turbulence model, it should be noted that this is, indeed, the limit that is reached as the wall is approached since, as noted, fluctuations normal to the wall vanish faster than those parallel to the surface. The constraint also helps ensure that the model is *realizable* [10,12], i.e. that impossible (as opposed to merely inaccurate) values of the Reynolds stresses cannot be returned – such as negative mean square velocity fluctuations. One might expect the satisfaction of the two-component limit would be especially helpful in predicting wall turbulence; yet, paradoxically, the model which it spawned led to marked improvements in the computation of *free* shear flows [5,6,13].

All model development requires guidance from a suitable data base and, like Shih and Mansour [2], we have adopted the DNS channel-flow data of Kim et al [3] as the most complete and reliable available. In order to decouple the problems of approximating the Reynolds stress from that of obtaining the turbulence energy dissipation rate we have, during the initial phases of development, prescribed from the direct simulation the distribution of the turbulent length scale $l \equiv k^{3/2}/\varepsilon$. In the second-moment computations the level of the energy dissipation rate was then obtained from:

$$\varepsilon = \frac{k^{3/2}}{l_{\rm dns}},$$

where the subscript dns serves as a reminder that the length scale is fixed from the direct simulations. The apparently more direct route of fixing ε itself is not viable for then the computed stress levels either collapse to zero or explode to infinity due to the absence of feedback from the sink (ε) to the source terms in the $\overline{u_i u_j}$ equations.

2.2 Modelling the Pressure-Strain Process

Adjustments to the Free Flow Form. The starting point for a model of ϕ_{ij} for near-wall flows is logically one for free flows. The version adopted here is basically that developed and used at UMIST for the calculation of a wide range of homogeneous and inhomogeneous free shear flows e.g. [5,6]:

$$\begin{aligned}\phi_{ij} = & -\tilde{c}_1\left[a_{ij} + c_1'\left(a_{ik}a_{kj} - {}^1\!/\!_3\delta_{ij}A_2\right)\right]\varepsilon \qquad \Big\}\ \phi_{ij1} \\ & \left.\begin{aligned} & -0.6\left(P_{ij} - {}^1\!/\!_3\delta_{ij}P_{kk}\right) + 0.3\varepsilon a_{ij}\left(P_{kk}/\varepsilon\right) \\ & -0.2\left[\overline{u_k u_j}\,\overline{u_l u_i}\left(\frac{\partial U_k}{\partial x_l} + \frac{\partial U_l}{\partial x_k}\right) - \overline{u_l u_k}\left(\overline{u_i u_k}\frac{\partial U_j}{\partial x_l} + \overline{u_j u_k}\frac{\partial U_i}{\partial x_l}\right)\right]/k \\ & -r\left[A_2\left(P_{ij} - D_{ij}\right) + 3a_{li}a_{kj}\left(P_{lk} - D_{lk}\right)\right]\end{aligned}\right\}\ \phi_{ij2}\end{aligned} \qquad (2)$$

Equation (2) follows broadly the direction proposed by Lumley [10] and Shih and Lumley [14] though, regarding mean-strain influences, ϕ_{ij2}, there are also some important philosophical and practical differences from that proposed in [14] (see [5]).

The earlier studies noted above included A as a factor in the numerator of $\tilde{c}_1$, a choice which would appear, from Fig. 1, to guarantee that the process ϕ_{ij1} vanished at the wall, as desired. However *consistency* of the model with the two-component limit does not unfortunately enforce compliance with it. As shown in [16], unless $\tilde{c}_1$ is independently forced to vanish at the wall, $\overline{u_2^2}/k$ remains finite there (x_2 being the direction normal to the surface) and thus neither A nor $\tilde{c}_1$ becomes zero. Only R_t seems capable of enforcing the desired limit, the form adopted being:

$$\tilde{c}_1 = 6.3 A F^{1/2} (1-f)$$
$$c_1' = 0.7\,;\quad f = \max(1 - R_t/140\,, 0)\,;\quad F = \min(0.6\,, A_2).$$

The model of the process ϕ_{ij2} has also been constructed so that it vanishes in the two-component limit. Nevertheless, like that for ϕ_{ij1}, it requires minor adaptation as the wall is approached. The problem stems from the cubic term in the shear stress equation ($\overline{u_1 u_2}$) for it leads to a sink of shear stress equal to:

$$r\left[(a_{11} - a_{22})\left(\tfrac{1}{2}A_2 - 3\,a_{33}^2\right)\right] k \frac{\partial U_1}{\partial x_2}.$$

Both $(a_{11} - a_{22})$ and A_2 soar rapidly as the wall is approached leading to the annihilation of $\overline{u_1 u_2}$. To limit the impact of the term, the free coefficient r is taken as:

$$r = \min(0.6, A).$$

The above proposals differ from the recommendations in [5] for large anisotropies ($A_2 > 0.6$; $A < 0.6$) but in practice such extreme levels of these invariants are only rarely encountered other than in the near-wall region.

Inhomogeneity Correction. The exact pressure-strain process may helpfully be expressed in a form from which the pressure has been eliminated, [17]:

$$\phi_{ij} = \frac{1}{4\pi}\int_{vol} \overline{\left(\frac{\partial^2 u_l u_m}{\partial x_l \partial x_m}\right)'\left(\frac{\partial u_i}{\partial x_j} + \frac{\partial u_j}{\partial x_i}\right)} \frac{d\,Vol}{|\underline{r}|}$$
$$+ \frac{1}{2\pi}\int_{vol}\left(\frac{\partial U_l}{\partial x_m}\right)' \overline{\left(\frac{\partial u_m}{\partial x_l}\right)'\left(\frac{\partial u_i}{\partial x_j} + \frac{\partial u_j}{\partial x_i}\right)} \frac{d\,Vol}{|\underline{r}|} + S_{ij}\,, \quad (3)$$

where the primes denote quantities evaluated at distance $|\underline{r}|$ from where ϕ_{ij} is evaluated and the integration is over all the flow region. S_{ij} refers to surface integrals arising from the fact that in wall flows pressure fluctuations do not vanish at the rigid boundary.

Equation (2) for ϕ_{ij1} and ϕ_{ij2} may, in free flows, be regarded as a model of the terms in (3) involving the volume integrals. In modelling near-wall turbulence, corrections are usually introduced to ϕ_{ij1} and ϕ_{ij2} notionally to account for the effects of the surface integrals (viz the next subsection). There is, however, a further assumption implicit in (2) that becomes highly questionable near the wall, namely the idea that the flow is *homogeneous*. For example, in arriving at the above model for ϕ_{ij2}, the assumption is made that $(\partial U_l/\partial x_m)'$ can be replaced by $(\partial U_l/\partial x_m)$ and thus taken through the integral in (3). The process ϕ_{ij2} is then represented as:

$$\phi_{ij2} = \frac{\partial U_l}{\partial x_m}\left(a_{lj}^{mi} + a_{li}^{mj}\right)\,, \quad (4)$$

where the fourth-rank tensor a_{ij}^{mi} is expressed as a series in ascending powers of a_{ij}, viz. [5,14].

Bradshaw et al [18] have used the data base provided by the direct simulations of channel flow [3] to explore how serious was the error in treating $(\partial U_l/\partial x_m)'$ as $(\partial U_l/\partial x_m)$. They concluded that the error was negligible *except for* x_2^+ *less than 30* where it became very large. Evidently, therefore, any closure attempting to span this near-wall sublayer ought to include some inhomogeneity correction. The practice adopted here is to replace $\partial U_l/\partial x_m$ in (4) by an *effective* velocity gradient $(\partial U_l/\partial x_m)_{\rm eff}$ where:

$$\left.\frac{\partial U_l}{\partial x_m}\right]_{\rm eff} = \frac{\partial U_l}{\partial x_m} + c_l\, l' \frac{\partial l'}{\partial x_k} \frac{\partial^2 U_l}{\partial x_k\, \partial x_m} , \tag{5}$$

where $l' \equiv (k/\varepsilon)(\overline{u_p u_q}\, n_p n_q)^{1/2}$ is a representative length scale normal to the wall. A discussion of the physical basis for this term is given in [7]. The reason for adopting l' as length scale rather than the more common l $(\equiv k^{3/2}/\varepsilon)$ is as much due to workability as physics: l increases very rapidly across the sublayer leading to a strong peak in the inhomogeneity term around $x_2^+ = 12$ and negligible values beyond $x_2^+ = 20$. By contrast the use of l' produces a more uniform influence over the range $10 < x_2^+ < 30$ which accords both with the findings of Bradshaw et al [18] and with our empirical discoveries of what is needed to bring accord with the data for the stress profiles. Results for two values of c_l are reported in §3.

If, as the direct simulations indicate, one needs to recognize inhomogeneities in the mean velocity gradient, one should probably also acknowledge those in the turbulent stress field (in the range $7 < x_2^+ < 20$, for example, the shear stress $\overline{u_1 u_2}$ varies at a rate comparable with $\partial U_1/\partial x_2$). For the present, however, no attempt has been made to do so.

Wall Reflection Effects. The contribution of the surface integrals S_{ij} in (3) is usually handled by way of wall corrections for both ϕ_{ij1} and ϕ_{ij2}. However, the processing of the direct-simulation data of homogeneous shear flow by Brasseur and Lee [19] showed that the volume integral arising from ϕ_{ij1} was associated with much finer-scale interactions than ϕ_{ij2}; thus, in wall turbulence we may expect that the latter process would "feel" the effects of the wall more than ϕ_{ij1}. In view of the fact that ϕ_{ij1} is already damped by the turbulent Reynolds number, we thus apply an explicit wall correction only to the process ϕ_{ij2}[1]. For present purposes we adopt broadly the version of Gibson & Launder [15]:

$$\phi_{ij2}^{w} = c_2^{w} \left(\phi_{km2}\, n_k n_m\, \delta_{ij} - \tfrac{3}{2}\phi_{ik2}\, n_k n_j - \tfrac{3}{2}\phi_{jk2}\, n_k n_i\right) \frac{l'}{x_n} , \tag{6}$$

where x_n is the normal distance from the wall. As discussed in §2.1 the use of wall distance is undesirable since near surfaces of very complex topography it is impossible to identify a physically meaningful distance. In mitigation, the importance of this process in the overall stress budget is here much less than in older closures such as [15]. Indeed, for the case of channel flow, at least, we show in §3 that it can be dropped entirely by suitably increasing the inhomogeneous correction discussed in the previous subsection.

[1] This choice is an oversimplification. Turbulent flow near a wall moving at the same speed as the mean flow is known to suffer wall proximity effects even though mean velocity gradients (and thus ϕ_{ij2}) are absent. In that case, therefore, the agency responsible must be ϕ_{ij1}.

(It seems rather doubtful, however, whether the process can be completely eliminated in the case of impinging wall flows such as that considered by Craft and Launder [20] elsewhere in these proceedings).

2.3 The Dissipation Tensor

The most popular route to modelling ε_{ij} is to assume local isotropy and imagine, following Lumley [10], that any anisotropies actually present are absorbed in the model for the process ϕ_{ij1}. In near-wall turbulence there are good reasons for *not* doing this, however: for one thing the limiting values of the relative component dissipation rates are different from and governed by different processes than ϕ_{ij}. Moreover, Kim et al [3] have processed their direct simulation data to provide values of ε_{ij} up to the wall thus providing values for direct comparison with computation.

The usual way of obtaining the limiting ratios of $\varepsilon_{ij}/\varepsilon$ at the wall is to expand the fluctuating velocities in a Taylor series and evaluate the gradient at $x_2 = 0$. This leads to the results [21]:

$$\frac{\varepsilon_{11}}{\varepsilon} = \frac{\overline{u_1 u_1}}{k} \quad ; \quad \frac{\varepsilon_{33}}{\varepsilon} = \frac{\overline{u_3 u_3}}{k} \quad ; \quad \frac{\varepsilon_{22}}{\varepsilon} = 4\frac{\overline{u_2 u_2}}{k} \quad ; \quad \frac{\varepsilon_{12}}{\varepsilon} = 2\frac{\overline{u_1 u_2}}{k} \quad . \tag{7}$$

The coefficients of ε_{12} and ε_{22} differ from unity because the leading term of the series for u_2 is taken proportional to x_2^2 to give $\partial u_2/\partial x_2 = 0$ at $x_2 = 0$. In fact, the use of a series containing only integer exponents of x_2 is not strictly valid if one seeks information about velocity derivatives at $x_2 = 0$. To recognize this one may note simply that $\partial u_2/\partial x_2 = 0$ at $x_2 = 0$ for any leading term $a_n x^{n+1}$ provided that n is greater than zero. Such a generalized form leads to:

$$\frac{\varepsilon_{22}}{\varepsilon}\frac{k}{\overline{u_2 u_2}} = (1+n)^2 \quad ; \quad \frac{\varepsilon_{12}}{\varepsilon}\frac{k}{\overline{u_1 u_2}} = (1+n) \ . \tag{8}$$

While there seems no obvious reason why the Navier Stokes equations should not support a value of n intermediate between zero and unity, as will be seen later the direct simulation data do strongly suggest $n = 1$. The appropriate limiting values may be achieved with the following model:

$$\varepsilon^*_{ij} = \frac{\varepsilon}{k}\left(\overline{u_i u_j} + \overline{u_i u_k}\, n_j n_k + \overline{u_j u_k}\, n_i n_k \right.$$
$$\left. + \overline{u_k u_l}\, n_k n_l n_i n_j\right) / \left(1 + {}^3\!/_2 \overline{u_p u_q}\, n_p n_q / k\right) \ . \tag{9}$$

We suppose that this highly non-isotropic limiting form gives way to a more isotropic behaviour away from the wall as the Reynolds number increases and the anisotropy of the stress field diminishes:

$$\varepsilon_{ij} = f_\varepsilon\, \varepsilon^*_{ij} + (1 - f_\varepsilon)\, \varepsilon'_{ij} \ . \tag{10}$$

In principle the quantity ε'_{ij} may be expected to be a Reynolds-number-dependent parameter ranging from the locally isotropic limit $\varepsilon'_{ij} = {}^2\!/_3 \delta_{ij}\varepsilon$ at very high Reynolds numbers to $\overline{u_i u_j}\,\varepsilon/k$ as R_t goes to zero. However, for the present we simply adopt the former irrespective of R_t. The weighting function f_ε that produces the changeover between ε^*_{ij} and ε'_{ij} may plausibly be modelled in terms of the stress anisotropy; indeed the parameter A offers a convenient way of doing this; we take:

$$f_\varepsilon = \exp\left(-20A^2\right) \ .$$

The question of a further amendment to the model of ε_{ij} is discussed in §3.

2.4 Stress Diffusion

Aside from viscous effects, our work has retained the model of Daly and Harlow [22] adopted for free shear flows. It is, however, interpreted as the *net* diffusion due to both velocity and pressure fluctuations rather than to velocity fluctuations alone:

$$d_{ij} = -c_s \frac{\partial}{\partial x_k}\left(\frac{k}{\varepsilon}\overline{u_k u_l}\frac{\partial \overline{u_i u_j}}{\partial x_l}\right) \ , \tag{11}$$

where c_s retains the value of 0.22 used in free shear flows.

As the wall is approached however, the turbulent triple products of velocity vanish faster than the pressure-velocity products (since, unlike the fluctuating velocities, pressure fluctuations do not vanish at the wall). It may be demonstrated that inclusion of a viscous-like pressure-diffusion process is essential if the equations for the stress components $\overline{u_1 u_2}$ and $\overline{u_2^2}$ are to balance at the wall. For example, with:

$$\begin{aligned} u_1 &= a_1 x_2 + a_2 x_2^2 + \ldots \\ u_2 &= \qquad\quad b_2 x_2^2 + \ldots \end{aligned}$$

we find $\overline{u_1 u_2} = \overline{a_1 b_2} x_2^3 + \ldots$ and $\varepsilon_{12} = 4\overline{a_1 b_2} x_2 + \ldots$. Thus, the budget equation for the shear stress $\overline{u_1 u_2}$ reduces to:

$$\left(6\nu\overline{a_1 b_2} - 4\nu\overline{a_1 b_2}\right) x_2 + \phi_{12} + \text{pressure diffusion} = 0 \ .$$

The model adopted for ϕ_{ij} implies at least a quadratic dependence of ϕ_{12} on x_2. Thus, in order that the $\overline{u_1 u_2}$ budget should balance for $x_2 \to 0$ we must accept a non-vanishing pressure diffusion. An analogous imbalance arises with the $\overline{u_2^2}$ budget also. Lumley [23] has shown that in the limit of weakly anisotropic turbulence, pressure transport opposes and makes more isotropic transport by triple velocity products. One may suppose an analogous process at work in the sublayer i.e. that due to turbulent pressure fluctuations, the effective viscous diffusion will be reduced and rendered more isotropic. Following this line of argument, the following model of the pressure transport due to viscous effects – denoted d_{ij}^p – is arrived at:

$$d_{ij}^p = -\tfrac{1}{3}\left(d_{ik}^\nu n_k n_j + d_{jk}^\nu n_k n_i\right) + \tfrac{1}{3}\delta_{ij} d_{kl}^\nu n_k n_l \tag{12}$$

and this is added to d_{ij}. It is readily verified that the above form correctly balances the $\overline{u_1 u_2}$ and $\overline{u_2^2}$ equations in the immediate wall vicinity while making a negligible contribution to the $\overline{u_1^2}$ and $\overline{u_3^2}$ budgets.

2.5 The Dissipation Rate Equation

The proposed equation for determining ε is similar to that adopted for free shear flows [5,6]:

$$\frac{D\varepsilon}{Dt}=\frac{\partial}{\partial x_k}\left[\left(c_\varepsilon\frac{\overline{u_k u_l}k}{\varepsilon}+\nu\,\delta_{kl}\right)\frac{\partial\varepsilon}{\partial x_l}\right]+\frac{1}{2}c_{\varepsilon 1}\frac{\varepsilon}{k}P_{kk}-c_{\varepsilon 2}\frac{\varepsilon^2}{k}\ , \tag{13}$$

where $c_\varepsilon = 0.18$, $c_{\varepsilon 1} = 1.0$ and $c_{\varepsilon 2} = 1.92/[1+0.63(AA_2)^{1/2}]$. Three modifications are introduced, however, to render it suitable for use in the viscous sublayer. Following [24], the sink term in (13) is modified to:

$$c_{\varepsilon 2}\frac{\varepsilon\tilde{\varepsilon}}{k}\quad;\quad \tilde{\varepsilon}\equiv\varepsilon-2\nu\left(\frac{\partial k^{1/2}}{\partial x_k}\right)^2$$

and a term involving the second derivative of the mean velocity is included on the right of (13):

$$2\,c_{\varepsilon 3}\,\nu\,\frac{k}{\varepsilon}\,\overline{u_k u_l}\,\frac{\partial^2 U_i}{\partial x_k\,\partial x_j}\,\frac{\partial^2 U_i}{\partial x_l\,\partial x_j}\ ,$$

where $c_{\varepsilon 3}$ takes the value 0.43. The third modification, proposed originally in [16] is the inclusion of a further term of the right side of (13), that is diffusive in character and which notionally represents the effects of pressure diffusion in the viscous sublayer:

$$d_\varepsilon^P=\frac{\partial}{\partial x_k}\left(2c_{\varepsilon 4}\,\nu\,\frac{\tilde{\varepsilon}}{k}\,\frac{\partial k}{\partial x_k}\right)\ , \tag{14}$$

where $c_{\varepsilon 4}$ takes the value 0.92. Note that since k varies as x_2^2 near the surface, the term tends to raise the level of ε there. Its effect, like those of the other amendments noted above, is insignificant outside of the sublayer region.

3. Application of the Model to Plane Channel Flows

The $\overline{u_i u_j}$, ε and momentum equations have been solved numerically for the case of fully-developed plane channel flow. One hundred nodes covered the region between the wall and the symmetry plane with about half of these concentrated in the region $x_2^+ < 40$. Figure 2 shows the distribution of the normal stresses normalized by k across the near-wall region for a Reynolds number of 5600; here $c_2^w = 0.2$ and $c_l = 0.3$. The DNS distributions are generally well captured by the model computations with $\overline{u_2^2}/k$ falling to zero as the wall is approached. However, for $x_2^+ < 10$, the simulation data show an interesting drop in $\overline{u_1^2}/k$ and a corresponding rise in $\overline{u_3^2}/k$ that is entirely missed by the computations.

The nature of the disagreement seems to suggest a pressure-induced transfer between x_1 and x_3. Such a transfer can be reproduced in the model by the inclusion of the following term in the $\overline{u_i u_j}$ equation:

$$-c_1^*\left[2\,a_{ij}+\left(a_{pq}\,n_p n_q\,\delta_{ij}-{}^3\!/\!_2 a_{pi}\,n_p n_j-{}^3\!/\!_2 a_{pj}\,n_p n_i\right)\right](\varepsilon-\tilde{\varepsilon})\ . \tag{15}$$

Note that $\tilde{\varepsilon}$ vanishes at the wall but is negligibly different from ε for $x_2^+ > 10$ so the influence of the term is automatically confined to the viscous sublayer, as desired.

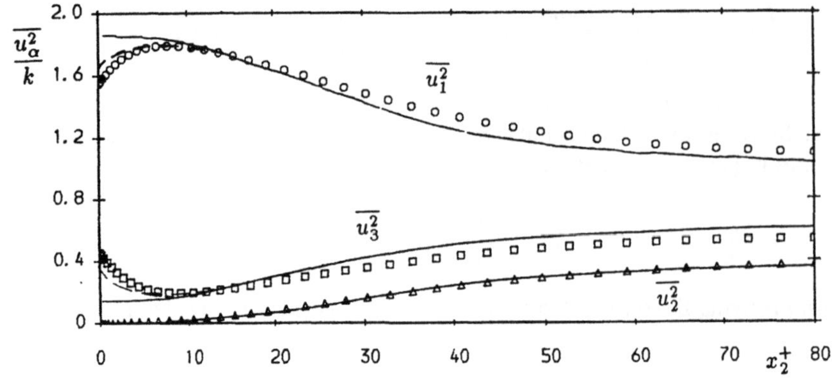

Fig. 2. Normalized profiles of normal stresses across sublayer at $Re = 5600$. Symbols: DNS data; ———— : present computations; — — — — — : present computations with inclusion of (15)

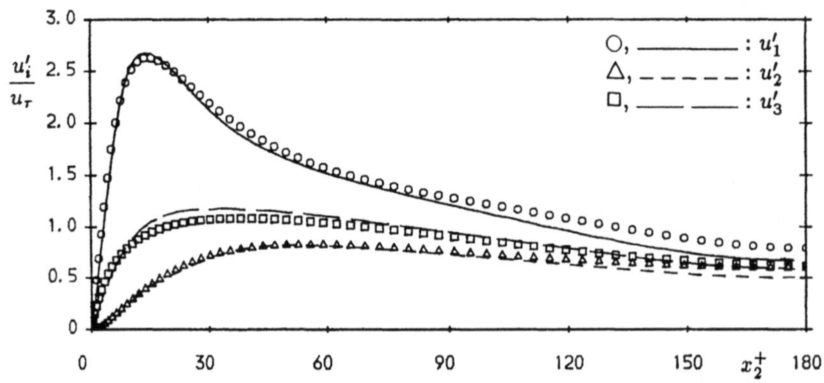

Fig. 3. Turbulence intensity profiles in plane channel at $Re = 5600$. Symbols: DNS data; Lines: present computations

With $c_1^* = 0.05$ a much improved distribution of $\overline{u_1^2}$ and $\overline{u_3^2}$ results. While the term is redistributive, it is unlikely that the phenomenon arises from ϕ_{ij} since, in the exact process, both ϕ_{11} and ϕ_{33} vanish at the wall. It seems more likely that the term is really providing a small correction to the model of ε_{ij}. Apart from the effect shown the term seems to be passive and it has thus been retained in all subsequent computations. Figures 3 and 4 show the computed rms velocity profiles and turbulent shear stress for this case; very close accord with the data is achieved. The corresponding profiles at the higher Reynolds number of 13750 appear in Figs. 5 and 6: agreement is again acceptably close though, by comparison with the previous case it is clear that the computations exhibit rather less sensitivity to Reynolds number than do the simulations. This is a little surprising since the present model seems to reproduce correctly the shift in the additive constant in the logarithmic law for the mean velocity.

In fact the best choice of c_l and c_2^w is hard to determine because both act in the same sense. Thus a reduction in one can be nearly offset by an increase in the other, at least in a simple shear flow. To illustrate this point, Fig. 7 shows the computed rms

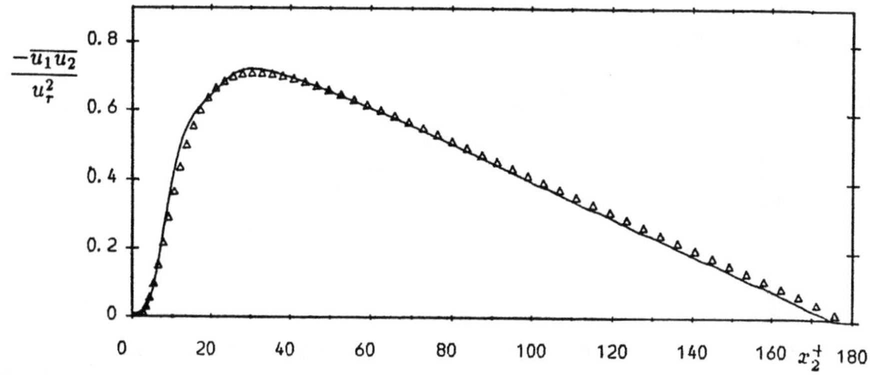

Fig. 4. Turbulent shear stress profile in plane channel at $Re = 5600$. △: DNS data; Line: present computations

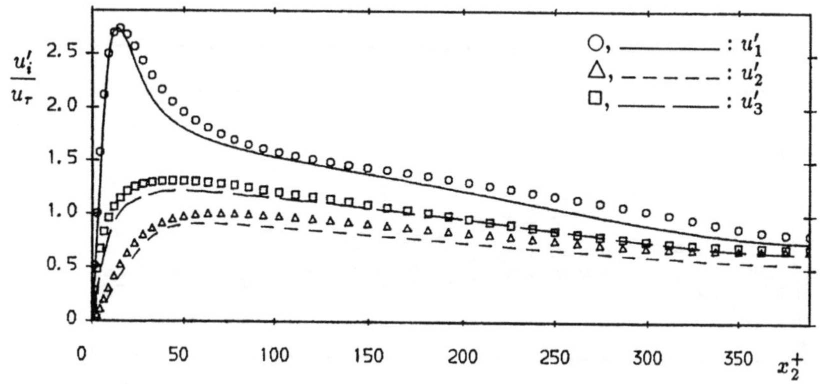

Fig. 5. Turbulence intensity profiles in plane channel at $Re = 13750$. Symbols: DNS data [11]; Lines: present computations

fluctuating velocity profiles that result from taking $c_l = 0.4$ and $c_2^w = 0$, i.e. *with the wall reflection term eliminated entirely*. The agreement with the DNS profiles is still very close though not quite as good as in Fig. 3. Nevertheless, the prospect that one *may* be able to devise a widely applicable 2nd moment closure that makes no reference to wall distance is certainly interesting from a practical point of view.

Let us now consider the "bad news" – the stress budgets. The profiles in Fig. 8 have all been obtained with the additional exchange process implied by (15) included; for book-keeping purposes it has been allocated to ε_{ij} – though it has only a very secondary effect on the budgets. For clarity the only modelled terms included are ε_{ij}, ϕ_{ij1} and ϕ_{ij2}; in the last of these we have included both the 'inhomogeneous' and the 'wall-reflection' contributions as these were not separated in the simulations. In both the $\overline{u_1^2}$ and $\overline{u_1 u_2}$ components the production term is reasonably well represented which is simply a consequence of the mean velocity and Reynolds stress profiles themselves being well represented.

In the budget for the streamwise component, the profiles are broadly in accord with the simulations – for example the fact that ϕ_{ij1} exceeds ϕ_{ij2} whereas in the free homogeneous

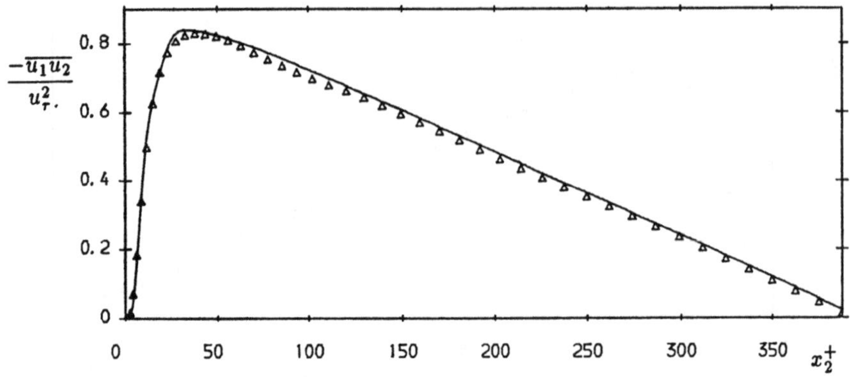

Fig. 6. Turbulent shear stress profile in plane channel at $Re = 13750$. △: DNS data [11]; Line: present computations

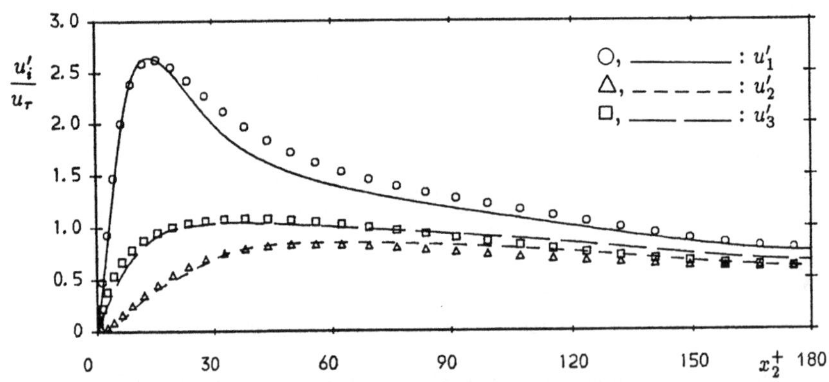

Fig. 7. Turbulence intensity profiles in plane channel at $Re = 5600$. Symbols: DNS data; Lines: present computations without wall reflection effects ($c_l = 0.4$, $c_2^w = 0$)

shear flow [19], the reverse is true. As with all other second-moment closures known to the authors the maximum in ε_{11} does not occur at the wall; in the present case the decrease in ε for values of $x_2^+ < 6$ is relatively small. It is difficult to say anything very encouraging about the other components. In the range $20 < x_2^+ < 60$ the magnitude of both ε_{ij} and ϕ_{ij1} is too high in the $\overline{u_2^2}$ and $\overline{u_3^2}$ components. The difficulties with ϕ_{ij2} are arguably even greater: in the $\overline{u_2^2}$ component the model gives values an order of magnitude too small while for $\overline{u_3^2}$ it is some 40-60% below the simulations except within the sublayer itself.

In view of the very poor agreement with the DNS stress budgets, it seemed worthwhile examining the simulation results in more detail to see whether in some respects they exhibited strong departures from conventional views on the turbulence structure. As an illustration, Fig. 9 shows the dimensionless dissipation/stress profiles, $\varepsilon_{\alpha\beta}\, k/\overline{u_\alpha u_\beta}\, \varepsilon$ (no summation), deducted from the direct simulation. We note first that at the wall the dimensionless dissipation ratios given by the direct simulations do indeed agree with

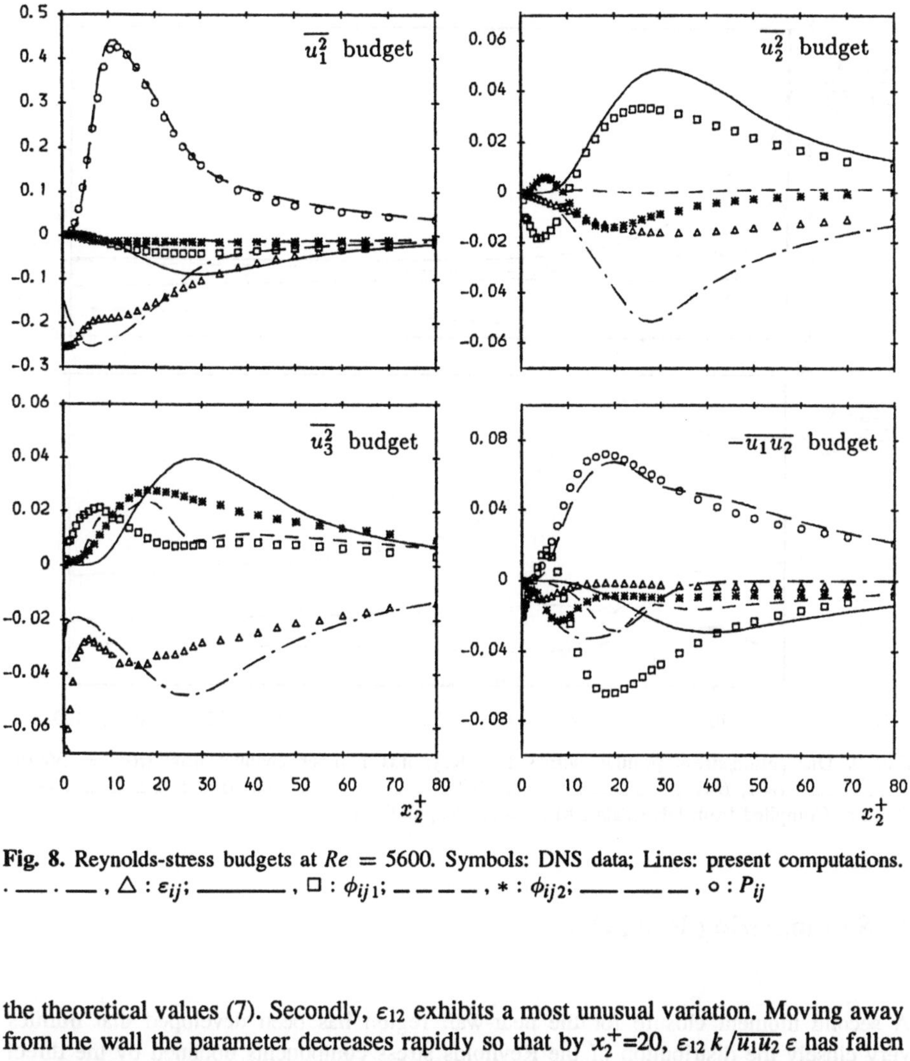

Fig. 8. Reynolds-stress budgets at $Re = 5600$. Symbols: DNS data; Lines: present computations. $\cdot$ — $\cdot$ —, $\triangle$: ε_{ij}; ———, $\square$: ϕ_{ij1}; – – – –, $*$: ϕ_{ij2}; —— ——, $\circ$: P_{ij}

the theoretical values (7). Secondly, ε_{12} exhibits a most unusual variation. Moving away from the wall the parameter decreases rapidly so that by $x_2^+=20$, $\varepsilon_{12}\, k/\overline{u_1u_2}\,\varepsilon$ has fallen to around 0.1 (in isotropic turbulence the parameter would be precisely zero). This result is perhaps surprising in view of the low Reynolds number, the large anisotropy of the stress field and the high strain rates present in that region. What is even less explicable is that, thereafter, the parameter should *rise* steadily towards the centreline. At the centre plane it is four times its minimum value yet there the stress field is relatively isotropic and the strain rate is zero – factors, which all conventional wisdom suggests, should favour a *reduction* in $\varepsilon_{12}\, k/\overline{u_1u_2}\,\varepsilon$.

Just what this indicates is not entirely clear at present. It may simply show that, yet again, real turbulence can easily outsmart attempts to pin it down in terms of mathematical or even conceptual models. There may, however, remain a slight query as to whether the dissipation components have been entirely resolved by the numerical simulation – that is whether the reported partitioning between ϕ_{ij1} and ε_{ij} is correct. The latter question is something that somebody will presumably resolve in a year or two by repeating the direct simulation with an order of magnitude more mesh points.

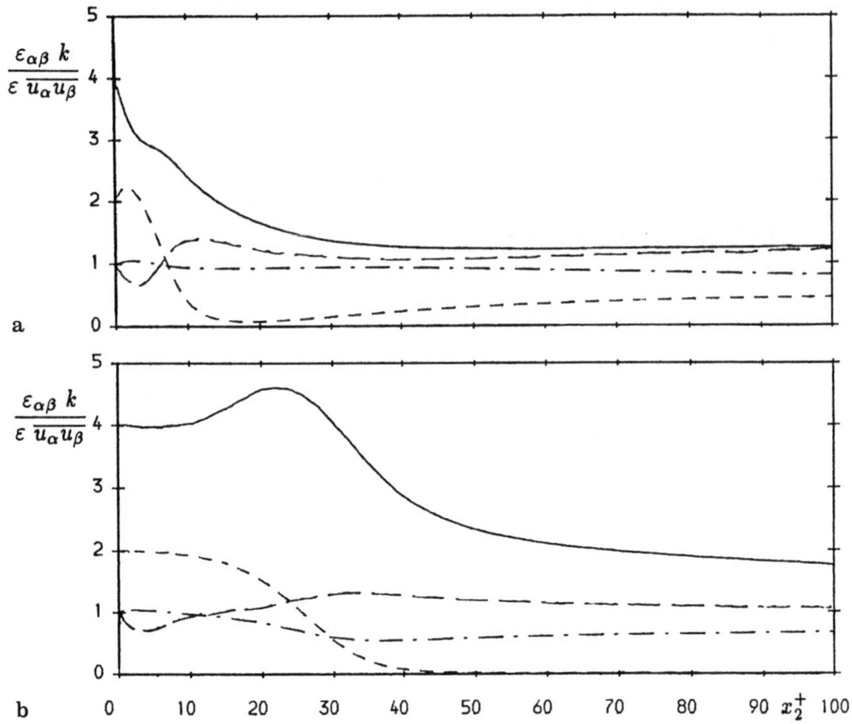

Fig. 9. Dissipation/stress profiles across low Reynolds number channel flow (Re = 5600). . —— . —— : α=1, β=1 ; ———— : α=2, β=2 ; —— —— — : α=3, β=3 ; — — — — : α=1, β=2 . a) Compiled from DNS data ; b) Present computations

4. Summarizing Remarks

A second moment closure for the near-wall region has been developed that mimics very closely the distribution of the Reynolds stress components obtained by the direct numerical simulations of flow through a plane channel.

New features of the model include:

- a reduced emphasis on the turbulent Reynolds number as a parameter for modifying the closure
- the inclusion of an *inhomogeneous* correction to ϕ_{ij} which exerts an important influence for levels of x_2^+ below about 50
- a much reduced importance of the model for wall-reflection effects on the fluctuating pressure field (indeed satisfactory behaviour can be obtained with the term eliminated entirely if the importance of the above noted inhomogeneity correction is increased)
- a process, tentatively attributed to ε_{ij}, that increases the relative importance of $\overline{u_3^2}$ relative to $\overline{u_1^2}$ for values of x_2^+ below 10 in accord with the DNS data
- pressure-diffusion model for $\overline{u_i u_j}$ in the viscous sublayer that is necessary to achieve the correct near-wall budget for $\overline{u_1 u_2}$ and $\overline{u_2^2}$.

Nevertheless, consideration of the profiles of ε_{ij} – and in particular the distribution of ε_{12} – suggests that considerably more effort is needed to arrive at a satisfactory modelling of the component processes. Indeed if the DNS data are correct a non-local model of the ε_{ij} would appear to be needed.

Acknowledgements

We gratefully acknowledge support of this research by the UK SERC under grant GR/D/8285.2. Our thanks also go to Mrs. J. Buckley for her help in preparing the camera-ready manuscript.

References

Authors are listed alphabetically.

1. Donaldson, C. du P., "A computer study of an analytical model of boundary layer transition", *AIAA Journal*, **7**, 271, 1969.
2. Shih, T.-H. and Mansour, N.N., "Modelling of near-wall turbulence", Proc. International Symposium on Engineering Turbulence Modelling and Measurements, Dubrovnik Yugoslavia, 13, 1990.
3. Kim, J., Moin, P. and Moser, R.D., "Turbulence statistics in fully developed channel flow at low Reynolds number", *J. Fluid Mechanics*, **177**, 133, 1987.
4. Spalart, P., "Direct simulation of a turbulent boundary layer up to $R_\theta = 1410$", *J. Fluid Mechanics*, **187**, 61, 1988.
5. Craft, T.J. and Launder, B.E., "A new model for the pressure/scalar-gradient correlation and its application to homogeneous and inhomogeneous free shear flows", Proc. 7th Symp. Turb. Shear Flows, Paper 17-1, Stanford University, 1989.
6. Cresswell, R.W., Haroutunian, V., Ince, N.Z., Launder, B.E. and Szczepura, R.T., "Measurement and modelling of buoyancy-modified, elliptic turbulent shear flows", Proc. 7th Symp. Turb. Shear Flows, Paper 12-4, Stanford University, 1989.
7. Launder, B.E. and Tselepidakis, D.P., "Directions in second-moment modelling of near-wall turbulence", AIAA Paper 91-0219, 1991.
8. Launder, B.E. and Tselepidakis, D.P., "Progress and paradoxes in modelling near-wall turbulence", Proc. 8th Symp. Turb. Shear Flows, Paper 29-1, Munich, 1991.
9. Lee, M.J. and Reynolds, W.C., "On the structure of homogeneous turbulence", Proc. 5th Symp. Turb. Shear Flows, p. 17.7, Cornell University, 1985.
10. Lumley, J.L., "Computational modelling of turbulent flows", *Adv. in Applied Mech.*, **18**, 123, 1978.
11. Kim, J., Personal communication, 1991.
12. Schumann, U., "Realizability of Reynolds stress turbulence models", *Phys. Fluids*, **20**, 721, 1977.
13. Launder, B.E., "Second-moment closure: present ... and future?", *Int. J. Heat and Fluid Flow*, **10**, 282, 1989.
14. Shih, T-H. and Lumley, J.L., "Modeling of Pressure Correlation Terms in Reynolds Stress and Scalar Flux Equations", Report FDA-85-3, Sibley School of Mech. and Aerospace Eng., Cornell University, 1985.
15. Gibson, M.M. and Launder, B.E., "Ground effects on pressure fluctuations in the atmospheric boundary layer", *J. Fluid Mechanics*, **86**, 491, 1978.

16. Launder, B.E. and Tselepidakis, D.P., "Contribution to the second-moment modelling of sublayer turbulent transport", Invited paper, Zoran Zarić Memorial, International Seminar on Near-Wall Turbulence, Dubrovnik, Yugoslavia, 1988.
17. Chou, P.-Y., "On velocity correlations and the solution of the equations of turbulent fluctuation", *Quart. Applied Mathematics*, **3**, 38, 1945.
18. Bradshaw, P., Mansour, N.N. and Piomelli, U., "On local approximations of the pressure-strain term in turbulence models", Proc. Summer Program, Center for Turbulence Research, Stanford University, 159, 1987.
19. Brasseur, J. and Lee, M.J., "Local structure of intercomponent energy transfer in homogeneous turbulent shear flow", Proc. Summer Program, Center for Turbulence Research, Stanford University, 165, 1987.
20. Craft, T.J. and Launder, B.E., "Computation of impinging flows using second-moment closures", Proc. 8th Symp. Turb. Shear Flows, Paper 8-5, Munich, 1991.
21. Launder, B.E. and Reynolds, W.C., "Asymptotic near-wall stress dissipation rates in a turbulent flow", *Phys. Fluids*, **26**, 1157, 1983.
22. Daly, B.J. and Harlow, F.H., "Transport equations in turbulence", *Phys. Fluids*, **13**, 2634, 1970.
23. Lumley, J.L., VKI Course "An Introduction to Turbulence Modelling", 1975.
24. Hanjalić, K. and Launder, B.E., "Contribution towards a Reynolds-stress closure for low-Reynolds-number turbulence", *J. Fluid Mechanics*, **74**, 593, 1976.

Direct Numerical Simulation of Low Prandtl Number Thermal Field in a Turbulent Channel Flow

N. Kasagi and Y. Ohtsubo

Department of Mechanical Engineering, The University of Tokyo, Hongo 7-3-1, Bunkyo-ku, Tokyo 113, Japan

Abstract

A direct numerical simulation of the fully developed thermal field in a two-dimensional turbulent channel flow was carried out. The simulation was made at a molecular Prandtl number of $Pr = 0.025$ to investigate low Prandtl number effects on the transport mechanism in wall turbulence. The computation was executed on about 1.6×10^6 grid points by using a spectral method. Mean field parameters as well as various thermal turbulence statistics including rms temperature fluctuations, turbulent heat fluxes and turbulent Prandtl numbers were obtained. Each term in the budget equations of temperature variance, its dissipation rate and turbulent heat fluxes was also calculated in order to establish a data base of convective heat transfer for thermal turbulence modeling. The results were compared mainly with those at $Pr = 0.71$ obtained by Kasagi et al. (1992). In addition, the computed thermal fields were visualized to investigate the role of the near-wall quasi-coherent structures in the turbulent transport mechanism.

Nomenclature

E^+_{aa}	dimensionless one-dimensional energy spectrum of fluctuating quantity a
h	heat transfer coefficient
k	turbulent kinetic energy, $\overline{u_i'u_i'}/2$
k_θ	variance of temperature fluctuation, $\overline{\theta'^2}/2$
k_x, k_z	wave numbers in the x- and z-directions
Nu	Nusselt number, $2h\delta/\lambda$
Pr	molecular Prandtl number, ν/α
Pr_t	turbulent Prandtl number, ν_t/α_t
R	timescale ratio, τ_θ/τ_u
R_{ij}	cross-correlation coefficient, $\overline{u_iu_j}/u_{i\,rms}u_{j\,rms}$
Re_m	Reynolds number, $2u_m\delta/\nu$
Re_τ	Reynolds number, $u_\tau\delta/\nu$
T	temperature
T_m	bulk mean temperature
T_τ	friction temperature
T^+	dimensionless temperature, $= (T - \langle T_w \rangle)/T_\tau = -\theta^+$

u_m	bulk mean velocity
u_τ	friction velocity
u, v, w	velocity components in the x-, y- and z-directions
x, y, z	streamwise, wall-normal and spanwise coordinates
α	thermal diffusivity
α_t	eddy-diffusivity for heat
δ	channel half width
ϵ	dissipation rate of k
ϵ_θ	dissipation rate of k_θ
θ	temperature difference, $= \langle T_w \rangle - T$
θ^+	dimensionless temperature, $= (\langle T_w \rangle - T)/T_\tau = -T^+$
θ_m	bulk mean temperature
λ	thermal conductivity
ν	kinematic viscosity
ν_t	eddy diffusivity
τ_u	velocity dissipation timescale, k/ϵ
τ_θ	temperature dissipation timescale, k_θ/ϵ_θ
$(\,)'$	fluctuating component
$(\,)^+$	normalized by the wall variables, u_τ, ν and T_τ
$\overline{(\,)}$	ensemble average over x-z plane and time
$\langle\,\rangle$	ensemble average over z-direction and time
$(\,)_{rms}$	root-mean-square value
$(\,)_w$	value at the wall

1. Introduction

The turbulent heat and mass transfer plays a key role in many engineering applications and will be of growing importance in resolving global environmental problems. The progress in turbulent heat transfer research, however, has been seemingly slow, although we have accumulated over the years a great deal of qualitative and quantitative knowledge on turbulent transport phenomena. This should be partly because a nonlinear nature common to turbulence phenomena is difficult to deal with and also because there has been serious lack of reliable experimental data with which proposed theories and turbulence closures are thoroughly evaluated.

In the meantime, with the advancement of large scale computers, direct numerical simulation (DNS, hereafter) of turbulent shear flows has become possible at moderate Reynolds numbers and now begun to provide valuable data bases (see, e.g., Moin & Spalart, 1989; Reynolds, 1990). DNSs are three-dimensional, time-dependent numerical solutions of the governing Navier-Stokes and energy equations that compute the evolution of flow and thermal fields without using any turbulence models. Recent DNSs of turbulent heat transfer in a channel flow of air (Kim & Moin, 1989; Kasagi et al., 1992) confirmed experimental observations such as a high correlation between streamwise velocity and temperature fluctuations (Perry and Hoffmann, 1976) and existence of the thermal streaky structures near a heated wall (Iritani et al., 1984). Furthermore, it was shown that the turbulent heat fluxes and the Reynolds stresses were also highly correlated and controlled by the same transport mechanism throughout the flow field when the Prandtl number was nearly equal to unity.

Although the turbulent heat transport mechanism in low Prandtl number fluids should be different from that in moderate Prandtl number fluids, the current knowledge is not satisfactory mainly owing to serious difficulties in experimental measurements, particularly in liquid metal flows (Reed, 1987). With the DNS, however, detailed information on the turbulent heat transfer over a wide Prandtl number range can probably be accumulated; the fundamental thermal turbulence statistics were obtained in channel flows at the Prandtl numbers from 0.1 to 2 by Kim & Moin (1989). The aim of this investigation is to offer new information on the turbulent heat transfer in a very low Prandtl number fluid, i.e., $Pr = 0.025$. For this, we simulate a fully-developed turbulent thermal field in a two-dimensional channel. Mean field parameters as well as various thermal turbulence statistics are calculated and compared with previous experimental and numerical results. In particular, extensive comparison is made with the DNS data at $Pr = 0.71$ obtained by Kasagi et al. (1992). The budgets of the temperature variance, its dissipation rate and the turbulent heat fluxes are discussed. In addition, instantaneous flow and thermal fields are visualized to explore the role of the quasi-coherent structures of wall turbulence in the turbulent transport mechanism.

2. Numerical Procedure

The flow geometry and the coordinate system are shown in Fig. 1. The Reynolds number Re_τ, based on the wall friction velocity u_τ and the channel half width δ, was set to be 150. The resultant bulk Reynolds number was 4580. Referring to the numerical procedure used by Kim et al. (1987), a fourth-order partial differential equation for v, a second-order partial differential equation for the wall-normal component of vorticity ω_y and the continuity equation was used to solve the flow field. A spectral method, i.e., Fourier series in the x- and z-directions and a Chebyshev polynomial expansion in the y-direction, was used. The computational periods were chosen to be $5\pi\delta$ and $2\pi\delta$ (2356 and 942 in wall units) in the x- and z-directions, respectively. 128×128 Fourier modes and Chebyshev polynomials up to the order 96 were used in order to resolve all essential turbulent scales on the computational grid. The collocation grid used to compute the nonlinear terms had 1.5 times finer resolution in each direction to remove aliasing errors. For time integration, the second-order Adams-Bashforth and Crank-Nicolson schemes were adopted for the nonlinear and viscous terms, respectively. The time increment was $0.12\nu/u_\tau^2$.

Once the velocity field was calculated at each time step, the scalar field was obtained by integrating the energy equation. This calculation was done by the spectral method with the same grid system used for the velocity field. Mercury was chosen as a test fluid, and the Prandtl number Pr was 0.025. Any buoyancy effect was neglected, and hence temperature was considered as a passive scalar. Since the flow and thermal fields were assumed to be fully developed, periodic boundary conditions were given in the x-

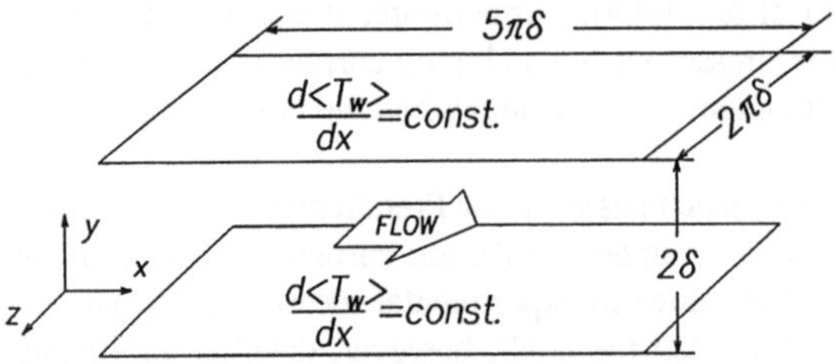

Fig. 1. Flow geometry and coordinate system

and z-directions. On the two walls, iso-flux heating boundary conditions were imposed so that the temperatures ensemble-averaged over the spanwise direction and time, $\langle T \rangle$, linearly increased in the streamwise direction, i.e.,

$$\frac{\partial \langle T \rangle}{\partial x} = \frac{\partial \langle T_m \rangle}{\partial x} = \frac{\partial \langle T_w \rangle}{\partial x} = \text{const.}, \tag{1}$$

while the temperature fluctuation on the wall is assumed to be zero. The dimensionless temperature difference θ^+ is defined as:

$$\theta^+ = \left(\langle T_w \rangle - T \right) / T_\tau . \tag{2}$$

Hence, the governing equation for the thermal field is given as follows:

$$\frac{\partial \theta^+}{\partial t} + u^+ \frac{\partial \theta^+}{\partial x} + v^+ \frac{\partial \theta^+}{\partial y} + w^+ \frac{\partial \theta^+}{\partial z} - \frac{2}{\int_{-1}^{1} \bar{u}^+ dy} u^+ = \frac{1}{Re_\tau Pr} \nabla^2 \theta^+, \tag{3}$$

where the velocities are non-dimensionalized by u_τ. The last term on the left hand side of Eq. (3) corresponds to $-u^+ \left(\partial \langle T_w \rangle^+ / \partial x \right)$.

As an initial condition, an instantaneous velocity field at the fully developed state was supplied by the numerical simulation of turbulent channel flow performed by Kuroda (1990). The initial mean temperature profile was given by the Kader's formula (1981), while a small constant value was given to all spectral modes of temperature fluctuations. A programm code, which was developed from the DNS code of Kuroda (1990) by Kasagi et al. (1992), was used. The time integration was repeated for about $800\nu/u_\tau^2$ until the thermal field was judged to be fully developed, and then the computation was further continued for about $2300\nu/u_\tau^2$ in order to calculate turbulence statistics as ensemble averages over space and time. The computation was carried out on a HITAC-S-820/80 supercomputer system at the Computer Center of the University of Tokyo. For one time step advancement, 5.26 second CPU time was required. Further details on the numerical procedure can be found in Kasagi et al. (1992).

3. Thermal Turbulence Statistics

The dimensionless mean velocity and temperature distributions are shown in Fig. 2. It has been confirmed that the velocity profile at $Re_\tau = 150$, which was obtained by

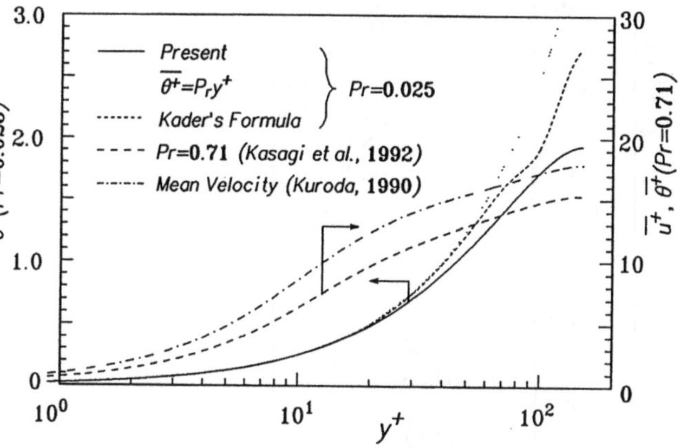

Fig. 2. Mean velocity and temperature profiles

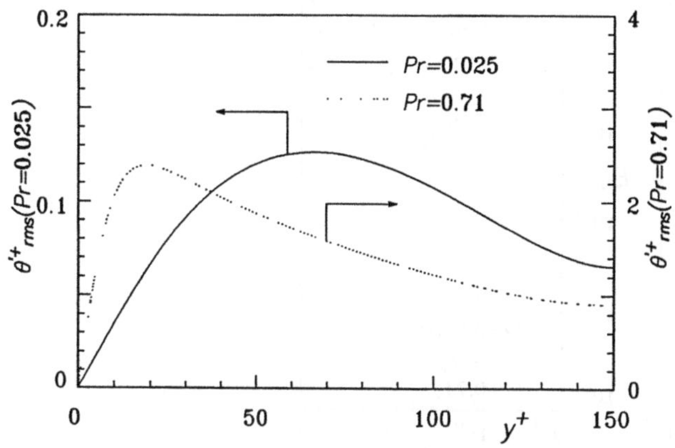

Fig. 3. RMS temperature fluctuation

Kuroda (1990), agrees well with that of Kim et al. (1987) at $Re_\tau = 180$, although there seems to be a slight difference in the logarithmic region owing to the low Reynolds number effect (see Kasagi et al., 1992). The thermal logarithmic region is visible in the temperature profile in the case of $Pr = 0.71$, but for $Pr = 0.025$ it does not exist owing to thickening of the conductive sublayer. Although it was found by Kasagi et al. (1992) that the temperature distribution agreed well with the Kader's empirical formula (1981) when $Pr = 0.71$, the present result is appreciably smaller. The Nusselt number calculated at $Pr = 0.025$ and $Re_m = 4580$ is 5.33, which compares reasonably well with the previous empirical correlations; according to Kays & Crawford (1980), the Nusselt numbers are 5.07 and 5.33 at $Re_m = 5000$, when $Pr = 0.01$ and 0.03, respectively, while that calculated by the Dwyer's correlation (1965) is 6.00 at the present condition.

The dimensionless root-mean-square temperature fluctuations θ'^+_{rms} at two Prandtl numbers are compared in Fig. 3. The rms temperature fluctuation at $Pr = 0.025$ is by an order of magnitude smaller than that at $Pr = 0.71$. It is also seen that the peak

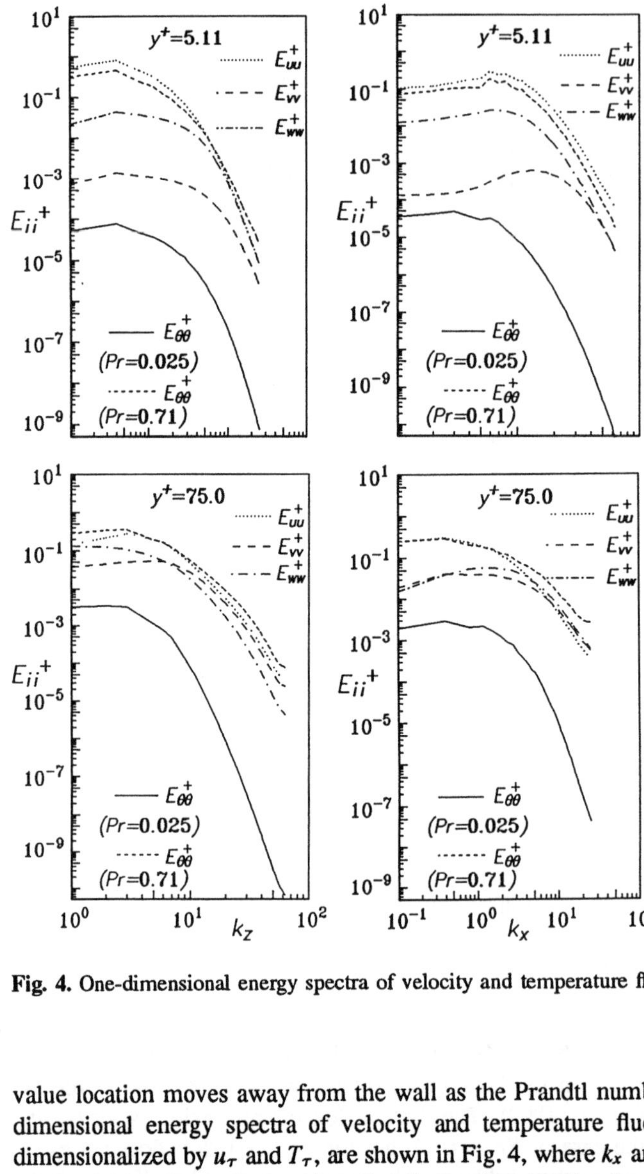

Fig. 4. One-dimensional energy spectra of velocity and temperature fluctuations

value location moves away from the wall as the Prandtl number is decreased. The one-dimensional energy spectra of velocity and temperature fluctuations, which are non-dimensionalized by u_τ and T_τ, are shown in Fig. 4, where k_x and k_z are the wavenumbers in the streamwise and spanwise directions, respectively. As expected, the temperature spectrum at $Pr = 0.71$ is similar to that of the streamwise velocity fluctuation. At $Pr = 0.025$, however, its decay at higher wave numbers is more rapid, particularly in the logarithmic region. Near the wall, a peak in the spanwise spectrum of $E_{\theta\theta}^+$, which is characteristic of the streaky structure discussed later, is shifted to a lower wave number.

Figure 5 shows the distributions of streamwise and wall-normal turbulent heat fluxes. As the Prandtl number is decreased, the turbulent heat fluxes also become very small owing to a vigorous conductive effect. When $Pr = 0.71$, the wall-normal heat flux is considerably smaller than the streamwise one over the channel cross section. Their difference, however, decreases as Pr is decreased and the anisotropy of turbulent heat

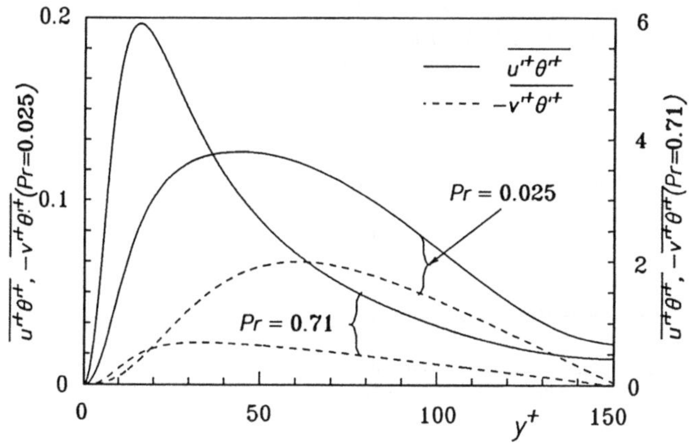

Fig. 5. Turbulent heat fluxes

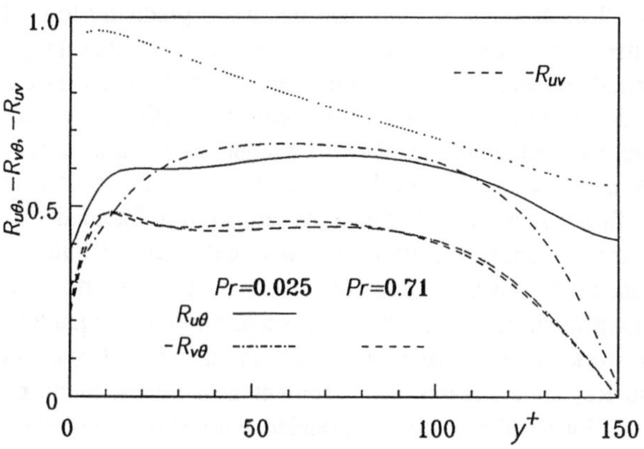

Fig. 6. Cross-correlation coefficients

fluxes is weakened. In addition, the maximum heat fluxes are obtained further away from the wall when the Prandtl number is small.

Several cross-correlation coefficients are shown in Fig. 6. In the case of $Pr = 0.71$, the correlation between the streamwise velocity fluctuation u' and the temperature fluctuation θ' is generally high and the maximum correlation coefficient $R_{u\theta}$ is as large as 0.96 near the wall. It is noted that the correlation coefficient $R_{v\theta}$ at $Pr = 0.71$ agrees quite well with R_{uv} and this suggests that the wall-normal turbulent heat flux and the Reynolds shear stress are generated by similar mechanisms. When the Prandtl number is low, the correlation coefficient $R_{u\theta}$ becomes smaller while $R_{v\theta}$ becomes larger, and they show roughly the same distributions in the flow field except the channel central region.

The turbulent Prandtl number Pr_t, which is defined as the ratio between the eddy diffusivities for momentum and heat, is shown in Fig. 7, together with the dissipation time scale ratio. For $Pr = 0.71$, Pr_t varies moderately across the channel and reaches a weak maximum of about 1.1 near the wall. For $Pr = 0.025$, however, it increases from

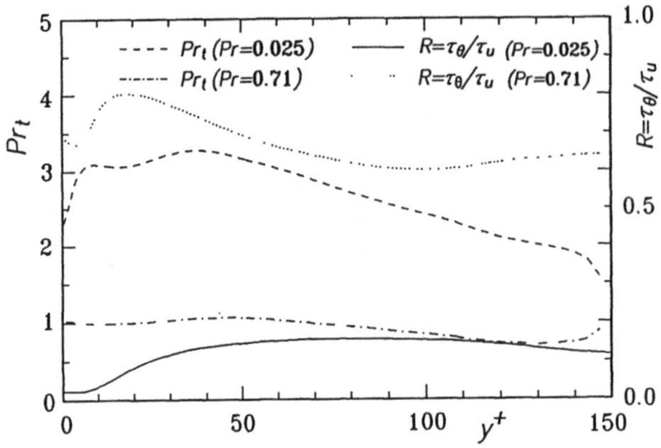

Fig. 7. Turbulent Prandtl number Pr_t and timescale ratio R

the wall to a maximum value of 3.27 at $y^+ = 37$ and then decreases gradually toward the channel center. The value itself and its change across the channel are much larger than those of the moderate Prandtl number fluid. The time scale ratio R, which is the ratio of temperature dissipation timescale τ_θ to velocity dissipation timescale τ_u, is often used in scalar flux modeling (see, e.g., Béguier et al., 1978; Launder, 1989). It is seen in Fig. 7 that the timescale ratio R at $Pr = 0.025$ is on the order of 0.1 and much smaller than that at $Pr = 0.71$, and this fact is reasonable because the temperature fluctuation is dissipated much faster due to a strong conductive effect in low Prandtl number fluids.

Current engineering turbulent heat transfer computations rely on various one-point closure turbulence models. Among them, those based on the second moment equations are of great interest for many workers (e.g., Launder, 1989). In the second moment equations such as those for the Reynolds stresses and turbulent heat fluxes, there are several terms of unknown correlations which must be modeled with known variables. Hence, the budget data obtained by DNSs should be useful for testing and improving existing closure models. Kasagi et al. (1992) obtained the budgets of the temperature variance, its dissipation rate and turbulent heat fluxes at $Pr = 0.71$. In the following, the results at $Pr = 0.025$ are presented.

In the fully developed thermal field in a two-dimensional channel, the budget equation for the temperature variance $k_\theta^+ = \overline{\theta'^+\theta'^+}/2$ leads to:

$$0 = \underbrace{\overline{u'^+\theta'^+}\frac{\partial \langle T\rangle^+}{\partial x^+} - \overline{v'^+\theta'^+}\frac{\partial \bar{\theta}^+}{\partial y^+}}_{\text{Production}} - \underbrace{\frac{1}{Pr}\overline{\frac{\partial \theta'^+}{\partial x_i^+}\frac{\partial \theta'^+}{\partial x_i^+}}}_{\text{Dissipation } \epsilon_\theta^+}$$

$$\underbrace{-\frac{1}{2}\frac{\partial}{\partial y^+}\overline{\theta'^{+2}v'^+}}_{\text{Turbulent Diffusion}} + \underbrace{\frac{1}{Pr}\frac{\partial^2 k_\theta^+}{\partial y^{+2}}}_{\text{Molecular Diffusion}} \quad . \tag{4}$$

Figure 8 shows each term included in Eq. (4). The production and dissipation rates are two major terms. Note that the first term of the production in Eq. (4) is negligibly

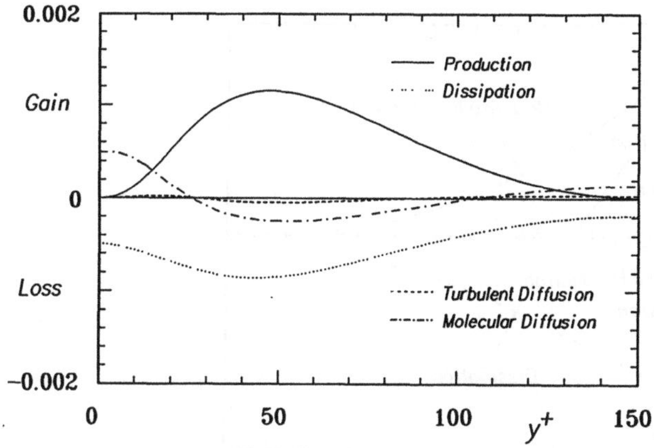

Fig. 8. Budget of the temperature variance k_θ^+

small compared with the second term, because the streamwise temperature gradient is much smaller than the wall-normal gradient. The molecular diffusion term is appreciable across the channel, and this implies that the influence of the thermal conduction can not be neglected throughout the flow field. On the other hand, the turbulent diffusion remains very small. The maximum of production rate is located at $y^+ = 48$, which is closer to the wall than the maximum temperature variance.

The transport equation of the dissipation rate of temperature variance ϵ_θ^+ is written as:

$$
\begin{aligned}
0 = & \underbrace{-\frac{2}{Pr}\overline{\frac{\partial\theta'^+}{\partial x_l^+}\frac{\partial u_j'^+}{\partial x_l^+}}\frac{\partial\bar{\theta}^+}{\partial x_j^+} - \frac{2}{Pr}\overline{\frac{\partial\theta'^+}{\partial x_l^+}\frac{\partial\theta'^+}{\partial x_j^+}}\frac{\partial\bar{u}_j^+}{\partial x_l^+}}_{\text{Mean Gradient Production}} \underbrace{- \frac{2}{Pr^2}\overline{\frac{\partial^2\theta'^+}{\partial x_j^+\partial x_l^+}\frac{\partial^2\theta'^+}{\partial x_j^+\partial x_l^+}}}_{\text{Dissipation}} \\
& \underbrace{-\frac{2}{Pr}\overline{\frac{\partial\theta'^+}{\partial x_l^+}\frac{\partial u_j'^+}{\partial x_l^+}\frac{\partial\theta'^+}{\partial x_j^+}}}_{\text{Turbulent Production}} \underbrace{-\frac{2}{Pr}\overline{u_j'^+\frac{\partial\theta'^+}{\partial x_l^+}}\frac{\partial^2\bar{\theta}^+}{\partial x_j^+\partial x_l^+}}_{\text{Gradient Production}} \\
& \underbrace{-\frac{1}{Pr}\frac{\partial}{\partial x_j^+}\overline{\left(\frac{\partial\theta'^+}{\partial x_l^+}\frac{\partial\theta'^+}{\partial x_l^+}u_j'^+\right)}}_{\text{Turbulent Diffusion}} \underbrace{+\frac{1}{Pr^2}\frac{\partial^2}{\partial x_j^+\partial x_j^+}\overline{\left(\frac{\partial\theta'^+}{\partial x_l^+}\frac{\partial\theta'^+}{\partial x_l^+}\right)}}_{\text{Molecular Diffusion}} \, .
\end{aligned}
\tag{5}
$$

Each term in Eq. (5) is shown in Fig. 9. It is seen that the budget is dominated by the mean gradient production and dissipation terms over the whole cross section. All the remaining terms are very small, but the molecular diffusion is appreciable near the wall. In contrast with this simple balance, the turbulent production term is the main source term away from the wall while the two mean gradient production terms become relatively large toward the wall when $Pr = 0.71$ (Kasagi et al., 1992).

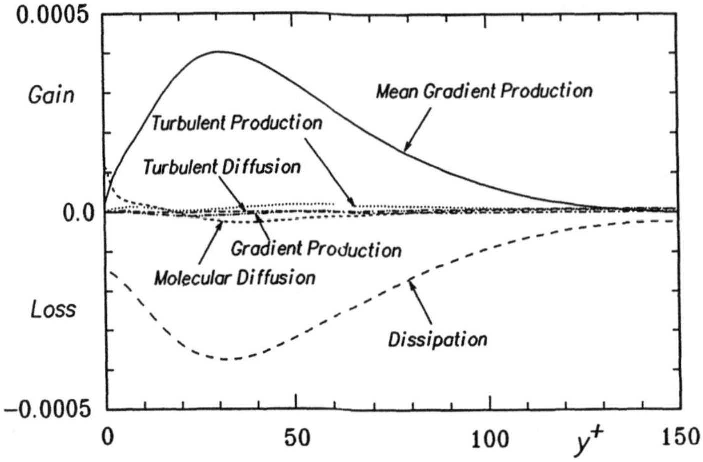

Fig. 9. Budget of the dissipation-rate of the temperature variance ϵ_θ^+

The transport equations of the streamwise and wall-normal turbulent heat fluxes, and of the Reynolds shear stress are respectively given as follows:

$$0 = \underbrace{\overline{u'^+u'^+}\frac{\partial \langle T\rangle^+}{\partial x^+} - \overline{u'^+v'^+}\frac{\partial \bar{\theta}^+}{\partial y^+} - \overline{v'^+\theta'^+}\frac{\partial \bar{u}^+}{\partial y^+}}_{\text{Production}} \underbrace{- \frac{1+Pr}{Pr}\overline{\frac{\partial \theta'^+}{\partial x_j^+}\frac{\partial u'^+}{\partial x_j^+}}}_{\text{Dissipation}} \underbrace{\overline{-\theta'^+\frac{\partial p'^+}{\partial x^+}}}_{\text{Temp. Press. Grad. Corr.}} - \frac{\partial}{\partial y^+}\left\{\underbrace{\overline{u'^+v'^+\theta'^+}}_{\text{Turbulent Diffusion}} - \underbrace{\left(\overline{\theta'^+\frac{\partial u'^+}{\partial y^+}} + \frac{1}{Pr}\overline{u'^+\frac{\partial \theta'^+}{\partial y^+}}\right)}_{\text{Molecular Diffusion}}\right\}. \tag{6}$$

$$0 = \underbrace{\overline{u'^+v'^+}\frac{\partial \langle T\rangle^+}{\partial x^+} - \overline{v'^+v'^+}\frac{\partial \bar{\theta}^+}{\partial y^+}}_{\text{Production}} \underbrace{- \frac{1+Pr}{Pr}\overline{\frac{\partial v'^+}{\partial x_j^+}\frac{\partial \theta'^+}{\partial x_j^+}}}_{\text{Dissipation}} \underbrace{\overline{-\theta'^+\frac{\partial p'^+}{\partial y^+}}}_{\text{Temp. Press. Grad. Corr.}} - \frac{\partial}{\partial y^+}\left\{\underbrace{\overline{v'^+v'^+\theta'^+}}_{\text{Turbulent Diffusion}} - \underbrace{\left(\overline{\theta'^+\frac{\partial v'^+}{\partial y^+}} + \frac{1}{Pr}\overline{v'^+\frac{\partial \theta'^+}{\partial y^+}}\right)}_{\text{Molecular Diffusion}}\right\}. \tag{7}$$

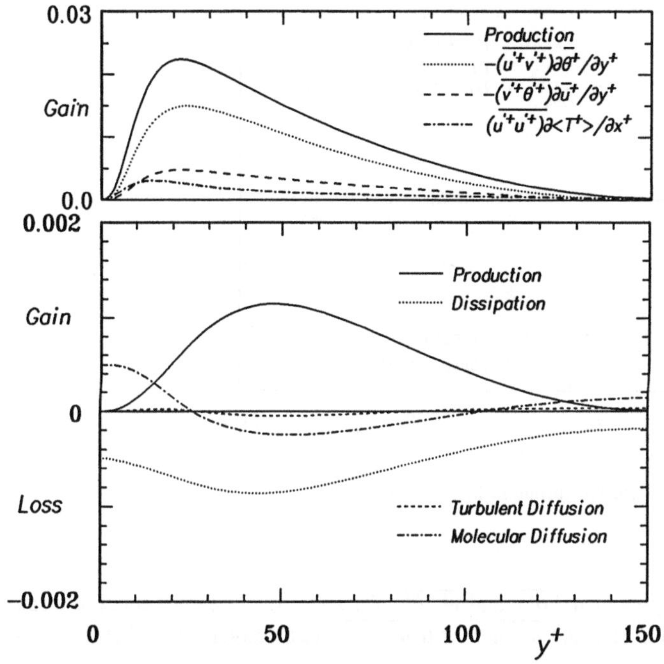

Fig. 10. Budget of the streamwise turbulent heat flux $\overline{u'^+\theta'^+}$

$$0 = \underbrace{-\overline{v'^+v'^+}\frac{\partial \bar{u}^+}{\partial y^+}}_{\text{Production}} \underbrace{-2\overline{\frac{\partial u'^+}{\partial x_j^+}\frac{\partial v'^+}{\partial x_j^+}}}_{\text{Dissipation}}$$

$$\underbrace{-\left(\overline{v'^+\frac{\partial p'^+}{\partial x^+}} + \overline{u'^+\frac{\partial p'^+}{\partial y^+}}\right)}_{\text{Vel. Press. Grad. Corr.}} \underbrace{-\frac{\partial}{\partial y^+}\overline{u'^+v'^{+2}}}_{\text{Turbulent Diffusion}} \underbrace{+\frac{\partial^2 \overline{u'^+v'^+}}{\partial y^{+2}}}_{\text{Molecular Diffusion}} . \quad (8)$$

Each term in Eq. (6) for the streamwise turbulent heat flux is shown in Fig. 10. The production and dissipation terms are large in the whole region, and besides them the molecular diffusion term becomes large toward the wall. The production term consists of the three terms in Eq. (6) and each of them is also shown in Fig. 10; among them the production by the Reynolds shear stress times the mean temperature gradient is the largest. The temperature pressure gradient correlation term remains small and always lies in the loss side. The turbulent diffusion term is very small as in the aforementioned budgets of k_θ^+ and ϵ_θ^+.

The calculated terms in the budget equation of the wall-normal turbulent heat flux $\overline{v'^+\theta'^+}$ are shown in Fig. 11. The production and dissipation terms are dominant in the whole flow region. The production by the mean streamwise temperature gradient in Eq. (7) is negligible. The temperature pressure gradient correlation term is again very small. It was shown by Kim (1988) and Kasagi et al. (1992) that the dominant sink term was the temperature pressure gradient correlation in the case of $Pr = 0.71$. In addition,

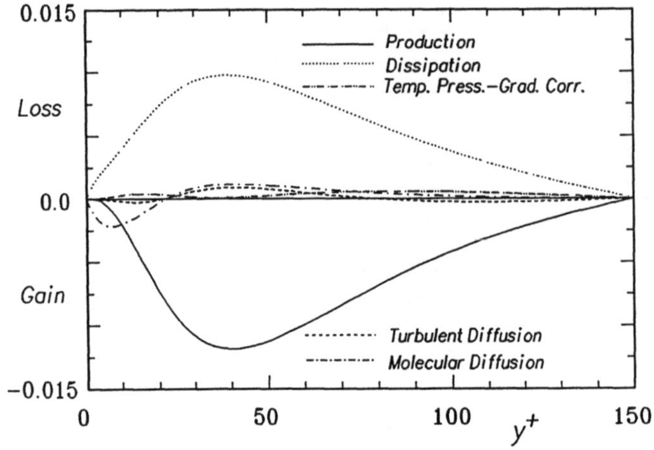

Fig. 11. Budget of the wall-normal turbulent heat flux $\overline{v'^+\theta'^+}$

they found that the budgets of $\overline{u'^+\theta'^+}$ and $\overline{v'^+\theta'^+}$ for $Pr = 0.71$ seemed much the same as those of the Reynolds stresses $\overline{u'^+u'^+}$ and $\overline{u'^+v'^+}$, respectively, but this is not the case for $Pr = 0.025$.

4. Near-Wall Turbulence Structure and Associated Heat Transfer Mechanism

Since the deterministic nature of wall turbulence was first explored in the 1950's, a number of studies have been made to further our understanding of its physics (see, e.g., Cantwell, 1981, Robinson, 1991a). The pioneering work made by Kline et al. (1967) and Kim et al. (1971) revealed that the near-wall region of turbulent boundary layers exhibits recognizable streaky structures which are fairly periodic in the spanwise direction. They also documented the deterministic behaviour of a fluid motion producing the Reynolds stress, which is labeled as bursting. Since the near-wall region is crucial for momentum, heat and mass transfer, understanding of these structures is of great importance from both scientific and engineering viewpoints. The present status of knowledge on the quasi-coherent structures of wall turbulence is summarized by Kline & Robinson (1990) and Robinson (1991a).

Visualization of structures in a numerically simulated turbulent boundary layer was extensively carried out by Robinson et al. (1990) and Robinson (1991b). They identified the structures experimentally observed, such as low- and high-speed streaks, arch-like vortical structures and local shear layers in the simulated boundary layer. They also found that the marked vortical structures, of which cores were identified as local low-pressure regions, were in close spatial association with the two major modes of turbulent motion generating the Reynolds shear stress, i.e., ejections and sweeps.

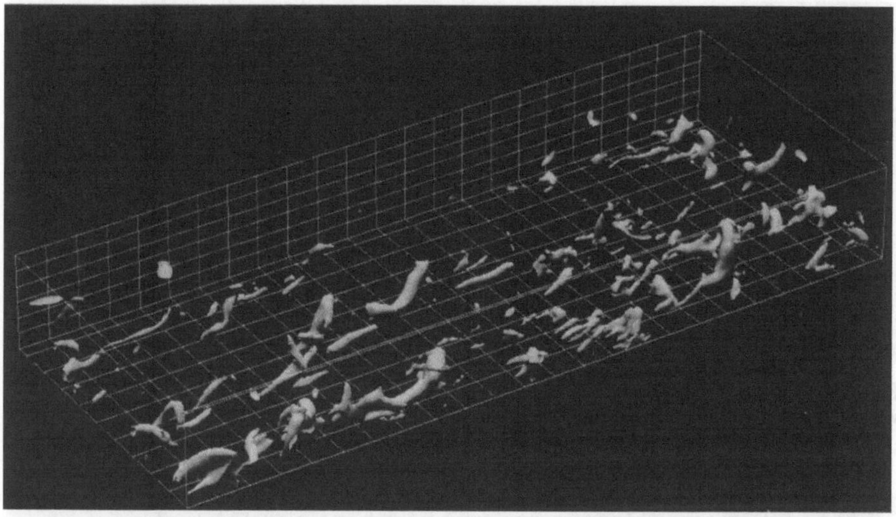

Fig. 12(a). Low pressure regions (white, $p'^{+} \leq -3.0$) in the lower half of the computational domain (2356 × 300 × 942 wall units)

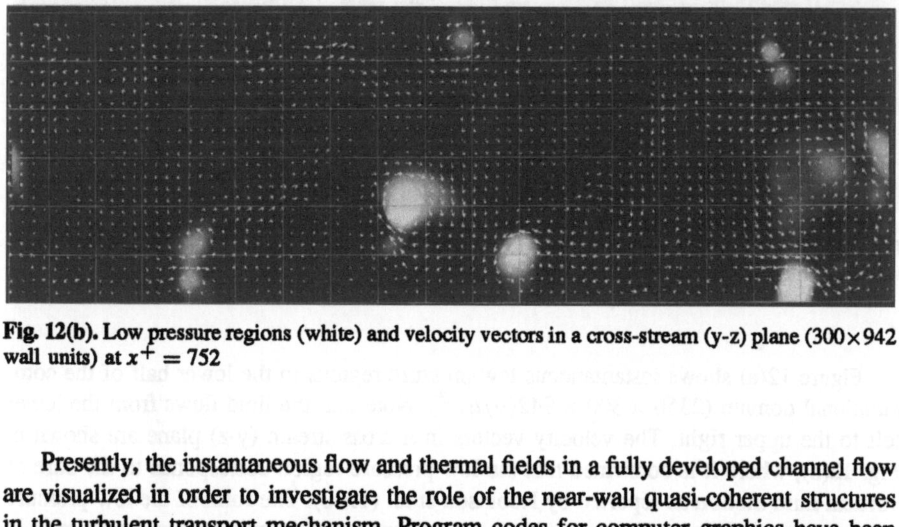

Fig. 12(b). Low pressure regions (white) and velocity vectors in a cross-stream (y-z) plane (300×942 wall units) at $x^{+} = 752$

Presently, the instantaneous flow and thermal fields in a fully developed channel flow are visualized in order to investigate the role of the near-wall quasi-coherent structures in the turbulent transport mechanism. Program codes for computer graphics have been developed to generate three-dimensional imagery of complex topology and probe the inter-relationship between various flow and thermal structures. The graphical analyses are carried out on a graphic workstation (Stardent Titan 3020), which is connected with the mainframe computer via a local area network (Ethernet). The data bases obtained by the present work and Kasagi et al. (1992) are used to generate graphical outputs, with which the Prandtl number effect is studied. The fully developed thermal fields at $Pr = 0.025$ and 0.71 are compared at the same instant which has been reached after starting the numerical computations with the same initial conditions. Further details on the hardware and software are found in Ohtsubo (1992).

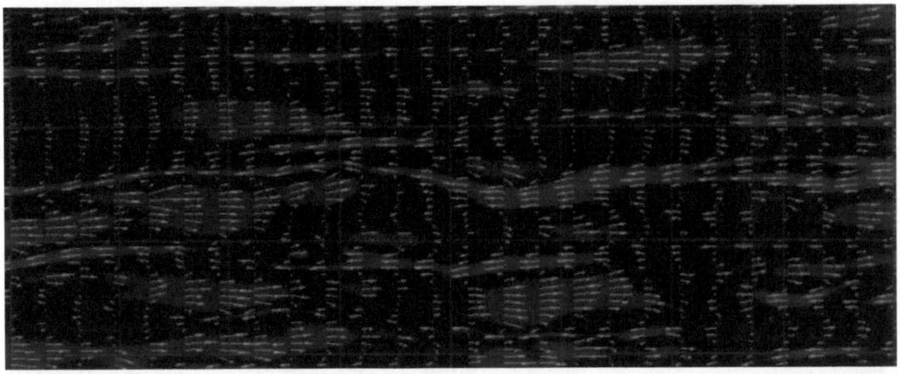

Fig. 13(a). Low and high temperature regions (blue, $T'^{+} \leq -1.0$; red, $T'^{+} \geq 1.0$) and velocity vectors in the x-z plane (1600 × 600 wall units) at $Pr = 0.71$

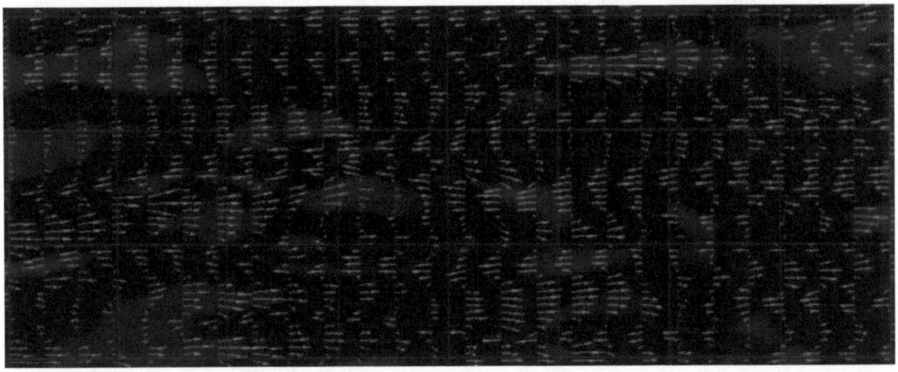

Fig. 13(b). Low and high temperature regions (blue, $T'^{+} \leq -0.03$; red, $T'^{+} \geq 0.03$) and velocity vectors in the x-z plane at $Pr = 0.025$

Figure 12(a) shows instantaneous low-pressure regions in the lower half of the computational domain ($2356 \times 300 \times 942(\nu/u_\tau)^3$). Note that the fluid flows from the lower left to the upper right. The velocity vectors in a cross-stream (y-z) plane are shown in Fig. 12(b), where it is confirmed that the low-pressure region corresponds to the core of vortical fluid motion as reported by Robinson et al. (1990). The form of the low-pressure region appears different from sample to sample, but a true hairpin vortex is very rare in the simulated flow. A hook-like asymmetric vortex relatively often exists and a single inclined streamwise vortex, which looks like a banana, appears most often. The influence of the threshold value on the low pressure region visualized seems to be limited to the diameter of the vortex core as reported by Robinson (1991b).

In figure 13 are typical vector-contour maps of the velocity and temperature fluctuations in the x-z plane near the wall ($y^+ = 8$). Hereafter, we use another dimensionless temperature $T^+ = -\theta^+$ to realize that the walls are heated and their temperature is higher than that of fluid. The velocity vectors in Fig. 13 clearly show the elongated streaky structures, of which spanwise spacing is roughly equal to 100 wall units in ac-

Fig. 14. Low and high temperature regions and low pressure regions (blue, $T'^{+} \leq -0.21$; red, $T'^{+} \geq 0.21$; white, $p'^{+} \leq -3.0$) (mesh: 100 × 50 × 100 wall units)

cordance with the experimental knowledge. Although not shown here, it is observed that a pair of these high and low speed regions exist mainly alongside the near-wall tail of the low pressure region. The thermal field at $Pr = 0.71$ shown in Fig. 13(a) also exhibits the streaky structures as reported by Iritani et al. (1984). As is deduced from the high correlation between u' and T', these thermal streaky structures show a strong resemblance to the momentum streaks. When $Pr = 0.025$, however, the thermal field appears much different as shown in Fig. 13(b). Even though correspondence between the momentum and thermal streaks is still confirmed in some degree, the latter is not so much elongated in the x-direction and its spanwise spacing seems to be considerably larger. The high- and low-temperature regions at $Pr = 0.025$ are shown in Fig. 14. Compared with those at $Pr = 0.71$ (figures not shown here), they seem larger and fatter, although there is an arbitrariness in specifying the threshold level of volume rendering. In addition, both high- and low-temperature regions appear away from the wall in accordance with the fact that the maximum rms temperature fluctuation is obtained around $y^{+} = 70$.

The ejection $u'v'(Q2)$ and the sweep $u'v'(Q4)$ are known to be the two main fluid motions contributing to instantaneous Reynolds shear stress generation. The former is outward movement of low-speed fluid, while the latter is wallward movement of high-speed fluid. Similarly, the turbulent heat flux can be decomposed into four quadrant components. Figures 15(a) and 15(b) show the contour surfaces of instantaneous turbulent heat flux at $Pr = 0.71$ and 0.025, respectively. $v'^+T'^+(Q1)$ represents the turbulent heat flux product in the first quadrant on the v'^+–T'^+ plane, i.e., the outward motion of high-temperature fluid, while $v'^+T'^+(Q3)$ is the third quadrant product, i.e., the wall-ward motion of low-temperature fluid. In Fig. 15(a), it is seen that $v'^+T'^+(Q1)$ and $v'^+T'^+(Q3)$ are located alongside the low pressure region. This represents upward- and downward-rotating motions around the vortex. In particular, $v'^+T'^+(Q1)$ is mostly associated with the vortical structures. However, fairly many regions of $v'^+T'^+(Q3)$ also appear in the vicinity of the wall. It is noted that the graphical view of the ejection and sweep in the flow field is almost indistinguishable from that in Fig. 15(a) if we interpret $v'^+T'^+(Q1)$ and $v'^+T'^+(Q3)$ as $u'^+v'^+(Q2)$ and $u'^+v'^+(Q4)$, respectively, although it is not shown here; at $Pr = 0.71$, even instantaneous momentum and heat fluxes are highly correlated in space. These features are in good agreement with the experimental observation that the ejection and sweep bear the most effective mechanism of turbulent heat transfer (e.g., Iritani et al., 1984; Nagano & Hishida, 1990). Thus, the heat and momentum transfer analogy holds when the Prandtl number nearly equals unity.

The high turbulent heat flux regions at $Pr = 0.025$ shown in Fig. 15(b) are not so much different from those at $Pr = 0.71$ in Fig. 15(a), although there is much difference in the momentum and thermal streaky structures as mentioned previously. There is again observed close association between the vortical structures and the heat flux products.

Fig. 15(a). High turbulent heat flux regions and low pressure regions ($Pr = 0.71$) (red, $v'^+T'^+(Q1) \geq 3.0$; blue, $v'^+T'^+(Q3) \geq 3.0$; white, $p'^+ \leq -3.0$)

Fig. 15(b). High turbulent heat flux regions and low pressure regions ($Pr = 0.025$) (red, $v'^+T'^+(Q1) \geq 0.23$; blue, $v'^+T'^+(Q3) \geq 0.23$; white, $p'^+ \leq -3.0$)

Thus, the vortical structures play an important role even in the low Prandtl number fluid. A close inspection reveals that, if compared with $v'^+T'^+(Q1)$, $v'^+T'^+(Q3)$ seems more influenced by the molecular conductive effect and grows more in size.

The major budget terms of $\overline{v'^+\theta'^+}$ in Eq. (7) are visualized together with the low-pressure regions in Figs. 16(a) and 16(b). In these figures, the spatial derivatives included in the dissipation and the temperature pressure gradient correlation are calculated in physical space by using central differences. The $\overline{v'^+\theta'^+}$ budget at $Pr = 0.71$ is quite similar to the $\overline{u'^+v'^+}$ budget given by Eq. (8); the temperature pressure gradient correlation acts as a main sink term as the velocity pressure gradient correlation (see, Kasagi et al., 1992). In Fig. 16(a), it is seen that the production takes large values on both sides of the inclined streamwise vortex, because the Reynolds stress product $v'^+v'^+$ associated with the ejection and sweep should also be large there. The temperature pressure gradient correlation also appears mainly around the vortex, but on its belly and back, where the heat flux vector turns to the spanwise direction. In addition, there seem to exist the high production regions in the vicinity of the wall owing to the sweep motions (see, $v'^+T'^+(Q3)$ in Fig. 15(a) instead), and they are associated with the temperature pressure gradient correlation as well as the dissipation. It is separately confirmed that the same features are found in the visualized spatial distributions of the corresponding $u'^+v'^+$ budget terms.

Fig. 16(a). Production and destruction regions of the wall-normal turbulent heat flux at $Pr = 0.71$ (red, production ≥ 0.2; blue, dissipation ≥ 0.08; purple, temperature pressure gradient correlation ≤ -0.3; white, low pressure region, $p'^{+} \leq -3.0$)

Figure 16(b) shows the high production and dissipation regions of $v'^{+}T'^{+}$, i.e., the two major terms in the budget at $Pr = 0.025$. In general, it seems that these regions tend to appear closely each other and a considerable number of them are again correlated with the vortical structures. For this Prandtl number, the temperature pressure gradient correlation is negligibly small. However, it is of great interest that the molecular dissipation takes large values where the temperature pressure gradient correlation is effective at $Pr = 0.71$ in Fig. 16(a).

Finally, the $\overline{u'^{+}\theta'^{+}}$ budget terms in Eq. (6) are visualized in Figs. 17(a) and 17(b). When $Pr = 0.71$, the production is large alongside the inclined vortex, because $v'^{+}\theta'^{+}$ (and also $u'^{+}v'^{+}$) is large there as shown in Fig. 15(a). However, since the mean gradient is larger toward the wall, the high production regions are located somewhat closer to the wall than the high heat flux regions. The temperature pressure gradient correlation is large again on the back and belly of the streamwise vortex. In particular, this sink term is remarkable upstream of the upper portion of the vortex, which seems to be located in the logarithmic region. Near the wall where the production balances mainly with the dissipation in the budget of $\overline{u'^{+}\theta'^{+}}$, they are closely associated in space and considerably elongated in the streamwise direction. Note that these spatial structures of the $\overline{u'^{+}\theta'^{+}}$ budget at $Pr = 0.71$ resembles very much with those of the $\overline{u'^{+}u'^{+}}$ budget.

Fig. 16(b). Production and destruction regions of the wall-normal turbulent heat flux at $Pr = 0.025$ (red, production ≥ 0.035; blue, dissipation ≥ 0.025; white, low pressure region, $p'^{+} \leq -3.0$)

The composition of the $\overline{u'^{+}\theta'^{+}}$ budget terms at $Pr = 0.025$ is qualitatively similar to that at $Pr = 0.71$ near the wall; i.e., the dissipation is the main sink term against the production. Hence, despite the Prandtl number difference the spatial structure of the $\overline{u'^{+}\theta'^{+}}$ budget shown in Fig. 17(b) is not very much different from that in Fig. 17(a). However, the dissipation extends to the region further away from the wall. The temperature pressure gradient correlation seems to be associated with the vortical structures in the logarithmic region, although its magnitude is relatively small in the budget as shown in Fig. 10.

5. Conclusions

A direct numerical simulation of a fully developed two-dimensional channel flow at $Re_m = 4580$ and $Pr = 0.025$ has been carried out. Various thermal turbulence statistics are calculated, and instantaneous flow and thermal fields are visualized. The following conclusions are derived:

Fig. 17(a). Production and destruction regions of the streamwise turbulent heat flux at $Pr = 0.71$ (red, production ≤ -0.9; blue, dissipation ≤ -0.5; purple, temperature pressure gradient correlation ≥ 0.5; white, low pressure region, $p'^{+} \leq -3.0$

(1) The effect of molecular conduction on the mean temperature profile is appreciable even in the channel central region and the thermal logarithmic region does not exist.
(2) As the Prandtl number becomes very low, the temperature variance and the turbulent heat fluxes are markedly decreased with their peak values moving away from the wall, and the anisotropy of turbulent heat fluxes is also decreased.
(3) The turbulent Prandtl number at $Pr = 0.025$ is much larger than that for moderate Prandtl number fluids and its maximum reaches about 3.27 at $y^{+} = 37$. The ratio of dissipation time scales is about 0.1, which is much smaller than that at $Pr = 0.71$.
(4) The terms included in the budget equations of the temperature variance, its dissipation rate and the turbulent heat fluxes are calculated. In general, the thermal conductive effect is considerable. The temperature pressure gradient correlation no longer acts as a primary sink mechanism for the turbulent heat fluxes when the Prandtl number is very low.
(5) A great difference is observed between the thermal streaky structures visualized at $Pr = 0.71$ and 0.025. Those in the low Pr fluid are thick in size and not so much elongated in the streamwise direction.
(6) The streamwise vortical structures, which are identified as local low-pressure regions, play a predominant role in the turbulent transport of momentum and heat in

Fig. 17(b). Production and destruction regions of the streamwise turbulent heat flux at $Pr = 0.025$ (red, production ≤ -0.08; blue, dissipation ≤ -0.08; purple, temperature pressure gradient correlation ≥ 0.02; white, low pressure region, $p'^{+} \leq -3.0$)

the channel flow; the production and destruction mechanisms for turbulent stresses and fluxes are closely associated with these structures. This is also the case for low Prandtl number fluids, although the molecular dissipation acts as a main sink mechanism instead of the temperature pressure gradient correlation. Thus, the turbulent transport mechanisms in wall turbulence are highly localized and intermittent in space.

Aknowledgements

The authors are grateful to A. Kuroda, Y. Suzuki and Y. Tomita for their cooperation during the course of this work and also to the financial support through the Grant-in-Aid for Cooperative Research (No. 02302043) by the Ministry of Education, Science and Culture.

References

Béguier, C., Dekeyser, I., Launder, B. E. (1978): "Ratio of Scalar and Velocity Dissipation Time Scales in Shear Flow Turbulence", Phys. Fluids, Vol. 21, pp. 307-310

Cantwell, B. J. (1981): "Organized Motion in Turbulent Flows", Ann. Rev. Fluid Mech., Vol. 13, pp. 457-515

Dwyer, O. E. (1965): "Heat Transfer to Liquid Metals Flowing Turbulently Between Parallel Plates", Nuclear Science and Engineering, Vol. 21, pp. 79-89

Iritani, Y., Kasagi, N., Hirata, M. (1984): "Heat Transfer Mechanism and Associated Turbulence Structure in a Near-Wall Region of a Turbulent Boundary Layer", in Turbulent Shear Flows IV, L. J. S. Bradbury et al., eds., Springer Verlag, Berlin, pp. 223-234

Kasagi, N., Tomita, Y., Kuroda, A. (1992): "Direct Numerical Simulation of Passive Scalar Field in a Turbulent Channel Flow", ASME J. Heat Transfer (in press)

Kader, B. A. (1981): "Temperature und Concentration Profiles in Fully Turbulent Boundary Layers", Int. J. Heat Mass Transfer, Vol. 24, pp. 1541-1544

Kays, W. M., Crawford, M. E. (1980): "Convective Heat and Mass Transfer", 2nd Ed., McGraw-Hill, New York

Kim, H.T., Kline, S. J., Reynolds, W. C. (1971): "Turbulence Near a Smooth Wall in a Turbulent Boundary Layer", J. Fluid Mech., Vol. 50, pp. 133-160

Kim, J., Moin, P. Moser, R. (1987): "Turbulence Statistics in Fully Developed Channel Flow at Low Reynolds Number", J. Fluid Mech., Vol. 177, pp. 133-166

Kim, J. (1988): "Investigation of Heat and Mass Transport in Turbulent Flows via Numerical Simulation", in Transport Phenomena in Turbulent Flows: Theory, Experiment and Numerical Simulation, M. Hirata and N. Kasagi, eds., Hemisphere, Washington, D.C., pp. 157-170

Kim, J., Moin, P., (1989): "Transport of Passive Scalars in a Turbulent Channel Flow", in Turbulent Shear Flows VI, J.-C. Andrè et al., eds., Springer Verlag, Berlin, pp. 85-96

Kline, S. J., Reynolds, W. C., Schraub, F. A., Runstadler, P. W. (1967): "The Structure of Turbulent Boundary Layers", J. Fluid Mech., Vol. 30, pp. 741-773

Kline, S. J., Robinson, S. K. (1990): "Quasi-Coherent Structures in the Turbulent Boundary Layer: Part I. Status Report on a Community-Wide Summary of the Data", in Near-Wall Turbulence, S. J. Kline and N. H. Afgan, eds., Hemisphere, Washington, D.C., pp. 200-217

Kuroda, A. (1990): "Direct Numerical Simulation of Couette-Poiseuille Turbulent Flows", (in Japanese), Dr. Eng. Thesis, Department of Mechanical Engineering, The University of Tokyo, Tokyo

Launder, B. E. (1989): "Second-Moment Closure: Present ... and Future?", Int. J. Heat & Fluid Flow, Vol. 10, pp.282-300

Moin, P., Spalart, P. R. (1989): "Contributions of Numerical Simulation Data Bases to the Physics, Modeling, and Measurement of Turbulence," in Advances in Turbulence, W. K. George and R. Arndt, eds., Hemisphere, Washington, D.C., pp. 11-38

Nagano, Y., Hishida, M. (1990): "Turbulent Heat Transfer Associated with Coherent Structures Near the Wall", in Near-Wall Turbulence, S. J. Kline and N. H. Afgan, eds., Hemisphere, Washington, D.C., pp. 568-581

Ohtsubo, Y. (1992): "Direct Numerical Simulation of Scalar Transport in Liquid Metal Turbulent Flow", (in Japanese), Department of Mechanical Engineering, The University of Tokyo, Tokyo

Perry, A. E., Hoffmann, P. H. (1976): "An Experimental Study of Turbulent Convective Heat Transfer from a Flat Plate," J. Fluid Mech., Vol. 77, pp. 355-368

Reed, C. B. (1987): "Convective Heat Transfer in Liquid Metals", in Handbook of Single-Phase Convective Heat Transfer, S. Kakaç et al., eds., John Wiley & Sons, New York, pp. 8.1-8.30

Reynolds, W. C. (1990): "The Potential and Limitations of Direct and Large Eddy Simulations", in Whither Turbulence? – Turbulence at the Crossroads, J. L. Lumley, ed., Springer Verlag, Berlin, pp. 313-342

Robinson, S. K., Kline, S. J., Spalart, P. R. (1990): "Quasi-Coherent Structures in the Turbulent Boundary Layer: Part II. Verification and New Information from a Numerically Simulated Flat-Plate Layer", in Near-Wall Turbulence, S. J. Kline and N. H. Afgan, eds., Hemisphere, Washington, D.C., pp. 218-247

Robinson, S. K. (1991a): "Coherent Motions in the Turbulent Boundary Layer", Ann. Rev. Fluid Mech., Vol. 23, pp. 601-639

Robinson, S. K. (1991b): "The Kinematics of Turbulent Boundary Layer Structure", NASA TM 103859

Part II

Separated Flows

Introductory Remarks

Roger L. Simpson

Department of Aerospace and Ocean Engineering, Virginia Polytechnic Institute and State University
Blacksburg, Virginia 24061, USA

The papers selected to appear in this section represent a cross-section of current research activities on separated flows: laser Doppler anemometer measurements of highly three-dimensional separations around and over obstacles and bodies that may have large mean vortical regions; large-eddy simulations (LES) with subgrid-scale modeling of separated flows dominated by the behavior of the larger structures; calculations using Reynolds-averaged equations and several closure models to assess their relative performance as compared to data for complex separated flows.

Using laser-Doppler-anemometer data, Larousse, Martinuzzi, and Tropea experimentally observed bimodal velocity probability histograms for the streamwise and normal to the wall velocity component in front of a cube mounted on the floor of a channel. Many flow structure similarities can be found between this case and the earlier work of Devenport and Simpson (1990), who examined the bimodal separated flow in front of 3:2 elliptic-nosed, NACA 0020 – tailed cylinder on the floor of a channel. Strong bimodal histograms exist between the primary separation that passes through the saddle point on the floor and a secondary separation closer to the cube, even out of the plane of symmetry.

An instantaneous unsteady four vortex model is suported by these data. Mode-averaged velocity vectors in the plane of symmetry show two modes that look much like those of Devenport and Simpson. The stronger backflow mode has a large elliptically-shaped vortex in front of the cube of the sense of the spanwise vorticity of the on-coming boundary layer; a much weaker elliptical vortical structure of the same sense is upstream and a very small counter-rotating structure lies on each side of the large structure. The weaker backflow mode has a large circularly-shaped vortex in front of the cube, a much weaker circularly-shaped structure further upstream, and very small counter-rotating structure on each side of the large structure. This pattern was observed to be dynamically chaotic in both investigations, with no preferred frequency of switching from one mode to the other.

While the four-vortex model has been widely mentioned in much earlier literature, e.g., Thwaites (1960), it has been previously assumed that the structures were *steady*. A major contribution of both of these investigation is the discovery of the *dynamically-unsteady* bimodal behavior in front of such low width-to-height obstacles.

Larousse et al. also found that no bimodal flow behavior existed for large width-to-height obstacles – the flow can simply go over the top rather than be forced around the sides of the obstacle. Ölçmen and Simpson (1992) also found no bimodal behavior in front of more streamlined cylindrical bodies such as a NACA 0012 wing. The bimodal behavior has important engineering implications; higher heat transfer and surface pressure fluctuations occur in the presence of such large-scale unsteadiness. Depending on the application, we can use available information to either avoid or exploit the phenomenon.

Werner and Wengle present a large-eddy simulation (LES) of the Larousse et al. flow. Although no direct comparisons with experimental data are made here, we can see the large eddy structure and the behavior of the helicity, enstrophy, and the $\vec{v} \times \vec{\omega}$, quantities that are difficult to measure. Hopefully, a future article will provide a detailed discussion of the similarities and differences with the experiment.

Arnal and Friedrich present a LES for a backstep flow for which experimental data are available. As in the Werner and Wengle calculation, this computational method provides simulation of the instantaneous flowfield and provides insights to the large eddy structure. While the calculated mean velocity profiles away from the wall are in good agreement with data, the Reynolds shear stress profiles are progressively poorer donwstream. Near the wall mean velocity profiles are poor. Improvements will require better subgrid-scale models.

Durao, Knittel, Pereira, and Rocha present LDA measurements and calculations using two turbulence models for the near wake recirculating flow around a disk with a central jet. The objective of the work is to investigate the performance of a second order closure model as compared to these new data and the k-ε model. The width of the flow around the disk, whose lip sloped away from the centerline, is approximately 1/3 of the disk radius. The central jet diameter is about 1/10 the disk diameter.

Since large mean velocity gradients are present for the recirculating flow between the flow around the disk and the jet, relatively fine spatial resolution is required for numerical accuracy. The second order closure scheme that is used for this pressure-driven flow produces much better agreement with the mean flow in the near wake then the standard k-ε model. Neither model produces good agreement with the turbulence quantities and with the far downstream flow.

As noted in the reviews of Simpson (1985, 1986), the k-ε model does not calculate recirculating flows very well, but does best for cases with comparable length scales for the forward and reversed flows. Although the discrepancies here are attributed to the k-ε model, the turbulence energy diffusion is also known to be important in recirculating flows (Simpson, 1985). The current work points to the need for improved models for these effects.

Lien and Leschziner address three issues associated with a non-orthogonal collocated finite-volume algorithm: wall-related pressure-strain terms, boundary conditions on curved walls, and iterative stability. Emphasis in the work is placed on numerical practices. The second-moment closure of Gibson and Launder is used in the calculation of a backward-facing step flow, a pipe constriction flow, and flow over a transonic bump.

For the backstep flow, there is some improvement of the mean velocity profiles near separation as compared to k-ε model results, although further downstream there is little difference. For the pipe constriction flow, there is some improvement of calculated surface pressures, although there still remain large discrepancies; velocities are not improved at all. The transonic bump shows quantitative improvements in surface pressure calculations and qualitative improvements in mean velocity profile shapes, although significant differences remain.

Franke and Rodi calculate vortex shedding past a square cylinder also using a k-ε model and a second-moment Reynolds stress closure. Comparisons of calculations using different near wall treatments indicate that the Reynolds-stress-equation models produce better agreement with the experimentally-measured phase-averaged velocities, while turbulence quantities are still poorly calculated. As in the Durao et al. and Lien and Leschziner works, further improvements in the second-order closure model are required for more accurate calculations.

References

Devenport, W.J. and Simpson, R.L. (1990) "Time-Dependent and Time-Averaged Turbulence Structure Near the Nose of a Wing-Body Junction", *J. Fluid Mech.*, *210*, pp. 23-55, 1990.

Ölçmen, S.M. and Simpson, R.L. (1992) "Influence of Wing Shapes on the Surface Pressure Fluctuations of a Wing-Body Junction", AIAA-92-0433, 30th AIAA Aerospace Sciences Meeting, Reno, NV., Jan. 6-9.

Simpson, R.L. (1985) "Two-Dimensional Turbulent Separated Flow", AGARD-AG-287-Vol.1, NATO-AGARD.

Simpson, R.L. (1986) "A Review of Two-Dimensional Turbulent Separated Flow Calculation Methods", *IUTAM Symposium on Boundary Layer Separation*, London, August 1986; proceeding, pp. 179-196, 1987, Springer-Verlag.

Thwaites, B. (1960) *Incompressible Aerodynamics*, Oxford Univ. Press; first Dover Edition, 1987.

Flow Around Surface-Mounted, Three-Dimensional Obstacles

A. Larousse, R. Martinuzzi and C. Tropea

Lehrstuhl für Strömungsmechanik der Universität Erlangen-Nürnberg, Cauerstraße 4, W-8520 Erlangen, Fed. Rep. of Germany

Abstract

The flow around surface-mounted obstacles placed in a turbulent channel flow ($Re = 10^5$) has been experimentally studied with the aim of determining separation and reattachment patterns and to investigate the major flow differences between two- and three-dimensional geometries. Obstacles of square cross-section and half channel height were studied while varying the cross-channel width between that of a cube and a two-dimensional rib. Surface flow patterns were obtained to study the flow topology. A two-component laser Doppler anemometer (LDA) was used to gain detailed information about the time-averaged and time-resolved turbulence parameters.

The results of this study lead to a picture of the flow pattern over a surface-mounted cube with quantitative data describing the mean and fluctuating velocity field. Major differences between the two- and three-dimensional flow patterns are identified, both upstream and downstream of the obstacle. Analysis of the time-dependent behaviour reveals a distinct unsteadiness of the upstream flow field for three-dimensional obstacles, which also leads to large contributions to the Reynolds shear stress.

Introduction

The study of the flow around sharp-edged obstacles placed in a channel contributes directly to the understanding of the flow mechanisms for complex two- and three-dimensional geometries. Past studies have often concentrated on two-dimensional geometries, for instance ribs or fences, because, as a diagnostic flow, the two-dimensional geometry is easier to handle experimentally and numerically. Hence, there exists a considerable amount of published data on two-dimensional, surface-mounted obstacles. However, little data exist on the flow development upstream of the obstacle. The present work indicates the importance of this flow region for understanding the rest of the flow field.

There has been very little work published on the flow around three-dimensional obstacles and no data has been found for such flows in a channel. Typically, the three-dimensional obstacle is studied as a perturbation to a boundary layer, for instance for purposes of artificial thickening or increasing overall drag (Castro (1979), Castro and Robins (1977)). Correspondingly, only few studies concentrated on describing the detailed flow patterns around the obstacle. Of the existing work, the theoretical analyses of Hunt *et al.* (1978), Perry and Horning (1984) and Fairlie (1980) address directly the expected separation and reattachment patterns around the obstacle. There exists, however, a severe lack of experimental data in this regard, a fact also recognized in a recent review of the flow over surface-mounted obstacles presented by Schofield and Logan (1990). The present work provides such data.

The transition between a two-dimensional obstacle flow and a three-dimensional one is not well defined. Work by de Brederode and Bradshaw (1972), in which the base pressure of a backstep flow was measured for various channel widths, is often cited as prescribing a 10:1 width-to-height ratio to ensure two-dimensionality in the recirculation region behind the step. In light of the complex nature of the obstacle flow as compared with a backstep flow, this ratio is not expected to be directly applicable in the present case. The nature of the two-dimensional to three-dimensional transition in flow pattern remains obscure. Therefore, a second focus of the present work was to investigate the effect of varying the obstacle cross-channel width on the resulting surface flow patterns. Detailed velocity data for two-dimensional geometries already exist (Dimaczek *et al.* (1989)) and can be used for comparison with the present measurements for three-dimensional obstacles.

Experimental Apparatus and Methods

The flow geometry to be studied and the nomenclature used are shown in Fig. 1. Some preliminary work, especially concerning the flow pattern on the front face of the obstacle, was performed in a water channel (550 mm (w) x 50 mm (h)). Most of the experiments were performed in a 600 mm (w) x 50 mm (h), blower-type air channel at a Reynolds number of 10^5 ($Re = U_B h/\nu$). Trip wires were placed at the inlet of the air channel flow. The obstacles were placed on the centre line at a downstream distance of 52 channel heights. The channel could be mounted either horizontally or vertically, as sketched in Fig. 2, to allow either the U–V (forward scatter) or the U–W (backscatter) velocity components to be acquired with a three-beam, two-component LDA (TSI 9100), also shown in Fig. 2. Beam expansion was used in the LDA system to achieve a measuring volume of 84 μm diameter and 610 μm length. Two counter processors (TSI 1980B) with a 5 μs coincidence acquisition window were connected to a PC for data reduction. Data rates of 500 Hz to 4000 Hz were typical. A di-ethylglycol/water mixture atomized by a

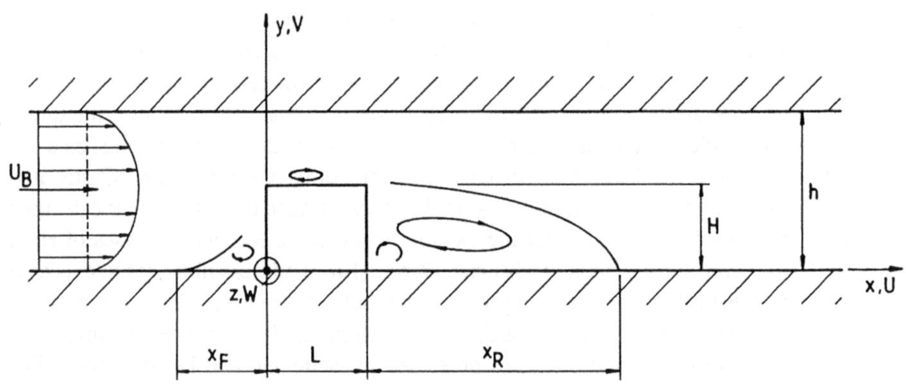

Fig. 1. Sketch of the obstacle geometry in a channel flow.

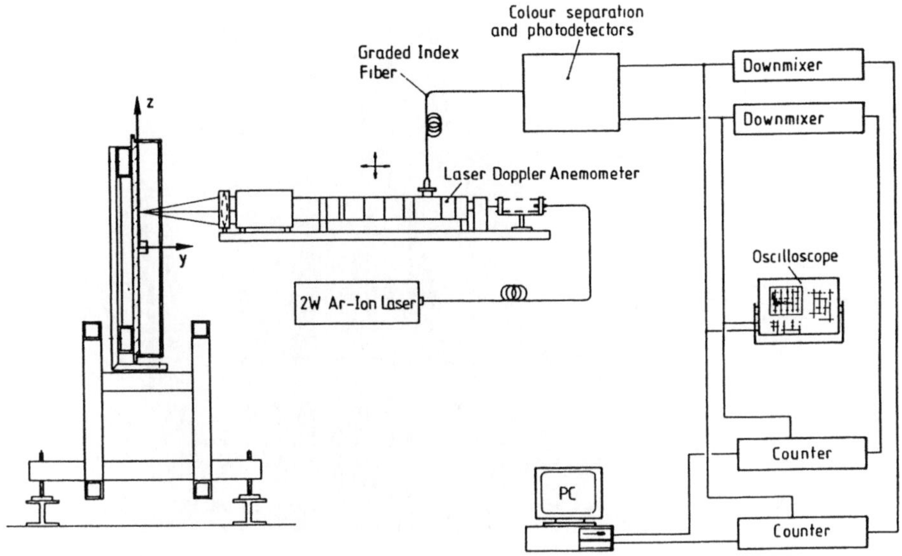

Fig. 2. Schematic representation of the test-rig.

medical nebulizer was used as seeding material. The particles were introduced upstream of the channel inlet in the settling chamber of the blower.

Surface flow pattern visualization was performed in the air channel using a mixture of kerosene, light transmission oil and carbon dust (toner). Visualization results were then photographed for analysis and documentation purposes. In the water channel, the crystal violet visualization technique (Dimaczek *et al.* (1988)) was used.

Experimental Results and Discussion

The obstacles used in this study were 25 mm in height (H) and length (L) and varied between 25 mm and 600 mm in width (W). Before mounting the obstacle, the channel flow was verified to be fully turbulent and uniform in mean velocity to within 2% over 85% of the channel width. With the cube obstacle mounted, the mean velocity about the z-axis was verified to be symmetric within 2%. Also, the measured quantities $\overline{W}$ and $\overline{uw}$ were acceptably close to zero on the z-plane of symmetry.

Flow Around the Cube

To begin with, the flow in front of the cube will be examined in more detail. Fig. 3 shows the mean streamlines corresponding to the measured flow patterns, surface oil patterns on

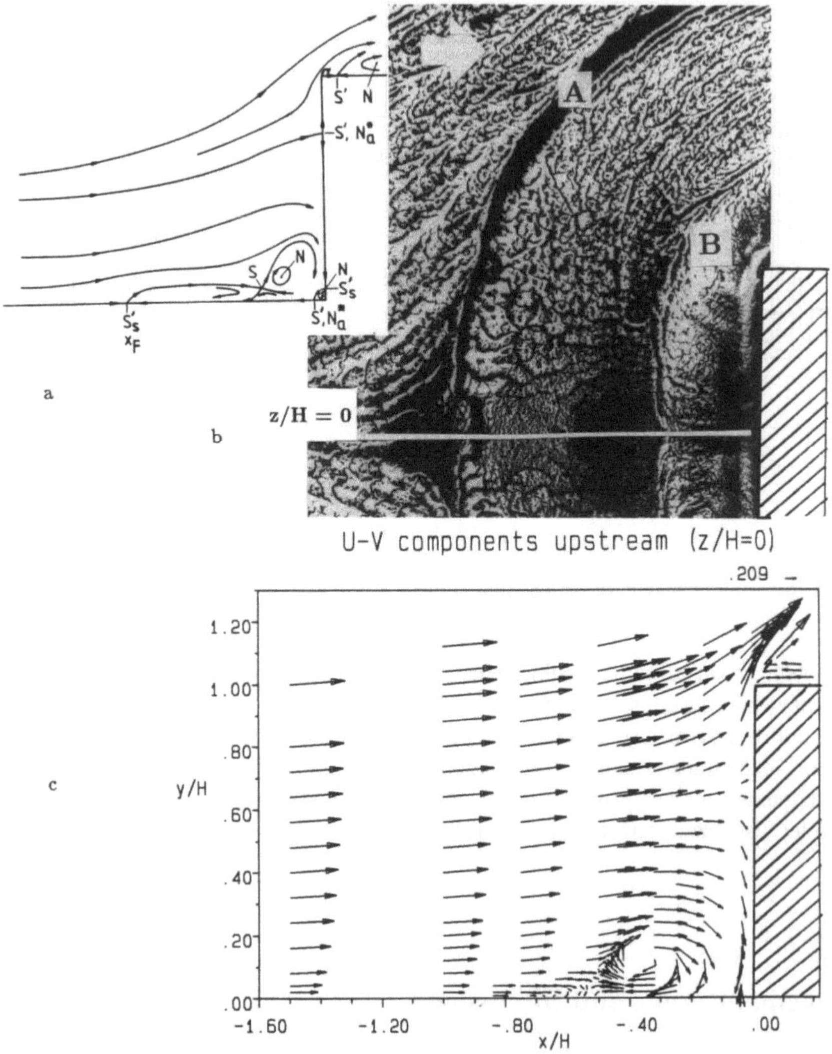

Fig. 3. Flow upstream of the cube: a) flow pattern b) oil-visualization c) measured time-averaged velocity vectors. Legend: S= saddle, N= node, '= half saddle or node,*= when point viewed perpendicularly to the wall; subscripts: s–separation and a–attachment.

the channel floor upstream of the obstacle and results of the mean velocity measurements in the z-plane upstream of the cube. The mean streamline pattern in this work correlates well with that of Hunt *et al.* (1978) except for the region upstream of the horseshoe vortex. No evidence could be found supporting the existence of two secondary vortices in this region.

Dark regions in Fig. 3b correspond to an accumulation of visualization pigment. Line **A** in Fig. 3b, for instance, is a half saddle point in the X-Y plane which is considered to correspond to the primary separation of the flow, marked x_F in Fig. 1, and is located

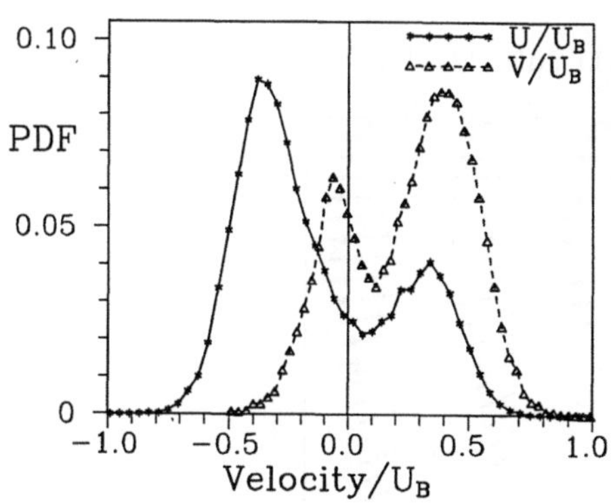

Fig. 4. Velocity PDF at x= −0.44H, y= 0.10H and z= 0.

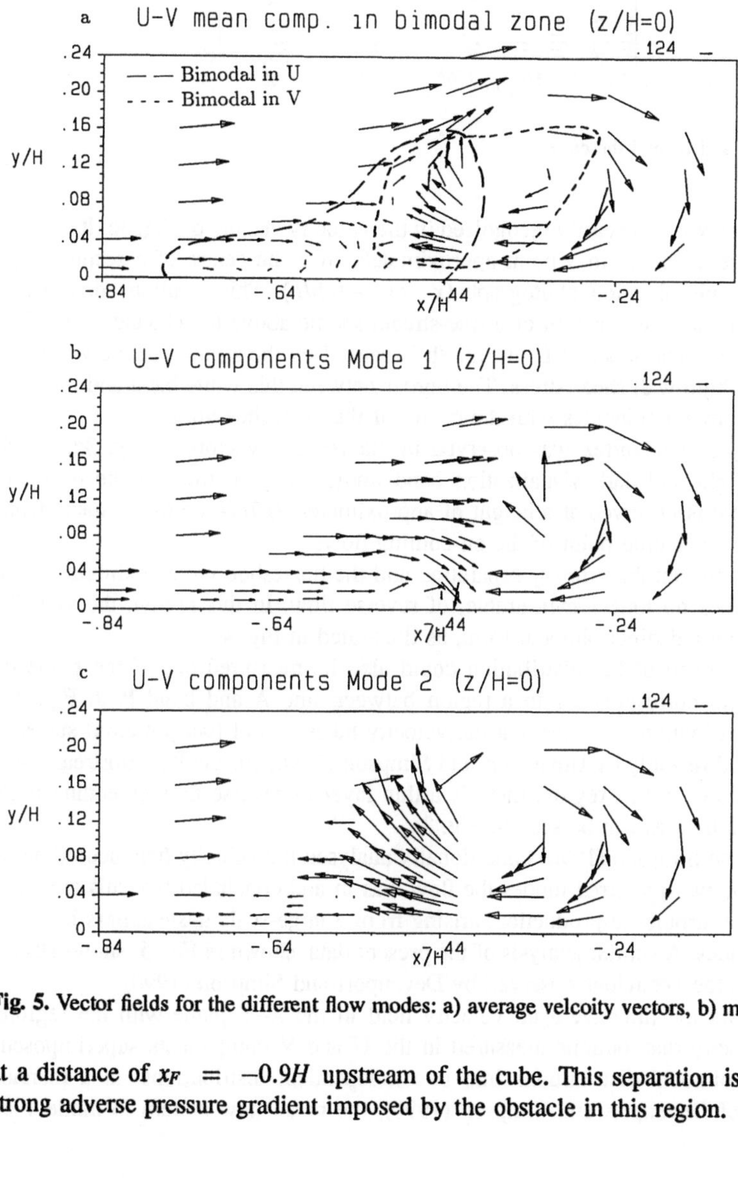

Fig. 5. Vector fields for the different flow modes: a) average velcoity vectors, b) mode 1 c) mode 2.

at a distance of $x_F = -0.9H$ upstream of the cube. This separation is caused by the strong adverse pressure gradient imposed by the obstacle in this region.

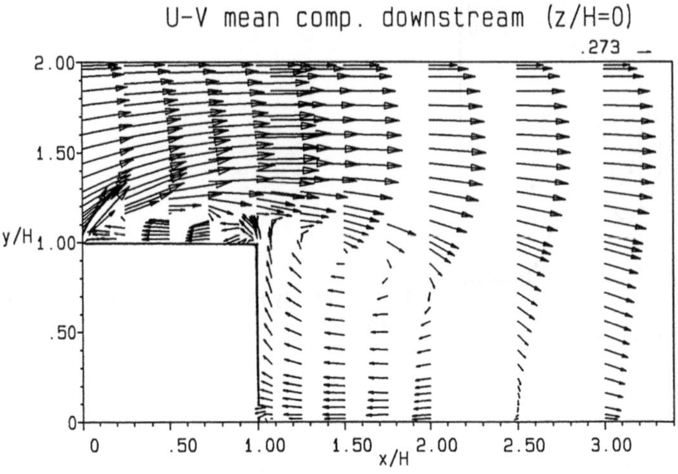

Fig. 6. Flow over and behind the cube.

The horseshoe vortex is easily recognized in the velocity vectors of Fig. 3c. It extends from the obstacle upstream to a point approximately over the centre of a second large dark area in the visualization photograph ($x \approx -0.5H$). This location corresponds roughly to the streamwise position of a free-stream saddle above the channel floor. The white visualization band labelled **B** in Fig. 3b is caused by the strong backflow and the correspondingly high wall shear stress. The border between this white band and the dark zone directly upstream coincides with the centre of the horseshoe vortex.

A secondary corner vortex was observed in the velocity vectors of Fig. 3c, which yields another light and dark visualization band immediately in front of the obstacle. Also a half saddle is observed at a height of approximately $0.76H$ on the obstacle front face. This is the stagnation point of the on-coming flow.

The region between the primary separation and the horseshoe vortex exhibits a very thin layer, at times no longer measurable, of reverse flow. In this region, the velocity distributions show a distinct bimodal form, as illustrated in Fig. 4.

The bimodal form of the distribution could also be measured out of the plane of symmetry, always corresponding to a region between line **A** and band **B** in Fig. 3b. This form of distribution indicates that the velocity takes one of two preferred states, a behaviour studied recently by Devenport and Simpson (1990) for the flow upstream of a wing-body junction. In this region, there is a thin layer of reverse flow extending back to the separation line, as can be seen in Fig. 3a.

Devenport and Simpson (1990) conditionally analysed the velocity field according to which one of the two preferred modes the flow was in and concluded that this area is a region of intense turbulence production, arising from a large-scale unsteadiness between the two flow modes. A similar analysis of the present data, shown in Fig. 5, shows strong similarities with the behaviour observed by Devenport and Simpson (1990).

Fig. 5a shows the time-averaged velocity field in the X–Y plane with the regions of bimodal velocity distributions measured in the U and V components superimposed. Separating the bimodal distributions into two independent distributions, in a manner similar to that of Devenport and Simpson (1990), the mean flow pattern in mode 1 or

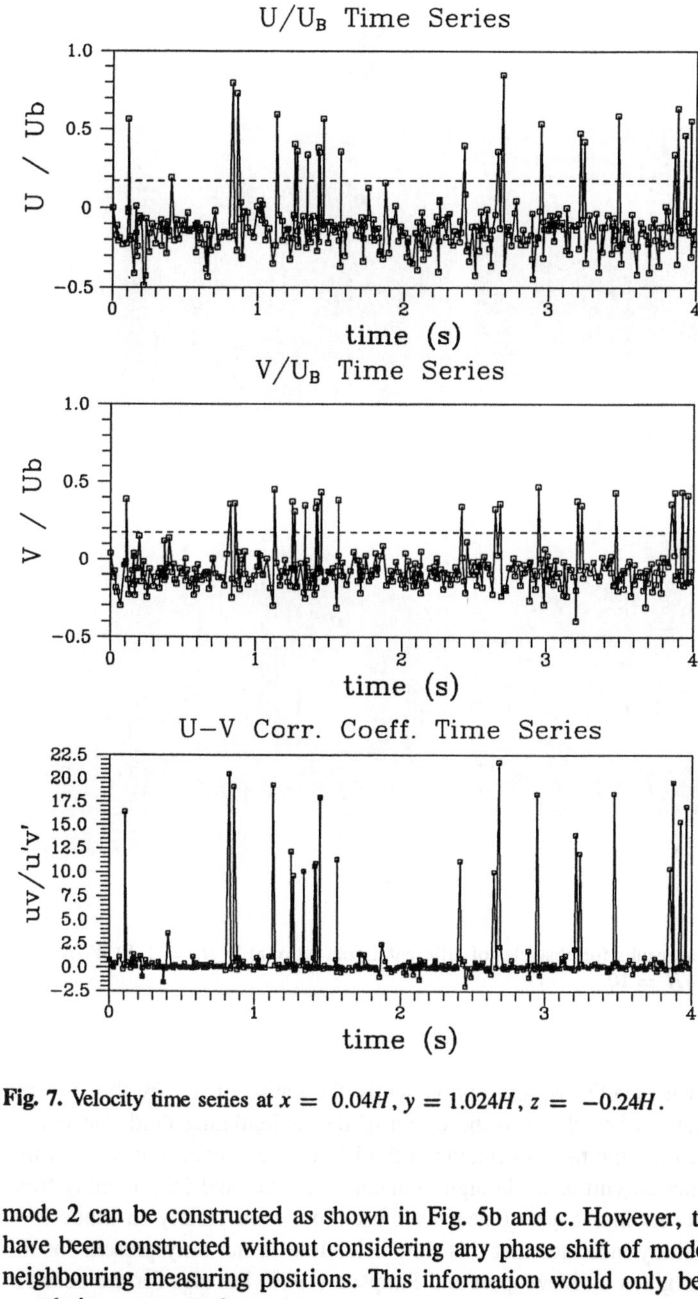

Fig. 7. Velocity time series at $x = 0.04H$, $y = 1.024H$, $z = -0.24H$.

mode 2 can be constructed as shown in Fig. 5b and c. However, these representations have been constructed without considering any phase shift of mode transition between neighbouring measuring positions. This information would only be available if spatial correlations were performed.

Further analysis of this data reveals extremely high values of the Reynolds shear stress in the region of measured bimodal velocity distributions suggesting that this production is associated with large-scale unsteadiness. Future analysis will investigate the cross spectral density of the U and V velocity fluctuations to determine the time scales of the $\overline{uv}$ contributions and their relative magnitude.

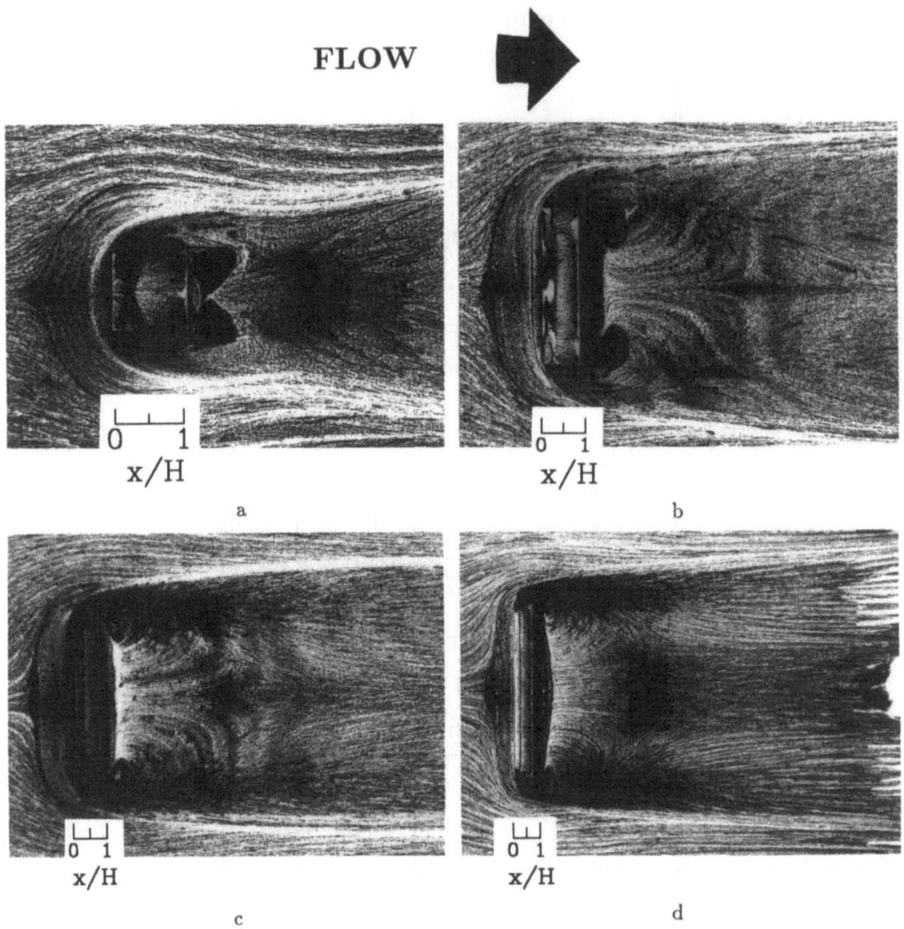

Fig. 8. Flow visualization results for obstacles of different aspect ratio: a) $W/H = 1$, b) $W/H = 2$, c) $W/H = 4$ and d) $W/H = 10$.

The cause of this unsteadiness, and consequently the cause of a very thin reverse flow layer, is postulated to be related to the origin of the recirculating fluid (Devenport and Simpson (1990)). Low momentum rotational fluid from the boundary layer is easily integrated into the junction vortex while high momentum irrotational fluid coming from the outer flow transforms into a jet of near-wall backflow seen clearly in mode 2 in Fig. 5c. The present study indicates that this flow behaviour is not only present for the wing-body junction as given by Devenport and Simpson (1990) but also for the obstacle placed in a channel flow. The major difference between these two geometries is that the flow can go over the top of the obstacle but can only go around a wing-body junction.

A further region studied in more detail is the flow separation at the upper leading edge of the obstacle, shown in Fig. 6. There exists a mean recirculation region over the top surface which also exhibits very strong velocity fluctuations and bimodal PDF's. Measurements of the U and V velocity components, taken just inside of the separating

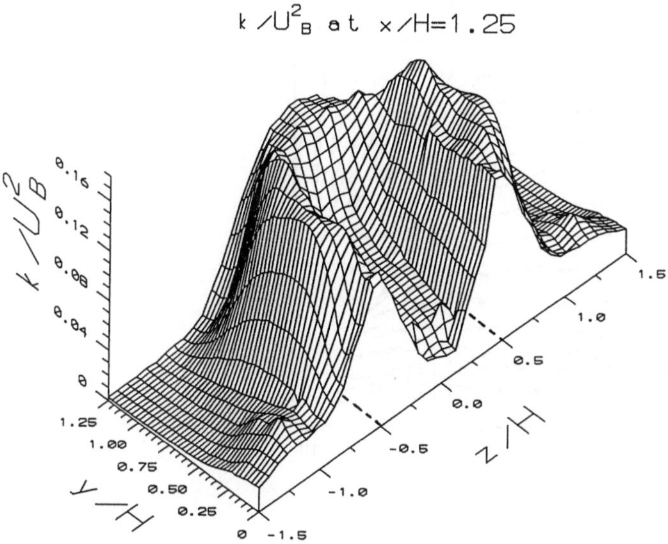

Fig. 9. Map of the turbulent kinetic energy at the plane $x = 1.25H$.

streamline above the obstacle, are shown in Fig. 7 as time traces. For the most part, the fluctuations are random in nature and contribute weakly to the Reynolds stress. At some instances, however, a very organized motion occurs in which the U and V velocity fluctuations become highly correlated. This would correspond to intense turbulence production and can be interpreted as a large-scale unsteadiness in the recirculation region similar to the behaviour seen in front of the obstacle.

The recirculation region behind the cube, also shown in Fig. 6, exhibits a structure similar to that described by Hunt *et al.* (1978). The two corner vortices visible on the floor behind the cube (as seen in the top view given in Fig. 8a) extend upwards to join in the plane of symmetry forming an arch along which strong velocity fluctuations and bimodal velocity distributions have been measured. As was the case for the regions of bimodal measurements previously described, this region also corresponds to one where the kinetic turbulent energy is very large. A map of the turbulent kinetic energy, $k = \frac{1}{2}(\overline{u^2}+\overline{v^2}+\overline{w^2})$, measured in the Y–Z plane immediately behind the obstacle, shows that large values of k clearly trace out the postulated arch vortex behind the cube (see Fig. 9).

Velocity measurements performed behind the obstacle show that the two arms of the horseshoe vortex downstream of the reattachment point remain adjacent to one another, with no further streamwise vortices occuring in between. This is in contradiction to recent flow patterns proposed by Schofield and Logan (1990). These results also show that the location of the horseshoe vortex centres change from $z \approx \pm 1.24H$ at $x = 1.75H$ to $z \approx \pm 0.95$ at $x = 5.3H$ (see Fig. 10). This narrowing is due to the influence of the recirculation region vortex which entrains inner portions of the horseshoe vortex and draws the arms towards the axis of symmetry.

Further analysis of the wall visualization results, with support from LDA velocity measurements, leads to a picture of the overall mean flow field as sketched in Fig. 11.

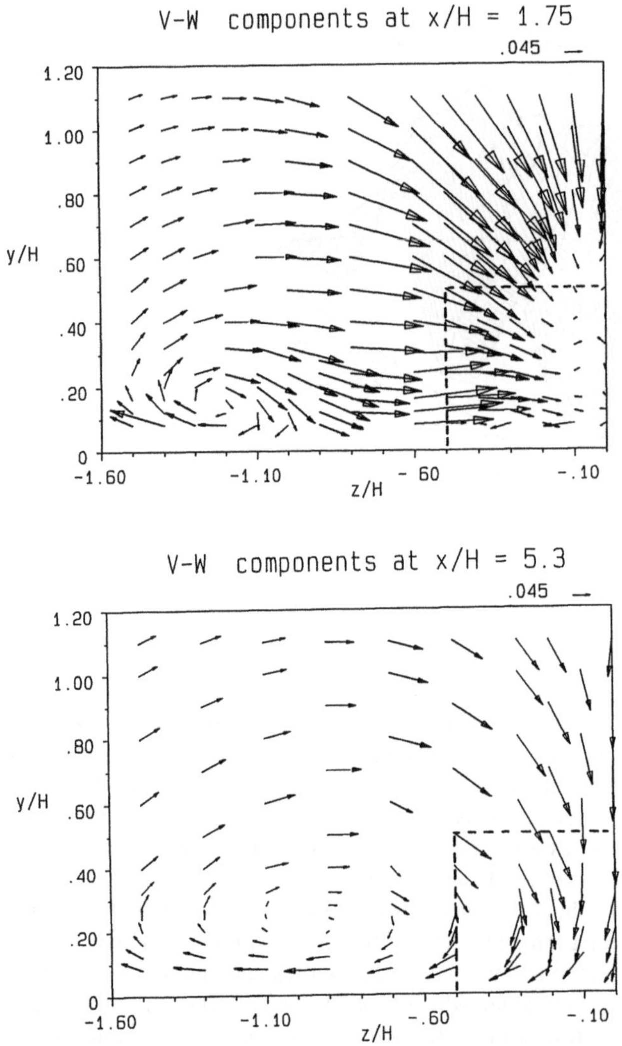

Fig. 10. Velocity vectors in the planes $x = 1.75H$ and $x = 5.3H$

Flow Around Obstacles of Various Aspect Ratios

The aspect ratio of the obstacle, defined as the width to height ratio (W/H), was varied between 1 (cube) and 24 (rib). The results of the visualization experiments are shown in Fig. 8 for four of the tested obstacles.

Observing first the dark line of primary separation in front of the obstacle, it is clear that at large obstacle aspect ratios ($W/H > 10$), it looses its curvature and becomes a separation line. The separation distance, x_F, shown in Fig. 12, increases with the aspect ratio to values of $W/H \approx 5$ and then becomes independent of the obstacle width. This result is in contradiction to the visualization measurements of König (1990), who

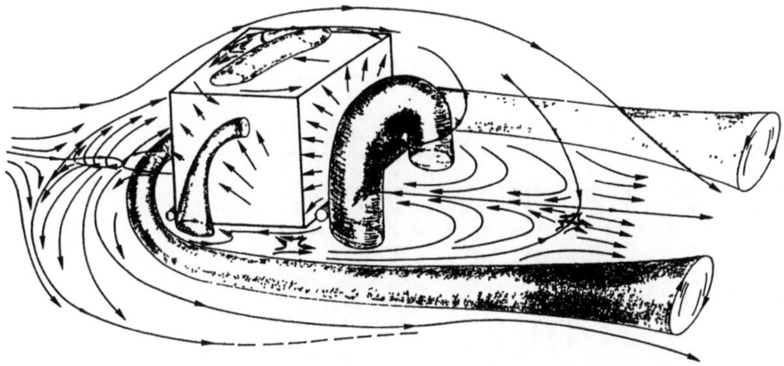

Fig. 11. Sketch of the flow around a cube.

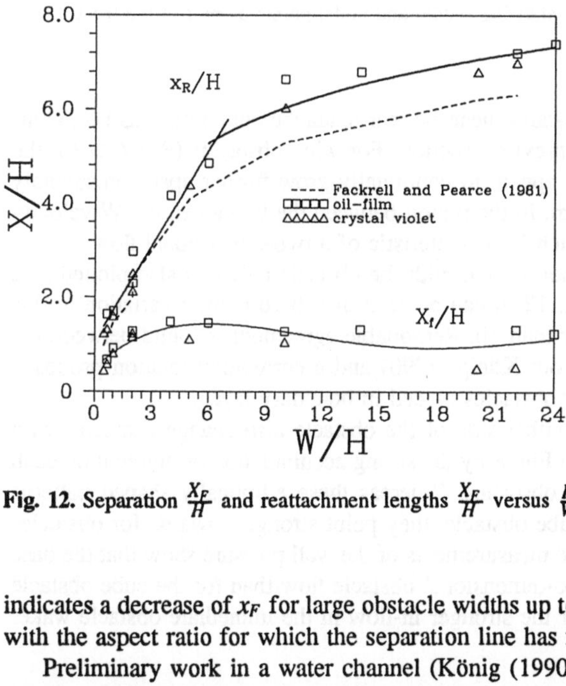

Fig. 12. Separation $\frac{X_F}{H}$ and reattachment lengths $\frac{X_F}{H}$ versus $\frac{H}{W}$.

indicates a decrease of x_F for large obstacle widths up to 20H. This latter value coincides with the aspect ratio for which the separation line has negligible curvature.

Preliminary work in a water channel (König (1990) and Theisinger (1990)) showed that for obstacle aspect ratios greater than 10, an alternating series of node and saddle points could be observed on the obstacle front face as shown by the crystal violet visualization in Fig. 13. Whereas the on-coming flow can easily be diverted around short obstacles, it must go over wider obstacles and the structure of the forward recirculation region apparently changes. In addition, no bimodal velocity distributions could be measured for the wider obstacles. Thus, the large-scale unsteadiness, which appears to be typical of three-dimensional obstacles, must be closely linked to the evacuation of recirculating fluid to the sides and around the obstacle, probably in the form of a horseshoe vortex.

Inspection of the series of photographs in Fig. 8 also reveals major differences downstream of the obstacles for different aspect ratios. For the cube obstacle, the horseshoe

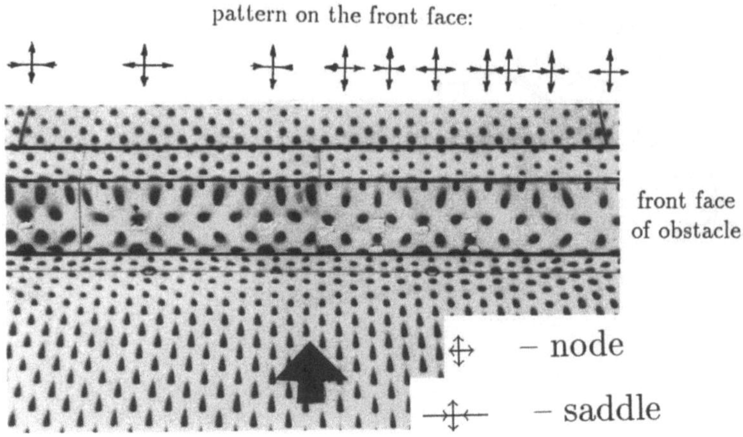

Fig. 13. Crystal violet results for a rib showing saddle and nodal points on part of the front face.

vortex narrows at a downstream distance near the rear reattachment point. This behaviour has been briefly described in the previous section. For wide obstacles ($W/H > 6$), the two arms of the horseshoe vortex appear to continually grow further apart, presumably with little interaction between them. In the region between the two arms, the W velocity component appears to be zero which is characteristic of a two-dimensional flow.

The reattachment length, x_R, normalized with the obstacle height is also plotted as a function of the aspect ratio in Fig. 12. It can be seen that there is little variation in this parameter for aspect ratios greater than 10. Reasonable agreement is found between the present results, previous results from König (1990) and a correlation relation proposed by Fackrell and Pearce (1981) for obstacles placed in boundary layers.

The trailing edge vortices on either side of the obstacle also change character with aspect ratio. These are indicated in Fig. 8 by the strong accumulation of pigment on each side and immediately behind the obstacles. Whereas these elliptically shaped patterns strongly point outwards for the cube obstacle, they point strongly inwards for obstacles with W/H greater than 6. Separate measurements of the wall pressure show that the base pressure is much lower for the two-dimensional obstacle flow than for the cube obstacle flow, which accounts therefore for the stronger in-flow in the immediate obstacle wake.

Conclusions

This experimental study has provided further details of the mean flow patterns around three-dimensional obstacles and a comparison of these patterns to those for two-dimensional obstacles. Major differences in the mean and fluctuating velocity field upstream of the obstacle between two- and three-dimensional obstacles were identified. In particular, a large scale unsteadiness was found in the case of the three-dimensional obstacle. This unsteadiness leads to very high values of the Reynolds shear stress. For wider obstacles, a flow of 'two-dimensional' character about the centre plane is only achieved for width-to-height ratios above 10. Even above these values, however, the flow upstream and downstream of the obstacle does exhibit spanwise variations, in particular an alternating saddle/node structure on the obstacle front face.

References

Castro, I.P. (1979): "Relaxing Wakes Behind Surface-Mounted Obstacles in Rough Wall Boundary Layers", J. Fluid Mech. **93** pp. 631-659.

Castro, I.P., Robins, A.G. (1977): "The Flow Around a Surface-Mounted Cube in Uniform and Turbulent Streams", Ann. Rev. Fluid Mech. **11** pp. 443-503

de Brederode, V., Bradshaw, P. (1972): "Three-Dimensional Flow in Nominally Two-Dimensional Separation Bubbles I: Flow Behind a Rearward-Facing Step", IC Aero Report 72-19, Imperial College, London

Devenport, W.J. and Simpson, R.L. (1990): "Time-Dependent and Time-Averaged Turbulence Structure Near the Nose of a Wing-Body Junction", J. Fluid Mech. **210** pp. 23-55

Dimaczek, G., Kessler, R., Martinuzzi, R. and Tropea, C. (1989): "The Flow Over Two-Dimensional, Surface-Mounted Obstacles at High Reynolds Numbers", 7th Symp. on Turb. Shear Flows, Stanford CA

Dimaczek, G., EH, C., Tropea, C. (1988): "Sichtbarmachung von Wasserströmungen mit Hilfe des Kristalviolettverfahrens", DLGR Workshop 2D-Meßtechnik, Marktdorf, DLGR-Bericht 88-04

Fackrell, J.E., Pearce, J.E. (1981): "Parameters Affecting Dispersion in the Near Wake of Buildings", CEGB Report No. RD/M/1179/N81

Fairlie, B.D. (1980): "Flow Separation on Bodies of Revolution at Incidence", Proc. 7th Australian Conf. on Hydr. and Fluid Mech., Brisbane.

Hunt, J.C.R., Abell, C.J., Peterka, J.A., Woo, H. (1978): "Kinematical Studies of the Flows Around Free or Surface- Mounted Obstacles; Applying Topology to Flow Visualization", J. Fluid Mech. **86** pp. 179-200

König, J. (1990): "Anwendung des Kristallviolettverfahrens zur Untersuchung der dreidimensionalen Hindernisströmung", Studienarbeit at the Lehrstuhl für Strömungsmechanik, University Erlangen–Nürnberg.

Perry A.E., Horning, H. (1984): "Some aspects of Three-Dimensional Separation", Z. Flugwiss. Weltraumforschung **8** Heft 3, pp. 155-160.

Schofield, W.H. and LOGAN, E. (1990): "Turbulent Shear Flow Over Surface Mounted Obstacles", J. of Fluid Engineering, Trans. of the ASME **112** pp. 376-385.

Theisinger,J (1990): "Untersuchung der Dreidimensionalität einer Rippenströmung im Flachkanal mittels LDA", Diplomarbeit at the Lehrstuhl für Strömungsmechanik, University Erlangen-Nürnberg.

Measurements and Modelling of the Turbulent Near Wake Flow of a Disk With a Central Jet

D.F.G. Durão, G. Knittel, J.C.F. Pereira and J.M.P. Rocha

Mechanical Engineering Department, Instituto Superior Técnico, Technical University of Lisbon, Av. Revisco Pais, 1096 Lisboa Cedex, Portugal

Abstract

LDV measurements of the velocity and turbulence characteristics and their comparison with numerical solutions are reported for the near wake recirculating flow around a disk with a central jet. Both the two-equation $k-\varepsilon$ eddy-viscosity model and the second-moment Reynolds stress transport closure (RSM) were used to close the mean-field equations. The 13-point Quadratic Weighted Upstream scheme was used for convection discretization and employed in all transport equations. It was found that the present flow configuration requires a very high computational resolution for numerical accuracy. Significant differences were noticed between the two levels of turbulence closure accordingly to previous disk flow prediction studies. The RSM provides a significantly better description of mean field compared with $k-\varepsilon$ eddy viscosity model. However, none of the models is able to predict accurately the measured turbulence characteristics.

Introduction

The velocity characteristics of the flow in the near wake of a disk play an important role in many engineering applications. Measurements of the flow in the disk near wake were reported previously e.g. Carmody (1964), Durão and Whitelaw (1978), Durão and Firmino (1983), Taylor and Whitelaw (1984) and Heitor *et al* (1987). The few numerical simulations of isothermal disk flows reported have shown that the numerical accuracy of flow computation is crucial to provide a meaningful assessement of turbulence models, see McGuirk *et al.* (1985a). Concerning turbulence modelling, McGuirk *et al.* (1985b) have applied two versions of a second-moment closure together with the $k-\varepsilon$ model for the prediction of the confined disk flow with comparison with $k-\varepsilon$ eddy viscosity model results. A noticeable feature of the results is that second-moment closures do not show notorious superiority over $k-\varepsilon$ model.

Related to the aforementioned studies, the present paper is concerned with the unconfined recirculating flow behind a disk with a central jet similarly to the bluff-body stabilized flame investigated by Schefer *et al.* (1987) and Li and Tankin (1987). The selected momentum ratio between the annular flow around the disk and central jet originates a complex recirculating flow structure. The objective of this study is to obtain measurements of mean and fluctuating velocity characteristics and to investigate the performance of a second-moment turbulence closure in this flow configuration.

Experimental Investigation

Experiments were conducted in the recirculation region behind a disk with a central orifice. Air was discharged at a velocity of 30 m/s at the nozzle surrounding the disk and also through the central disk jet. Air was supplied independently to the inner pipe and discharged through the disk central orifice. The Reynolds numbers of the two incoming streams were respectively $Re_a = 1.2 \times 10^5$ and $Re_j = 10^4$ based on disk diameter and disk orifice diameter respectively.

Velocity measurements were obtained with one-component dual beam LDV in forward scattering mode. The optical system consisted of an 0.8 Argon-Ion laser using the green line and OEI transmitting optics. The signal was processed by a counter (TSI 1980B) interfaced with a computer which provided the velocity moments. The air in both streams was seeded with nominal 1mm diameter TiO_2 particles. Mean axial and radial variances were calculated from 10 to 50 thousand samples. The maximum uncertainties in the mean and variances velocity at 95% confidence interval were 2% and 4% respectively. Detailed tests were carried out regarding flow axisymmetry, influence of seeding conditions, etc. and excellent long-term repeatability of the measurements was obtained. Details concerning the LDV system and data averaging procedure may be found in Durão *et al.* (1990).

Mathematical Model

The continuity and momentum equations for a steady, two-dimensional, incompressible fluid flow at high Reynolds number are

$$\frac{\partial U_j}{\partial x_j} = 0 \, , \tag{1}$$

$$\frac{\partial U_j U_i}{\partial x_j} = -\frac{1}{\rho}\frac{\partial P}{\partial x_i} - \frac{\partial <u_i u_j>}{\partial x_j} \, . \tag{2}$$

Upper and lower cases relate to mean and turbulence quantities, respectively; <> denotes temporal average of the variable; P represents pressure variable.

To close the above equations either a second-moment model and the standard $k - \varepsilon$ eddy viscosity (EVM) model were considered. Regarding RSM, the transport equations for the components of Reynolds stress tensor can be written as

$$\frac{\partial U_k <u_i u_j>}{\partial x_k} = \mathcal{D}_{ij} + \mathcal{P}_{ij} + \Phi_{ij} - \varepsilon_{ij} \tag{3}$$

where the terms in RHS of Eq. (3) represent diffusion, production, pressure redistribution and dissipation rate of $<u_i u_j>$. Apart from $\mathcal{P}_{ij}$ term all the remaining ones in RHS of (3) need to be modelled.

Diffusion term is represented by a generalised gradient diffusion model (Daly and Harlow, 1970):

$$\mathcal{D}_{ij} = \frac{\partial}{\partial x_k}\left[C_s \frac{k}{\varepsilon} <u_k u_l> \frac{\partial <u_i u_j>}{\partial x_l}\right] . \tag{4}$$

Pressure-redistribution term is modelled by Rotta (1951) and Naot *et al.* (1970) "slow" and "rapid" models:

$$\Phi_{ij} = \Phi_{ij,1} + \Phi_{ij,2} \tag{5}$$

$$\Phi_{ij,1} = -C_1 \frac{\varepsilon}{k}\left[<u_i u_j> - \frac{\delta_{ij}}{3} <u_k u_k>\right] , \tag{6}$$

$$\Phi_{ij,2} = -C_2\left[\mathcal{P}_{ij} - \frac{\delta_{ij}}{3}\mathcal{P}_{kk}\right] , \tag{7}$$

where the production term is given by

$$\mathcal{P}_{ij} \equiv - <u_i u_k> \frac{\partial U_j}{\partial x_k} - <u_j u_k> \frac{\partial U_i}{\partial x_k} . \tag{8}$$

For the basic model, the dissipation rate of $<u_i u_j>$ is represented by an isotropic tensor as $\varepsilon_{ij} = 2/3\varepsilon\delta_{ij}$. The scalar dissipation rate is computed from solving a transport equation in the form

$$\frac{\partial U_j \varepsilon}{\partial x_j} = \frac{\partial}{\partial x_i}\left(C_\varepsilon \frac{k}{\varepsilon} <u_i u_k> \frac{\partial \varepsilon}{\partial x_k}\right) + C_{\varepsilon 1}\frac{\varepsilon}{k}\mathcal{P}_k - C_{\varepsilon 2}\frac{\varepsilon^2}{k} \tag{9}$$

where $\mathcal{P}_k \equiv \mathcal{P}_{kk}/2$. The value of model constants is taken as previous studies and set out below, see Table 1.

Together with the above model the standard form of the $k-\varepsilon$ eddy-viscosity model was considered (Launder and Spalding, 1974).

Table 1. Model Constants.

C_1	C_2	C_s	C_ε	$C_{\varepsilon 1}$	$C_{\varepsilon 2}$
1.8	0.6	0.24	0.15	1.44	1.92

Numerical Model

A finite-volume method using a staggered variable arrangement is used for solving the equations system. The pressure and mean velocity fields are coupled with SIMPLE algorithm (Patankar and Spalding, 1972). The 13-point quadratic upstream weighted scheme

(QUDS; Leonard, 1979) was used for convection discretization in all transport equations. At each control volume face, the dependent variable ϕ was interpolated by the quadratic surface yielding e.g. for $\phi_{i-1/2,j} > 0$ the following expression when $U_{i-1/2,j} > 0$, $V_{i-1/2,j} > 0$,

$$\phi_{i-1/2,j} = a_0 + a_1 \frac{\Delta x_i}{2} + a_2 \frac{\Delta x_i^2}{4} - a_4 \frac{1}{3} \frac{(C_y^3 + 1)}{(C_y + 1)^3} \left(\frac{\Delta y_i + \Delta y_{i+1}}{2} \right)^2 \tag{10}$$

where $C_y = \Delta y_j / \Delta y_{j-1}$ takes into account the grid non-uniformity and a_i denotes coefficients that contain the dependent variable ϕ in the surrounding grid nodes (see for further details Durão and Pereira, 1987). The strongly implicit method, (Stone, 1968; Azevedo *et al.*, 1988) was used to solve the system of algebraic equations including only 5 diagonals in the coefficients matrix. The other coefficients were appropriately incorporated in the source term.

The stabilizing practices suggested by Huang and Leschziner (1985) were used in the numerical solution of the second moment closure. The under-relaxation form of the $< u_i u_j >$ finite-difference equations was used for the splitting procedure of $< u_i u_j >$ of apparent diffusion and source terms included in the momentum equations. The use of the under-relaxation form prevents the occurence of large variations of the value of the variables during the iterative solution procedure.

The solution domain was extended to cover 2.7 disk diameters in the axial direction and 1.4 diameters in the radial direction. This computational domain was shown to be adequate to compute the recirculating flow region up to 1.2 diameters.

Boundary conditions for U, V and the Reynolds stresses along the jet-exit plane were taken from experimental values. The inlet unknown turbulence-energy dissipation ε was prescribed according to $\varepsilon = k^{3/2}/0.3D$. Along the entrainment boundary, the pressure was prescribed and the radial momentum equation solved with the assumption of zero axial velocity at this boundary and arbitrary very small values were assigned to all turbulent quantities. At the exit plane the grid was refined and a zero second derivative was used for the axial velocity component. Special attention was taken to evaluate the radial velocity at the entrainment boundary outside the domain to ensure mass conservation at the exit plane.

Calculations with EVM and RSM closures were made with non-uniform meshes consisting of 30×23, 60×46 and 100×94 control volumes; the last grid is shown in Fig. 1. Solution convergence problems were encountered for both eddy viscosity and second moment closure computations due to high negative coefficients in the finite difference equations and also to rapidly varying gradients of dependent variables. Three approaches have been investigated to overcome this problem:

i) A mixed quadratic upstream (QUDS) and first order upwind scheme which shifts from QUDS to first order upwind on the basis of local cell Peclet number, similar to that used by McGuirk *et al.* (1985b);
ii) A blending flux correction of the form:

$$^{L}\{A\}^{n+1}[\phi]^{n+1} = {}^{H}[S]^{n} + \beta \left({}^{L}\{A\}^{n}[\phi]^{n} - {}^{H}\{A\}^{n}[\phi]^{n} \right) \tag{11}$$

where L and H denote lower and higher order convection discretization schemes and n the iteration counter. If $\beta = 1$ the converged solution corresponds to the higher order scheme;

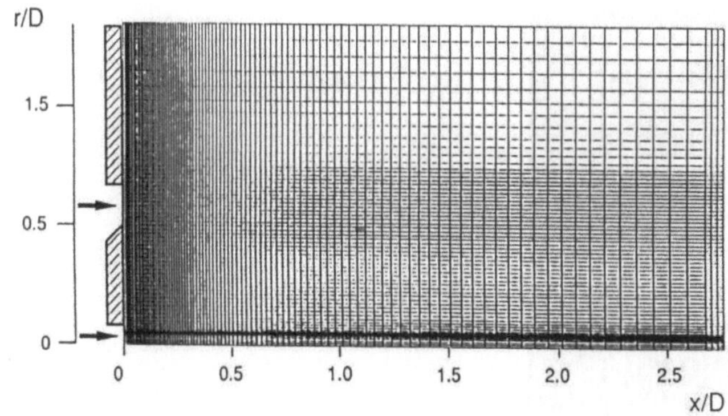

Fig. 1. Computational Grid, 100 × 66 nodes.

iii) A mixed procedure, with option i) adopted for momentum equations and option ii) for the other transport equations.

Option i) was applied with the shift from QUDS to first order Upwind at Peclet greater than 10^3. For eddy-viscosity calculations, more than 80% of the computational domain was calculated with QUDS. In this region, the negative coefficients were appropriately incorporated in the source term. For RSM calculations, the shift Peclet number had to be reduced in turbulence variables equations. To avoid this reduction, option ii) was tried and surprisingly, it was possible to obtain converged solutions either with central differences, quadratic upstream or high-order upwind depending on the correct choice of β. However, for $\beta > 0.05$ solution wiggles were found in almost all quantities and mainly in steep gradient regions. The best compromise was provided by option iii). For turbulence variables equations, β was assigned to 0.05 and option i) was used in momentum equations with the limit Peclet number $Pe = 550$. This procedure was required as first order Upwind results obtained with 60 × 48 and 100 × 94 control volumes were far from being grid independent.

Regarding the separated flow region, solution grid independence was not found even with the finer grid. Deviations up to 10% in the location of the first stagnation point were encountered for RSM calculations using the two finer grids. However, it was assumed that the grid comprising 100 × 66 control volumes was sufficiently fine to permit the response of turbulence-model variations to be interpreted on the assumption that the solutions were not biased by numerical errors.

Results

Fig. 2 shows the flow visualization results. The measured mean flow velocities are presented in Fig. 3 in the form of vector plots and streamlines. The figures show that the annular air jet converges towards the centreline downstream of the disk and forms a wake region which extends approximately 1.23 diameters. A toroidal recirculating flow

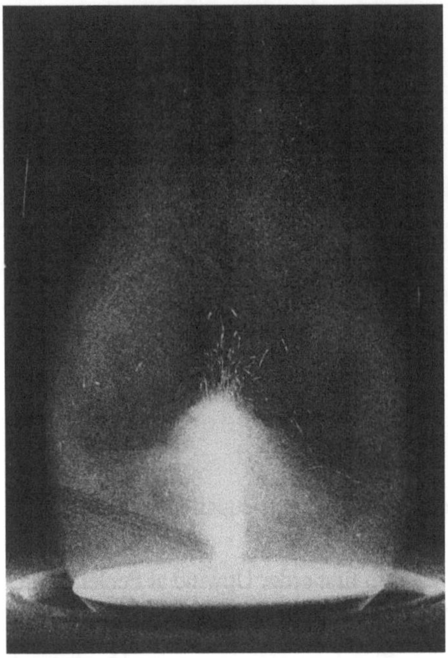

Fig. 2. Flow Visualization. (Durão *et al.*, 1990).

region is formed within the wake region by the interaction of the outer annular air with the central jet flow. The central jet stagnates and is turned upstream and outward by the opposing reverse flow of the annular air. The stagnation point location is dictated by both high shear stresses generated along central jet boundary and by the adverse pressure gradient.

Fig. 4a) and b) presents computed streamlines for RSM and EVM, respectively. Broadly, the main features of the flow are captured in the computations. The two-stagnation points located along the centreline are also reproduced in the predictions, but their locus vary depending on the turbulence model used, specially the first stagnation point.

The centreline U-velocity evolution is presented in Fig. 5 for EVM and RSM models. The size and minimum flow velocity within recirculation zones is correctly predicted by RSM, while EVM underpredicts the location of the first stagnation point by almost 50%. The centreline velocity decay is also rather differently predicted by the two models.

Fig. 6 shows radial profiles of U-velocity. Notably, RSM predictions follow quite well the experimental data across all axial stations. Only small discrepancies are observed, mainly within the outer region bounding the annular flow, where experimental values are larger than the predicted ones. This is probably due to flow intermittency in this zone, a feature previously identified by Schefer *et al.* (1987) and that none of the present turbulence models include. As a consequence of erroneous prediction of the first stagnation point, large discrepancies with the measurements are displayed in radial profiles obtained with EVM.

The radial profiles of V-velocity component are shown in Fig. 7. RSM predicted profiles are close to measurements, denoting that the flow entrainment rate is correctly

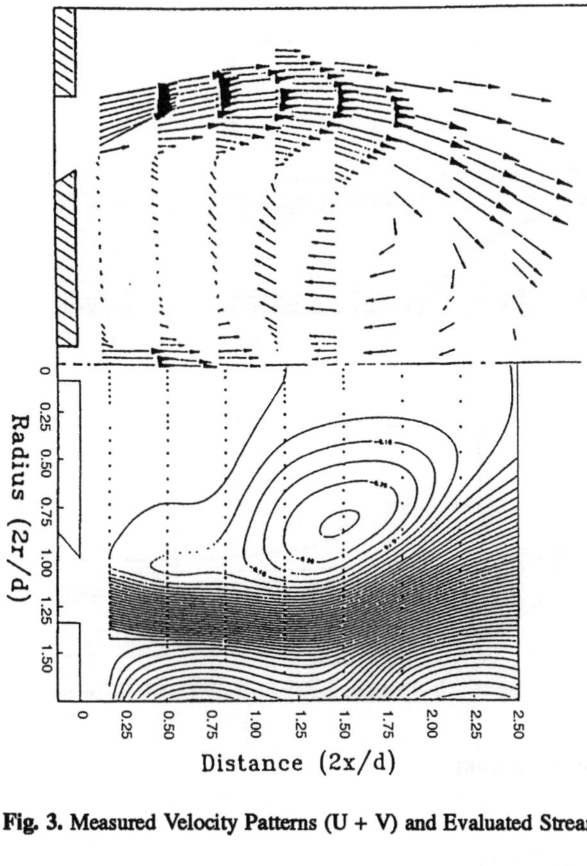

Fig. 3. Measured Velocity Patterns (U + V) and Evaluated Stream Functions. (Durão *et al.*, 1990).

predicted. EVM predictions are in severe error in the zone upstream of the first stagnation point inside the central vortex. In this zone, the transition of an inward flow (negative V-velocities) to a outward flow (positive V-velocities) observed near the first stagnation is correctly captured by RSM, while EVM is in error both in magnitude and in sign of the V-velocity.

Figs. 8, 9 and 10 show the radial development of Reynolds stress tensor components $<u^2>$, $<v^2>$ and $<uv>$ near the stagnation points, respectively. A striking feature, already observed in the computations of McGuirk *et al.* (1985b), is the strong underprediction of stresses levels in the core region of the flow, and specially at centreline.

This major defect of both RSM and EVM can not be attributable to an erroneous computation of mean-field, which if true, would yield biasing effects in Reynolds stresses production rates via mean-strain. However, this is not the case as RSM predictions of mean-flow field are quite close to measurements. It was speculated by McGuirk *et al.* (1985b) that the problem could arise from a deficient representation of pressure-strain process by the present model in the prediction of Reynolds stress anisotropy near stagnation points. Although this argument can be valid concerning the modelling of the intercomponent energy transfer, it is believed that is not enough to justify the systematic underprediction of values of turbulence energy components observed, as pressure strain process mainly represents an energy redistribution mechanism. Previous studies (We-

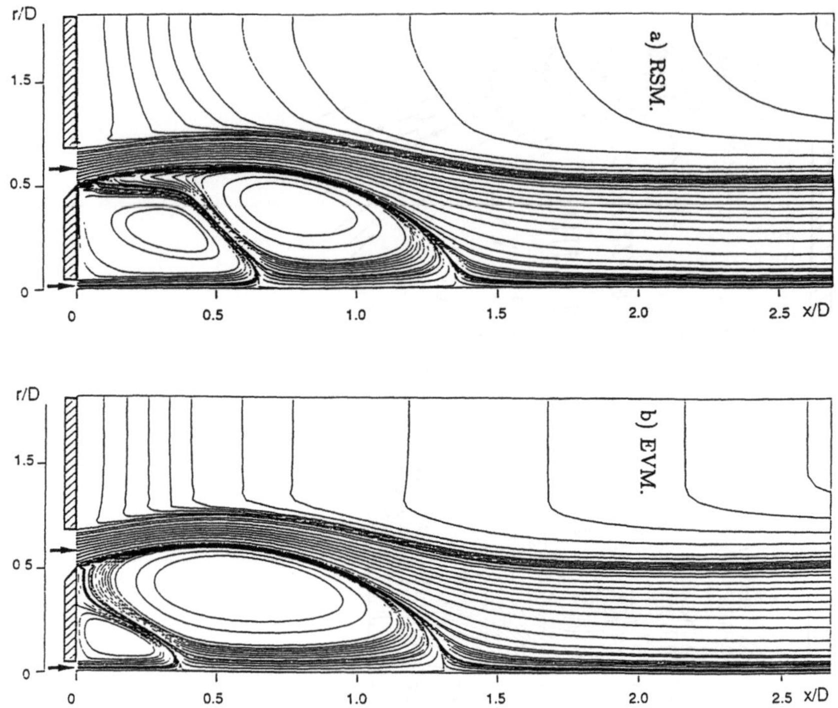

Fig. 4. Computed Streamlines. a) RSM. b) EVM.

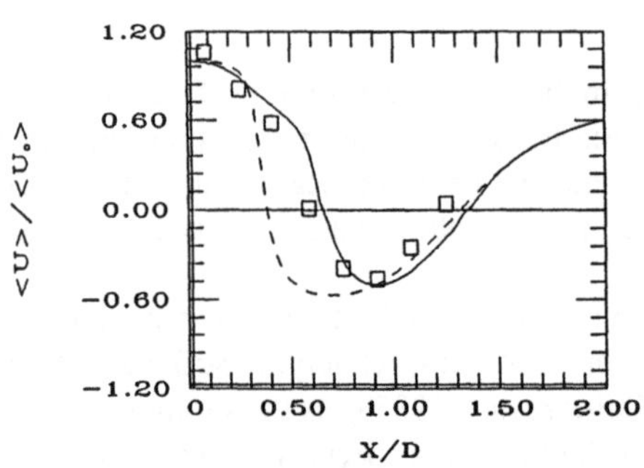

Fig. 5. Centreline Velocity Development. Predictions: — RSM; , - - - EVM; Measurements: Symbols.

instock and Burk, 1985, Sreenivasan, 1985; Daly and Harlow, 1970) have attempted to explore a functional dependence of pressure-strain model coefficients C_1 and C_2 on mean strain. After some initial trials, this route was pursued no further in the present study as, one believes, the major weakness stem from ε-equation. Moreover, EVM predictions show the same discrepancies also.

However, due to the complexity of the flow and variety of effects present, it is not an easy task to identify which dominant mechanisms are not represented by ε-equation in its standard form and in which flow zones their influence is dominant, and consequently to devise any reasonably correction term. Two agencies are likely to be responsible: The

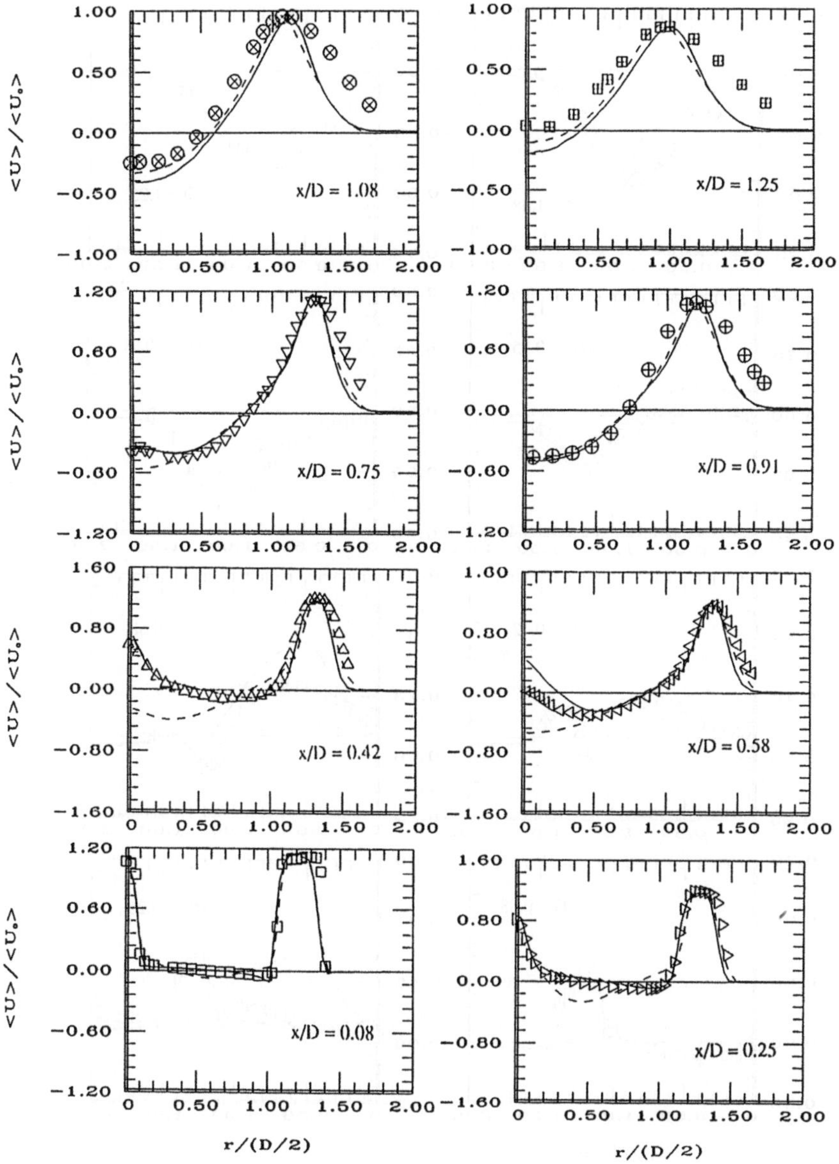

Fig. 6. U-velocity Radial Development. Predictions: — RSM; , - - - EVM; Measurements: Symbols.

first is the strong pressure gradient, which drives the central jet flow and establishes the first stagnation point together with shear stresses generated in the border of the jet; the second is the curved shear layer developing within the annular flow which is responsible for the formation of the second stagnation point. This shear layer dictates the levels of turbulence energy across the annular flow and by transport the values of k at the second stagnation region. However, and this point must be emphasized, experimental measurements of probably density distribution functions show clearly a bimodality in

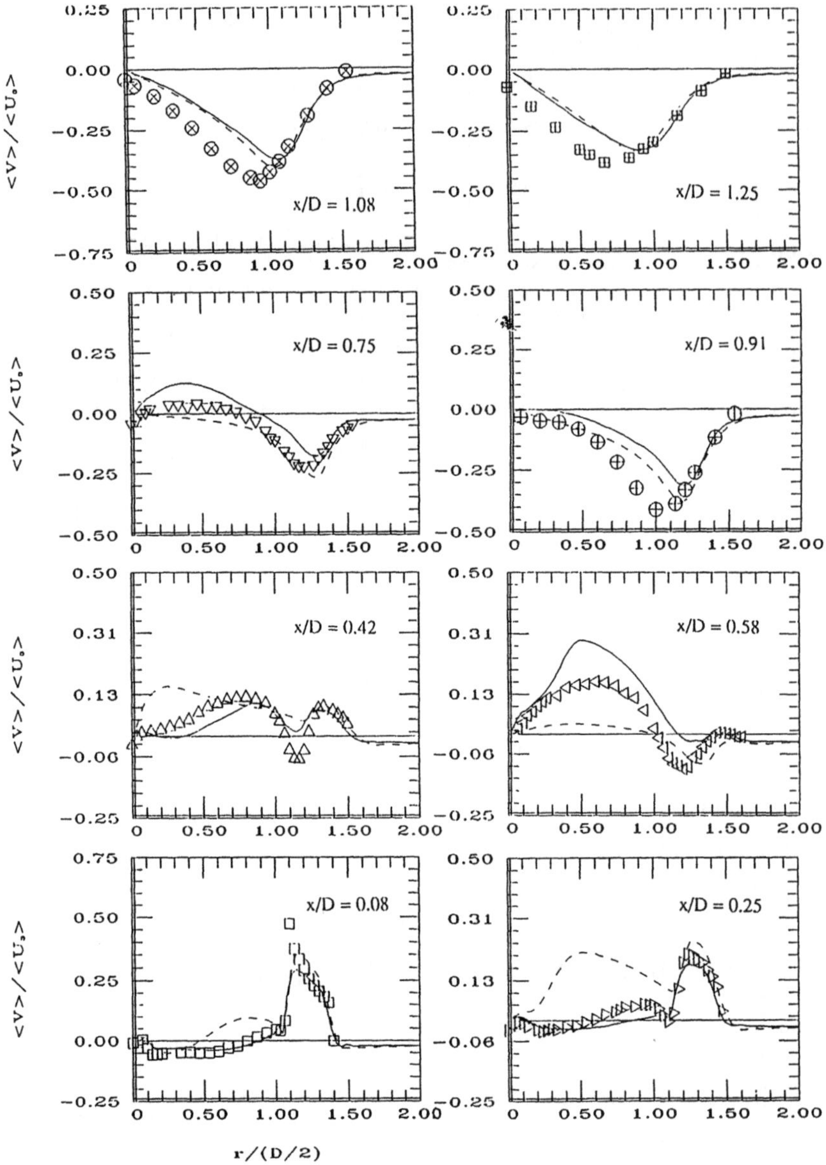

Fig. 7. V-velocity Radial Development. Predictions: — RSM; , - - - EVM; Measurements: Symbols.

U-velocity components at the first stagnation point and in V-velocity component near the second stagnation point. No bimodality was identified within the annular flow bordering shear layer. This may indicate that high levels of normal stresses found near stagnation regions are due also to the contribution to turbulence energy budjet from oscillating behaviour of the flow. This speculation may be supported by the comparison of present measurements with those reported previously by Schefer *et al.* (1987).

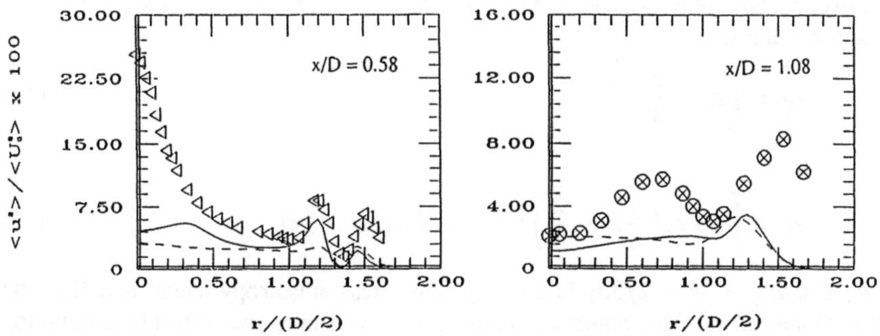

Fig. 8. Normal Axial Reynolds Stress, Radial Development. Predictions: — RSM; - - - EVM; Measurements: Symbols.

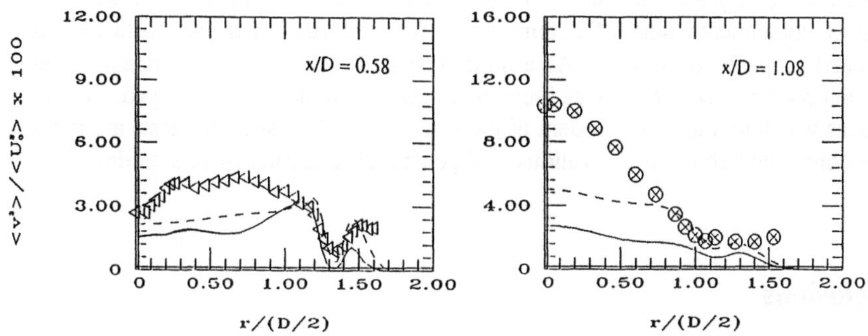

Fig. 9. Normal Radial Reynolds Stress, Radial Development. Predictions: — RSM; - - - EVM; Measurements: Symbols.

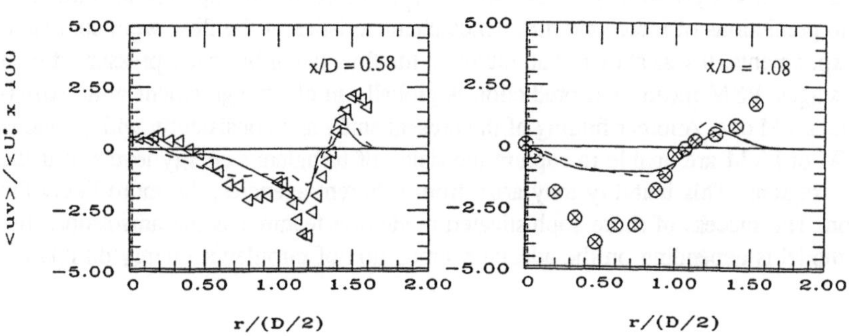

Fig. 10. Reynolds Shear Stress, Radial Development. Predictions: — RSM; - - - EVM; Measurements: Symbols.

Recent experiments of Hallbäck *et al.* (1990) have shown that small scale anisotropy may be significant particulary in flows with large strain rates, and that effects of anisotropy at small scale may be comparable to those of energy redistribution. Based on this evidence, these authors have proposed an algebraic model for dissipation rate of Reynolds stress tensor, ε_{ij}, aiming to include small scale anisotropy in modelling $<u_i u_j>$-equation. The model has the form

$$\varepsilon_{ij} = \left(e_{ij} + \frac{2}{3}\delta_{ij} \right) \varepsilon \tag{12}$$

where

$$e_{ij} = \left[1 + \alpha \left(\frac{1}{2}\mathbf{II} - \frac{2}{3} \right) \right] b_{ij} - \alpha \left(b_{ik} b_{kj} - \frac{1}{3}\mathbf{II}\delta_{ij} \right). \tag{13}$$

where $b_{ij} \equiv <u_i u_j> /k - 2/3\delta_{ij}$ is the Reynolds stress anisotropy tensor and $\mathbf{II}$ is the second invariant of b_{ij}. This algebraic model basically relaxes the isotropic assumption for ε_{ij}, allowing that non-normal components ($< uv >$ in the present case) having a dissipating process other than pressure-strain. In order to explore the importance of this modelling feature in the present flow, the model was incorporated in the RSM closure and predictions of the flow were obtained for several values of α, the only model constant. The computations (not presented) have not shown an improvement in the maximum values of energy components, since the same model for ε-equation was used with the above mentioned supposed drawbacks. As a noticeable effect, the locus of maximum normal stress component at centreline was found to change if the model is employed. Moreover, this locus was found to be dependent of the value of α. More stringing tests are required before concluding about of the validity and general applicability of this model.

Conclusions

In the present paper a numerical and experimental study of an unconfined wake recirculating flow behind a disk with a central jet was presented. A second-moment closure as well as $k - \varepsilon$ eddy-viscosity model were incorporated in an appropriate numerical solver and predictions of mean and turbulence characteristics of the flow were compared against experiments. It was shown that, although the flow being basically pressure-driven at initial stages, RSM mean field prediction is globally in closer agreement with experiments than EVM one. Another finding of the present study and consistently with previous ones, RSM or EVM are unable to capture the rising of turbulence energy levels near the stagnation regions. This inability may arise from different agencies, the more likely the ε-equation. The success of more sophisticated modelling features as the anisotropic dissipation model is depending on the adequate simulation of turbulence energy dissipation rate.

Acknowledgements

The authors wish to acknowledge the financial support of EEC contract No. PL890926 under the JOULE Programme.

References

Azevedo, J.L.T., Durst F. & Pereira J.C.F. (1988): Comparison of strongly implicit procedures for the solution of the fluid flow equations in finite-difference form, App. Math. Modelling, **12**, pp. 51-62.

Carmody, T., (1964): Establishment of the wake behind a disk, Trans. ASME D., J. Basic Engng. 86, pp. 869-882.

Daly, B.J. & Harlow, F.H., (1970): Transport Equations in turbulence, Phys. Fluids, **13**, No. 11, pp. 2634-2649.

Durão, D.F.G. & Firmino, F.C. (1983): "Velocity field around a disk", 4th Symposium on Turbulent Shear Flows.

Durão, D.F.G., Knittel, G. & Pereira, J.C.F., (1990): "Velocity characteristics of a central jet impingement in an unconfined wake behind a disk", 5th Int. Symp. on Applications of Laser Techniques to Fluid Mechanics.

Durão, D.F.G. & Pereira, J.C.F., (1987): "A Numerical-Experimental study of confined unsteady laminar flow around a square obstacle", 5th Int. Conference on Numerical Methods and Turbulent Flow, pp. 261-272.

Durão, D.F.G. & Whitelaw, J.H. (1978): Velocity characteristics of the flows in the near wake of a disk, J. Fluid Mech., **85**, pp. 369-385.

Hallbäck, M., Groth, J. & Johansson, A.V. (1990): An algebraic model for nonisotropic turbulent dissipation rate in Reynolds stress closures, Phys. Fluids, A2 (10), pp. 1859-1866.

Heitor, M.V., Taylor, A.M.K.P. & Whitelaw J.H. (1987): The interaction of turbulence and pressure gradients in a baffle-stabilised premixed flame, J. Fluid Mech., 181, pp. 387-413.

Huang, P.G. & Leschziner, (1985):" Stabilization of recirculating flow computations performed with second-moment closures and third-order discritization", Proc. 5th Symp. Turbulent Shear Flows, Cornell University, 20.7-20.12.

Launder, B.E. & Spalding D.B. (1974): The Numerical Computation of turbulent flows, Comp. Meth. Appl. Mech. Eng., **3**, pp. 269-289.

LI, X. & R.S. Tankin, (1987): A Study of cold and combusting flow around bluff-body combustors, Combust. Sci and Tech., **52**, pp. 173-206

Leonard, B.P. (1979): A Stable and accurate convective modelling procedure based on quadratic upstream interpolation, Comp. Meths. Appl. Mech. Engng. **19** p. 59-98.

Mcguirk, J.J., Taylor, A.M.K.P. & Whitelaw, J.H. (1985a): "The assessment of numerical diffusion in upwind difference calculations of turbulent recirculating flows", Turbulent Shear Flows - 3, Springer Verlag, pp. 206-224.

Mcguirk, J.J., Papadimitriou, C. & Taylor, A.M.K.P. (1985b): "Reynolds Stress model calculations of two-dimensional plane and axisymmetric recirculating flows", 5th Symposium Turbulent Shear Flows.

Naot, D. A. Shavit & Wolfshtein M. (1970): Interaction between components of the turbulent velocity correlation tensor, Israel Journal of Technology, **88**, pp. 259.

Patankar, S.V. & Spalding D.B. (1972): A calculation procedure for heat, mass and momentum transfer in three-dimensional parabolic flow, Int. J. Heat Mass Trans., **15**, pp. 1787-1806.

Rotta, J.C. (1951): Statistische Theorie Nichthomogener Turbulenz, Z. Phys., **129**, pp. 547-572.

Schefer, R.W., Namazian, N. & Kelly, J. (1987): Velocity characteristics in the turbulent near wakes of confined axisymmetric bluff bodies, Combust. Sci and Tech., **56**, pp. 101-138.

Screenivasan, K.R., (1985): The effect of contraction on a homogeneous turbulent shear flow, J. Fluid Mech., **154**, pp. 187-213.

Stone, H.L. (1968): Iterative solution of implicit approximations of multi-dimensional partial differential equations, SIAM J. Numer. Anal. **5**, No. 3, pp. 530-558.

Taylor, A.M.K.P. & Whitelaw, J. (1984): Velocity characteristics in the turbulent near wakes of confined axisymmetric bluff bodies, J. Fluid Mech., **139**, pp. 391-416.

Weinstock, J. & Burk, S., (1985): Theoretical pressure-strain term, experimental comparison, and resistance to large anisotropy, J. Fluid Mech., **154**, pp. 429-443.

Large-Eddy Simulation of Turbulent Flow Over and Around a Cube in a Plate Channel

H.Werner and H.Wengle[1]

Institut für Strömungsmechanik und Aerodynamik, LRT, WE7, Universität der Bundeswehr München, D-8014 Neubiberg, Germany

Abstract

The solution concept of large-eddy simulation (LES) has been applied to simulate turbulent flow over and around a single cube mounted on the bottom of a plate channel for a Reynolds number of 50000 (based on the incoming mean bulk velocity and obstacle height). Here we present, as a first part of the evaluation of the data, a few interesting views of the time-dependent fields and results for the three-dimensional mean fields (velocity, vorticity, Reynolds stress, enstrophy, helicity). Having engineering application in mind, this flow problem represents an ideal case, which is very well suited for testing and validating numerical simulation techniques and turbulence models.

1. Introduction

The purpose of our work is to apply the solution concept of large-eddy simulation (LES) to turbulent flow over and around flow obstacles with sharp edges and corners. The work is motivated by the need for applications of LES to engineering problems involving (a) high Reynolds number flow, (b) three-dimensionality of the *mean* flow and (c) non-periodic boundary conditions in the main flow direction. In this paper, we present results from a numerical simulation of turbulent flow over and around a cube mounted on the bottom of a plate channel. For the example presented here, we selected (a) a Reynolds number of about 50000, based on mean bulk velocity and obstacle height (which is equal to the channel half width), (b) a three-dimensional obstacle, creating a flow field without any homogeneous direction, making it necessary to provide the statistics by time-averaging only, and (c) a *single* obstacle, i.e. not allowing the use of periodic boundary conditions in the main flow direction.

Earlier related work on LES of turbulent flow over a periodic arrangement of cubes in a simulated atmospheric boundary layer has been published by Murakami, Mochida & Hibi (1987), results for the flow over a periodic arrangement of square ribs in a channel have been presented by Kobayashi, Kano & Ishihara, (1985), and results from our own earlier work on flow over a single square rib on the bottom of a plate channel can be found in Werner & Wengle (1988, 1989).

---

[1] to whom correspondence should be sent

2. Mathematical Models, Solution Technique and Evaluation of the Statistics

The governing equations describing the resolvable flow quantities (grid scale, GS) are derived from the integral conservation equations for mass and momentum applied to a finite grid volume $\Delta V = \Delta x * \Delta y * \Delta z$ following the 'volume balance method' of Schumann (1975). From this it follows that GS quantities of the three velocity components $U_\alpha(\alpha = 1,2,3)$ are defined as averages over the corresponding surfaces, $\Delta A_\alpha(\alpha = 1,2,3)$ of the grid volumes, ΔV.

The subgrid scale (SGS) stresses, arising from the nonlinear convection terms, are evaluated by the Smagorinsky-Lilly model (with $c_1 = 0.1$) which relates the SGS stresses to the GS velocity field via an eddy viscosity model. In grid volumes next to walls, we used for the mixing length the smaller value of $\kappa * x_\alpha$ and $0.1 * (\Delta x * \Delta y * \Delta z)^{1/3}$ respectively (x_α is the distance normal to a wall).

The geometry of the computational domain and of the flow obstacle is evident from figure 1. Measured in units of the reference height H (height of the cube) of the problem, the dimensions of the computational domain are $(X,Y,Z) = (10,7,2)$ and, for the results presented here, we used $(NX,NY,NZ) = (144,92,58)$ grid points.

The governing equations are solved numerically on a staggered and non-uniform grid using second order finite-differencing in time and space (explicit leap-frog for time discretization, central differencing for convection terms and time-lagged diffusion terms). The problem of pressure-velocity coupling is solved iteratively (point-by-point relaxation).

The direct results from LES are the time-dependent and three-dimensional fields for the GS quantities of the three velocity components and the pressure. Figure 2a shows a sample of the instantaneous flow field in a vertical plane through the center of the cube. By time-averaging of the instantaneous flow field, the mean flow field (fig. 2b) is obtained, and as soon as the mean velocity field has reached stable (i.e. time-independent) values, the fluctuating velocity field can be evaluated as the difference of the two fields (fig. 2c). This procedure leads to the classical Reynolds-decomposition of a turbulent flow field. Finally, from the fluctuating fields, the root-mean-square values, e.g. for velocity, vorticity and pressure fluctuations can be calculated, as well as the Reynolds-stresses and other statistics desired.

3. Inflow and Wall Boundary Conditions

At the inflow section, we used at each time step the instantaneous flow field of a LES result of the corresponding (fully developed) channel flow (see figure 1). Boundary conditions at horizontal and vertical walls were specified by assuming that at the grid points (P) closest to the wall, (a) the instantaneous velocity components tangential to the wall (u_P, v_P) are in phase with the instantaneous wall shear stress components (τ_{ub}, τ_{vb}) and (b) the instantaneous velocity distribution is assumed to follow the linear law-of-

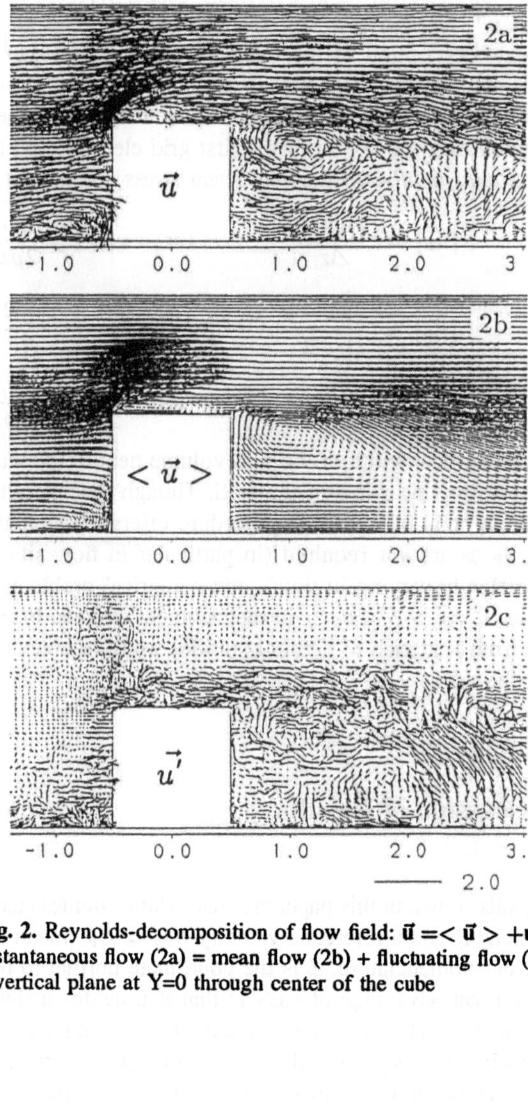

Fig. 2. Reynolds-decomposition of flow field: $\vec{u} = <\vec{u}> + \vec{u}'$ instantaneous flow (2a) = mean flow (2b) + fluctuating flow (2c) in a vertical plane at Y=0 through center of the cube

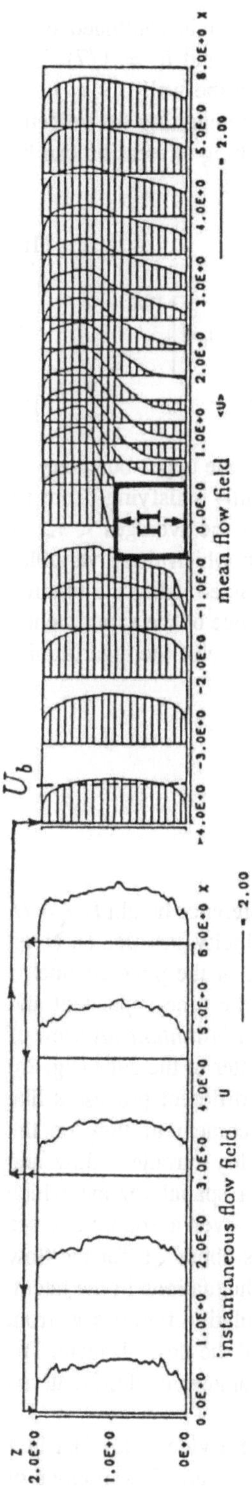

Fig. 1. Geometry of computational domain and inflow boundary conditions
left: channel flow $NX * NY * NZ = 54 * 92 * 58$ grid points
right: flow over cube $NX * NY * NZ = 144 * 92 * 58$ grid points

the-wall $u^+ = z^+$ for $z^+ \leq 11.81$, and for $z^+ = z_m > 11.81$ it is continued by a power-law description of the form $u^+ = A(z^+)^B$ (with $A = 8.3$ and $B = 1/7$). The velocity components tangential to a wall at the grid point next to the wall (u_P, v_P) can be related to the corresponding wall shear stress components by integrating the velocity distribution over the height of the first grid element, and the resulting expression can be resolved *analytically* for the wall shear stress component, i.e.

$$|\tau_{ub}| = \frac{2\mu|u_P|}{\Delta z} \qquad \text{for} \qquad |u_P| \leq \frac{\mu}{2\rho\Delta z}A^{\frac{2}{1-B}} \tag{1a}$$

$$|\tau_{ub}| = \rho\left[\frac{1-B}{2}A^{\frac{1+B}{1-B}}\left(\frac{\mu}{\rho\Delta z}\right)^{1+B} + \frac{1+B}{A}\left(\frac{\mu}{\rho\Delta z}\right)^{B}|u_P|\right]^{\frac{2}{1+B}}$$

$$\text{for} \qquad |u_P| > \frac{\mu}{2\rho\Delta z}A^{\frac{2}{1-B}} \tag{1b}$$

Δz is the vertical width of the grid volume next to the wall, z_m is the upper boundary of validity of the linear-law-of-the-wall. Though theoretically not fully satisfying, from an application point of view this procedure offers the advantage that the averages $< \tau_{ub} >$ and $< u_P >$ are *not* required (in particular in flow situations in which these variables may be slowly varying in time), and numerical problems are avoided using the relation $\tau_{ub} = u_P < \tau_{ub} > / < u_P >$ in reattachment regions. In consequence of the experimental results from Ruderich & Fernholz (1986) we have abandoned the use of the logarithmic-law-of-the-wall.

4. Discussion of Results

The results shown in this paper are made dimensionless using a reference height $L_{ref} = H$ (cube height) and a reference velocity $U_{ref} = U_b$ (mean bulk velocity, see fig. 1). Note, that in our nomenclature, Z is the coordinate normal to the walls of the plate channel.

The great advantage of LES is that it provides insight into the time-dependent and three-dimensional large-scale structure of a turbulent flow field. An *instantaneous* view of the velocity vectors projected onto a vertical plane through the center of the cube (fig. 2a) exhibits an extremly complicated flow field. The interaction of different processes like the development of a three-dimensional shear layer, the reattachment of flow on the bottom plate behind the flow obstacle, the recirculation of highly turbulent flow and its reentrainment into the free shear layer takes place within a spatial regime which is significantly smaller compared to the case of turbulent flow over a square-rib, see Werner & Wengle (1988, 1989): the mean recirculation length is about 2.0 for the flow over a cube, and about 7.0 for the square-rib. Strong horizontal fluctuations in the lateral direction close to the walls can be observed in the small recirculation regimes in front and on top of the cube and in the large recirculation regime behind the flow obstacle. The creation of strong horizontal fluctuations by splashing down of tongues of fluid material can be seen in figure 3.

An ideal way of presenting time-dependent results would be a video film, but in a printed medium the only way is to present a series of snapshots from flow regions of

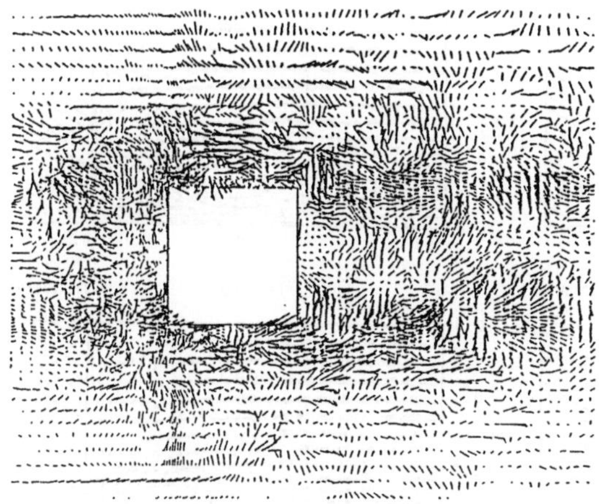

Fig. 3. Fluctuating flow field in a horizontal plane close to the bottom plate (Z = 0.01)

particular interest. Figure 4 shows ten snapshots of the flow in a vertical plane in front of the cube. The time difference between the snapshots is $0.4 * T_{ref}$. *One* reference time T_{ref} is equivalent to the time a tracer particle needs to travel with bulk velocity U_b (see fig. 1) over a distance of *one* obstacle height H. One or somtimes two large vortices (rotating in the same direction) can be observed in front of the cube: they are pushed toward the front face, and later pushed back again by a very violent stream of fluid from above. In front of the horse-shoe vortex and close to the bottom plate a very calm region in the flow field can be observed. Figure 5 shows snapshots close to the top face of the cube, demonstrating the complicated separation process which not always starts immediately at the sharp leading edge. Finally, figure 6 presents a few snapshots from the three components of the time-dependent vorticity field $(\omega_x, \omega_y, \omega_z)$ in the vertical center plane, illuminating the recirculating region behind the cube which is bounded by a free shear layer reattaching again on the bottom plate (the mean reattachment length is about 2.25 measured from the center of the cube).

The *mean* structure of the flow field (e.g. mean velocity, mean vorticity, second-order statistics, mean enstrophy, mean helicity) can be provided by time-averaging only. For the results shown here, we started averaging for the first-order statistics after 60 reference times T_{ref}; the first-order statistics was sufficiently stable after about 100 reference times (taking one sample at every 40th time step), samples for the second-order statistics have been collected over the latest 140 reference times. Figure 1 shows vertical profiles of the mean U-velocity component; the profiles are quite smooth, indicating that a sufficient number of samples has been taken for the first-order statistics. From distributions of the mean U-velocity component close to the bottom plate ($Z = 0.01$ =location of first vertical grid point) and close to the top face of the cube ($Z = 1.01$), the different mean recirculation lenghts can be approximately determined to $XF = 0.9$, $XR1 = 1.75$, $XR2 = 0.2$ and $XT = 0.75$ (fig. 7a,b).

In a vertical cut normal to the main flow direction at a dimensionless position $X = 0.88$ behind the cube (fig. 7d), two pairs of counter-rotating mean vortices can be observed (one pair very close to the side walls of the cube, and the other one at $Y = +1.5$ and $Y = -1.5$ from the center of the cube). In that vertical plane, the lateral shape of the mean shear layer also becomes visible (fig. 7d).

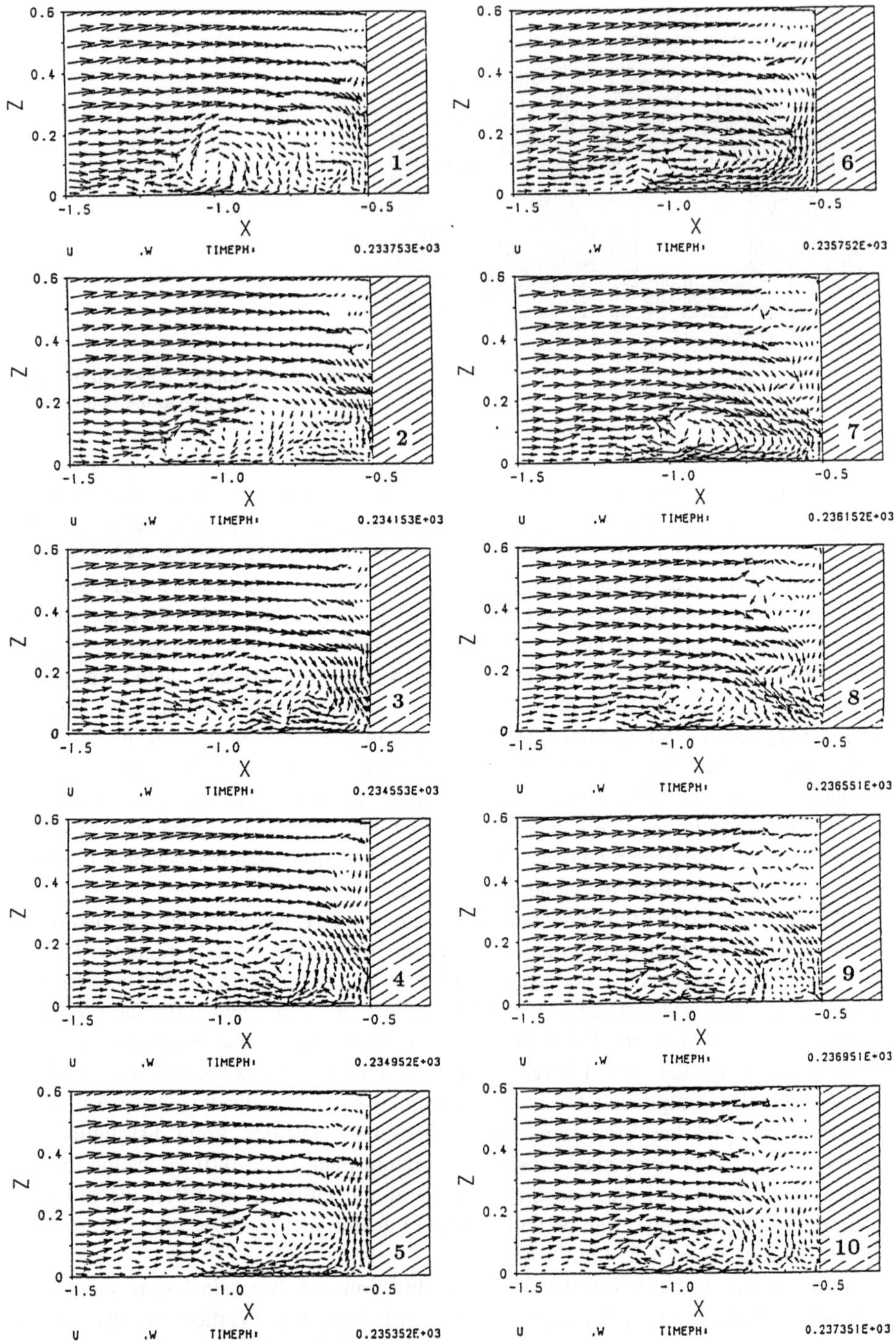

Fig. 4. Snapshots of instantaneous velocity field ($\Delta t = 0.4 * T_{ref}$), in front of the cube, in a vertical plane through the centerline of the cube (Y = 0)

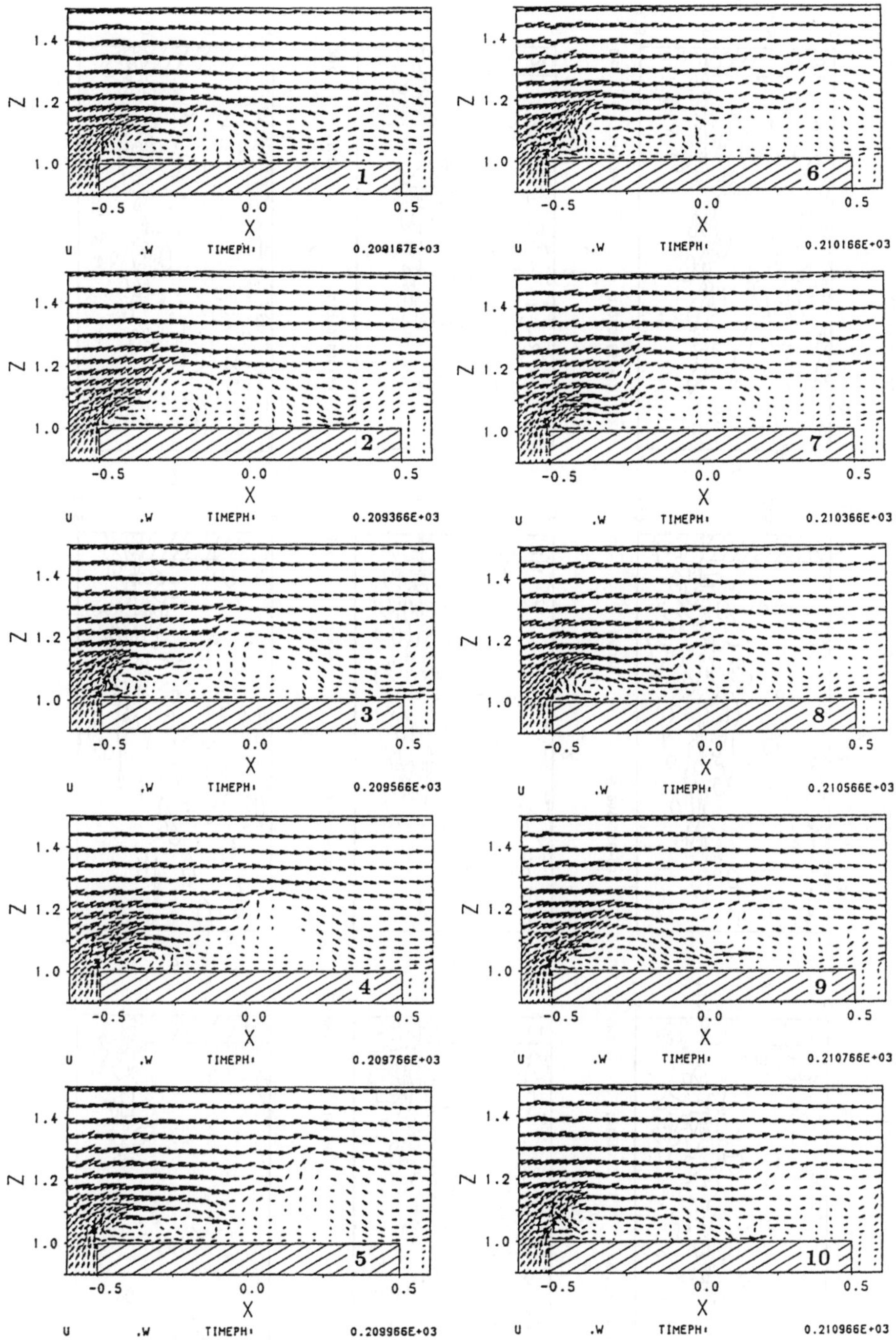

Fig. 5. Snapshots of instantaneous velocity field ($\Delta t = 0.2 * T_{ref}$), at the top face of the cube, in a vertical plane through the centerline of the cube (Y = 0)

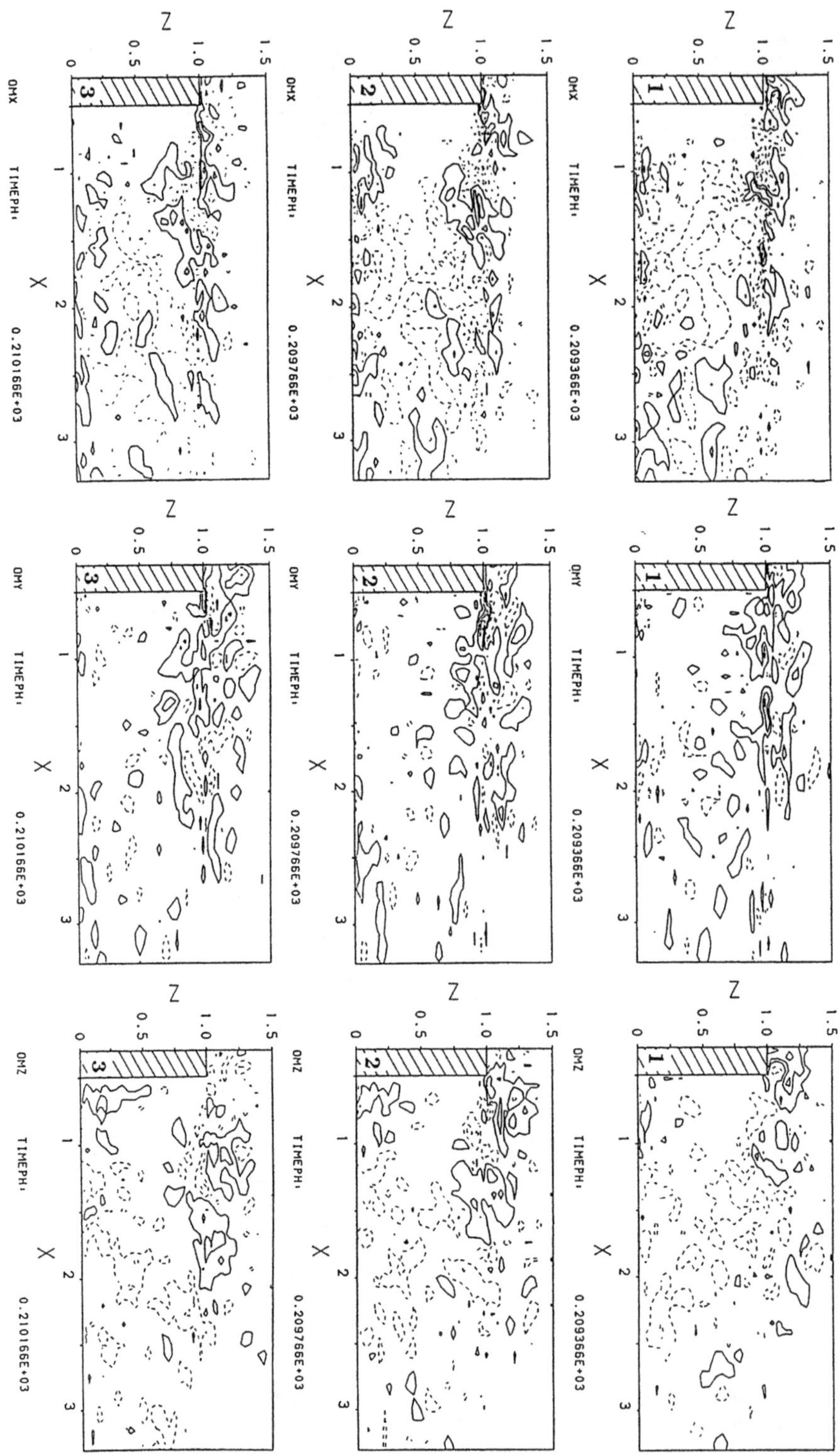

Fig. 6. Snapshots of instantaneous vorticity field ($\Delta t = 0.4 * T_{ref}$), in the recirculation regime behind the cube in a vertical plane through the centerline (Y = 0), dashed lines: negative values, full lines: positive values, column on the left: ω_x, column in the middle: ω_y, column on the right: ω_z.

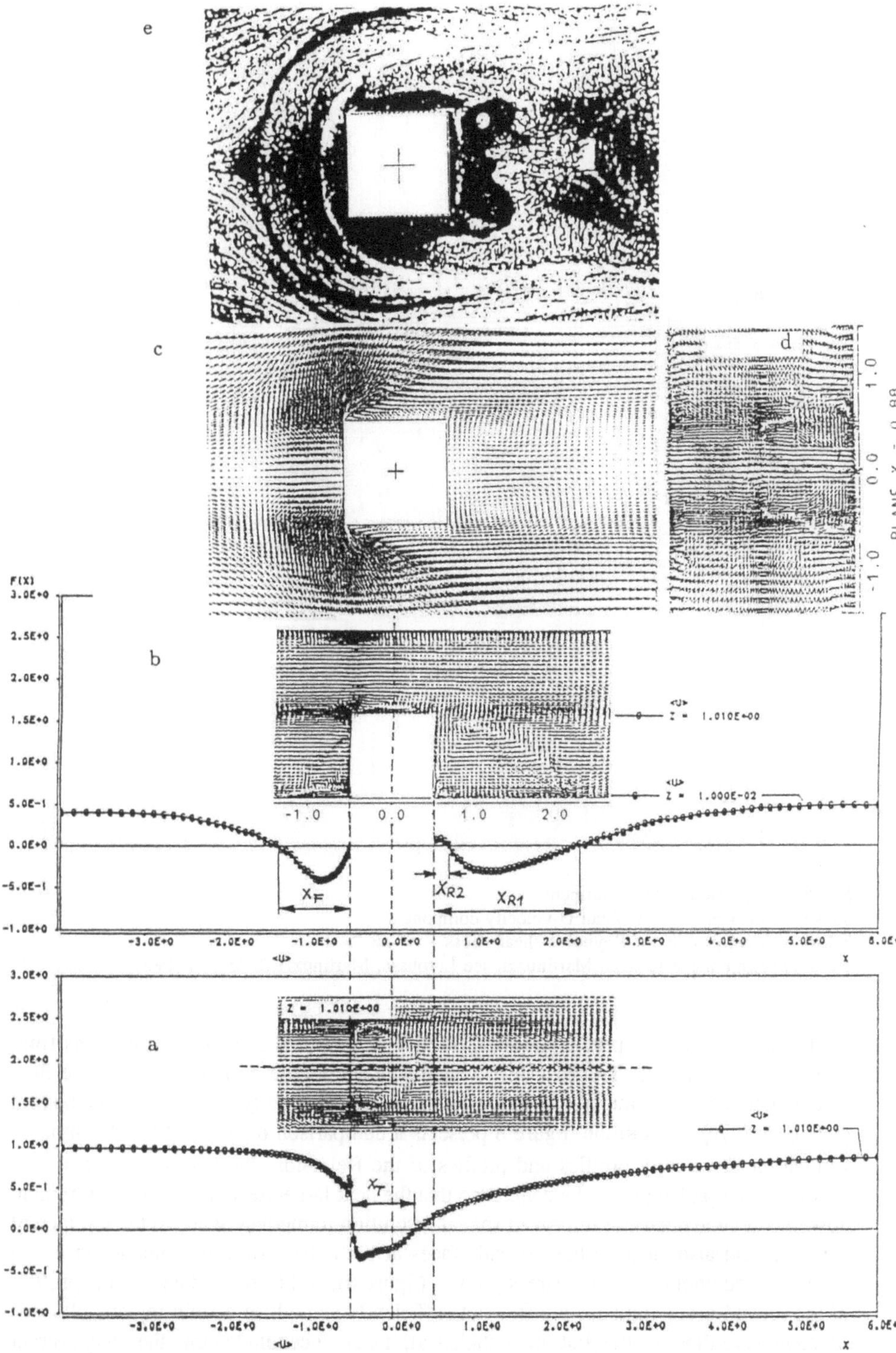

Fig. 7. Determination of mean recirculation lengths, Fig. 7a,b: from profiles of mean velocity at heights Z = 0.01 and Z = 1.01 at Y = 0, Fig. 7c: mean velocity vectors at Z = 0.01, Fig. 7d: mean velocity field (directions of velocity vectors) at X = 0.88Fig. 7e: surface flow pattern from the experiment of Martinuzzi, see Larousse, Martinuzzi & Tropea (1991)

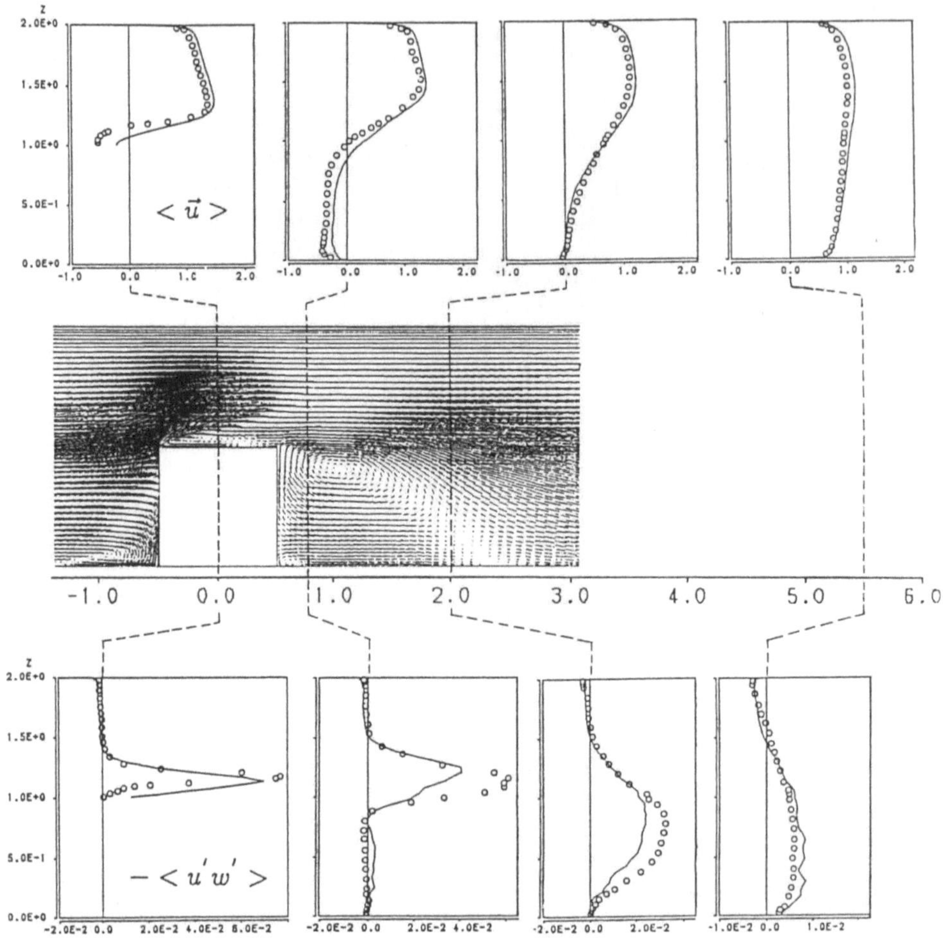

Fig. 8. Comparison with experiment
above: vertical profiles of Mean U-velocity component
below: vertical profiles of Reynolds shear stress $- \langle uw \rangle$
o o o: experimental data from Martinuzzi, see Larousse, Martinuzzi & Tropea (1991)

In the meantime, experimental data are available for the same flow conditions from Larousse, Matinuzzi & Tropea (1991). Figure 7e shows the experimental surface flow pattern which can be compared with the computed mean velocity field close to the bottom plate (figure 7c). In addition, figure 8 presents a comparison of measured and computed vertical mean velocity profiles and profiles of the Reynolds shear stress $- \langle u'w' \rangle$. From that comparison it can be concluded that the next large-eddy simulation of turbulent flow over a cube needs an improved spatial resolution immediately above the top face of the cube, and also close to the two side faces of the cube. From the comparison of the measured and computed shear stress profiles (figure 8), and from the fact that the profiles of the second-order statistics are still not sufficiently smooth in certain regions of space (in contrast to those of the first-order statistics), it can be concluded that the second-order statistics is very slowly converging in particular flow regions. One of these critical flow

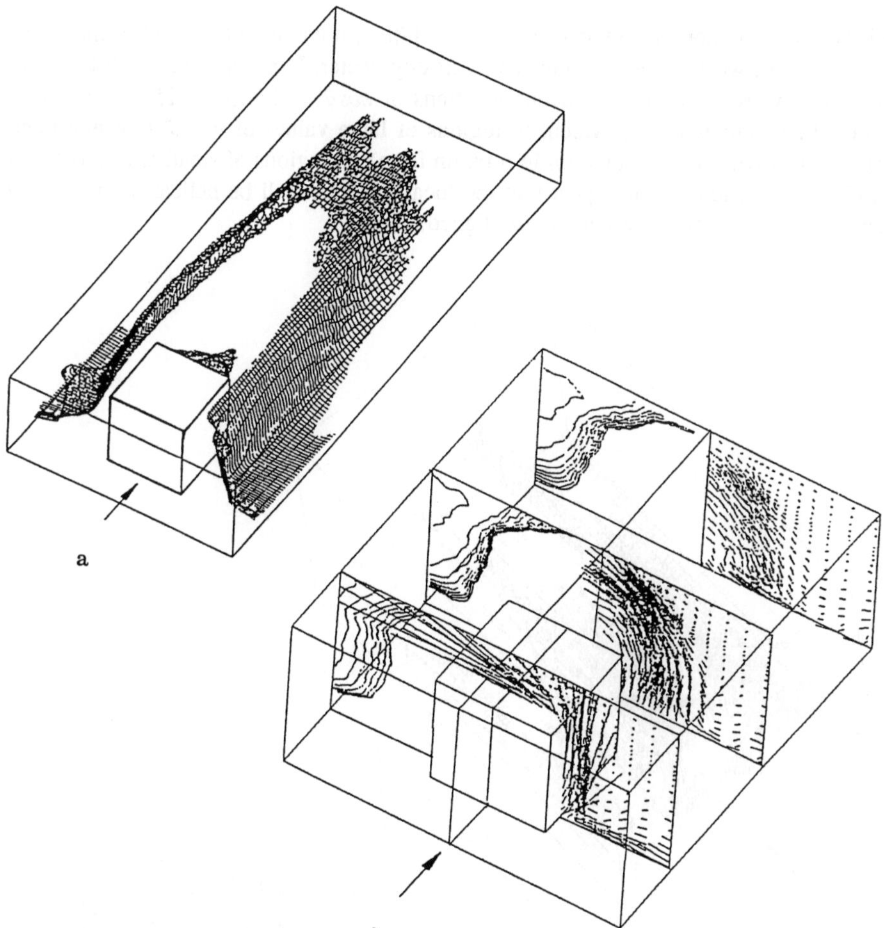

Fig. 9. Mean vorticity field
Fig. 9a: right half: vorticity vectors projected onto vertical planes at X = -0.3, +1.0, +2.5, left half: isolines of total mean enstrophy $< \omega_x^2 + \omega_y^2 + \omega_z^2 >$
Fig. 9b: isosurface of total mean enstrophy (isosurface cut open at Z = 1.4)

regions can be identified immediately behind the flow obstacle (at about $X = 1.0$), and another one is the reattachment region (at about $X = 2.0$).

The evaluation of the *mean* vorticity field in figure 9 gives an indication of the geometry of the mean shear layer behind the cube. Figure 9a shows, on the right hand side, the mean vorticity vectors projected onto vertical planes normal to the main flow direction (the length of a vorticity vector is proportional to the strength of the local mean vorticity) and, on the left hand side, isolines of the mean total enstrophy. An interesting perspective view of the shape of the 3D mean recirculation zone is given in fig. 9b showing an isosurface of the mean total enstrophy $< {\omega_x}^2 + {\omega_y}^2 + {\omega_z}^2 >$ (for an arbitrarily chosen value of 20.0).

Finally, we evaluated the so-called helicity $\vec{v} \cdot \vec{\omega}$ of the flow field. Fig. 10a shows a perspective view of a *mean* helicity iso-surface for a value of 0.25. If the *instantaneous* helicity density is normalized as $cos\theta = \vec{v} \cdot \vec{\omega}/|\vec{v}||\vec{\omega}|$, it can be interpreted as the cosine of the angle between velocity vector and vorticity vector. For example, fig. 10b gives a perspective view of the instantaneous locations of $cos\theta = 0.90(\theta = 25^o)$. Regions of small helicity values are equivalent to regions of large values of $\vec{v} \times \vec{\omega}$ (the non-linear term of the Navier-Stokes equation in rotation form), in regions of small angles between $\vec{v}$ and $\vec{\omega}$ it is reasonable to expect that the local structure will be helical (an ideal case would be $cos\theta = 1.0$, i.e. $\vec{v}$ and $\vec{\omega}$ are aligned).

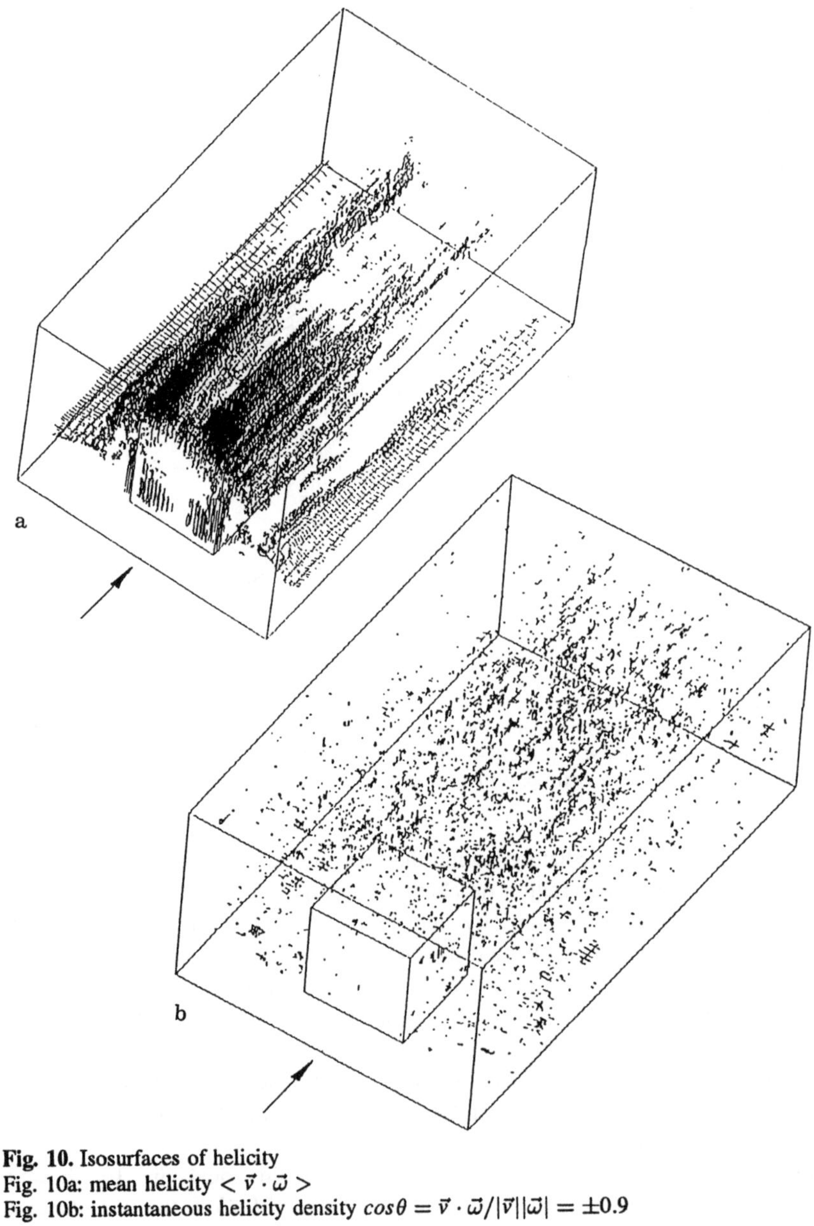

Fig. 10. Isosurfaces of helicity
Fig. 10a: mean helicity $< \vec{v} \cdot \vec{\omega} >$
Fig. 10b: instantaneous helicity density $cos\theta = \vec{v} \cdot \vec{\omega}/|\vec{v}||\vec{\omega}| = \pm 0.9$

5. Conclusions

The most valuable result from a large-eddy simulation is the time-dependent and three-dimensional large-scale structure of a turbulent flow field, and a lot of additional effort seems to be necessary to present such data to the research community in a proper way. It is always very helpful and, at least for some time in the future, it will be necessary to compare LES data with corresponding experimental data for the first-order and, most important, for the second-order statistics. Then, the two methods will be able to supplement each other, and LES will develop to a valuable tool for a sufficiently accurate prediction of turbulent flows of engineering interest. For a flow case without any homogeneous direction, as presented in this paper, about 300 CPU hours on a CRAY/Y-MP must be spent to provide sufficiently stable second-order statistics, and the number of grid points used here should be considered to be a minimum to reach satisfying agreement with the experiment. The case of turbulent flow over a cube in a plate channel at high Reynolds number represents a flow problem which is very well suited for testing and validating numerical simulation techniques and turbulence models.

Acknowledgements

This work is currently supported by the German Research Society, Priority Research Program, Project No. Ro 497/5 (Römer/Wengle). We also gratefully acknowledge the support by the computing center of the Universität der Bundeswehr München, by the Leibniz Computing Center of the Bavarian Academy of Sciences, and by Convex Computer GmbH, GS Bayern.

References

Baetke, F., Werner, H. Wengle, H., (1987): "Computation of turbulent flow around a cube on a vector computer", in Proc.6th Symposium on Turbulent Shear Flows, Sept. 7-9, Toulouse, France.

Kobayashi,T., Kano,M., Ishihara,T. (1985): "Prediction of turbulent flow in two-dimensional channel with turbulence promoters", Bull. JSME,Vol. 28, No. 246, 2948-2953.

Larousse, A., Martinuzzi, R., Tropea, C. (1991): "Flow around surface-mounted three-dimensional obstacles", in Proc.8th Symposium on Turbulent Shear Flows, Sept. 9-11, Technical University of Munich, see also this volume of selected papers from TSF8.

Murakami, S., Mochida, A., Hibi, K. (1987): "Three-dimensional numerical simulation of air flow around a cubic model by means of large eddy simulation", J.Wind Engng. and Ind. Aerodynamics, 25, 291-305.

Ruderich, R., Fernholz, H.H. (1986): "An experimental investigation of a turbulent shear flow with separation, reverse flow, and reattachment", J.Fluid Mech. 163, 283-322.

Schumann, U., (1975): "Subgrid scale model for finite difference simulations of turbulent flows in plane channels and annuli", J.Comp.Phys.18, 376-404.

Werner, H., Wengle, H. (1988, 1989): "Large-eddy simulation of turbulent flow over a square rib in a channel", in Proc.2nd European Turbulence Conference, August 30 – September 2, 1988, Berlin, and in: Advances in Turbulence 2, Springer., Proc. 7th Symposium on Turbulent Shear Flows, August 21-23, 1989, Stanford University, USA.

Large-Eddy Simulation of a Turbulent Flow with Separation

Michel Arnal and Rainer Friedrich

Lehrstuhl für Fluidmechanik, Technische Universität München, D-8000 München 2, Germany

Abstract

Large-eddy simulations of the fully-developed, turbulent channel flow over a rearward-facing step are presented and discussed. Predicted mean-flow quantities of the high Reynolds number flow compare well with the results of experimental investigations of the same flow. A momentum balance including all the terms appearing in the Reynolds-averaged Navier-Stokes equations demonstrates the importance of the stress terms in the separation and reattachment regions of the flow. Spectra taken from time series of the fluctuating pressure reveal the presence of low-frequency motion in the recirculation zone. Similar behavior has been observed in experimental studies of separated flows occurring in a variety of geometries. Finally, the instantaneous flow field is analyzed for the presence of topological flow patterns typical for turbulent shear flows. To what extent such events contribute to the statistical properties of the flow is discussed.

1. Introduction

The present study is part of a program to develop the large-eddy simulation technique as an engineering tool for the investigation of flows of industrial interest. In addition to the backward-facing step geometry with which we are concerned, other recent studies have focussed on fully-developed pipe flow (Unger and Friedrich 1991), channel flow both with and without large-eddy break-up (LEBU) devices (Klein and Friedrich 1990) and the flow in straight ducts (Su and Friedrich 1991).

In the present study the rearward-facing step geometry was chosen because it contains regions of flow separation, recirculation and reattachment. Such features are common occurrences in flows of engineering importance. Because of this, its relatively simple geometry and a well-defined separation line it has also served as a benchmark test-case for the turbulence modelling and simulation communities.

The separated flow is characterized by the low-frequency unsteadiness of the shear layer and recirculation zone. Experimental investigations by Eaton and Johnston (1981) and, more recently by Driver *et al.* (1987), report a flapping motion of the reattaching shear layer and oscillation of the instantaneous reattachment point about the mean value. The reattachment point was found to drift slowly up- and downstream over a range of approximately two step-heights. This low-frequency motion has been observed in a number of experimental studies of separated flows in various geometries (Kiya and

Sasaki 1983; Dianat and Castro 1991), and is thought to be inherent to such flows. The unsteadiness occurs at time-scales much longer than those typical of the turbulent eddies present in the separated shear layer. It has been identified as one of the reasons why statistical models consistently fail to reproduce the mean-flow features of turbulent separated flows.

The large-eddy simulation (LES) technique presents an attractive alternative to the statistical-turbulence modelling approach of predicting high Reynolds number flows. In applying this technique, the large, energy-rich structures in the flow are directly simulated while the smallest, dissipative scales are modelled. Since the time-dependent, three-dimensional flow field is simulated, instantaneous distributions of the resolvable flow quantities can be obtained. In addition, the availability of the time-varying flow field makes the determination of hard-to-measure quantities (such as the pressure-strain correlations) possible.

In the present paper, we emphasize the range of results which may be obtained from a large-eddy simulation of a separating and reattaching turbulent flow. Following a summary of the numerical methodology employed in the present computations, a discussion of the predicted mean flow field is given. Subsequently, the presence of the experimentally observed low-frequency unsteadiness in our simulations is considered. Instantaneous contour plots shed light on the sequence of events leading to the low-frequency motion in the flow. Finally, the instantaneous flow field is analyzed for the presence of typical streamline patterns which have been found to arise frequently in turbulent shear flows. Such an analysis is meant to identify characteristic or dominant topological features which contribute to the statistical properties of such flows. The important conclusions are summarized to close the study.

2. Numerical Methodology

2.1 Governing Equations

The governing equations are derived by applying a spatial, lowpass filter to the Navier-Stokes equations for an incompressible, constant-viscosity fluid. In the present investigations, we employ Schumann's (1975) "volume balance procedure". In this case, filtering is accomplished by integration of the governing equations over a control volume ΔV, of an equidistant, Cartesian, staggered grid. This yields a spatially second-order accurate difference scheme which conserves mass exactly and leads to a set of ordinary differential equations in time, describing the resolvable (grid-scale, GS) flow quantities. The resulting equations have the following non-dimensional form:

$$\delta_j \overline{u_j} = 0\,, \tag{1}$$

$$\frac{\partial}{\partial t}\left(\overline{u_i}\right) + \delta_j\left(\overline{u_j u_i} + \overline{u_j' u_i'} + \overline{p}\delta_{ji} - \overline{\tau_{ji}}\right) = 0\,, \tag{2}$$

where $\overline{\tau_{ji}}$ is the viscous stress tensor, $\overline{p}$ the pressure and $\overline{u_j}$ the velocity component in one of the three coordinate directions. The central finite-difference operator is represented by δ_j.

Spatial deviations from the filtered velocity components result in an additional transport of momentum through the mesh-cell surfaces in the form of subgrid-scale (SGS) stresses, $-\overline{u_j'u_i'}$. These stresses arise from the non-linear convection terms in the momentum equations and must be modelled in terms of known GS quantities in order to close the equations. In the present study the SGS model due to Schumann (1975) is adopted:

$$-\overline{u_j'u_i'} + \frac{1}{3}\delta_{ji}\overline{u_k'u_k'} = 2\mu_{iso}\left(\overline{S_{ji}} - \left\langle\overline{S_{ji}}\right\rangle\right) + 2\mu_{inh}\left\langle\overline{S_{ji}}\right\rangle \ . \tag{3}$$

This two-part eddy-viscosity model relates the SGS-stresses to the fluctuating and statistically-averaged portions of the filtered deformation tensor. In (3) the filtered deformation tensor is defined as $\overline{S}_{ji} = 1/2(\delta_j\overline{u_i} + \delta_i\overline{u_j})$ and the angular brackets ($\langle\rangle$) denote statistical averages. The fluctuating (isotropic) part of the model vanishes in the mean while the statistical (inhomogeneous) part is important in flow regions with strong mean shear. The eddy-viscosity coefficients μ_{iso} and μ_{inh} consist of length scales related to the mesh cell size and velocity scales related to the filtered, fluctuating and mean deformation tensors, respectively. Further details concerning the model are given in Friedrich (1988).

2.2 Boundary Conditions

Inflow boundary conditions are determined from a separate simulation of the fully-developed, turbulent channel flow with the identical Reynolds number, spatial resolution, and time-step size as in the backward-facing step flow. At each time step, the instantaneous GS velocity vector on a plane perpendicular to the streamwise direction is stored. The data is then used to specify the inlet conditions for the step flow at a plane upstream of the step.

At the *outlet* plane the concept of "frozen" turbulence successfully applied by Richter *et. al.* (1987) is employed. This leads to a simplified convection equation for the instantaneous velocity vector, where the local-mean, streamwise velocity $\langle\overline{u}\rangle$, serves as the convection velocity:

$$\frac{\partial}{\partial t}\left(\overline{u_i}\right) + \langle\overline{u}\rangle\,\delta_x\left(\overline{u_i}\right) = \langle\overline{u}\rangle\,\delta_x\left\langle\overline{u_i}\right\rangle \ . \tag{4}$$

The mean streamwise velocity gradient is obtained by linear extrapolation. A constant extrapolation is used for the remaining two mean velocity components.

Since the flow is statistically stationary and two-dimensional, periodic conditions are employed in the *spanwise* direction.

For high Reynolds number flows with engineering applications approximate boundary conditions for the *solid-wall* boundaries help to reduce the necessary number of grid points in the direction normal to solid walls. In this case the wall layer is not resolved and the approximate boundary conditions model its influence on the turbulent core of the flow. In the staggered-grid arrangement the wall-normal velocity component and two tangential shear stress components must be specified at the boundaries. As suggested by Schumann (1975) we assume that the streamwise wall shear-stress is proportional to and in perfect phase with the tangential velocity at the grid point nearest to the wall. The model for the spanwise stress has been modified according to the proposal of Piomelli *et al.* (1989). In this case the proportionality factor for the spanwise component is taken to be that computed for the streamwise direction. The normal velocity component is defined at the wall boundary and set to zero there.

Initial values for the three velocity components are needed in the entire computational domain at the start of the simulations. These are obtained from a separate two-dimensional, steady-state calculation of the flow using the standard k-ε model of turbulence. The instantaneous velocity field is determined by adding a fluctuating component to each of the three mean velocities. These fluctuating components are specified from random numbers weighted with $(2k/3)^{1/2}$. The resulting initial velocity field is then made divergence free by applying the Poisson equation solver for the pressure field once and correcting the velocities.

2.3 Solution Algorithm

The momentum equations are integrated in time using the second-order leapfrog scheme with an initial Euler-step to start the simulations. The Poisson equation for the pressure field is computed exactly using the projection method (Chorin 1968) and a fast elliptic solver. A Fourier transformation in the spanwise direction reduces the three-dimensional problem to a set of decoupled, two-dimensional Helmholtz problems. These are solved in parallel employing the cyclic reduction technique for computational domains which are simply connected (box-like). In order to apply the direct solver to flows in multiply-connected domains, such as the step flow with an inlet channel, the capacitance matrix technique is utilized. This essentially leads to the need to solve a three-dimensional Poisson equation twice at each time step. The Poisson equations can be evaluated by employing techniques for simply-connected regions.

2.4 Geometry and Flow Parameters

The backward-facing step geometry consists of a 1:2 ratio sudden expansion with solid-wall boundaries above and below the step. The inlet plane is located 4 step-heights (h) upstream of the step and the outlet plane $20h$ downstream. The computational domain is $8h$ in width. Including the separate channel flow for the inlet conditions, the computed domain is illustrated in Fig. 1. The step-flow simulations are performed on a mesh of $192\times64\times32$ cells in the x-y-z coordinate directions, respectively. The Reynolds number, based on the step-height h and the maximum inlet velocity U_0, is $Re_0 = 1.55 \times 10^5$.

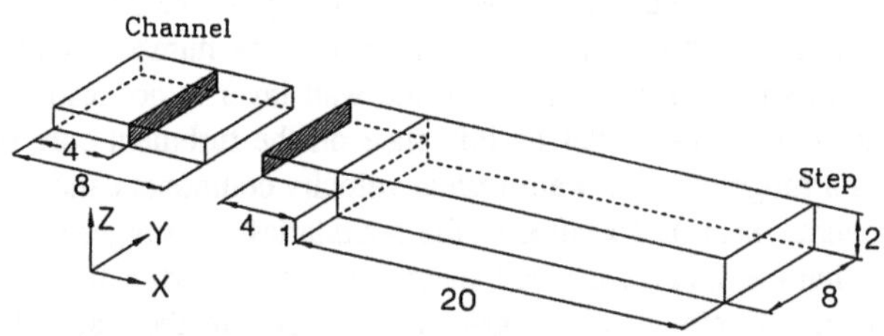

Fig. 1. Geometry and coordinate system foı the computational domain

If we define a characteristic time for the flow as $T_0 = L/U_0$ where L is the streamwise length of the computational domain, then the entire time simulated is $100T_0$. The total simulation time is sufficient for the flow to reach a statistically stationary state in the mean which is independent of the initial conditions. Time averaging begins after $20T_0$ and correlations are calculated after $40T_0$ of elapsed time. The entire simulation requires 125 hours of CPU time on the CRAY Y-MP 4/464.

3. Results and Discussion

3.1 Mean Flow Variables

The most common parameter characterizing a separated, reattaching flow is the time-mean position of reattachment. The mean reattachment length X_r, is often used as a criterion to judge the quality of computations and/or turbulence models. It can be defined in terms of several mean-flow quantities, including the dividing streamline $\langle\psi\rangle = 0$, the velocity contour line $\langle u\rangle = 0$, and the wall shear-stress distribution along the reattachment wall, $\langle\tau_w\rangle = 0$. In our calculations we have determined X_r based on the dividing streamline as well as from the wall shear-stress variation and have found differences of less than one percent. The calculated value $X_r = 7.32h$ is approximately 14% below the experimentally determined value of $X_r = 8.5h$ (Durst and Schmitt 1985). The latter authors used the LDA technique to investigate the backward-facing step flow in a closed channel with the same expansion ratio as in our calculations and at a comparable Reynolds number of $Re_0 = 1.13 \times 10^5$.

There are several possible explanations for the discrepancy between the measured and predicted values. These include variations in the inlet conditions between the experimental and numerical studies, three-dimensional effects in the measurements, as well as the possible influence of the SGS model or incomplete grid resolution in the simulations.

In a previous study we have investigated the influence of the grid resolution and SGS model on the predictions (Arnal and Friedrich 1991b). There we have shown that doubling the number of grid points in all three coordinate directions compared with the present case led to changes in the predicted mean flow field of less than 2%. Separate simulations of the flow on the same grid with the basic Smagorinsky-Lilly SGS-model have resulted in a reattachment length of $X_r = 7.17h$, less than 2% below that in the present simulations. Of course, other models may yield better results. Of particular interest is the dynamic SGS model of Germano (1991) which has been applied with success to both compressible and incompressible flows, Moin *et al.* (1991).

Several experimental studies have shown that the mean flow field characteristics are remarkably sensitive to small changes in the inlet conditions. For example, Isomoto and Honami (1989) have demonstrated that a 20% increase in the turbulence intensity upstream of the step can lead to a reduction in the reattachment length of as much as two step-heights. Studies by Adams and Johnston (1988) have also shown that the mean reattachment length is sensitive to the boundary-layer thickness in the case of boundary layer separation at the step. Thus, differences in the inlet conditions between the calulations and measurements could be responsible for discrepancies in the reattachment length. While the flow in the calculations upstream of the step is fully-developed (Sect. 2.2), the measured Reynolds-stress profiles in the upstream channel indicate that the flow is still developing in the experimental study (Durst and Schmitt 1985).

Finally, the differences between the measured and predicted reattachment lengths could also be a result of three-dimensional effects in the experimental study. A recent study (McCluskey *et al.* 1991) has called into question the assumption of two-dimensional behavior in many previous experimental studies of separated flows in channels. They have shown that spanwise dimensions of 2-3X_r are necessary to guarantee that the flow behavior is two-dimensional. The aspect ratio of the channel used in Durst and Schmitt's study was $12h$, or approximately $1.5X_r$, which indicates that three-dimensionality could be important.

We have also investigated the influence of the spanwise dimension of our computational domain on the predicted mean flow characteristics (Arnal and Friedrich 1991b). We found that for spanwise dimensions of 8 step-heights or more, periodic boundary conditions can successfully be applied and have no influence on the predicted value of X_r. However, for dimensions less than 4 step-heights, the mean reattachment length was as much as 30% longer than the measured value.

Because of the sensitivity of the reattachment length to geometry and flow parameters, Roshko and Lau (1965) proposed normalizing the backward-facing step data using the inlet-channel centerline velocity U_0, and the mean reattachment lenght, X_r. The authors found that experimental data obtained under a variety of inlet and freestream conditions collapsed together when normalized in this way. In the comparisons of our predictions with the measurements (Durst and Schmitt 1985) we will employ the same normalization scales in the discussion which follows.

Figure 2 shows the distribution of the predicted mean streamlines for the flow. The dividing streamline ($\langle\psi\rangle = 0$) defines the mean spatial extent of the primary recirculation zone. Of interest is the smaller, secondary recirculation region in the corner of the step, which was also observed in the measurements. Because it is much smaller and has a weaker recirculation than the primary one, it is often not observed in simulations of the flow. Although its predicted size ($0.040X_r \times 0.020X_r$ in the x-z directions) is smaller than that observed in the measurements ($0.11X_r \times 0.04X_r$), this is almost certainly connected with the relatively coarse resolution in the present simulations.

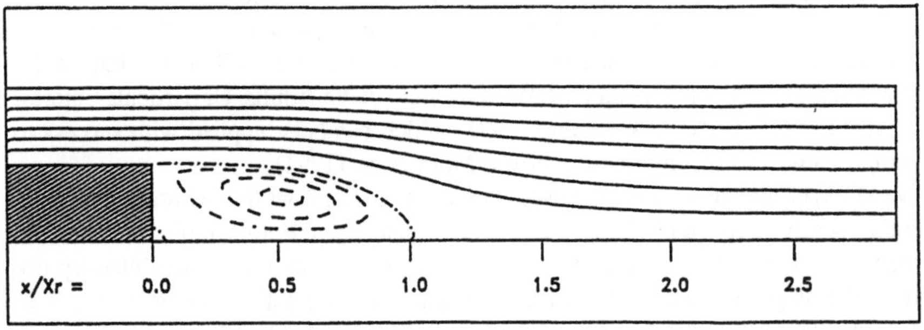

Fig. 2. Predicted mean streamlines, Contours of $\phi < 0$: - - -, $\Delta\phi = 0.025$; $\phi > 0$: —, $\Delta\phi = 0.13$

Note that although this secondary recirculation region is present in the mean velocity field, observations of the instantaneous velocity field as shown in Fig. 3 often fail to reflect its existence. Instead, the presence of this secondary zone in the mean flow is a result of small three-dimensional bubbles occurring in this region. These bubbles result from the entrainment of fluid with positive streamwise velocity from the incoming channel flow. An additional influence is the positive circulation in the step corner induced by the neighboring, primary separation zone.

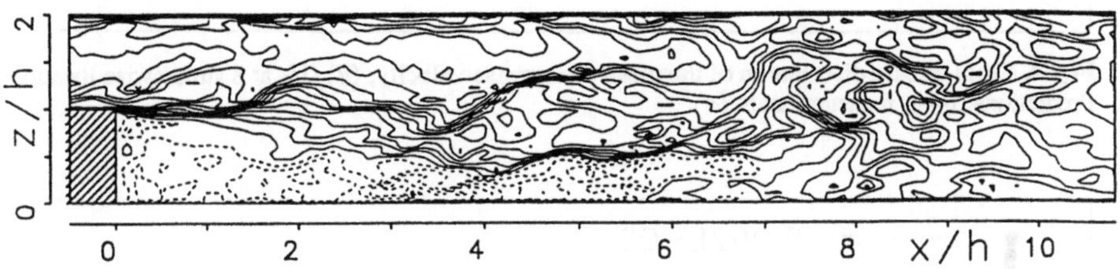

Fig. 3. Contours of the instantaneous streamwise velocity, $-0.48 < u/U_0 < 1.14$. Contours: - - - < 0; — > 0

In Fig. 2 the stream function has its minimum at the point in the flow where the mean streamwise and cross-stream velocities are zero, $\langle u \rangle = \langle w \rangle = 0$. The ratio of the minimum value of the stream function to the mean mass flow, $\langle \psi \rangle_{min}/(U_b h)$, gives an indication of the relative mass flux in the recirculation zone compared with that in the downstream direction. The predicted value and position of $\langle \psi \rangle_{min}$ compare well with the measured values (Durst and Schmitt 1985) as shown in Table 1.

Table 1. Value and position of $\langle \psi_{min} \rangle$

Source	$\langle \psi \rangle_{min}/(U_b \cdot h)$	x_{min}/X_r
Present (calcs.)	−0.09	0.48
Durst/Schmitt (expts.)	−0.1	0.47

In Figs. 4 and 5 the calculated mean velocity field is compared with the measurements (Durst and Schmitt 1985). The mean streamwise velocity profiles are in excellent agreement with the measurements (Fig. 4) both upstream and downstream of the mean reattachment point. The negative velocities near the step are slightly underpredicted, however, the maximum reverse flow velocity of $\langle u \rangle/U_0 = -0.216$ which occurs at $x/X_r = 0.47$ is in good agreement with the measured value ($\langle u \rangle/U_0 = -0.218$) at this position.

Figure 5 compares the cross-stream velocity distribution with the measured profiles. At first glance the discrepancies are quite large, however, note that the velocity magnitudes are much smaller than those shown in Fig. 4. In addition, we note that there is a fair amount of scatter in the measured profiles, particularly in the region near the step ($x/X_r = 0.12, 0.24$).

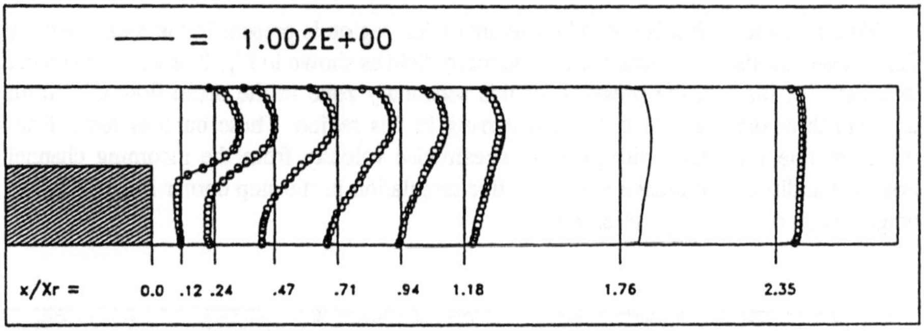

Fig. 4. Cross-stream distribution of the mean streamwise velocity, $\langle\overline{u}\rangle/U_0$ at 8 streamwise locations. Predictions: (—); measurements (o)

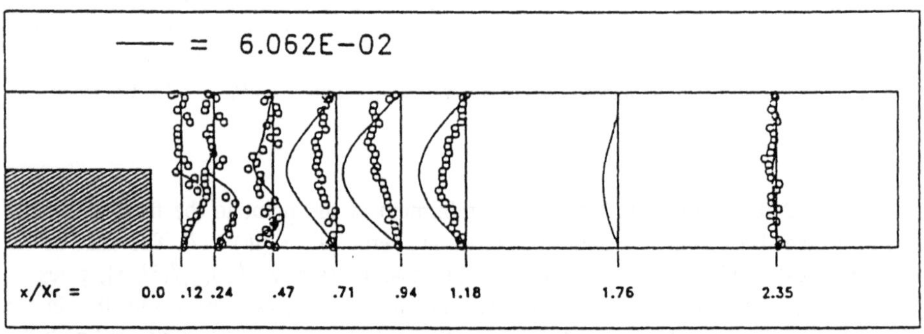

Fig. 5. Cross-stream distribution of the mean cross-stream velocity, $\langle\overline{w}\rangle/U_0$ at 8 streamwise locations. Predictions: (—); measurements (o)

Figure 6 illustrates the predicted Reynolds shear-stress profiles at several streamwise locations. Here, the overall agreement is quite satisfactory. The predicted stress levels in the recirculation area are in good agreement with the measured values. We note however, that the cross-stream, z-position of the maximum shear stress does not appear to fall as rapidly in the predictions as in the experiments. As discussed above, there are various possible reasons for the differences between the computations and measurements. In addition to those cited, the discrepancies may be due to inaccuracies in the measurements.

3.2 Momentum Balances

The Reynolds-averaged momentum equations in the x- and z-directions are given by:

$$\langle u\rangle\frac{\partial}{\partial x}\langle u\rangle + \langle w\rangle\frac{\partial}{\partial z}\langle u\rangle = -\frac{1}{\rho}\frac{\partial}{\partial x}\langle p\rangle - \frac{\partial}{\partial z}\left\langle u''w''\right\rangle - \frac{\partial}{\partial x}\left\langle u''u''\right\rangle, \quad (5)$$

$$\langle u\rangle\frac{\partial}{\partial x}\langle w\rangle + \langle w\rangle\frac{\partial}{\partial z}\langle w\rangle = -\frac{1}{\rho}\frac{\partial}{\partial z}\langle p\rangle - \frac{\partial}{\partial x}\left\langle u''w''\right\rangle - \frac{\partial}{\partial z}\left\langle w''w''\right\rangle, \quad (6)$$

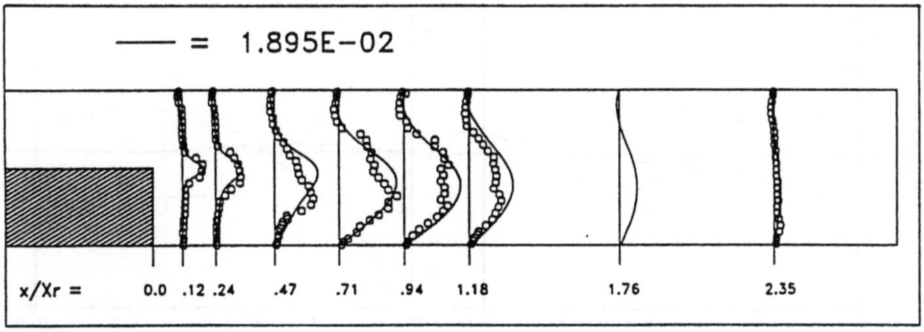

Fig. 6. Cross-stream distribution of the Reynolds stress, $\langle u''w''\rangle/U_0^2$ at 8 streamwise locations. Predictions: (—); measurements (o)

where the double prime ($''$) indicates fluctuations in Reynolds' sense. Note that the viscous terms have been neglected since these are small except for the near-wall regions. Each of these terms can be evaluated from the simulations which have been performed. In order to account for the effects of the SGS model, we have made the following approximations:

$$\langle u''w''\rangle \approx \langle(\overline{u} - \langle\overline{u}\rangle)(\overline{w} - \langle\overline{w}\rangle)\rangle + \tau_{xz,\mathrm{inh}} \tag{7}$$

$$\langle u''u''\rangle \approx \langle(\overline{u} - \langle\overline{u}\rangle)^2\rangle + \frac{2}{3}\langle E'\rangle \tag{8}$$

$$\langle w''w''\rangle \approx \langle(\overline{w} - \langle\overline{w}\rangle)^2\rangle + \frac{2}{3}\langle E'\rangle \tag{9}$$

These approximations are based on the assumption that the total Reynolds stresses are composed of grid-scale (GS) and subgrid-scale (SGS) components. In (7)-(9), the SGS energy is given by E' and the inhomogeneous part of the SGS shear-stress is given by $\tau_{xz,\mathrm{inh}} = 2\mu_{\mathrm{inh}}\langle\overline{S}_{xz}\rangle$. Recall that the time-average of the isotropic portion of the SGS stresses is zero, since it is proportional to the fluctuating deformation tensor, which vanishes in the mean.

Typical cross-stream, z-direction momentum balances for the flow are presented in Fig. 7. The corresponding streamwise, x-direction momentum balances are shown in Fig. 8. Streamwise locations correspond to the recirculation, reattachment and relaxation regions of the flow.

Turning first to Fig. 7, we note that the cross-stream transport due to convective terms is only important in the mixing layer region and above, Fig. 7a. In contrast, the transport of z-momentum due to the Reynolds shear-stress term $\partial\langle u''w''\rangle/\partial x$ is relatively unimportant throughout the flow. In the backflow region and in the entire channel downstream of reattachment (Figs. 7b,c) the cross-stream momentum balance reduces to

$$-\frac{1}{\rho}\frac{\partial}{\partial z}\langle p\rangle = \frac{\partial}{\partial z}\langle w''w''\rangle\,. \tag{10}$$

For these regions we may integrate (10) as discussed by Simpson *et al.* (1980) to yield:

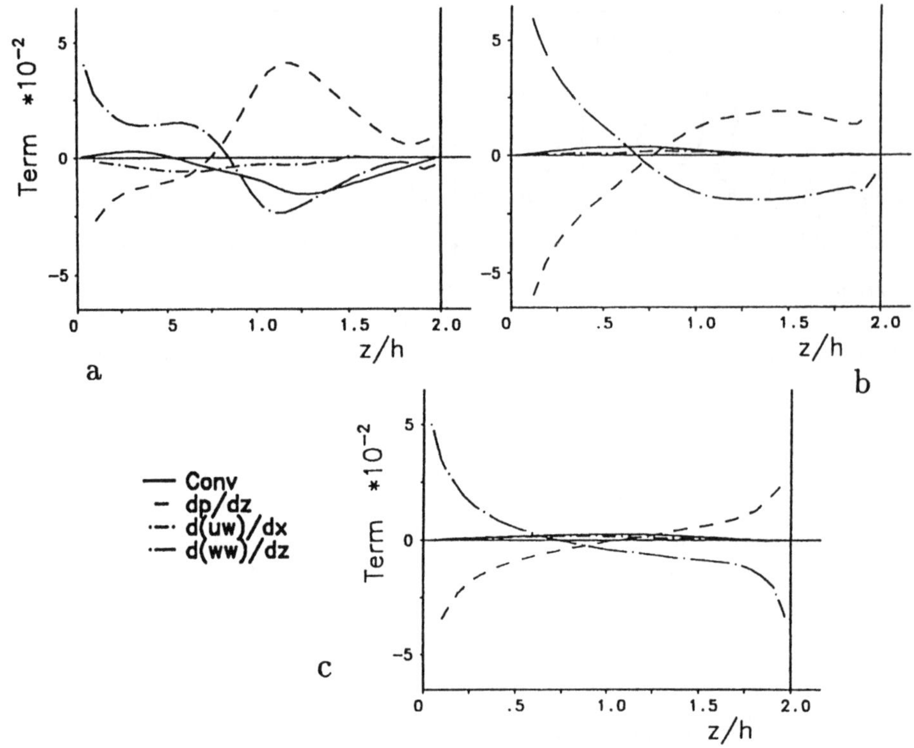

Fig. 7. Cross-stream z-momentum balances at (a) $x/X_r = 0.47$; (b) $x/X_r = 0.99$; (c) $x/X_r = 1.7$. All terms are normalized using U_0 and h

$$\langle p(x,z)\rangle = \langle p_0(x)\rangle - \rho\left\langle w''w''\right\rangle \tag{11}$$

where $\langle p_0(x)\rangle$ is the local wall mean static pressure. Differentiating this equation with respect to x yields:

$$-\frac{1}{\rho}\frac{\partial}{\partial x}\langle p\rangle = -\frac{1}{\rho}\frac{d}{dx}\langle p_0\rangle + \frac{\partial}{\partial x}\left\langle w''w''\right\rangle . \tag{12}$$

For those regions where $\partial\langle w''w''\rangle/\partial x$ is much smaller than $d\langle p_0\rangle/dx$, the mean pressure gradient will be nearly independent of z. We found this to be the case in the region downstream of reattachment, Figs. 8b,c.

In contrast to the cross-stream momentum balance, the transport due to the Reynolds shear stress has a more important role in the streamwise momentum balance. Note that the term $\partial\langle u''w''\rangle/\partial z$ is significant in the recirculation zone (Fig: 8a) near reattachment and downstream (Fig. 8b,c). In addition, the transport due to the normal stress $\partial\langle u''u''\rangle/\partial x$ is important in the shear layer and recirculation zone (Fig. 8a). Downstream of reattachment this term plays a less important role.

Only within the reattachment zone is it possible to make a momentum balance simplification similar to that leading to (10). In this region convection is unimportant and (5) reduces to

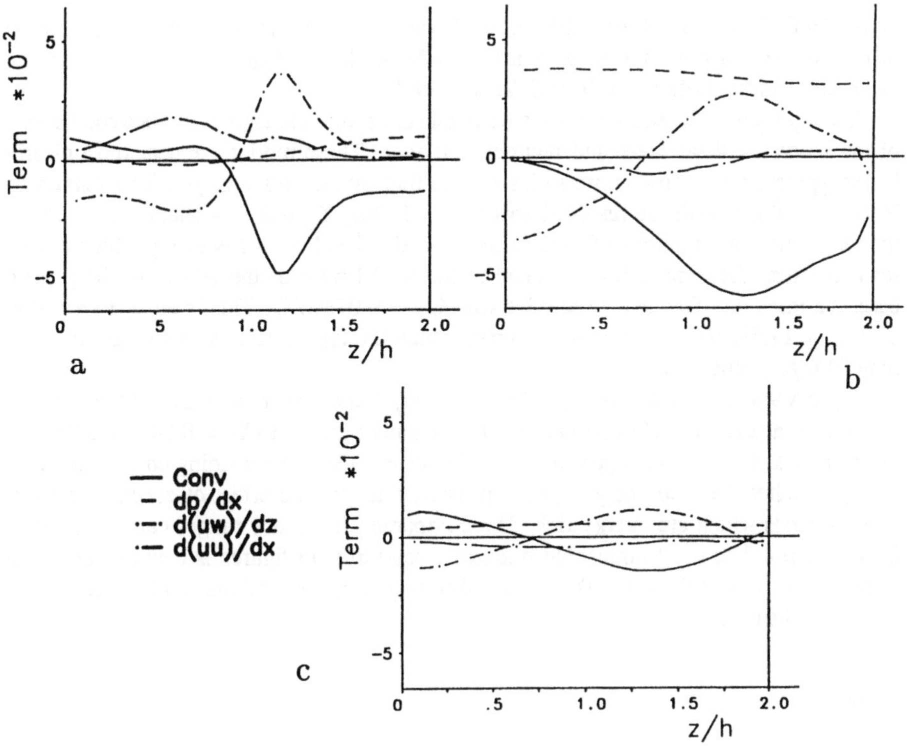

Fig. 8. Streamwise x-momentum balances at (a) $x/X_r = 0.47$; (b) $x/X_r = 0.99$; (c) $x/X_r = 1.7$. All terms are normalized using U_0 and h

$$-\frac{1}{\rho}\frac{\partial}{\partial x}\langle p\rangle = \frac{\partial}{\partial z}\left\langle u''w''\right\rangle, \tag{13}$$

which indicates a balance between the mean pressure gradient and the cross-stream gradient of the shear stress.

The representative momentum balances shown in Figs. 7 and 8 demonstrate the importance of developing accurate models for the Reynolds stresses. Momentum transport due to both normal and shear stresses is significant in the separated and shearing regions of the flow. Only in the relaxation region downstream of reattachment does the streamwise normal stress begin to lose importance in the overall balance of momentum. Simple models which neglect one or more of these terms cannot be expected to perform well in predictions of complex separated flows.

3.3 Low-frequency Time Variations

The experimental study of the flow behind a backward-facing step (Eaton and Johnston 1981) revealed the presence of low-frequency motions in the shear layer and reattachment region. Several later studies of separated and reattaching flows in different geometries have confirmed the observations of a low-frequency flapping of the shear layer (Kiya and

Sasaki 1983; Dianat and Castro 1991). In addition, a recent three-dimensional simulation of the flow past a blunt flat plate at a Reynolds number of $Re_0 = 1000$ has shown the presence of these motions (Tafti and Vanka 1991b).

In the present simulations we have been able to capture the large timescale oscillations of the separated shear layer and recirculation zone. These motions appear most clearly in the spectra taken from time-series of the fluctuating pressure, p''. The maximum frequency which could be resolved was $f_{max} = 160U_0/X_r$, this was adequate to resolve the long-time scale motions of interest in the study. The lowest frequency which can be resolved from the time-series is related to the total length of the series. In the present case, the minimum frequency recorded was $f_{min} = 0.07U_0/X_r$. This is equal to a period of $T_{max} = 13.5X_r/U_0$ which is much smaller than the typical time series length of more than $33X_r/U_0$ time units.

Figures 9 and 10 illustrate typical spectra for p'' taken near the wall in the recirculation region at $x/X_r = 0.47$ and near the reattachment line at $x/X_r = 0.94$. Note that the ordinate axis is given as the product of the frequency and corresponding energy spectrum $f \cdot S(f)$, rather than the energy spectrum itself. This has the effect of accentuating frequency bands which are energy rich. Three important frequency ranges can be observed in the figures. The most noticeable frequency band in both figures is centered around a frequency of $f = 2.0U_0/X_r$. This is the shedding frequency of the large eddies in the separated shear layer.

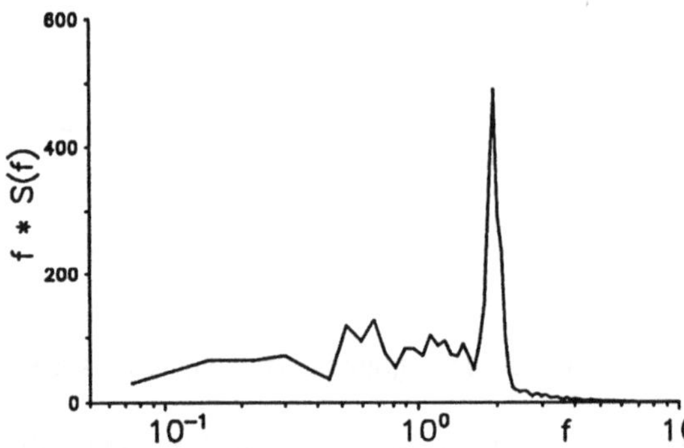

Fig. 9. Spectrum of the wall pressure fluctuations, p'' in the recirculation zone, $x/X_r = 0.47$

A second band of frequencies in the range of $0.5 < fX_r/U_0 < 0.7$ can also be seen. This frequency range is more prevalent in the reattachment region (Fig. 10) and has been identified with the oscillation of the instantaneous reattachment point. This value compares well with the peak frequency of $f \approx 0.52U_0/X_r$, observed by Eaton and Johnston (1981) in this region. The latter authors identified this frequency with vortex shedding from the recirculation bubble.

The third band of frequencies which appear in the spectra is centered around $f \approx 0.07-0.3U_0/X_r$. This is the lowest range which was resolved in the current investigations and is more important in the region upstream of reattachment, close to the step, Fig. 9. This range compares fairly well with the frequency of the flapping shear-layer ($f \approx 0.12U_0/X_r$) in investigations of the flow past a blunt flat plate (Kiya and Sasaki 1983).

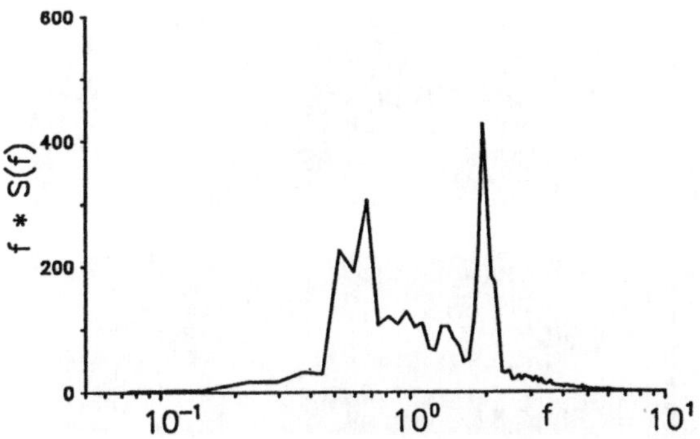

Fig. 10. Spectrum of the wall pressure fluctuations, p'' in the reattachment region, $x/X_r = 0.99$

The flapping of the shear layer behind the backward-facing step has been hypothesized (Eaton and Johnston 1981) to be two-dimensional in nature and responsible for the wandering of the reattachment point in the up- and downstream directions. The authors further conjecture that the long timescale motions are caused by an instantaneous imbalance between shear layer entrainment and reinjection of fluid into the separation bubble near reattachment.

Our numerical observations of the instantaneous flow field lead us to believe that the growth and shrinkage of the separation bubble is indeed a result of variations in the entrainment of the shear layer and fluid injection near reattachment. However, the process is almost certainly three-dimensional in nature. If the effect were two-dimensional, one would expect variations in the predicted entrainment in two-dimensional simulations of the separated flows. In fact, no such observations were made in the two-dimensional simulations of the flow past a blunt flat plate by Tafti and Vanka (1991a). On the other hand, the same authors found the low-frequency motions in their three-dimensional simulations of the same flow made at the same Reynolds number, Tafti and Vanka (1991b). In the present calulations, we find that both the eddies which develop in the shear layer and the entrainment process which they influence become three-dimensional a short distance downstream of the step, Figure 11.

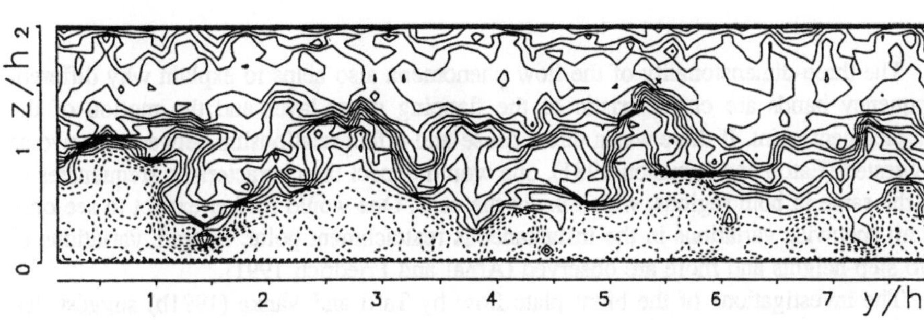

Fig. 11. Three-dimensional structure of the shear layer. Contours of the instantaneous streamwise velocity in a cross-stream plane, $x/X_r = 0.6$. Flow is out of the page. $-0.52 < u/U_0 < 1.12$. Contours: - - - < 0; — > 0

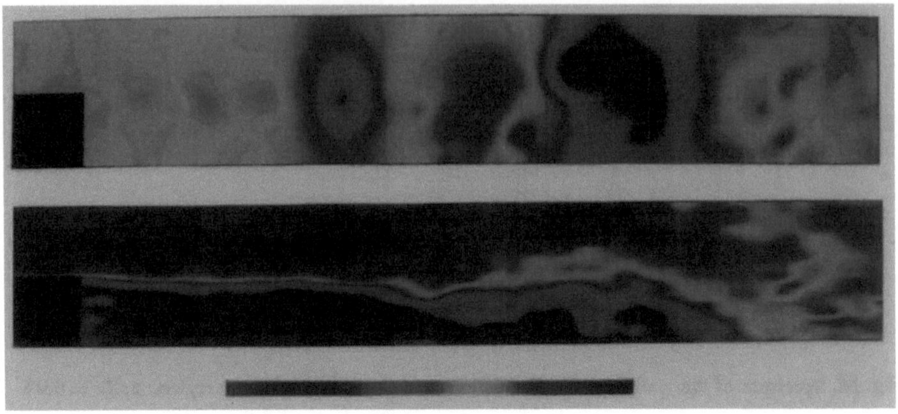

Fig. 12. Contours of (top) fluctuating pressure, p'' and (bottom) instantaneous streamwise velocity, $\overline{u}$ with the recirculation zone in an expanding state. Scale values increase from left to right

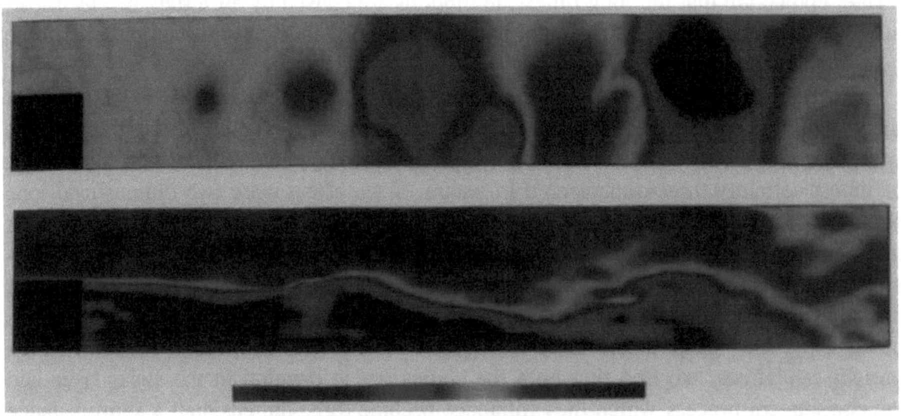

Fig. 13. Contours of (top) fluctuating pressure, p'' and (bottom) instantaneous streamwise velocity, $\overline{u}$ with the recirculation zone in a state just prior to reduction. Scale values increase from left to right

The three-dimensionality of the flow phenomena also helps to explain why different frequency bands are characteristic of the flapping shear layer and the motion of the reattachment point. If there was a direct cause-and-effect relationship between the two as suggested (Eaton and Johnston 1981), one would expect the characteristic frequencies to be the same in both regions, which is not the case. One would further expect to see only small spanwise variations in the instantaneous reattachment point. Instead, variations of two step-heights and more are observed (Arnal and Friedrich 1991).

The investigations of the blunt plate flow by Tafti and Vanka (1991b) suggest that variations in the shear layer entrainment are connected with the shedding of vortices of different strength from the separated shear layer. We have found evidence that this is the case in our simulations of the step flow.

Figures 12 and 13 show color contour plots of the streamwise velocity and fluctuating pressure field at two different times. Typical eddy (or, vortical) structures in the shear layer are shown in the form of local, instantaneous pressure minima. In Fig. 12, the recirculation zone is in an expanding state. The eddies in the shear layer are shown as regions of blue-green alternating with the high pressure regions in red.

In Fig. 13, two of the eddies begin to merge in the shear layer. At the instant shown, the entrainment of fluid *from* the bubble prevails over the *reinjection* of fluid in the reattachment region. On the upstream side of the structure, fluid is being entrained from the backflow region into the shear layer. On the downstream side, positive velocity fluid from above the shear layer is being forced down into the recirculation region. This leads to a pinching of the separation bubble and shedding of fluid from the recirculation zone. Following this, the separation region becomes more contracted in size.

The two eddies which have merged in Fig. 13 are shed from the shear layer and move downstream. The recirculation zone then gradually increases in size as the pressure within it increases. This expansion of the separation region leads to downstream movement of the instantaneous shear layer impingement point and causes the shear layer itself to move away from the recirculation zone.

The entire unsteady behavior of the separation bubble is very complex and three-dimensional. Further work remains to be done in order to understand the exact sequence of events which take place during the expansion and contraction of the recirculation region.

3.4 Topological Flow Features

In the previous section we discussed the motion of eddies in the shear layer and identified vortical structures with local regions having an instantaneous pressure minimum. This is just one of many means of identifying predominant structures or flow regions which occur in a turbulent flow. These topological features are studied because they are associated with relatively large contributions to certain mean flow features, such as the Reynolds stresses or the turbulent kinetic energy. The vortical structures which occur in separated shear layers also appear to play an important role in the low-frequency dynamics of such flows.

In a recent study Hunt *et al.* (1988) have analyzed direct numerical simulation (DNS) data for the presence of three distinct streamline patterns. These included flow regions where the instantaneous streamlines converge (C-regions), vortical eddies (E-regions) and streaming S-regions. These flow regions were characterized in terms of local values of the turbulent kinetic energy and fluctuating pressure fields. We have applied the topological analysis of Hunt *et al.* in a modified way to the instantaneous flow fields in our simulations to determine to what extent such flow pattern occur. Some of the ideas we employ in our modified analysis are due to Kleiser and Sandham (1991).

Figure 14 shows three different views of the instantaneous flow field downstream of the step. In Fig. 14a the fluctuating pressure field and contours of the velocity product, $-u''w''$. A comparison of the three figures indicates that there are several regions which are correlated with one another.

In Figs. 14a,b there is a close correspondence between regions exhibiting a local pressure minimum and vector pattern typical for a vortex (circulating streamlines). We have identified these as the eddy structures (E-regions) discussed by Hunt *et al.* Note that regions with high levels of vorticity need not necessarily be identified with these vortical structures. High levels of vorticity may also arise due to simple shear.

Alternating with the low-pressure regions in the shear layer (Fig. 14b) are high-pressure areas which can be associated with a local stagnation-type flow pattern in the vector field (Fig. 14a). These structures correspond to the convergence (C) regions of the flow.

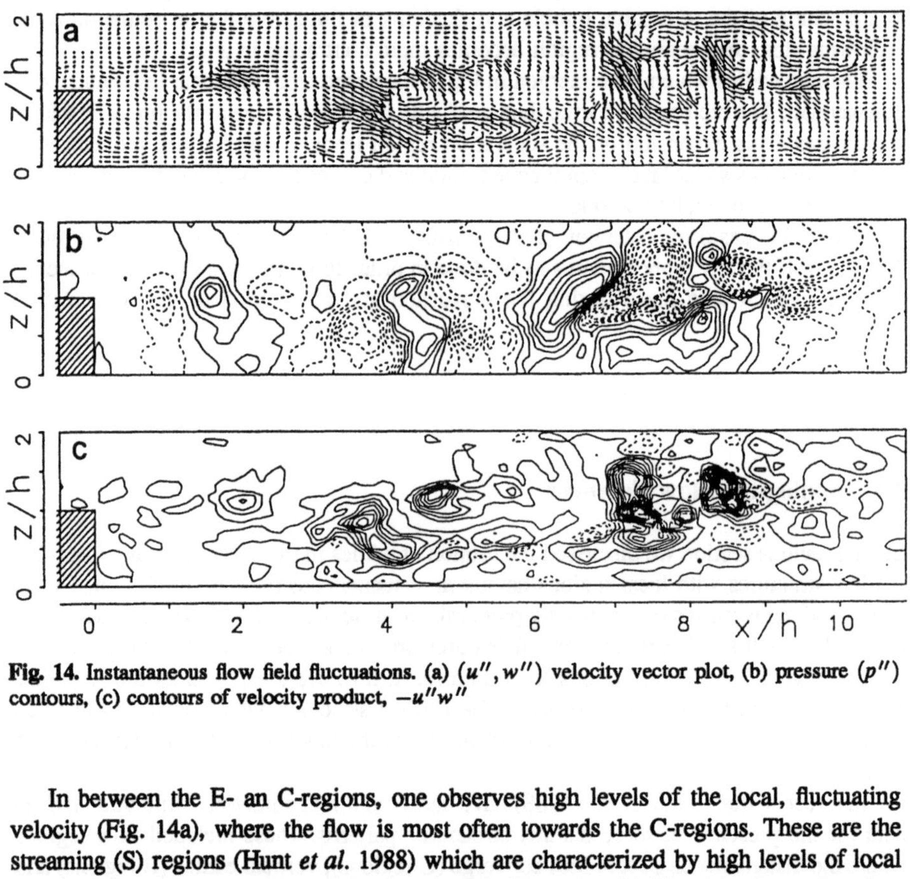

Fig. 14. Instantaneous flow field fluctuations. (a) (u'', w'') velocity vector plot, (b) pressure (p'') contours, (c) contours of velocity product, $-u''w''$

In between the E- an C-regions, one observes high levels of the local, fluctuating velocity (Fig. 14a), where the flow is most often towards the C-regions. These are the streaming (S) regions (Hunt *et al.* 1988) which are characterized by high levels of local kinetic energy. A comparison of these regions in Fig. 14a with Fig. 14c shows that they correspond closely with high levels of the velocity product, $-u''w''$. Instantaneous values of $-u''w''$ occur which are an order of magnitude larger than the local Reynolds shear stress $\langle -u''w'' \rangle$.

Once the various structures can be identified, one may investigate to what extent they contribute to the structure of the mean flow field. For example, it seems likely that the streaming regions are responsible for a significant portion of the local Reynolds shear stress. An analysis of time series recorded in the separated shear layer has shown this to be the case. While fluctuating velocities which are twice as large as the local rms value occur only 10 percent of the time, the resulting velocity product, $-u''w''$ contributes more than 30 percent of the local mean Reynolds shear stress. This confirms the assumption that such structures play an important role not only in the instantaneous dynamics of the flow but also in determining the mean flow field structure.

4. Conclusions

The LES technique has been used to investigate the separated and reattaching flow behind a backward-facing step. Predictions of mean flow quantities compare fairly well with measurements (Durst and Schmitt 1985). Momentum balances in the recirculation and reattachment regions have shown the importance of transport due to the Reynolds stress terms. At the same time, there are regions in the flow where simplifications can be made to the Reynolds-averaged momentum equations. These are undoubtedly of significance in developing new turbulence models for turbulent, separated flows.

Time variations in the flow field have been analyzed for the presence of the experimentally-observed, low-frequency motions. Both the low-frequency flapping of the shear layer and the vortex shedding in the reattachment region are observed. The value of the low-frequency band ($0.07 < fX_r/U_0 < 0.3$) compares well with the experimentally observed values of $f \approx 0.15U_0/X_r$. Dominant frequencies in the reattachment region are also in good agreement with the observed experimental values (Eaton and Johnston 1981). The large-time scale motions have been linked to variations in the shear-layer entrainment level caused by vortex interaction.

Finally, the instantaneous flow field has been analyzed for the presence of typical large-scale structures. Eddy, convergence and streaming zone discussed by Hunt *et al.* (1988) were observed. The streaming zones were linked to high levels of the fluctuating velocity product $-u''w''$. These large instantaneous values of $-u''w''$ make a large contribution to the local Reynolds shear stress, $\langle -u''w'' \rangle$. Further investigation into the role which the eddy, convergence and streaming zones play in the mean flow characteristics and instantaneous flow dynamics seems to be warranted.

Acknowledgements

The authors are grateful to Dr.L. Schmitt for his work on the original code development and for many useful discussions. Figures 12 and 13 were produced using the NCSA software developed at the University of Illinois at Urbana-Champaign. The work has been supported by the German Research Society (DFG).

References

Adams, E.W. & Johnston, J.P (1988): "Effects of the separating shear layer on the reattachment flow structure, part 2: reattachment length and wall shear stress", Exp. Fluids, Vol. 6, pp. 493-499.

Arnal, M. and Friedrich, R. (1991a): "Investigation of the Pressure and Velocity Fields in a Turbulent Separated Flow Using the LES Technique", AIAA Paper No. 91-0251, 29th Aerospace Sciences Meeting, Reno, USA, Jan. 7-10.

Arnal, M. & Friedrich, R. (1991b): "On the effects of spatial resolution and subgridscale modeling in the large eddy simulation of a recirculating flow", *to appear in* Proc. of th 9th GAMM-Conf. on Numerical Methods in Fluid Mechanics, (Vieweg Verlag, Braunschweig).

Chorin, A.J. (1968): "Numerical solution of the Navier-Stokes equations", Math. of Comp., Vol. 22, pp. 745-762.

Driver, D.M., Seegmiller, H.L. & Marvin, J.G. (1987): "Time-dependent behavior of a reattaching shear layer", AIAA Journal, Vol. 25, No. 7, pp. 914-919.

Dianat, M. & Castro, I.P. (1991): "Turbulence in a separated boundary layer", J. Fluid Mech., Vol. 226, pp. 91-123.

Durst, F. & Schmitt, F. (1985): "Experimental study of high Reynolds number backward-facing step flow", in Proc. 5th Symp. on Turbulent Shear Flows, Aug. 7.-9 (Cornell University, Ithaca, NY).

Eaton, J.K. & Johnston, J.P. (1981): "Low frequency unsteadiness of a reattaching turbulent shear layer"; in Turbulent Shear Flows 3, ed. L.J.S. Bradbury, F. Durst, B.E. Launder, F.W. Schmitt & J.H. Whitelaw (Springer-Verlag, Berlin) pp. 162-170.

Friedrich, R. (1988): "Simulation of turbulent flows", in Computational Methods in Flow Analysis, ed. by H. Niki and M. Kawahara (Okayama University of Science, Japan), Vol. 2, pp. 833-843.

Germano, M., Piomelli, U., Moin, P. & Cabot, W.H. (1991): "A dynamic subgridscale eddy viscosity model", Phys. Fluids A, Vol. 3, No. 7, pp. 1760-1765.

Hunt, J.C.R., Wray, A.A. & Moin, P. (1988): "Eddies, streams and convergence zones in turbulent flows", in Proc. of the Summer Program 1988, Report CTR-S88 (Center for Turbulence Research, Stanford, CA) pp. 193-208.

Isomoto, K. & Honami, S. (1989): "The effect of inlet turbulence intensity on the reattachment process over a backward-facing step", J. Fluids Eng., Vol. 111, pp. 87-92.

Kiya, M. & Sasaki, K. (1983): "Structure of a turbulent separation bubble", J. Fluid Mech., Vol. 137, pp. 83-113.

Klein, H. & Friedrich, R. (1990): "Large eddy simulation of manipulated boundary layer and channel flows", in Turbulence Control by Passive Means, ed. by E. Coustols (Kluwer Academic Publishers, Dordrecht) pp. 41-65.

McCluskey, F., Hancock, P.E. & Castro, I.P (1991): "Three-dimensional separated flows", in Proc. 8th Symp. on Turbulent Shear Flows, Sept. 9-11 (Technical University of Munich, Germany) pp. 9-5-1-6.

Moin, P., Squires, K., Cabot, W. & Lee, S. (1991): "A dynamic subgrid-scale model for compressible turbulence and scalar transport", Phys. Fluids A, Vol. 3, No. 11, pp. 2746-2757.

Piomelli, U., Ferziger, J., Moin, P. & Kim, J. (1989): " New approximate boundary conditions for large-eddy simulations of wall-bounded flow", Phys. Fluids A, Vol. 1, No. 6, pp. 1061-1068.

Richter, K., Friedrich, R. & Schmitt, L. (1987): "Large-eddy simulation of turbulent wall boundary layers with pressure gradient", in Proc. 6th Symp. on Turbulent Shear Flows (Paul Sabatier University, Toulouse) pp. 22.3.1-7.

Roshko, A. & Lau, J.C. (1965): "Some observations on transition and reattachment of a free shear layer in incompressible flow", in Proc. Heat Transfer and Fluid Mechanics Institute, ed. A.F. Charawat (Stanford University Press, Standord, CA) pp. 157-167.

Sandham, N.D, & Kleiser, L. (1991): "Vortex formation in the late stages of transition to turbulence", in Proc. of the Boundary Layer Transition and Control Conf., April 8-12 (Royal Aeronautical Society, Cambridge, UK) pp. 26.1-12.

Schumann, U. (1975): "Subgrid scale model for finite difference simulations of turbulent flows in plane channels and annuli", J. Comp. Phys., Vol. 18, pp. 376-404.

Simpson, R.L., Chew, Y.-T. & Shivaprasad, B.G. (1981): "The structure of a separating turbulent boundary layer. part 2: Higher-order turbulence results", J. Fluid Mech., Vol. 113, pp. 53-73.
Su, M.D. & Friedrich, R. (1991): "Large eddy simulation of fully-developed turbulent flow in a straigth duct", in Proc. 8th Symp. on Turbulent Shear Flows, Sept. 9-11 (Technical University of Munich, Germany) pp. II-19-1-2.
Tafti, D.K. & Vanka, S.P. (1991a): "A numerical study of flow separation and reattachment on a blunt plate", Phys. Fluids A, Vol. 3, No. 7, pp. 1749-1759.
Tafti, D.K. & Vanka, S.P. (1991b): "A three-dimensional numerical study of flow separation and reattachment on a blunt plate", Phys. Fluids A, Vol. 3, No. 12, pp. 2887-2909.
Unger, F. & Friedrich, R. (1991): "Large eddy simulation of fully-developed turbulent pipe flow", in Proc. 8th Symp. on Turbulent Shear Flows, Sept. 9.11 (Technical University of Munich, Germany) pp. 19-3-1-6.

Calculation of Vortex Shedding Past a Square Cylinder with Various Turbulence Models

R. Franke[1] and W. Rodi

Sonderforschungsbereich 210, University of Karlsruhe, Kaiserstr. 12, 7500 Karlsruhe, Germany
[1] present address: Dornier GmbH, 7990 Friedrichshafen, Germany

Abstract

The vortex-shedding flow past a square cylinder at Re = 22.000 was calculated with various turbulence models. The 2D periodic shedding motion was resolved in an unsteady calculation, and the superimposed stochastic turbulent fluctuations were simulated both with the $k-\varepsilon$ eddy-viscocity model and with a Reynolds-stress equation model. For both models, the viscosity-affected near-wall region was either bridged by wall functions or was resolved with a simpler one-equation model using a prescribed length-scale distribution. The $k-\varepsilon$ model with wall functions does not yield unsteady vortex motion while the other model variants do. The two-layer $k-\varepsilon$ model underpredicts severely the periodic fluctuations and also the Strouhal number and drag coefficient. The Reynolds-stress-equation models yield considerably better agreement with experiments, but tend to overpredict the periodic fluctuating motion and also miss some other details of the flow behaviour.

1. Introduction

The flow past slender, bluff structures is frequently associated with periodic vortex shedding causing dynamic loading on the structures. For the design of such structures, the unsteady loading forces must be known and hence methods for predicting the flow and the forces are of great practical importance. For situations with low Reynolds numbers, successful calculations have been obtained for square and circular cylinders by solving numerically the unsteady Navier-Stokes equations, e.g. Lecointe and Piquet (1989), Franke et al. (1990). For situations with high Reynolds numbers, which usually occur in practice, stochastic three-dimensional turbulent fluctuations are superimposed on the periodic vortex-shedding motion. A resolution of these motions in a direct simulation is not feasible at present. Tamura et al. (1990) have reported such calculations without a turbulence model, but for the higher Reynolds numbers considered by them, the small-scale turbulent motion could certainly not be resolved by the numerical grids used. They employed a third-order upwind differencing scheme which introduces numerical damping and basically takes over the role of a subgrid-scale model in a large-eddy simulation. The results obtained by Tamura et al. (1990) are certainly very interesting and surprisingly successful, but the approach of relying on numerical damping instead of a subgrid-scale model requires further examination. The work has shown, however, that large-eddy simulations are possible for vortex-shedding flow at high Reynolds numbers, but such calculations are very costly. Hence, there is still a need for more economic calculation methods based on the use of a turbulence model for simulating the influence of the stochastic fluctuations on the periodic vortex-shedding motion. So far, successsful applications of a turbulence model to vortex-shedding flow have not been reported. It should be mentioned here that steady calculations for the time-mean field using standard turbulence models were not successful (e.g. Majumdar and Rodi, 1985) since the strong momentum exchange by the periodic vortex motion is not accounted for in this approach.

The aim of the research reported here was to test various established turbulence models in calculations of high-Reynolds-number vortex-shedding flows. One candidate was the $k-\varepsilon$ turbulence model as this is presently the most widely used model in practice. The preliminary study of Franke et al. (1989) evaluating the detailed experiments of Cantwell and Coles (1983) has shown already that this eddy-viscosity model is unlikely to yield satisfactory results since it does not take proper account of history and transport effects which are of eminent importance in unsteady vortex-shedding flows. A Reynolds-stress-equation model was therefore considered necessary to account for the transport and history effects, and hence the model of this type due to Launder, Reece and Rodi (1975) was also tested. Two approaches were used for the treatment of near-wall regions, namely the wall-function approach and the two-layer approach in which the viscosity-affected near-wall region is resolved with a simpler one-equation model. The test calculations were carried out for two-dimensional vortex-shedding flow past a square cylinder. For this, detailed measurements have recently become available for a Reynolds number of Re = 22.000 (Lyn, 1990).

2. Calculation Procedure

2.1 Mean-Flow Equations

In flows with periodic unsteadiness, an instantaneous quantity f can be separated into (see Fig. 1)

$$f = \underbrace{\bar{f} + \tilde{f}}_{<f>} + f' \tag{1}$$

where $\bar{f}$ is the time-mean value, $<f>$ the ensemble-average value, $\tilde{f}$ the periodic fluctuation and f' the stochastic turbulent fluctuation. In the calculations presented here, equations governing the temporal and spatial variation of the ensemble-averaged velocity components and pressure representing the mean flow are solved. These are the ensemble-averaged continuity and momentum equations, which in tensor notation read as follows:

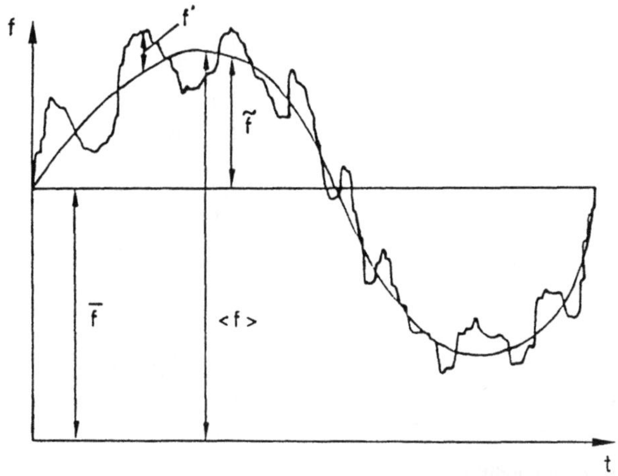

Fig. 1. Definition of mean and fluctuating values in periodic unsteady motion

$$\frac{\partial <u_i>}{\partial x_i} = 0 \tag{2}$$

$$\frac{\partial <u_i>}{\partial t} + <u_j>\frac{\partial <u_i>}{\partial x_j} = -\frac{1}{\rho}\frac{\partial <p>}{\partial x_i} + \frac{\partial}{\partial x_j}\left[\nu\frac{\partial <u_i>}{\partial x_j} - <u_i'u_j'>\right] \tag{3}$$

In the present 2D calculations, only the velocity components $<u>$ and $<v>$ are present and all quantities depend only on x and y (see Fig. 2 for coordinate system). In the momentum equations, the correlations $<u_i'u_j'>$ between the fluctuating velocities appear which represent the Reynold stresses and need to be simulated by a statistical turbulence model. In laminar vortex-shedding flow at low Reynolds numbers, these terms are absent and the only stresses acting are the viscous ones. In the present work, the calculation procedure of Franke et al. (1990) for laminar flows was extended by incorporating various turbulence models into the numerical solution procedure.

2.2 Turbulence Models

Before the turbulence models used are introduced, the problem areas associated with the challenging vortex-shedding flow considered are discussed briefly with respect to turbulence modelling. Three sub-areas of the flow can be identified:

(i) Stagnation Flow

In front of a bluff body, a stagnation flow exists which behaves virtually like potential flow. Except very near the wall, stresses (viscous and turbulent) have very little influence on the mean-flow behaviour. However, any turbulence and associated stresses generated in this region influence the flow behaviour downstream, e.g. the separated shear layer on the upper and lower surface of a square cylinder. If turbulent fluctuations are present,

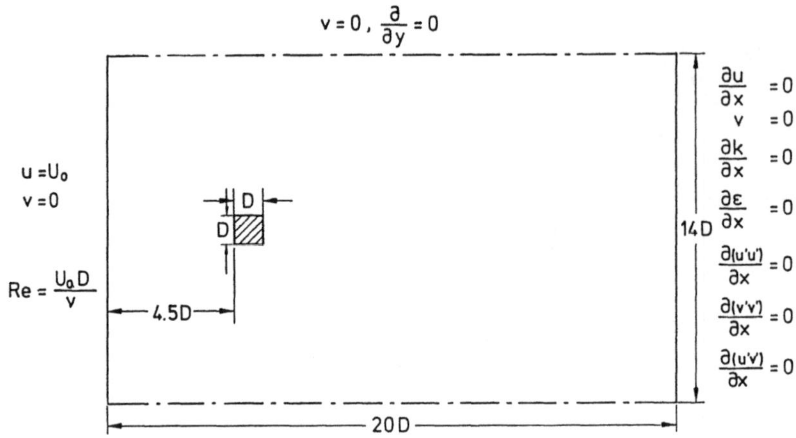

Fig. 2. Calculation domain and boundary conditions

there is no shear-stress production on the stagnation line but only normal-stress production which governs the turbulence level in this region. The normal-stress production depends on the separation between the normal stresses $<u'^2>$ and $<v'^2>$ which is grossly over-predicted by an isotropic eddy-viscosity model and can only be simulated realistically with a Reynolds-stress-equation model (Taulbee and Tran, 1988).

(ii) Attached and Detached Shear Layers on Side Walls

In the vicinity of the side walls of a square cylinder, a complex flow field develops with partly attached and partly detached shear layers, possible reattachment on the side wall, as well as laminar-turbulent transition. Here, inertial forces, viscous and turbulent stresses interplay, and the region is of considerable importance for the development of the vortex-shedding flow as here the origin of the vortex instability is located. The details of the flow mechanisms in this region are not well understood, and the phenomena are difficult to describe by a turbulence model. Because of the influence of viscous forces, wall functions are unlikely to be very suitable.

(iii) Wake Region

In the wake past the cylinder, convection and turbulent stresses dominate and the viscous forces are negligible. The vortices generated in the region discussed under (ii) are convected downstream forming a vortex street. The periodic vortex motion is damped by turbulent stresses. The modelling of turbulence in this region mainly affects the vortex frequency but not so much the amplitude of the forces acting on the body. History and transport effects are particularly important in this region so that an eddy-viscosity model is unlikely to do justice to this region (see Franke et al., 1989).

Two different models for simulating the turbulent stresses in the bulk of the flow away from walls were tested and are briefly introduced in the following. The two approaches used for treating the near-wall region are described thereafter. The actual turbulence-model equations are not given since only models were used which are fairly standard by now and have been described elsewhere (e.g. Rodi, 1980). For all equations, the ensemble-average forms were employed.

$k-\varepsilon$ model. The standard form of the $k-\varepsilon$ model has been applied to many steady flows and had some success also for calculating separated flows, even though the accuracy of the predictions was not as high as for shear-layer flows. Here, the standard $k-\varepsilon$ model is simply adopted for ensemble-averaged quantities. The Reynolds stresses are calculated from the eddy-viscosity relation

$$-<u_i'u_j'> = <\nu_t> \left(\frac{\partial <u_i>}{\partial x_j} + \frac{\partial <u_j>}{\partial x_i} \right) - \frac{2}{3} <k> \delta_{ij} \tag{4}$$

The eddy viscosity $<\nu_t>$ is related to the kinetic energy $<k>$ and its dissipation rate $<\varepsilon>$and the spatial and temporal distribution of these quantities is determined from model transport equations. The standard values were adopted for the empirical constants (see Rodi, 1980).

Reynolds-Stress Equation (RSE) Model. In order to account better for the important transport and history effects in vortex-shedding flows, a second-order closure model was tested which does not employ the eddy-viscosity hypothesis but determines the individual turbulent stresses $<u_i'u_j'>$ from transport equations. The standard model of Launder, Reece and Rodi (1975) was adopted, with wall corrections to the pressure-strain terms due to Gibson and Launder (1978). The wall corrections effect damping of the normal fluctuations near the wall and enhance the fluctuations parallel to the wall. Again, standard values were taken for the empirical constants (see Franke, 1991).

Near-Wall Treatment. For both turbulence models described above, two approaches were tested for handling the near-wall region. Wall functions are still used in most practical calculations and were therefore tested also here. In this approach, the viscous sublayer is not resolved but the first grid point in a numerical calculation is located outside this layer. Based on the assumption of the logarithmic velocity distribution and of local equilibrium of turbulence (production = dissipation) the flow quantities at this grid point with wall distance y_c are related to the friction velocity $u_\tau = (\tau_{\text{wall}}/\rho)^{1/2}$ by

$$u_c = \frac{u_\tau}{\kappa} \ln(y^+ E) \qquad k_c = \frac{u_\tau^2}{\sqrt{c_\mu}} \qquad \epsilon_c = \frac{u_\tau^3}{\kappa y_c} \tag{5}$$

Here, $y^+ = u_\tau y/\nu$ is a dimensionless wall distance, κ is the von Karman constant and E is a roughness parameter. In connection with the RSE model, the shear stress at point c is equal to u_τ^2, and invoking local equilibrium conditions on the stress equations yields the normal stresses at the first grid point.

The assumptions of a logarithmic velocity distribution and of local equilibrium of turbulence are violated in separated flows (see e.g. Rodi, 1991), especially near separation and reattachment regions. One way around this problem would be to use low-Reynolds-number versions of the models employed away from the wall. For the $k-\varepsilon$ model, a variety of such versions is available, but they have virtually all been tested only for attached boundary layers. For RSE models, low-Re versions are still under development and not yet ready for application to a complex flow. In both cases, such model versions require very fine numerical resolution near the wall which has an adverse effect on the computational effort required. For these reasons, a two-layer approach was adopted here in which the viscosity-affected near-wall region is resolved by a simpler one-equation model, which does not solve a dissipation-rate equation but uses a prescribed length-scale distribution and hence requires less numerical resolution near the wall. The one-equation model is then matched to the model used away from the wall. In calculations with both the $k-\varepsilon$ model and the RSE model away from the wall, the one-equation model due to Norris-Reynolds (1975) was adopted near the wall. The $k-\varepsilon$-based two-layer model was already tested quite extensively for steady flows, including separated flows (see Rodi, 1991). The Norris-Reynolds model employs the eddy-viscosity hypothesis and calculates the eddy viscosity from:

$$<\nu_t^{NR}> = f_\mu c_\mu \sqrt{<k>} L, \qquad f_\mu = 1 - e^{(-.0198\sqrt{<k>} y_w/\nu)} \tag{6}$$

where f_μ is a damping function similar to the van Driest function accounting for the reduction of turbulent momentum exchange very near the wall. The kinetic energy $<k>$ is obtained from the same $<k>$-equation as used in the $k-\varepsilon$ model, while the length scale L is assumed to vary linearly with distance from the wall as $L = \kappa y_w / c_\mu^{3/4}$ in the very thin near-wall layer where the model is applied ($\kappa = 0.4187, c_\mu = 0.09$). The dissipation rate $<\varepsilon>$ appearing as source term in the $<k>$-equation is not obtained from an equation but calculated from the following algebraic relation:

$$<\epsilon> = \frac{<k>^{3/2}}{L}\left[1 + \frac{2\kappa^2 \nu}{c_\mu^{1,5}\sqrt{<k>} L}\right] \tag{7}$$

When used in combination with the RSE model, only the shear stress in the viscous sublayer is calculated with the aid of an eddy viscosity relation according to (6). The normal stresses appearing in the momentum equations are determined by assuming $<u_i' u_j'>/<k>$ = constant across the viscous sublayer. The ratio taken is the one resulting from the RSE model at the matching point between this model and the one-equation model, while $<k>$ is taken from the k-equation. This approach is certainly very crude and does not account for the damping of the normal fluctuations very near the wall, but in the thin region where the one-equation model is employed the flow has boundary-layer character and the normal stresses are small compared with the shear stress so that they do not need to be determined very accurately.

The matching between the one-equation model and the model used further away from the wall should be effected near the edge of the viscous sublayer where viscosity effects have become small. This is the case when the damping function f_μ in (6) has a value near unity, and in the calculations presented the matching was done where $f_\mu = 0.95$. For fully turbulent boundary layers this corresponds to $y^+ \sim 83$. For good near-wall resolution, the first grid point should be located at $y^+ \sim 1$. When the viscous sublayer is resolved with the one-equation model, the no-slip conditions are used as boundary conditions, i.e. both velocity components and the turbulent kinetic energy $<k>$ are zero.

2.3 Numerical Solution Procedure

The differential equations governing the mean-flow and the turbulence field were solved numerically with an extended version of the program TEACH developed by Gosman and Pun (1973). This employs a finite-volume method for solving the equations in primitive variables on a two-dimensional staggered grid. The coupling between continuity and momentum equations was achieved with the SIMPLEC predictor-corrector algorithm of van Doormal and Raithby (1984), which is an improved version of the SIMPLE algorithm incorporated in the original TEACH program. The central/upwind hybrid spatial discretization scheme in the original TEACH program was replaced by the QUICK scheme (Quadratic Upwind Interpolation for Convective Kinematics) proposed by Leonard (1979). This scheme combines the high accuracy of a third-order scheme with the stabilising effect of upwind weighting. A disadvantage of the scheme is its unboundedness, which may cause over- and undershoots. It should be mentioned that central/upwind hybrid differencing introduced too much damping so that no unsteady vortex shedding resulted. For time discretization, the fully implicit first-order Euler scheme was chosen. It provides high stability but requires small time steps in order to obtain accurate solutions (more than 100 time steps per period were used). The resulting system of linear difference equations was solved by the strongly implicit method of Stone (1968). A more detailed description of the numerical method is given in Franke (1991). The computational domain in which the equations were solved and the outer boundary conditions are shown in Fig. 2. In connection with wall functions, a 70 x 64 grid was used with $\Delta y_c/D = 0.022$. For calculations with the two-layer approach, in which the viscosity-affected sublayers are resolved, a 186 x 156 grid was used with the first point at a distance from the wall of $\Delta y_c/D = 0.00125$. On an IBM 3090, the calculation time to cover one shedding period was typically 1/2 hour for the RSE model with wall functions and 5 hours for the two-layer RSE-model.

3. Calculation Results and Comparison with Experiments

3.1 Experimental Data

The availability of experimental data is crucial for the testing of turbulence models. In the case of unsteady vortex-shedding flows, a proper testing of the models can only be done when results of time-resolved measurements are available which give information on the temporal variation of ensemble-averaged quantities and separate the fluctuations into periodic and turbulent ones. Such detailed data on flows past bluff bodies are rare, but recently Lyn (1990) has completed time-resolved two-channel LDA measurements in the wake past a square cylinder at $Re = 22.000$. In these experiments, the shedding period was divided into 20 phase angles, for which the velocity-component and Reynolds-stress distributions were reported as ensemble-averaged values in a plane perpendicular to the cylinder. The turbulence level in the oncoming flow was 2%. The only other fairly detailed measurements of the flow past a square cylinder known to the authors are those of Durao et al. (1988) for $Re = 14.000$. However, these authors obtained only time-averaged data and did not separate the fluctuations into periodic and turbulent components. The flow situation investigated by Lyn (1990) was therefore taken as test case.

3.2 Integral Parameters and Time-Mean Distributions

The calculations with the $k-\varepsilon$ model using wall functions yielded a steady solution and no vortex shedding. Apparently this combination of model and near-wall treatment introduces too strong damping so that a periodic vortex-shedding motion cannot persist. Periodic vortex shedding was obtained with the other three model variants, and a sequence of streamlines covering approximately one period is shown in Fig. 5 below. Values of various integral parameters predicted with the turbulence-model variants tested are compared in Table 1 with experimental values. The parameters considered are the dimensionless shedding frequency (Strouhal number $St = fD/U_0$), the time-averaged drag coefficient $\overline{c_D}$ and the amplitudes of the oscillations in drag coefficient c_D and lift coefficient c_L. The $k-\varepsilon$ model yields too low shedding frequency and drag coefficient, the latter however being somewhat higher than that obtained in the stationary solution with wall functions. The drag coefficient does not oscillate, but there is a small oscillation in the lift coefficient. Results obtained with a modified two-layer $k-\varepsilon$ model are also included in Table 1. In this model, the unrealistic normal-stress production of k in the stagnation region due to the use of an isotropic eddy-viscosity model, which leads to excessive k-levels (see Fig. 5 below) is suppressed in front of the body. This measure reduces significantly the k-level in the front part and influences mainly the vortex formation regions near the side walls of the cylinder. The prediction of the integral parameters is improved somewhat by this measure. Further improvement is achieved by use of the RSE models. In combination with wall functions, the predicted Strouhal number and drag coefficient are now in good agreement with the measured ones. When the two-layer variant is used, these quantities are predicted too high; by switching from wall functions to a two-layer approach, the shedding frequency increases while the amplitude of the oscillating forces acting on the body is reduced. This detail of the calculation is not well understood at the present time. Consistent with the increase in $\overline{c_D}$ when moving from the standard $k-\varepsilon$ model to the two-layer RSE model is a reduction of the base pressure. The distribution of the time-mean pressure along the cylinder walls is compared in Franke (1991) with various experiments carried out at different Reynolds numbers and free-stream turbulence levels. The experimental results themselves differ by about 15 to 20%. The base pressure predicted by the RSE model with wall functions falls between the measurement results, while the $k-\varepsilon$ model predictions yield too high and the predictions with the two-layer RSE model too low base pressures.

Table 1. Integral Parameters

	Turbulence models					Experiments
	k-ε with wall functions	two-layer $k-\varepsilon$	mod. two-layer $k-\varepsilon$	RSE with wall functions	two-layer RSE	
St	stationary	0.124	0.129	0.136	0.159	Lyn 0.135 Durao 0.139
$\overline{c}_D$	1.55	1.79	1.89	2.15	2.43	2.05-2.23
$\tilde{c}_D$	-	0.0	0.0	0.383	0.079	-
$\tilde{c}_L$	-	0.323	0.883	2.11	1.84	-

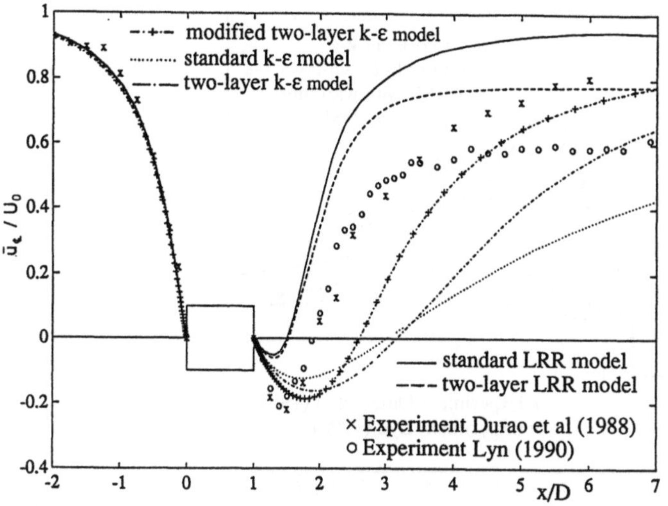

Fig. 3. Time-mean velocity $\bar{u}$ along a centre line

Fig. 3 displays the distribution of the time-mean velocity $\overline{u}$ along the centre-line and gives information on the length of the time-averaged separation zone behind the cylinder. Experimental data from Lyn's (1990) and Durao et al.'s (1988) experimental study are included. The data agree fairly well in the near-cylinder region, but the centreline velocity measured by Lyn approaches the free-stream velocity much slower than that measured by Durao et al. The reasons for this difference and for the very slow increase of the centreline velocity in Lyn's experiments are not clear. While, as expected, the velocity distribution in front of the cylinder is influenced very little by the turbulence model used, there are fairly large differences in the wake region, and in particular regarding the length of the separation zone. The $k-\varepsilon$ model variants overpredict the length of this zone considerably, indicating that there is not enough momentum exchange introduced by these models. Since in vortex-shedding flows the momentum exchange is mainly due to the periodic vortex-shedding motion (see e.g. Cantwell and Coles 1983, Franke 1991), this also means that the periodic fluctuations are underpredicted significantly, as will be confirmed shortly. The modified $k-\varepsilon$ model predicts a somewhat shorter separation zone, pointing to a more realistic simulation of the periodic vortex-shedding motion. Both RSE model variants predict on the other hand too short separation bubbles, and there is little difference between the results obtained with the two versions near the cylinder; there are larger differences further downstream but these should not be taken too seriously since here there may be an influence of the different numerical grids used, which are quite coarse in this downstream region.

The distribution of the total (periodic plus turbulent) fluctuating kinetic energy along the centreline shown in Fig. 4 gives a clue to the results discussed so far and in particular to the differences between the various turbulence models. The two-layer $k-\varepsilon$ model can be seen to underpredict severely the fluctuation level behind the cylinder, while the RSE models give approximately the correct level and distribution of the total fluctuations. The differences between the RSE calculations using wall functions and the two-layer

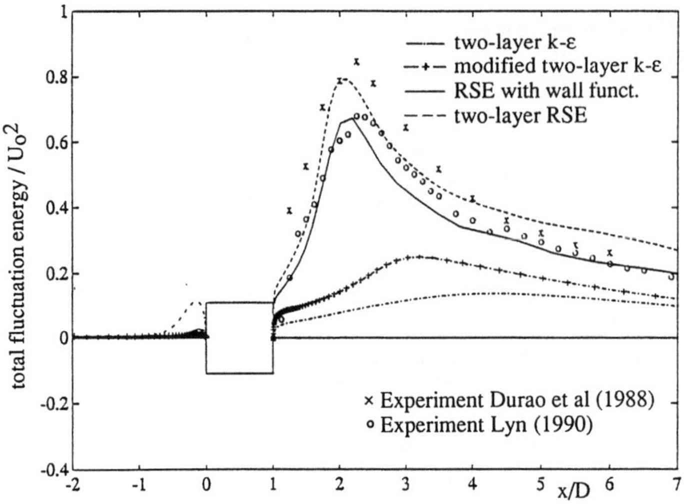

Fig. 4. Total kinetic energy of fluctuations (periodic + turbulent) along centre line

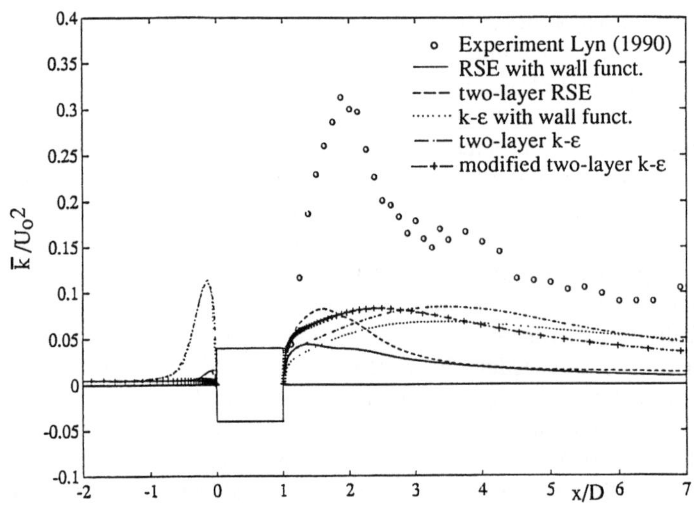

Fig. 5. Time-mean turbulent kinetic energy $\bar{k}$ along centre line

approach are of the same order as the differences between the experimental values of Lyn (1990) and Durao et al. (1988), but the two-layer version yields a higher fluctuation level, which is consistent with the higher Strouhal number and drag coefficient predicted. The modified $k-\varepsilon$ model, in which the k-production in front of the body is suppressed, yields a somewhat improved fluctuation level which, however, is still far below the measured one. This shows that the problems with the $k-\varepsilon$ model originate only to a small extent from the excessive k-production by this model in the stagnation region.

The effect of this excessive production can be seen best in Fig. 5, which displays the distribution of the time-mean turbulent kinetic energy $\overline{k}$ along the centreline. Fairly large k-levels are produced right in front of the cylinder which are unrealistic. When the RSE models are used, much lower k-levels result in the stagnation region. Behind the cylinder, all models can be seen to underpredict considerably the level of turbulence energy on the centreline and particularly so in and near the separation region. Since the RSE models predict the correct level of total fluctuation energy, this means that they overpredict the periodic fluctuations. The $k - \varepsilon$ models yield higher turbulent fluctuation levels in the wake than the RSE models and this may cause a damping of the periodic fluctuating motion and in turn the observed underprediction of this motion. The use of the two-layer approach in connection with the RSE model increases the turbulent fluctuations in the separation region. This may cause the reduced amplitudes in the drag and lift forces on the cylinder, but apparently on the whole the periodic fluctuations are increased rather than reduced as can be seen from Fig. 4 and can also be inferred from the increased Strouhal number and mean drag coefficient.

3.3 Phase-Dependent Results

Since the Reynolds-stress models gave much more realistic fluctuation levels than the $k - \varepsilon$ model, phase-dependent results are only presented for the calculations with the RSE models. Fig. 6 shows the streamlines predicted with the RSE model using wall functions at three phases; at two of the phases experimentally determined streamlines are available and are also given for comparison. The streamlines predicted with the two-layer RSE model are not significantly different. The alternating vortex shedding can be seen clearly, and there is reasonable agreement about the streamline development with the experiments. However, certain differences can be noted. At some phases, the predictions do not show a negative $\langle u \rangle$-velocity on the leeward side of the cylinder while the experiments indicate that the velocity is negative at all times (for further evidence see Franke, 1991). Also, the model predicts temporary reattachment of the separated flow on the side walls near the rear corner (e.g. phase 1, topside) which was not found in the experiments. This can be seen more clearly in Fig. 7 where the $\langle u \rangle$-velocity distribution at the x-position of the rear cylinder wall is compared with measurements. Here, in addition to Lyn's (1990) two-channel measurements also Lyn's (1989) one-channel measurements extending closer to the wall are included. At the two phases considered (1 and 9) the calculations do not show negative velocities on one of the side walls while in the experiments the velocity very near the wall is always negative. Considering the results at all phases (which could not be included here) the two-layer RSE model gives slightly better agreement with the measurements than the model using wall functions. However, the discrepancies point to a weakness of the model in underpredicting the size of the separation zone near the side walls. Fig. 8 presents the variation of the ensemble-averaged lateral velocity $\langle v \rangle$ along the centreline for two phases. The agreement with the measurements is fairly good, both with respect to the maximum values and to the zero crossings, the latter being a measure of the size of the shed vortices. There is no significant difference here between the model

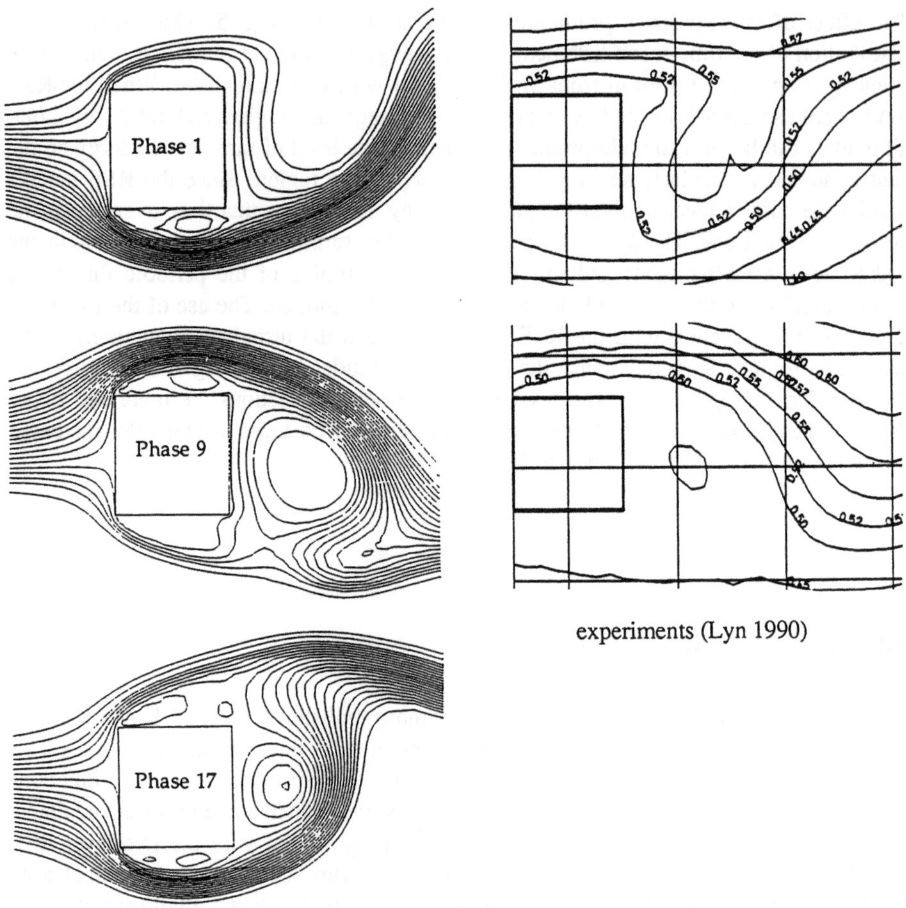

Fig. 6. Streamlines at 3 phases (Phase 1=1/20 T, Phase 9=9/20 T, Phase 17=17/20 T)

using wall functions and the two-layer approach. Similar distributions for the velocity component $<u>$ as given in Franke (1991) show inferior agreement with experiments, firstly because in the downstream region the time-mean u-velocity is overpredicted, as can be seen from Fig. 3, but also because near the rear cylinder wall the velocity is not always negative as it was found in the experiments.

Finally, Fig. 9 compares calculated and measured $<k>$-contours at one phase. The differences between calculations obtained with the two RSE variants are not very significant compared with the differences to the experiments. Of course, the asymmetry in the contour lines is basically the same in the calculations and experiments, as must be expected from the similarity of the streamlines, but the maxima of $<k>$ in the experimental data are located much closer to the centreline than in the calculations. This points to a weakness of the present calculation model which appears to predict the centres of the vortices considerably further from the centreline than the experiments indicate. This is supported by a comparison of the RMS values of the periodic fluctuations ($\tilde{u}$) on the

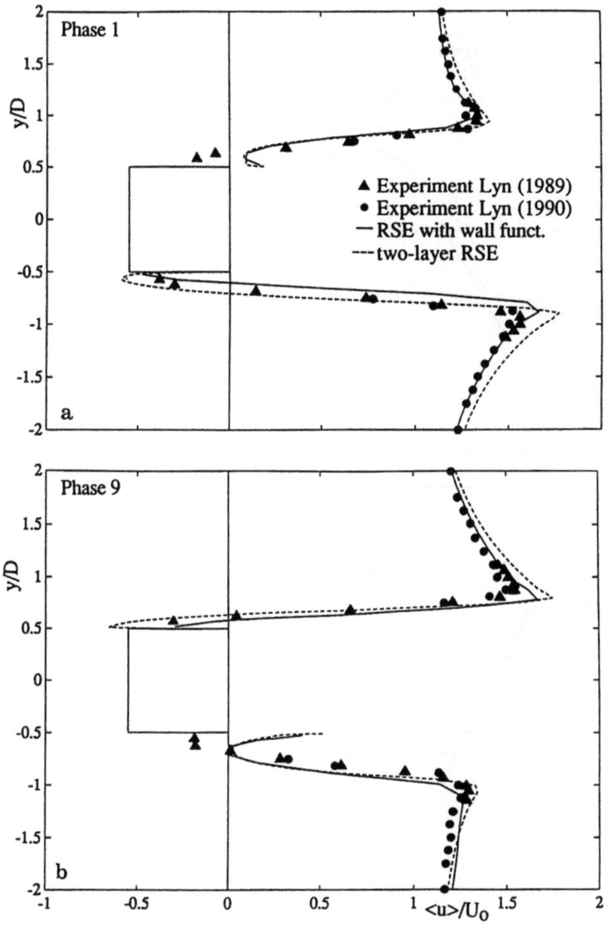

Fig. 7. Profiles of $<u>$ at x-location of rear cylinder wall for 2 phases

centreline, which are virtually zero in the experiment and take on a sizeable value in the predictions (Franke, 1991).

4. Conclusions

Time-dependent calculations of vortex-shedding flow past a square cylinder were carried out with various turbulence models and near-wall treatments. The $k-\varepsilon$ eddy-viscosity model combined with wall functions introduces too much damping so that no vortex shedding is predicted. With the two-layer approach, the model does predict vortex shedding but the periodic motion is considerably too small so that there is too little momentum

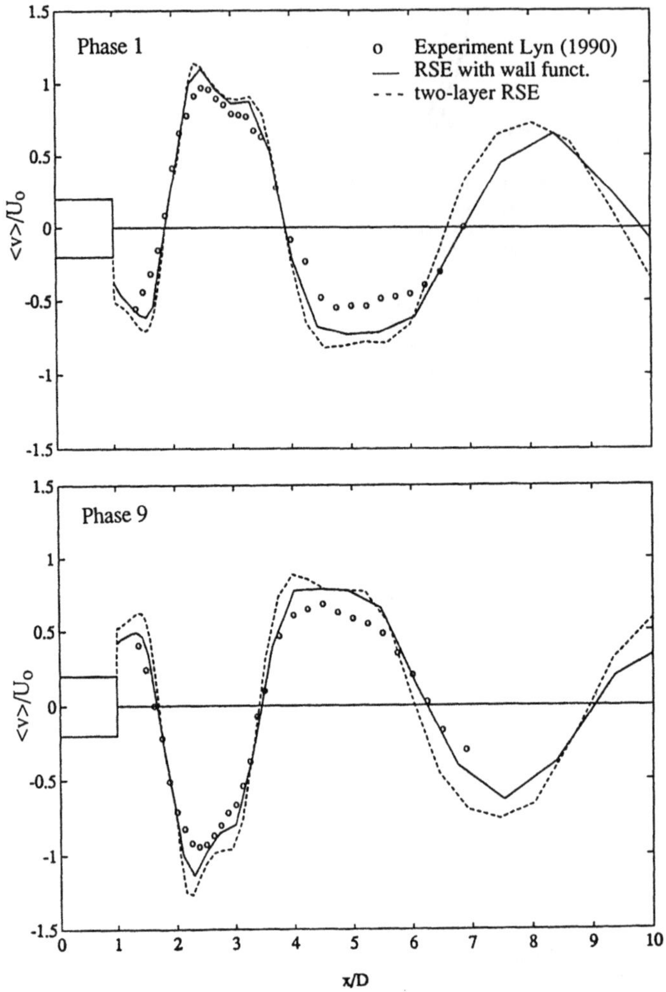

Fig. 8. Lateral velocity $<v>$ along centre line at 2 phases

transport resulting in a too long time-mean separation zone and too low values of the Strouhal number and drag coefficient. The RSE models predict the correct level of total fluctuations, but overpredict the periodic component while they considerably undepredict the level of turbulence fluctuations in the wake. Concerning the Strouhal number and mean drag coefficient, the RSE model with wall functions yields the best agreement with the experiments while the two-layer RSE model produces too high values. Concerning other flow features there is not too much difference in the calculations obtained with the two model variants. Both versions yield a reattachment on the side walls and positive u-velocity near the rear wall at certain times during one period which was not observed in the experiments. These predicted features are probably due to an overprediction of the periodic fluctuations. Also, the vortex centres are calculated too far from the centre line

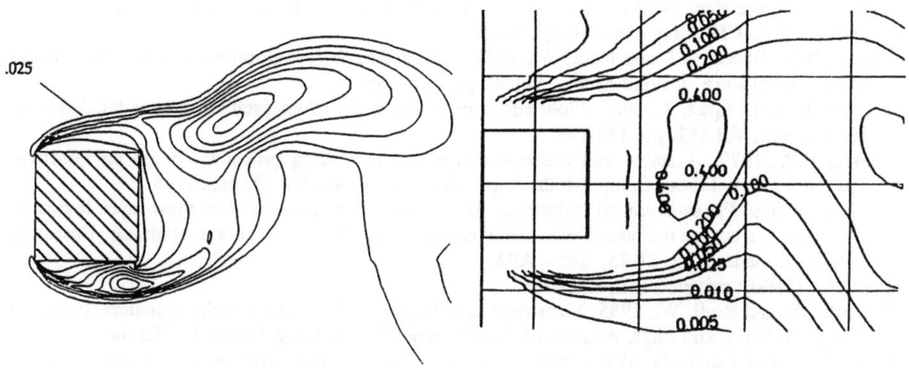

calculations, 2-layer RSE

experiments (Lyn 1990)

Fig. 9. <K> contours at phase 1

of the cylinder. A refinement of the RSE model making it more responsive to unsteady effects may improve the calculations. However, the limits of conventional turbulence models are probably reached in this flow with large-scale vortex structure and a better reproduction of the details probably calls for a large-eddy simulation.

Acknowledgements

The work reported here was sponsored by the Deutsche Forschungsgemeinschaft through the Sonderforschungsbereich 210. The calculations were carried out on the IBM 3090 computer of the University of Karlsruhe. The authors are grateful to Mrs. R. Zschernitz for her expert typing of the manuscript.

References

Cantwell, B. and Coles, D., 1983, An experimental study of entrainment and transport in the turbulent near-wake of a circular cylinder. J. Fluid Mech., Vol.**136**, pp. 321-374.

Durao, D.F.G., Heitor,M.V.,Pereira,J.C.F., 1988: Measurements of turbulent and periodic flows around a square cross-section cylinder. Experiments in Fluids, Vol.**6**, pp.298-304.

Franke, R., 1991, Numerische Berechnung der instationären Wirbelablösung hinter zylindrischen Körpern, Ph.D. Thesis, University of Karlsruhe.

Franke, R., Rodi, W. and Schönung, B., 1989, Analysis of experimental vortex-shedding data with respect to turbulence modelling, Proc. 7th Turbulent Shear Flow Symposium, Stanford, USA.

Franke, R., Rodi, W. and Schönung, B., 1990, Numerical calculation of laminar vortex-shedding flow past cylinders, J. Wind Engineering and Industrial Aerodynamics, Vol.**35**, pp. 237-257.

Gibson, M.M. and Launder, B.E., 1978, Ground effects on pressure fluctuations in the atmospheric boundary layer, J. Fluid Mech., Vol.**86**, p- 491.

Gosman, A.D. and Pun, W.M., 1973, Calculation of recirculating flows, Lecture Notes for Course entitled "TEACH", Imperial College, London.

Launder, B.E., Reece, G.J. and Rodi, W., 1975, Progress in the development of a Reynolds-stress turbulence closure, J. Fluid Mech., Vol.**68**, pp. 537-566.

Lecointe, Y. and Piquet, J., 1989, Flow structure in the wake of an oscillating cylinder, J. Fluids Engineering, Vol.**111**, pp. 139-148.

Leonard, B.P., 1979, A stable and accurate convective modelling procedure based on quadratic upstream interpolation, Comp. Meth. Appl. Mech. and Eng., Vol.**19**, pp. 59-98.

Lyn, D.A., (1989):Phase-averaged turbulence measurements in the separated shear-layer region of flow around a square cylinder. Proc. 23rd congress of the Int. Association f. Hydraulic Research, Ottawa, Canada, Aug. 21-25, A85 - A92.

Lyn, D.A., (1990): private communication.

Majumdar, S. and Rodi, W., 1985, Numerical calculations of flow past circular cylinders, Proc. 3rd Symp. on Num. and Phys. Aspects of Aerodynamic Flows, Long Beach, California.

Norris, H.L. and Reynolds, W.C., 1975, Turbulent channel flow with moving wavy boundary, Department Mech. Eng. Rept. FM-10, Stanford University.

Rodi, W., 1980, Turbulence Models and their Application in Hydraulics, Int. Association for Hydraulic Research, Delft, The Netherlands.

Rodi, W., 1991, Experience with two-layer turbulence models combining the $k - \varepsilon$ model with a one-equation model near the wall, AIAA paper 91-0216.

Stone, H.L., 1968, Iterative solution of implicit approximations of multidimensional partial differential equations, SIAM J. Num. Anal., Vol.3, No. 3, pp. 530-558.

Tamura, T., Ohta, I. and Kuwahara, K., 1990, On the reliability of two-dimension simulation for unsteady flows around a cylinder-type structure, J. Wind Engineering and Industrial Aerodynamics, Vol.**35**, pp. 275-298.

Taulbee, D.E. and Tran, L, 1988, Stagnation streamline turbulence, AIAA J., Vol.**26**, pp. 1011-1013.

van Doormal, J.P. and Raithby, R.D., 1984, Enhancement of the SIMPLE method for predicting incompressible fluid flows, Num. Heat Transfer, Vol.**7**, pp. 147-163.

Second-Moment Modelling of Recirculating Flow with a Non-Orthogonal Collocated Finite-Volume Algorithm

F.-S. Lien, M.A. Leschziner

Department of Mechanical Engineering, University of Manchester, Institute of Science and Technology, Manchester, UK

Abstract

The incorporation of Reynolds-stress closure into a non-orthogonal, collocated finite-volume framework in which the discretisation of convection is non-diffusive, presents a number of algorithmic problems not encountered in more traditional schemes employing staggered, rectilinear volume arrangements. Three issues requiring special consideration are: the tensorially correct incorporation of the wall-related pressure-strain terms which are important fragments in the stress closure, boundary conditions at curved walls, and iterative stability. The first issue, in particular, arises because the wall-related terms are tied to the orientation of the wall relative to the directions of the Reynolds stresses. The paper reports practices which address all three problem areas. Four complex applications are presented, among them the flow through a sinusoidal pipe constriction and shock-induced separation over a channel bump.

1. Introduction

Over the past few years, CFD for general turbulent flow has advanced along two major fronts without significant cross-links. Numerically, there has been a dramatic shift away from simple rectilinear geometries towards complex domains requiring the use of body-fitted meshes. A number of strategies have been pursued, but the one which seems to emerge as offering the best compromise between flexibility and algorithmic simplicity adopts a Cartesian or cylindrical-polar velocity decomposition within a non-orthogonal finite-volume framework. Such schemes have been developed, for example, by Peric (1985), Shyy & Braaten (1986) and Majumdar et al (1989). Moreover, a number of commercial packages – for example, FLOW3D and FLUENT/BFC – are based on this approach. Physically, there has been a similarly strong shift away from eddy-viscosity models towards second-moment closure. This shift has been motivated, principally, by the realisation that only the latter type of model is able to capture, in any fundamentally meaningful sense, the interaction between curvature-related strain and turbulence transport.

The importance of curvature in all recirculating and swirling flows naturally favours the combination of second-moment models with elliptic-flow solvers. While this turns out to be a non-trivial task, such models have been successfully incorporated into a number of recirculating-flow algorithms [e.g. Jones & Marquis (1985), Gaskell & Lau (1987), Jones & Manners (1988) and Lin & Leschziner (1989)]. However, with very few exceptions, such schemes are applicable to Cartesian or cylindrical-polar meshes only, and most adopt a staggered finite-volume arrangement. One procedure based on the collocated approach is that of Obi et al (1989), but this applies to regular grids only, although Obi (1991) has recently presented an extension of the method to irregular grids. Finally, a scheme by Sebag & Laurence (1990) adopts a non-orthogonal, semi-staggered formulation.

Combining second-moment closure with a non-orthogonal, collocated scheme is a particularly challenging task. Apart from 'ordinary' complexities associated with non-orthogonality, the main difficulties arise from boundary conditions, the awkward tensorial nature of the highly influential wall-reflection terms in the pressure-strain components of the Reynolds-stress equations, and from the fact that storage of all stresses at the same spatial location is a strong source of numerical instability or physically unrealistic oscillatory behaviour.

This paper presents practices directed towards all three problem areas, realised within a recently developed non-orthogonal-grid methodology combining second-moment closure, non-diffusive discretisation and multigrid convergence acceleration. That same methodology has also been extended to compressible flow, including transonic conditions, and one of the applications presented below relates to shock-induced separation from a channel bump.

2. Basic Numerical Framework

The discretisation process is preceded by a transformation of the Cartesian forms of the flow-governing equations to forms involving spatial derivatives along the non-orthogonal co-ordinates and the Jacobian transformation matrix. Because the velocity components are chosen to remain Cartesian (or, more generally, cylindrical-polar, toroidal or adapting to an arbitrary, user-defined datum line), they can be treated as scalars, a practice circumventing the complexities associated with fully covariant differentiation of vectors and tensors.

With ξ and η representing the non-orthogonal directions and r being the radial co-ordinate, the transport equation for any scalar property ϕ can be written in the following form:

$$\frac{\partial}{\partial \xi}[\rho r U \phi - \frac{r^2}{(rJ)} q_1^{\phi} \frac{\partial \phi}{\partial \xi}] + \frac{\partial}{\partial \eta}[\rho r V \phi - \frac{r^2}{(rJ)} q_2^{\phi} \frac{\partial \phi}{\partial \eta}] = rJS^{\phi}, \tag{1}$$

where U and V are contravariant velocity components,

$$U = u y_\eta - v x_\eta, \qquad V = v x_\xi - u y_\xi, \tag{2}$$

the subscripts ξ and η denote partial differentiation with respect to the associated directions, J is the Jacobian of the transformation, rJ is the ratio of corresponding volumes (areas) in physical and transformed (ξ, η) space, and S^{ϕ} contains diffusive terms involving mixed derivatives, sources and sinks, including pressure terms in the case of momentum. For isotropic diffusivity Γ_{ϕ}, the q^{ϕ}-terms are:

$$q_1^\phi = \Gamma_\phi(y_\eta^2 + x_\eta^2), \qquad q_2^\phi = \Gamma_\phi(y_\xi^2 + x_\xi^2). \tag{3}$$

The above terms will be redefined later in the context of anisotropic turbulence transport.

Next, the equation is integrated over the volume shown in Fig. 1 to yield, after application of the Gauβ Divergence Theorem, a balance of face fluxes and volume-integrated net source:

$$\begin{aligned}(\rho rU)_e\phi_e - (\rho rU)_w\phi_w + (\rho rV)_n\phi_n - (\rho rV)_s\phi_s = rJS^\phi \\ +[(\frac{rq_1^\phi}{J})_e + (\frac{rq_1^\phi}{J})_w + (\frac{rq_2^\phi}{J})_n + (\frac{rq_2^\phi}{J})_s]\phi_P \\ -[(\frac{rq_1^\phi}{J})_e\phi_E + (\frac{rq_1^\phi}{J})_w\phi_W + (\frac{rq_2^\phi}{J})_n\phi_N + (\frac{rq_2^\phi}{J})_s\phi_S]\end{aligned} \tag{4}$$

The convection fluxes are approximated either by the 'Hybrid' (upwind/central-) differencing scheme of Patankar (1980) or by the quadratic scheme QUICK of Leonard (1979) or by van Leer's TVD scheme MUSCL (1979). The last scheme is also used for all turbulence-model equations.

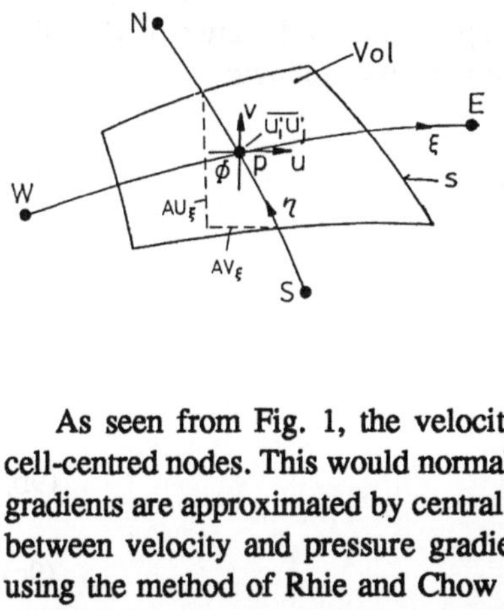

Fig. 1. Finite volume and storage arrangement

As seen from Fig. 1, the velocity components and pressure are stored together at cell-centred nodes. This would normally lead to *chequerboard oscillations* when pressure gradients are approximated by central differencing, reflecting a computational decoupling between velocity and pressure gradient. The difficulty may be overcome, however, by using the method of Rhie and Chow (1983) which, essentially, introduces a measure of artificial fourth-order pressure smoothing. For the east and north faces of the cell shown in Fig. 1, the Rhie and Chow interpolation may, for plane flow, be written as:

$$\begin{aligned}U_e = [(\overline{u - DU_\xi\,\delta_\xi p})_e + \overline{DU_{\xi,e}}(p_P - p_E)]y_\eta \\ -[(\overline{v - DV_\xi\,\delta_\xi p})_e + \overline{DV_{\xi,e}}(p_P - p_E)]x_\eta\end{aligned} \tag{5}$$

$$\begin{aligned}V_n = [(\overline{v - DV_\eta\,\delta_\eta p})_n + \overline{DV_{\eta,n}}(p_P - p_N)]x_\xi \\ -[(\overline{u - DU_\eta\,\delta_\eta p})_n + \overline{DU_{\eta,n}}(p_P - p_N)]y_\xi\end{aligned} \tag{6}$$

where

$$\delta_\xi p_P = (p_w - p_e); \qquad \delta_\eta p_P = (p_s - p_n); \tag{7}$$

overbars denote averages, evaluated at faces 'e' and 'n', respectively, from values pertaining to volumes which abut on the face in question; DU_ξ, DV_ξ – say at point P – are projected areas, AU_ξ and AV_ξ, identified in Fig. 1, divided by the coefficient multiplying the velocity u_P in the final form of the finite-volume momentum equation (4) (i.e. for $\phi = u_P$).

The pressure itself is determined by using the SIMPLE pressure-correction algorithm of Patankar (1980). This is modified to include a *density-retardation* methodology, leaning on a *pressure retardation* practice by McGuirk & Page (1989) and permitting an accurate capturing of shocks [Lien & Leschziner (1992)]. One example presented later has been computed with this methodology.

3. Implementation of Second Moment Closure

3.1 Present Closure Form

The closure adopted herein is a variation of the high-Reynolds-number Reynolds-stress model of Gibson and Launder (1978), the point of difference relative to the "standard" form being the manner in which stress diffusion is approximated. In terms of Cartesian tensor notation, for plane flows, this closure may be written as follows:

$$\frac{\partial \rho u_k \overline{u_i' u_j'}}{\partial x_k} = D_{ij} + P_{ij} - \frac{2}{3}\delta_{ij}\rho\epsilon + \Phi_{ij}, \tag{8}$$

in which

$$D_{ij} = \frac{\partial}{\partial x_k}\left[\frac{c_\mu \rho k^2}{\sigma_k \epsilon}\frac{\partial \overline{u_i' u_j'}}{\partial x_k}\right], \tag{9a}$$

$$P_{ij} = -\rho\overline{u_i' u_k'}\frac{\partial u_j}{\partial x_k} - \rho\overline{u_j' u_k'}\frac{\partial u_i}{\partial x_k}, \tag{9b}$$

$$\Phi_{ij} = \Phi_{ij1} + \Phi_{ij2} + \Phi_{ijw1} + \Phi_{ijw2}, \tag{9c}$$

$$\Phi_{ij1} = -c_1\rho\frac{\epsilon}{k}\left[\overline{u_i' u_j'} - \frac{\delta_{ij}}{3}\overline{u_k' u_k'}\right], \tag{9d}$$

$$\Phi_{ij2} = -c_2\rho\left[P_{ij} - \frac{\delta_{ij}}{3}P_{kk}\right], \tag{9e}$$

$$\Phi_{ijw1} = c_1'\rho\frac{\epsilon}{k}\left[\overline{u_k' u_m'}n_k n_m \delta_{ij} - \frac{3}{2}\overline{u_k' u_i'}n_k n_j - \frac{3}{2}\overline{u_k' u_j'}n_k n_i\right]f \tag{9f}$$

$$\Phi_{ijw2} = c_2'\left[\Phi_{km2}n_k n_m \delta_{ij} - \frac{3}{2}\Phi_{ik2}n_k n_j - \frac{3}{2}\Phi_{jk2}n_k n_i\right]f \tag{9g}$$

In the last two contributions, n_i is the wall-normal unit vector in the direction i and $f = C_\mu^{0.75}k^{1.5}/\epsilon\kappa\Delta n$, with Δn being the wall-normal distance. It is noted that the model (9a) differs from the usual Daly & Harlow "generalised gradient diffusion hypothesis" (1970). While the latter form has been used by the present authors, the isotropic variant has been found to yield consistently better agreement with experimental data, in particular as regards the details of the reattachment process in separated conditions. With the simplified form (9a) chosen, the value of σ_k was adjusted subject to the constraint that this form should give an identical effect to that arising from the Daly & Harlow term in homogeneous shear. This requirement led to $\sigma_k = 0.81$. Consistent with form (9a) is the use of an isotropic diffusivity in the turbulence-dissipation equation, which then involves the Prandtl/Schmidt number σ_ϵ. To ensure compatibility with the state of turbulence-energy equilibrium in near-wall flow, σ_ϵ was finally adjusted to the value 1.1.

In principle, the equations within set (8) are discretised in precisely the same manner as any other for any transported property ϕ, i.e. the equations are transformed, as expressed by form (4), and then integrated over the finite volume in Fig. 1. However, in the course of the discretisation and solution process, three issues require special consideration, namely stability, boundary conditions and the tensorially correct incorporation of the wall-reflection terms Φ_{ijw}. The last issue arises because the wall-reflection approximations are clearly tied to the orientation of the wall, via f. Each issue is considered separately below.

3.2 Boundary Conditions

In the context of using high-Reynolds-number transport models of turbulence within finite-volume schemes, it is standard practice to adopt log-law-based relations to bridge the semi-viscous near-wall region. One feature of this treatment is the solution of the turbulence energy or stress equations over the near-wall cells, with cell-integrated productions and dissipation terms evaluated from the log law.

The treatment adopted here is different and involves a prescription of all stresses at the near-wall nodes. Values for the stresses in the log-law region may be derived by focusing on the stress equations applicable to local energy equilibrium, $P_{kk} = 2\rho\epsilon$, and setting the wall-distance function $f = 1$. Adopting the wall-oriented Cartesian frame of reference shown in Fig. 2, this route leads to:

$$\widehat{\overline{u'^2}} = \frac{k\left(4c_1 + 2c_1^2 - 4c_1c_2 + 2c_1c_2c_2' + 6c_1' + 6c_1c_1' - 6c_1'c_2\right)}{3c_1\left(c_1 + 2c_1'\right)}$$

$$= 1.098k, \qquad (10a)$$

$$\widehat{\overline{v'^2}} = \frac{2\left(-1 + c_1 + c_2 - 2c_2c_2'\right)k}{3\left(c_1 + 2c_1'\right)} = 0.247k, \qquad (10b)$$

$$\widehat{\overline{w'^2}} = 2k - \widehat{\overline{u'^2}} - \widehat{\overline{v'^2}} = 0.655k, \qquad (10c)$$

$$-\widehat{\overline{u'v'}} = \sqrt{\left[\frac{1 - c_2 + 1.5c_2c_2'}{c_1 + 1.5c_1'}\right]\left[\frac{2\left(-1 + c_1 + c_2 - 2c_2c_2'\right)}{3\left(c_1 + 2c_1'\right)}\right]}\,k$$

$$= 0.255k, \qquad (10d)$$

with the last obtained with the aid of

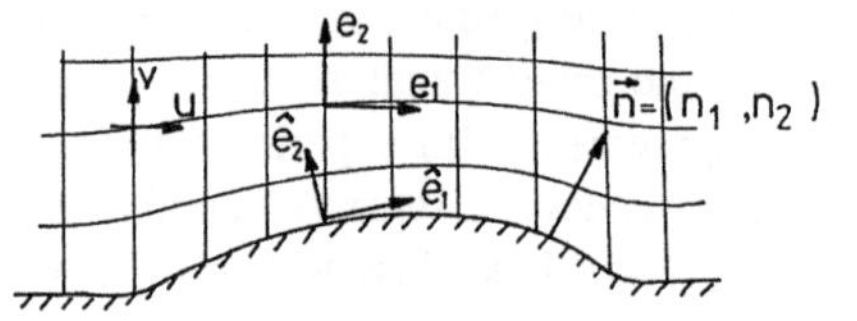

Fig. 2. Wall- and mesh-oriented coordinate systems

$$\epsilon \simeq \frac{u_\tau^3}{\kappa \Delta n}. \tag{11}$$

Next, the above stresses in wall-oriented co-ordinates must be transformed to the primary Cartesian system in which the velocity-vector is decomposed within the numerical scheme. This can be achieved by:

$$[\sigma_{ij}] = A^T \, [\hat{\sigma}_{ij}] \, A, \tag{12}$$

where $[\hat{\sigma}_{ij}]$ is the wall-oriented stress system and $[\sigma_{ij}]$ is the primary Cartesian one. With the wall-oriented unit vectors denoted by $\hat{e}_i$ and decomposed into components associated with the primary Cartesian system,

$$\hat{e}_1 = (t_1, t_2, 0), \tag{13a}$$

$$\hat{e}_2 = (n_1, n_2, 0), \tag{13b}$$

$$\hat{e}_3 = (0, 0, 1), \tag{13c}$$

as shown in Fig. 2, A arises as:

$$A = \begin{bmatrix} t_1 & t_2 & 0 \\ n_1 & n_2 & 0 \\ 0 & 0 & 1 \end{bmatrix}, \tag{14}$$

and A^T is the transpose of A. Hence, the mesh-oriented stresses may be evaluated as:

$$\overline{u'^2} = \widehat{\overline{u'^2}}\, t_1^2 + \widehat{\overline{v'^2}}\, n_1^2 + 2\widehat{\overline{u'v'}}\, t_1 n_1, \tag{15a}$$

$$\overline{v'^2} = \widehat{\overline{u'^2}}\, t_2^2 + \widehat{\overline{v'^2}}\, n_2^2 + 2\widehat{\overline{u'v'}}\, t_2 n_2, \tag{15b}$$

$$\overline{u'v'} = \widehat{\overline{u'^2}}\, t_1 t_2 + \widehat{\overline{v'^2}}\, n_1 n_2 + \widehat{\overline{u'v'}}(t_1 n_2 + t_2 n_1). \tag{15c}$$

The above stresses are used as boundary conditions for the inner field.

3.3 Wall-Reflection Terms

The task at hand may be illustrated by focusing on any single curved wall, as shown in Fig. 2, with the wall-normal vector being (n_1, n_2). The wall-distance function f may then be related to Cartesian, velocity-oriented components by:

$$f_x = n_1^2 f, \quad f_y = n_2^2 f, \quad f_{xy} = n_1 n_2 f. \tag{16}$$

With the above transformed wall-distance-function components, it can be shown that the stress equations (for 2D plane cases) may be written as:

$$\begin{aligned} C_{ij} - D_{ij} = \; & \alpha_1 P_{11} + \alpha_2 P_{22} + \alpha_3 P_{12} + 0.5\alpha_4 P_{kk} \\ & + \rho\frac{\epsilon}{k}[\alpha_5 \overline{u'^2} + \alpha_6 \overline{v'^2} + \alpha_7 \overline{u'v'}] + \alpha_8 \rho\epsilon \end{aligned} \tag{17}$$

with the α-coefficients given in Table 1 for $\overline{u'^2}$, $\overline{v'^2}$ and $\overline{u'v'}$.

Table 1. Coefficients in (17)

$\overline{u_i'u_j'} = \overline{u'^2}$	
$\alpha_1 = 1 - c_2 + 2c_2c_2'f_x$	$\alpha_2 = -c_2c_2'f_y$
$\alpha_3 = c_2c_2'f_{xy}$	$\alpha_4 = \frac{2}{3}(c_2 - 2c_2c_2'f_x + c_2c_2'f_y)$
$\alpha_5 = -(c_1 + 2c_1'f_x)$	$\alpha_6 = c_1'f_y$
$\alpha_7 = -c_1'f_{xy}$	$\alpha_8 = \frac{2}{3}(c_1 - 1)$
$\overline{u_i'u_j'} = \overline{v'^2}$	
$\alpha_1 = -c_2c_2'f_x$	$\alpha_2 = 1 - c_2 + 2c_2c_2'f_y$
$\alpha_3 = c_2c_2'f_{xy}$	$\alpha_4 = \frac{2}{3}(c_2 - 2c_2c_2'f_y + c_2c_2'f_x)$
$\alpha_5 = c_1'f_x$	$\alpha_6 = -(c_1 + 2c_1'f_y)$
$\alpha_7 = -c_1'f_{xy}$	$\alpha_8 = \frac{2}{3}(c_1 - 1)$
$\overline{u_i'u_j'} = \overline{u'v'}$	
$\alpha_1 = 1.5c_2c_2'f_{xy}$	$\alpha_2 = 1.5c_2c_2'f_{xy}$
$\alpha_3 = 1 - c_2 + 1.5c_2c_2'(f_x + f_y)$	$\alpha_4 = -2c_2c_2'f_{xy}$
$\alpha_5 = -1.5c_1'f_{xy}$	$\alpha_6 = -1.5c_1'f_{xy}$
$\alpha_7 = -[c_1 + 1.5c_1'(f_x + f_y)]$	$\alpha_8 = 0$

The above relationships have been derived via two entirely different routes, one involving the direct insertion of the wall-distance transformations into the wall-reflection terms of equations (9f) and (9g), and the other proceeding via a rigorous transformation of all terms in Φ_{ijw}. Both routes are found to lead to identical results (details may be found in Lien (1992)).

3.4 Stability

Low iterative stability is a serious difficulty in elliptic solvers, particularly in those based on a collocated arrangement and using non-diffusive discretisation. In essence, the difficulty arises from the absence of eddy-viscosity terms and a numerical decoupling of stresses from related strains. The problem has been addressed here by extracting apparent viscosities from the (ξ, η)-transformed partial differential equations governing the stresses. It can be shown that terms appearing naturally in the stress equations allow, for $i \neq j$, the interpretation:

$$\nu_{ij} = \frac{1 - c_2 + 1.5 c_2 c_2' (f_i + f_j)}{c_1 + 1.5 c_1' (f_i + f_j)} \frac{k\overline{u_j'^2}}{\epsilon} \tag{18}$$

$\longrightarrow$ which is associated with the strain $(\frac{\partial u_i}{\partial \xi_j})$

while for $i = j$ (no summation)

$$\nu_{ii} = \frac{2 - \frac{4}{3} c_2 + \frac{2}{3} c_2 c_2' (4 f_i + f_j)}{c_1 + 2 c_1' f_i} \frac{k\overline{u_i'^2}}{\epsilon} \tag{19}$$

$\longrightarrow$ which is associated with the strain $(\frac{\partial u_i}{\partial \xi_i})$

where f_i and f_j are either f_x or f_y, as defined by equation (16). Insertion of the above viscosities into the u-momentum equation, for example, leads to:

$$q_1^u = \mu_{11} y_\eta^2 + \mu_{12} x_\eta^2, \quad q_2^u = \mu_{11} y_\xi^2 + \mu_{12} x_\xi^2, \tag{20}$$

$$\begin{aligned} S^u = -\{ & \frac{\partial}{\partial \xi}(r y_\eta) p^{u\xi} + \frac{\partial}{\partial \eta}(-r y_\xi) p^{u\eta} \\ & + \frac{\partial}{\partial \xi}(-r x_\eta) \widetilde{\rho \overline{u'v'}}^{u\xi} + \frac{\partial}{\partial \eta}(r x_\xi) \widetilde{\rho \overline{u'v'}}^{u\eta} \} \end{aligned} \tag{21}$$

with

$$p^{u\xi} = p + \rho \overline{u'^2} + (\frac{\mu_{11} y_\eta}{J}) \frac{\partial u}{\partial \xi}, \quad p^{u\eta} = p + \rho \overline{u'^2} - (\frac{\mu_{11} y_\xi}{J}) \frac{\partial u}{\partial \eta}, \tag{22a}$$

$$\widetilde{\rho \overline{u'v'}}^{u\xi} = \rho \overline{u'v'} - (\frac{\mu_{12} x_\eta}{J}) \frac{\partial u}{\partial \xi}, \quad \widetilde{\rho \overline{u'v'}}^{u\eta} = \rho \overline{u'v'} + (\frac{\mu_{12} x_\xi}{J}) \frac{\partial u}{\partial \eta}, \tag{22b}$$

The addition of normal stresses to the pressure, as done in relations (22a) which are inserted into equation (21), is of considerable significance to stability in the context of the Rhie & Chow interpolations (5) and (6). It should be evident that the interpretation of normal stresses as pressure fragments means that any cell-face velocity is not merely sensitised to the pressure differences centred on that face but also to the normal-stress difference. In effect, this practice introduces (artificial) fourth-order *normal-stress diffusion* in addition to fourth-order *pressure diffusion*.

4. Results

The capabilities and performance of the above methodology are illustrated by four examples: two flows behind backward-facing steps – one in an expanding channel [Driver & Seegmiller (1982)], and the other in a constant-area channel [Durst & Schmitt (1985)] but computed with a deliberately distorted mesh, a flow through a sinusoidal pipe constriction [Deshpande & Giddens (1980)] and a transonic flow over a channel bump 'Case B' of Delery (1983)] in which the shock induces boundary-layer separation. All cases are two-dimensional, although the code (in conjunction with eddy-viscosity modelling) applies to and has been used for three-dimensional flows [Lien & Leschziner (1991)]. All flows contain regions of separation, and thus offer the opportunity to examine the response of the turbulence models to curvature in the intense shear layer bordering the separated regions. The last two flows present the added challenge of separation taking place from a continuous, curved surface.

Figs. 3 to 5 compare $k - \epsilon$ and Reynolds-stress-model calculations performed with a 110×60 grid for Driver & Seegmiller's geometry in which the backward-facing step is followed by a channel expanding at an angle of 6 degrees. QUICK was used to approximate the convection of momentum, while convection of turbulence quantities was discretised by means of the TVD-MUSCL scheme. With these high-accuracy numerical approximations, the above mesh –the finest in a sequence of three examined – effectively supports a grid-independent solution. The economy of the computations was considerably enhanced by the use of 3-level multigrid method, which yielded speed-up ratios of order 5-8, depending on model and grid.

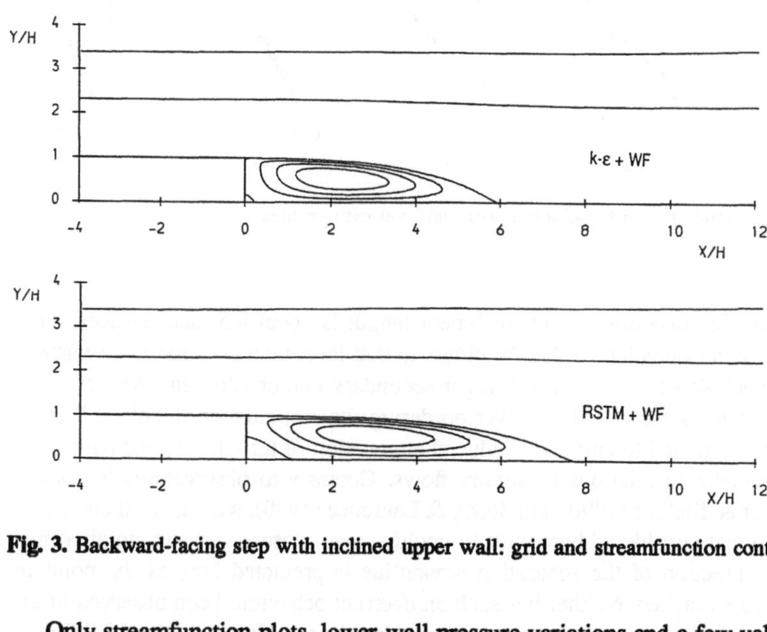

Fig. 3. Backward-facing step with inclined upper wall: grid and streamfunction contours

Only streamfunction plots, lower-wall pressure variations and a few velocity profiles have been included here, although extensive comparisons with experimental data have

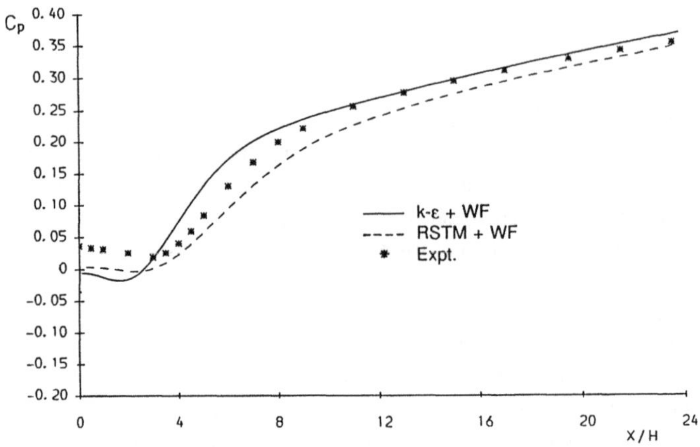

Fig. 4. Backward-facing step with inclined upper wall: pressure variation along lower wall

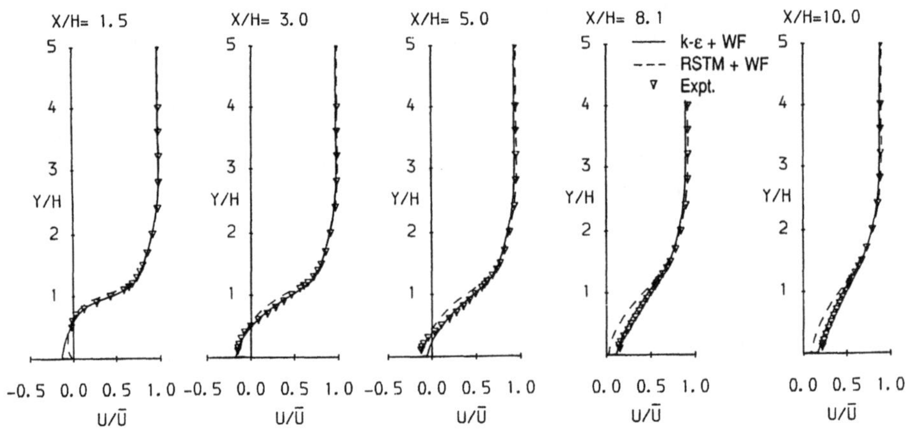

Fig. 5. Backward-facing step with inclined upper wall: velocity profiles

been performed. The experimental reattachment length is about 8.5, and the Reynolds-stress model is seen to predict a value far closer to that level than does the $k-\epsilon$ variant. The former model also returns a much larger secondary corner eddy and steeper shear strains in the curved, separated shear layer bordering the recirculation bubble. All these features may be attributed to curvature-induced attenuation of turbulence, and have been observed in previous calculations of similar flows. Contrary to observations by Obi et al (1989), Lasher & Taulbee (1990) and Sebag & Laurence (1990), who have all computed flows similar to that considered here with Reynolds-stress closure variants, no abnormal reversal in the direction of the separation streamline is predicted here as the point of reattachment is approached. Neither has such an aberrant behaviour been observed in the second backward-facing step geometry to be considered next. However, the third flow – in which separation is provoked by a continuous constriction in a pipe – does display

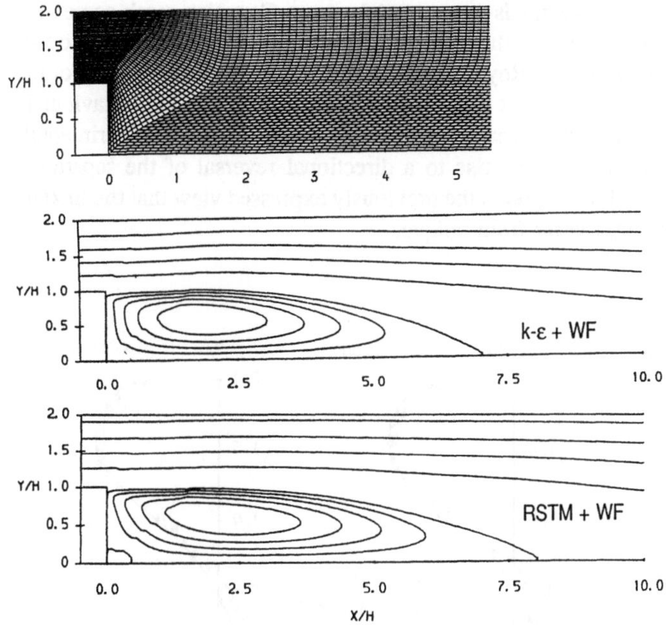

Fig. 6. Backward-facing step: grid and streamfunction contours

this abnormal feature, and this suggests that its occurrance is case-dependent. In fact, the feature is also sensitive to the precise details of the near-wall treatment: in some instances where a reversal of the separation streamline was returned, seemingly minor variations in the treatment of the boundary condition for shear stress were observed to materially modify the behaviour in the immediate vicinity of the reattachment point. The above anomaly has been attributed to an inappropriate amplification, rather than attenuation, of the wall-normal turbulence intensity by the wall-related contribution to the pressure-strain model, Φ_{ijw}, at impingement or reattachment points. Clearly, however, the consequences of this undisputed model weakness vary considerably, and the behaviour at the reattachment point depends upon a number of issues associated with near-wall modelling, both inviscid and viscous.

Figs. 6 and 7 provide a comparison of turbulence-model performance for Durst & Schmitt's step. This geometry is rectilinear, but has been computed with a deliberately distorted grid in order to examine numerical weaknesses associated with grid skewness and discontinuity. The approximation schemes employed for convection were identical to those used in the previous case. Here again, grid dependence was investigated, and revealed that the precise position of reattachment was fairly sensitive to the degree of mesh skewness in the vicinity of this position. The present results have been obtained with a 200 × 50 grid, which gives a grid-independent solution, except for a residual uncertaintly in the critical reattachment region. Closer inspection of the streamlines in Fig. 6 also reveals minor 'kinks' at locations coincident with the boundary between two zonal grids generated separately by transfinite interpolation; at this boundary, grid smoothness is evidently poor. Confidence in the present results is strengthened by a test calculation performed with the Reynolds-stress model over a Cartesian mesh somewhat coarser than the present 200×50 non-orthogonal one. This yielded a solution close to that shown in Fig. 6. Significantly, neither Reynolds-stress solution displayed a reversal of the separation streamline at the reattachment point, which is consistent with the behaviour in the previous case. Interestingly, however, a replacement of the rounded (experimental) inlet profile by a uniform one did give rise to a directional reversal of the separation streamline - an observation which supports the previously expressed view that the linkage between this anomaly and Φ_{ijw} is far from unique.

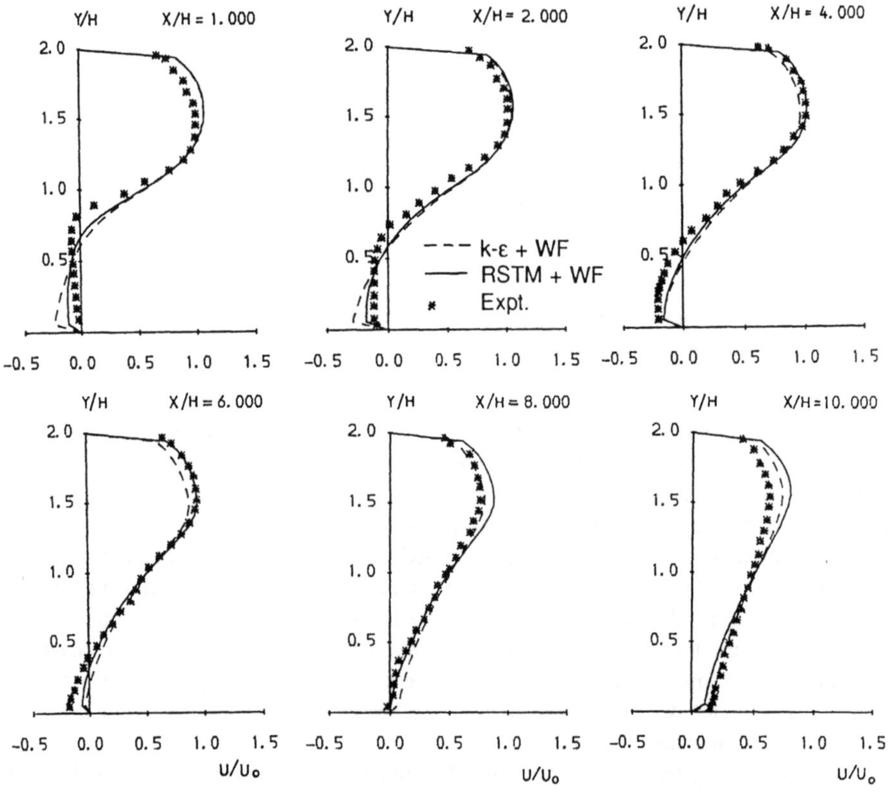

Fig. 7. Backward-facing step: velocity profiles

The results given in Figs. 6 and 7 show features well in line with the previous case: the Reynolds-stress model returns a longer recirculation zone, a more pronounced secondary eddy and steeper gradients in the shear layer bordering the recirculation bubble. Reattachment is predicted by the stress model to occur at about 8 step heights downstream of the step, a value marginally below that recorded experimentally. Here again, the superior characteristics of the stress model are rooted, principally, in its ability to account for curvature-induced turbulence attenuation in the recirculation zone.

Figs. 8 to 11 show solutions obtained with the "standard" high-Reynolds-number $k-\epsilon$ model, with a low-Reynolds-number $k-\epsilon$[1] variant of Lien & Leschziner (1990) and with the stress-model for the flow through Deshpande & Giddens' sinusoidal pipe constriction. As in the previous cases, a combination of QUICK and MUSCL was used for approximating convection, here over grids varying between 120 × 24 and 120 × 40 lines.

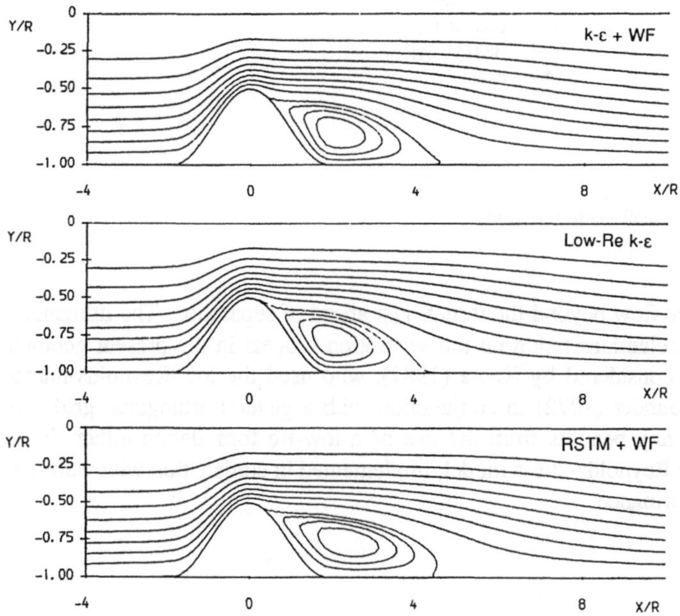

Fig. 8. Pipe constriction: streamfunction contours

The comparison of streamfunction plots given in Fig. 8 reveals some trends which are consistent with those observed in the previous case. For example, the stress closure predicts, here too, steeper velocity gradients in the shear layer bordering the separation zone. However, there are also marked differences. Thus, here, the tendency of the separation streamline to double up at the point of reattachment is evident, a feature which may have been enhanced by the nearly uniform velocity profile ahead of separation. None of the models returns the correct behaviour of the centreline velocity and the wall pressure, but the stress model clearly responds more sensitively to the adverse pressure gradient beyond the throat, inducing earlier separation and diminishing the consistent tendency towards an excessive recovery of pressure. Interestingly, the comparison conveys the message that any relaminarisation of the boundary layer during the acceleration phase ahead of the throat does not seem to play a material role in the separation process. Implicit in this statement is, however, the assumption that the low-Reynolds-number $k-\epsilon$ model gives a realistic representation of the reversion to turbulent conditions following

[1] This model has been specifically constructed to yield a near-wall length scale in accord with that used in Wolfshtein's one-equation model (1969).

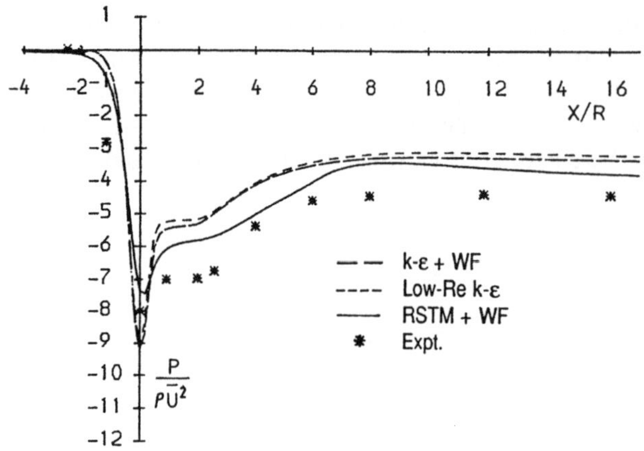

Fig. 9. Pipe constriction: wall static pressure

relaminarisation in the flow beyond the throat, just ahead of separation. The influence of acceleration-induced relaminarisation on the separation process in the present geometry had previously been considered by Kadja (1987), who used the low-Reynolds-number model of Jones & Launder (1972) in conjunction with a general-orthogonal grid. That study did not reveal any benefits from the use of a low-Re formulation either. It also demonstrated that the Reynolds-stress model, implemented in conjunction with wall laws, yielded the best performance.

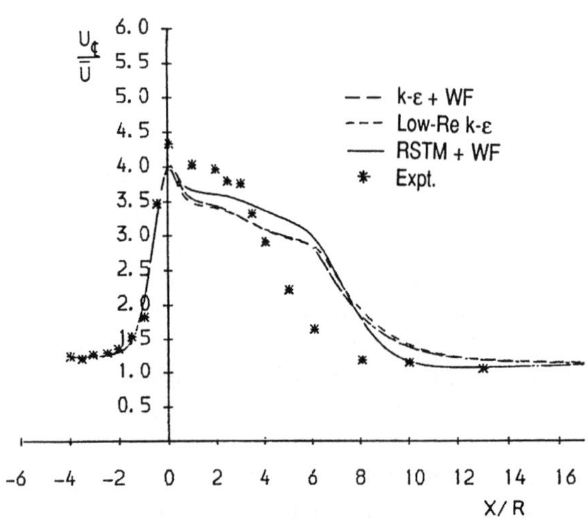

Fig. 10. Pipe constriction: centreline velocity

Finally, Figs. 12 to 14 give results obtained with the high-Reynolds-number $k - \epsilon$ model and the stress closure for the transonic bump flow of Delery. For the latter model, solutions are reported for two grids, one containing 95×35 and the other 130×50 lines.

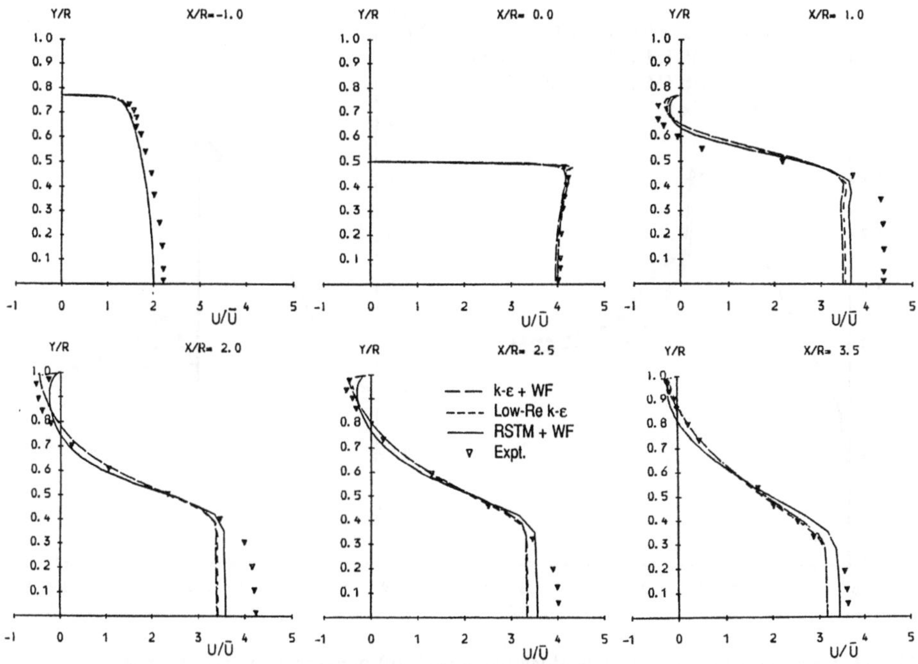

Fig. 11. Pipe constriction: velocity profiles

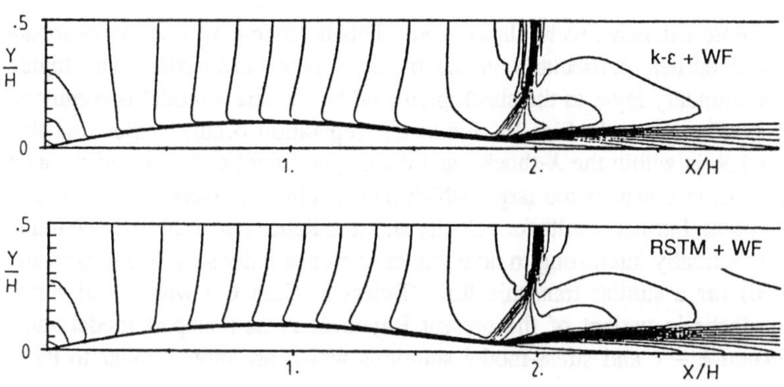

Fig. 12. Transonic bump: Mach-number contours

Fig. 12 gives an overall view of the flow field in terms of Mach-number contours, and the plots reveal the characteristic λ-shock structure arising from the interaction of the normal shock with the wall boundary layer. Fig. 13 compares the predicted pressure variations along the bump wall and the duct-centre plane with Delery's experimental data. The centreline variation has been included to convey the shock-capturing capability of the procedure; this is not well brought out in Fig. 12 in which the shock appears contaminated by minor oscillations. The characteristic pressure plateau in Fig. 13 is a reflection of the separation zone, and the comparison gives a clear indication that the Reynolds-stress model is considerably more sensitive to the shock, yielding earlier

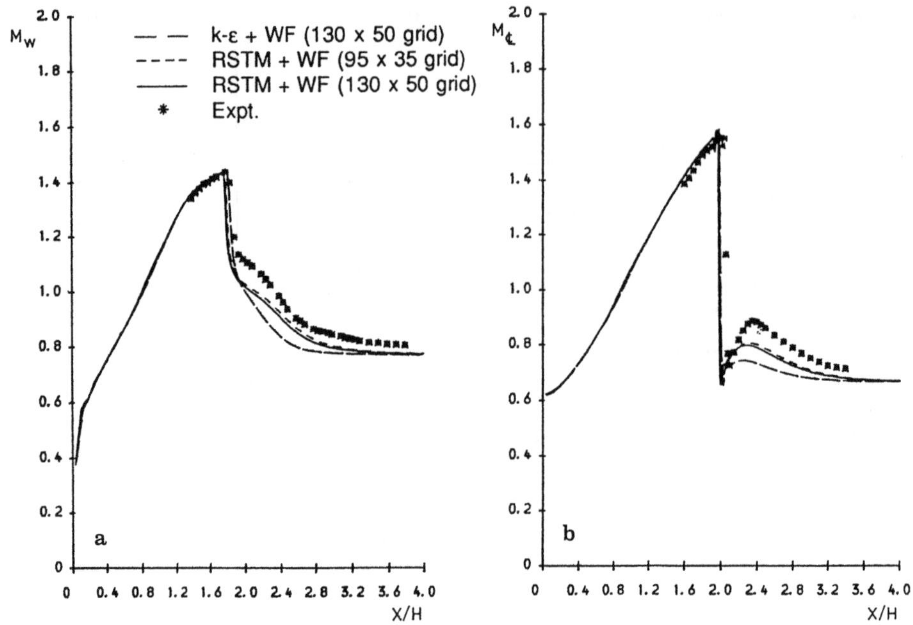

Fig. 13. Transonic bump: (a) pressure along bump wall; (b) pressure along centre plane

separation and a more extensive recirculation zone. This behaviour is broadly consonant with that observed earlier, particularly in the previous pipe constriction. The higher sensitivity of the boundary layer to the shock predicted by the stress model is confirmed by the velocity profiles given in Fig. 14. Evidently, separation occurs earlier (the first location, $x/H = 1.9$, is within the λ-shock), and the displacement of the boundary layer is larger; indeed, displacement is too large, which then inhibits recovery downstream of the reattachment point. Further results for velocity and turbulence quantities may be found in Lien (1992). It is finally interesting to note that earlier calculations by Dimitriadis and Leschziner (1990) for a similar transonic flow (Delery's 'Case C') with a cell-vertex scheme and an algebraic variant of the present Reynolds-stress-transport model show differences between $k-\epsilon$ and stress-model solutions which are quite similar to those reported here.

5. Conclusions

The paper reported the successful inclusion of full second-moment closure into a general, non-orthogonal, collocated finite-volume scheme in which convection is approximated by essentially non-diffusive convection schemes. Particular emphasis was put on numerical practices pertaining to stability, wall-boundary conditions and the wall-related fragments of the pressure-strain model. The application examples reported, whilst necessarily giving

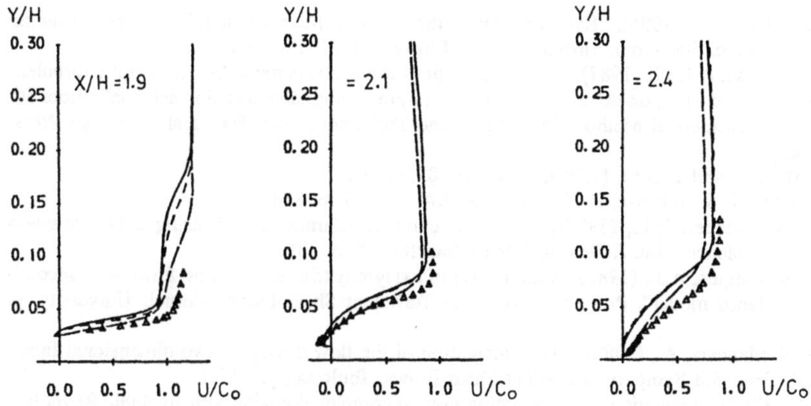

Fig. 14. Transonic bump: velocity profiles (Key as in Fig. 13)

a rather superficial and incomplete view, serve to reinforce earlier observations that, for separated flow, second-moment closure offers benefits relative to the eddy-viscosity framework. They also confirm previously reported weaknesses, however. Thus, there is evidence of an insufficient rate of recovery following reattachment, and of an erroneous representation of the reattachment process itself – at least in some flow conditions. The latter may be alleviated, but not eliminated, by modifications to the wall-function-based near-wall treatment, and this points to the need for new proposals for modelling the influence of wall-induced pressure reflections on the pressure-strain process.

Acknowledgement

The authors are grateful to the UK Science and Engineering Research Council which supported the present work via grants GR/E26808 and GR/F86816.

References

Daly, B.J. & Harlow, F.H., (1970), Phys. Fluids, 13, p. 2364.

Delery, J., (1983), J. Aiaa, Vol. 21, p. 180.

Deshpande, M.D. & Giddens, D.P., (1980), JFM, Vol. 97(1), p. 65.

Dimitriadis, K.P. & Leschziner, M.A., (1990), "Modelling shock/boundary layer interaction with a cell-vertex scheme and second-moment closure"., Proc. 12th Int. Conf. on Numerical Methods on Fluid Dynamics, Oxford, *Lecture Notes in Physics* (K.W. Morton, ed.), Springer Verlag, p. 371.

Driver, D.M. & Seegmiller, H.L., (1982), "Features of a reattaching turbulent shear layer", AIAA Paper 82-1029.

Durst, F. & Schmitt, F., (1985), "Experimental studies of high Reynolds number backward-facing step flows", Proc. 5th Symp. Turbulent Shear Flows, Cornell University, p. 5.19.

Gaskell, P.H. & Lau, A.K.C., (1987), "An assessment of direct stress modelling for elliptic turbulent flows with the aid of a non-diffusive, boundedness-preserving, discretisation scheme", Proc. 5th Int. Conf. on Numerical Methods in Laminar and Turbulent Flows, Montreal, Pineridge Press, Swansea, p. 351.

Gibson, M.M. & Launder, B.E., (1978), JFM, Vol. 85, p. 491.

Hafez, M., South, J. & Murman, E., (1979), J. AIAA, Vol. 17, p. 838.

Jones, W.P. & Launder, B.E., (1972), "The prediction of relaminarisation using a two-equation model of turbulence", Int. J. Heat and Mass transfer, 15, p. 301.

Jones, W.P, & Marquis, A.J., (1985), "Calculation of axisymmetric recirculating flows with second-order turbulence model", Proc. 5th Symp. on Turbulent Shear Flows, Cornell University, p. 20.1.

Jones, W.P. & Manners, A., (1988), "The calculation of the flow through a two-dimensional faired diffuser", Proc. 6th Symp. on Turbulent Shear Flows, Toulouse, p. 17.7.1.

Kadja, M., (1987), "Computation of recirculating flow in complex domains with algebraic Reynolds-stress closure and body-fitted meshes", Ph.D. Thesis, University of Manchester.

Lasher, W.C. & Taulbee, D.B., (1990), *Engineering Turbulence Modelling and Experiments*, Elsevier, (eds. Rodi and Ganic), p. 195.

Van Leer, B., (1979), JCP, Vol. 32, p. 101.

Leonard, B.P., (1979), Comp. Meths. Appl. Mech. Engng., Vol. 19, p. 59.

Lien, F.S. & Leschziner, M.A. (1990), "Modelling variable-area curved duct flow with a 3D non-orthogonal collocated FV method", Proc. 4th UMIST Colloquium on Computational Fluid Dynamics, Dept. of Mech. Engng., UMIST, Manchester, p. 6.4.

Lien, F.S., (1992), "Computational modelling of 3D flow in complex ducts and passages", Ph.D. Thesis, University of Manchester.

Lien, F.S. & Leschziner, M.A., (1991), "Multigrid convergence acceleration for complex flow including turbulenceß", *International Series of Numerical Mathematics*, Birkhauser Verlag Basel, Vol. 98, p. 277.

Lien, F.S. & Leschziner, M.A., (1992), "Modelling shock/turbulent-boundary-layer interaction with second-moment closure within a pressure-velocity strategy", Proc. 13th Int. Conf. on Numerical Methods in Fluid Dynamics, Rome (to be published).

Lin, C.A. & Leschziner, M.A., (1989), " Computation of three-dimensional injection into swirling flow with second-moment closure", Proc. 6th Int. Conf. on Numerical Methods in Laminar and Turbulent Flows, Swansea, Pineridge Press, Swansea, p. 1711.

Majumdar, S., Rodi, W. I & Schonung, B., (1989), *Finite Approximations in Fluid Mechanics II, Notes on Numerical Fluid Mechanics*, Vieweg Verlag.

Mcguirk, J.J. & Page, J.G., (1989), "Shock capturing using a pressure-correction method", AIAA 27th Aerospace Sciences Meeting, Reno, Nevada.

Obi, S. (1991), "Berechnung komplexer turbulenter Strömungen mit einem Reynolds-Spannungs-Modell", Doctoral Dissertation, University of Erlangen-Nürnberg.

Obi, S., Peric, M. & Scheuerer, G., (1989), "A finite-volume calculation procedure for turbulent flows with second-order closure and collocated variable arrangement", Proc. 7th Symp. Turbulent Shear Flows, Stanford University, p. 17.4.

Patankar, S.V., (1980), *Numerical Heat Transfer and Fluid Flow*, McGraw-Hill, New York.

Peric, M. (1985), "A finite volume method for the prediction of three dimensional fluid flow in complex ducts", Ph.D. Thesis, University of London.

Rhie, C.M. & Chow, W.L., (1983), J. AIAA, Vol. 21, p. 1525.

Sebag, S. & Laurence, D., (1990), *Engineering Turbulence Modelling and Experiments*, Elsevier, (eds. Rodi and Ganic), p. 175.

Shyy, W. & Braaten, M.E., (1986), Int. J. Numerical Methods Fluids, Vol. 6, p. 861.

Wolfshtein, M.W., (1969), Int. J. Heat and Mass Transfer, Vol. 12, p. 301

Part III

Compressibility Effects

Introductory Remarks

A.J. Smits

Department of Mechanical and Aerospace Engineering, Princeton University, Princeton, N.J. 08544, USA

The history of research in turbulent compressible flows is somewhat checkered, in that shifts in national and international priorities have had a large impact on the continuity of effort. The high level of activity that lasted from the end of the Second World War to about 1965 was largely driven by the wish to fly at supersonic speeds, and to solve the problems associated with re-entry. Once these problems were "solved" (in an engineering sense), the funding dried up, and further efforts became severely restricted. The recent upsurge in activity has been mainly driven by a new set of priorities such as the projected need for a low-cost supersonic transport aircraft, and the desire to fly at hypersonic speeds. However, much of the fundamental knowledge required to attain hypersonic flight is not yet available, and "generic" hypersonic research is now receiving considerable attention. In a similar way, the next generation of supersonic transports will need to be highly fuel efficient, cause minimum ozone depletion and minimum noise levels. The understanding of compressibility effects on turbulence will undoubtedly have a significant impact on the development of engineering solutions to these problems.

In this Chapter, five papers are presented which consider fundamental aspects of the effects of compressibility on turbulence. To understand their significance, we note that in a turbulent shear layer mean density gradients will exist in addition to mean velocity gradients, and the turbulent field consists of pressure, density and velocity fluctuations. Energy is continually transferred among these three modes, and the transport mechanisms are therefore more complex than those encountered in constant property flows. In some parts of the flow, the relative speed of adjacent turbulent motions may be transonic or supersonic, and it is possible that local compression waves and "shocklets" could affect the turbulence evolution. Vorticity can be produced through baroclinic torques, since the density is not just a function of temperature, and energy may be dissipated by sound radiation, in addition to the usual viscous effects. All these processes can influence the production, dissipation and advection of the turbulence, and they may have significant effect on the mixing by turbulence. The most obvious example of these direct compressibility effects is the reduced growth rate of free shear layers observed in flows with high convective Mach numbers (Bogdanoff 1983, Papamoschou and Roshko, 1986). It seems reasonable to expect that near-wall motions in supersonic and hypersonic boundary layers, where the fluctuating Mach number takes the highest values, may also be affected. In addition, some of the most notable differences between subsonic and supersonic shear layers may be due to the large variation in fluid properties that can occur across the layer. For example, in an adiabatic wall boundary layer at Mach 3, there is a six-fold variation in kinematic viscosity can occur across the layer.

When turbulence enters a region of strong pressure gradients, or when it interacts with shock waves, dramatic compressibility effects can occur. Longitudinal pressure gradients lead to the compression or dilatation of vortex tubes, enhancing or reducing turbulent velocity and pressure fluctuations. When shock waves are present, the interaction with the turbulence can become extremely complicated. In wall bounded flows, there will be the possibility of separation caused by the occurrence of severe pressure gradients (a phenomenon that can be understood from subsonic experience), but even in the absence of separation there will be a strong coupling between the shock and the turbulence, and the resultant distortions of the shock sheet and its unsteady motion have been widely observed (see, for example, Muck and Andreopoulos, 1986). As far as the turbulence is concerned, the overall pressure rise and turning angle seems to be more important than the rate at which the perturbation occurs (Jayaram, Taylor and Smits 1987), presumably as long as the total temperature or entropy changes are small over the region of perturbation. Interestingly, the unsteady shock motion does not appear to play a significant role in turbulence modification (Selig and Smits 1989).

The five papers presented here provide some very basic insight into the physics of compressible turbulence. The response of homogeneous compressible turbulence to isotropic (spherical) compression is presented by G.N. Coleman and N.N. Mansour "Simulation and modeling of homogeneous compressible turbulence under isotropic mean compression." Low Mach number cases, such as those experienced in internal combustion applications, and high Mach number cases (fluctuating Mach numbers up to 0.5) were studied. Since this flow may be seen as a limiting case, the results may be used as a means of "tuning" turbulence models for a wide variety of compressible flows. Here, the DNS code of Blaisdell, Mansour and Reynolds (1991) was used to generate the results, which were then used to test some proposed closures for two-equation turbulence models. For the low Mach number cases, where the compressibility only affects the mean density, the primary conclusion is that the variation of the kinematic viscosity due to the compression must be taken into account precisely in developing two-equation models. For the higher Mach number cases it was found that, unlike shear, spherical compression does not drive the acoustical field to a unique state, since the only coupling between the solenoidal and acoustic fields is due to nonlinear interactions which are slow compared to the turbulence time scale at reasonable Reynolds numbers. Hence, models for the energy dissipation due to dilatational (acoustic) velocity fluctuations and the pressure-dilatation term must account for the initial acoustic conditions.

The paper by O. Zeman and G.N. Coleman "Compressible turbulence subjected to shear and rapid compression." is a useful companion to the previous paper. Zeman and Coleman first present a review of the physics and modeling of compressibility effects (in terms of two-equation models) in homogeneous decaying, and shear-driven turbulence. This may be a useful starting point for readers of this chapter who are new to the field of compressible turbulence. The authors also consider the differences between rapid spherical and one-dimensional compression on initially isotropic turbulence, and show that in the case of one-dimensional rapid compression the pressure-dilatation term cannot be neglected, even for very low fluctuating Mach numbers. This observation is in contrast to the observation made by Coleman and Mansour in the case of spherical compression, and belies the common wisdom that the significance of the compressibility effects can be measured solely by the level of the fluctuating Mach number. Apparently, when turbulence is subjected to a rapid anisotropic compression, high negative values of the pressure-dilatation correlation results, which lead to a considerable transfer of turbulent kinetic energy to pressure fluctuations. The effects are illustrated by reference to the model predictions of the amplification of turbulence through a normal shock (upstream Mach number of 1.18), as compared to the results of a direct numerical simulation of the same flow by Lee (1991).

In the paper by S. Sarkar, G. Erlebacher and M.Y. Hussaini "Compressible homogeneous shear: Simulation and modeling," direct numerical simulation is used to study the effects of compressibility on the evolution of a homogeneous shear flow. By solving the equations in a frame of reference moving at the mean velocity (first suggested by Rogallo 1981), periodic boundary conditions can be used in the direction of shear. Taylor microscale Reynolds numbers of about 40 were achieved. Two cases with slightly different speeds of sound were computed, so that the maximum fluctuating Mach number varied from 0.46 to 0.67. By the end of the two computations, the non-dimensional strain rate $\frac{\partial u}{\partial y}\frac{k}{\epsilon}$ was 7 and 9 respectively, which may be compared to a typical value of about 10 to 12 found in the logarithmic region of a turbulent boundary layer (Sreenivasan 1985). The results indicated a significant attenuation of the kinetic energy growth rate due to compressibility, specifically due to the pressure-dilatation and compressible dissipation terms which act as sinks for the kinetic energy (and express the explicit influence of the non-solenoidal part of the fluctuating velocity field), and a reduced level of the Reynolds shear stress. It was also found that the dilatational velocity component was strongly anisotropic, and therefore contributes predominantly to the transverse velocity intensity. This has the effect of reducing the transverse mixing, since the dilatational component is a relatively poor mixer of scalar compared to the vortical component.

J.H. Chen, in her paper "The effect of compressibility on conserved scalar entrainment in a plane free shear layer," describes a direct numerical simulation of a temporally evolving mixing layer, and it is therefore the only paper in this chapter to consider a flow other than initially homogeneous turbulence. The numerical method was developed by Sandham and Reynolds (1991), and periodic boundary conditions were used in the streamwise and spanwise directions, and non-reflecting boundary conditions in the third direction. The primary motivation for these computations was to study the evolution of a conserved scalar field and its relation to the structure of the hydrodynamic fields. The highest Reynolds number considered was 1600. Two convective Mach numbers, 0.4 and 0.8, were studied. At the lower value, compressibility effects were found to be small, as may be expected, but at the higher value, significant compressibility effects were observed. In terms of the turbulence structure, compressibility appears to enhance the three-dimensionality of the flow. What is particularly interesting is the effect on the scalar concentration distributions. At the lower convective Mach number the spanwise rollers bring the fluid from the high- and low-speed sides into the center, and the concentration profiles show a well-defined peak at this location. At the higher convective Mach number the presence of three-dimensional lambda vortices leads to local upwash and downwash regions which create pockets of partially mixed fluid near the tips of the vortices. As a result, the instantaneous concentration profiles show two preferred mixture fractions at a given position in the layer.

In the only experimental paper in this chapter, L. Jacquin, E. Blin and P. Geffroy discuss "An experiment on free turbulence/shock wave interaction." A grid was placed in the wind tunnel test section where the flow was supersonic, and a normal shock was generated at a distance 36 mesh sizes downstream. The Mach number upstream of the shock was about 1.4, so the results should be comparable to the direct numerical simulation by Lee (1991) where the upstream Mach number was 1.18 (Lee's results were also used by Zeman and Coleman in their discussion of turbulence modeling for compressible turbulence). However, the experimental results were quite unexpected, in that no net amplification of the turbulent kinetic energy was observed downstream of the shock. This is in contrast to previous experimental and numerical work in similar flows where considerable amplification has been noted. The experiments also indicated an increase in the decay rate across the shock, which is in agreement with previous work, and perhaps the decay rate was increased sufficiently to mask the expected rise in turbulence intensity level.

In any case, the papers in this chapter make a very useful contribution to our understanding of the effects of compressibility on turbulence. As is so often the case, however, the authors also raise more questions than they answer, indicating that there is considerable need for further work in this area.

References

Andreopoulos, J. and Muck, K.C. (1987) Some new aspects of the shock wave boundary layer interaction in compression ramp flows. *J. of Fluid Mech.*, Vol 180, pp. 405-428.

Blaisdell, G.A., Mansour, N.N. and Reynolds, W.C. (1991) Numerical simulation of compressible homogeneous turbulence. Department of Mechanical Engineering, Stanford University, Thermosciences Division Report F-50.

Bogdanoff, D.W. (1983) Compressibility effects in turbulent shear layers. *AIAA Journal*, Vol. 21, pp. 926-927.

Jayaram, M., Taylor, M.W., and Smits, A.J. (1987) The response of a compressible turbulent boundary layer to short regions of concave surface curvature. *J. of Fluid Mech.* Vol. 175, pp. 343-362.

Lee, S. (1991) Direct numerical simulations of compressible turbulence. Ph.D. Thesis, Mechanical Engineering Department, Stanford University, Stanford, CA.

Rogallo, R.S. (1981) Numerical experiments in homogeneous turbulence. NASA TM 81315.

Sandham, N.D. and Reynolds, W.C. (1991) Three dimensional simulations of large eddies in the compressible mixing layer. *J. Fluid Mech.* Vol. 224, pp. 133-159.

Selig, M.S. and Smits, A.J. (1991) Effect of periodic blowing on attached and separated supersonic turbulent boundary layers. *AIAA Journal*, Vol. 29, pp. 1651-1658.

Sreenivasan, K.R. (1985) The effect of contraction on a homogeneous turbulent shear flow. *J. Fluid Mech.*, Vol. 154, pp. 187-213.

An Experiment on Free Turbulence/Shock Wave Interaction

L. Jacquin, E. Blin and P. Geffroy

Office National d'Etudes et de Recherches Aérospatiales, 92 320 Châtillon, France

Abstract

Grid generated turbulence impinging on a standing normal shock wave is probed by LDV. The results show that the weak turbulence coming from the grid suffers no net amplification when traversing the shock, and that it decays more rapidly downstream. LDV accuracy in the vicinity of the shock is discussed. The results are analysed in the framework of available linear theories.

1. Introduction

Supersonic turbulent flows accelerated by expansion waves or decelerated by compression or shock waves involve different types of compressibility effects. The first effect is related to a rapid adaptation of turbulence to the variation in the mean density of the flow. Other effects are related to the compressible nature of the fluctuating field itself. Small fluctuations of the thermodynamic variables, whose effect could be neglected in equilibrium flows (e.g. in flat plate supersonic boundary layers at moderate Mach number), may influence the turbulent kinetic energy when they are coupled with large mean field gradients.

The majority of published experimental results in this field concern the shock/boundary layer interaction, a crucial configuration in supersonic and transonic aerodynamics (see [1] for a review). However, the complexity of this flow makes the physics of the interaction difficult to sort out. Modeling this interaction is not feasible with the basic low-speed turbulence models. The direction to be followed to improve these models has not been clearly identified, mainly because the new physics involved in this process need further clarifications (see e.g. [2]).

The aim of the present study is precisely to produce a much simpler flow which could help to scrutinize basic aspects of this problem. Indeed, the grid generated turbulence appears as a good candidate. Experiments on grid flow/shock interaction have been already performed in shock tubes (see [3], [4]) but in the present work, the experiment is conducted in a continuous wind tunnel which enables to perform classical measurements on longer time and length scales.

Apart from wall bounded flows, supersonic combustion systems may constitute an important application domain for shock interactions. While it is now well admitted that effects induced by compressibility inhibit the turbulent mixing, it has been observed that shocks can enhance shear flow turbulence. In particular, the various shocks which characterize the inviscid structure of non-adapted jets may contribute for a large part to the mixing properties of these flows and manipulation of supersonic free flows by impinging shocks constitutes an interesting topic for combustion applications.

This last subject has motivated another experiment which is presently still under investigation. This experiment is devoted to an interaction between a supersonic jet and a normal shock. First results were presented during the symposium, but they are not included in the present paper. This inhomogeneous test case, of a much more complex nature than the grid flow, will be the object of a future publication.

The results presented in sections 2 and 3 essentially concern mean and fluctuating velocities obtained with a two component Laser Doppler Anemometric system. A discussion is given in section 3 of the accuracy of this technique with regards to the problem of the particle behaviour across the shock. Section 4 is devoted to a review of theoretical results and a concluding discussion is given in section 5.

2. Experimental Results

The experiments have been conducted in one of the supersonic continuous wind tunnel of the Fluid Mechanics Laboratory at ONERA Meudon. The stagnation temperature and pressure are respectively 300 K and 10^5 Pa.

Figure 1 shows a schematic view of the test duct. The turbulence comes from a grid placed in the entrance section of a $120 \times 100 \times 450$ mm straight duct. The grid, a simple wire netting whose mesh size is $m = 7$ mm, serves as a sonic throat and as a turbulence generator, at the same time. It generates a supersonic flow whose Mach number is equal to about 1.65 with low but measurable turbulence fluctuations.

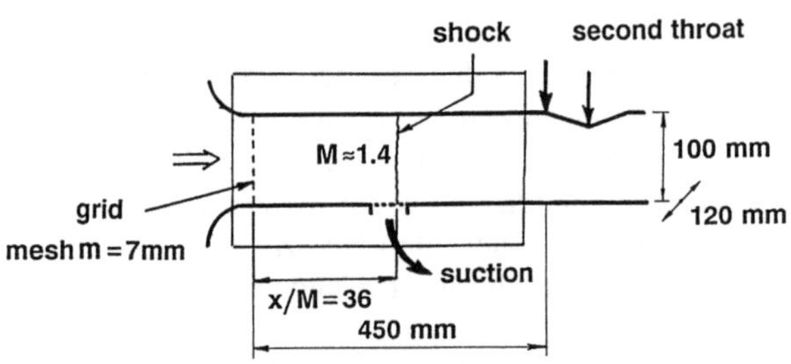

Fig. 1. Schematic view of the test section

A straight shock wave was controlled with a second adjustable throat located downstream. In the absence of a smooth nozzle profile, the stabilisation of the shock has been difficult to realize. After having tried different systems, we have been able to insure a satisfactory control of the shock by means of an aspiration of the bottom wall boundary layer. The shock fixes itself at the end of the suction part of the bottom wall, at a distance from the grid equal to $x/m = 36$. However, in spite of the suction, shock instabilities still arise from the interactions with the side wall boundary layers. A decrease of the Mach number, by means of a slight convergence of the top wall, has been also necessary to fully stabilize the shock.

Figure 2 shows a schlieren picture of the flow. One sees that the flow is perturbed by compression waves issued from the grid. These waves are dampened downstream, except those issued from the grid sides. The suppression of this phenomenon is hard and we had to endure it. The convergence of the upper wall together with the compression wave effects make the Mach number decreasing from about $M = 1.65$ at the grid level down to about $M = 1.39$ just upstream of the shock.

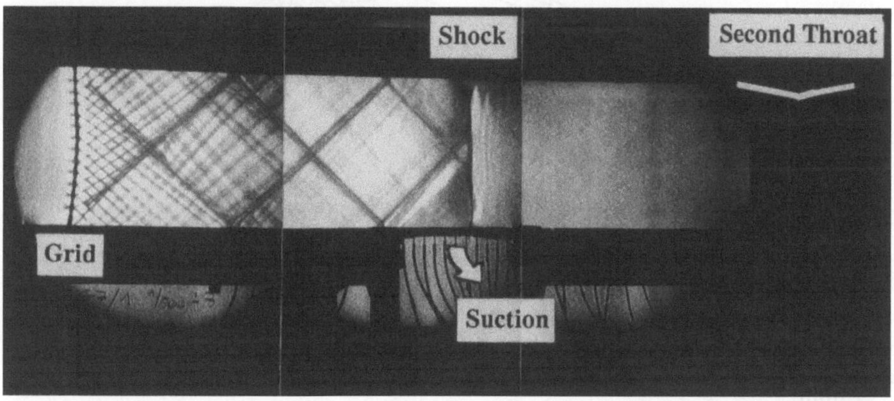

Fig. 2. Schlieren picture of the flow

The axial and vertical components of the mean and fluctuating velocities have been measured by LDV on a fine vertical mesh in the center of the duct ($y = 0$). The probing mesh size is $(\Delta x, \Delta z) = (10\ \text{mm}, 5\ \text{mm})$ upstream and downstream of the shock, and $(\Delta x, \Delta z) = (2\ \text{mm}, 5\ \text{mm})$ at the level of the shock. Incense smoke particles have been used to seed the flow. Each velocity value corresponds to a sample of 2000 instantaneous values. Some corrections have been tested in order to minimize the various usual bias errors. They have been found to be negligible. Another source of errors is related to the response of the particles to the shock induced deceleration. This problem will be discussed in section 3. But the main weakness of this experiment lies in the low signal-to-noise ratio obtained by LDV in such a weakly turbulent flow.

Figure 3(a) shows a map of the mean axial velocity U. The figure at the bottom shows the corresponding velocity contours. The mean velocity is highly perturbed by the compression waves issued from the grid (see Fig. 2) and the mark of the diamond-shaped compression waves can be identified on the contours. Just behind the shock, an abrupt thickening of the wall boundary layers leads to a slight local re-acceleration of the flow. The velocity then becomes nearly uniform.

Figure 3(b) shows the corresponding evolution of (twice) the turbulent kinetic energy defined as $q^2 = u^2 + 2w^2$. The energy amplitude is visualized for $-45 \leq z \leq 25$ mm so that only the bottom boundary layer is depicted. As we will see in section 3, the large amplitude peaks are due to the spatial intermittency of the shock but also to particle

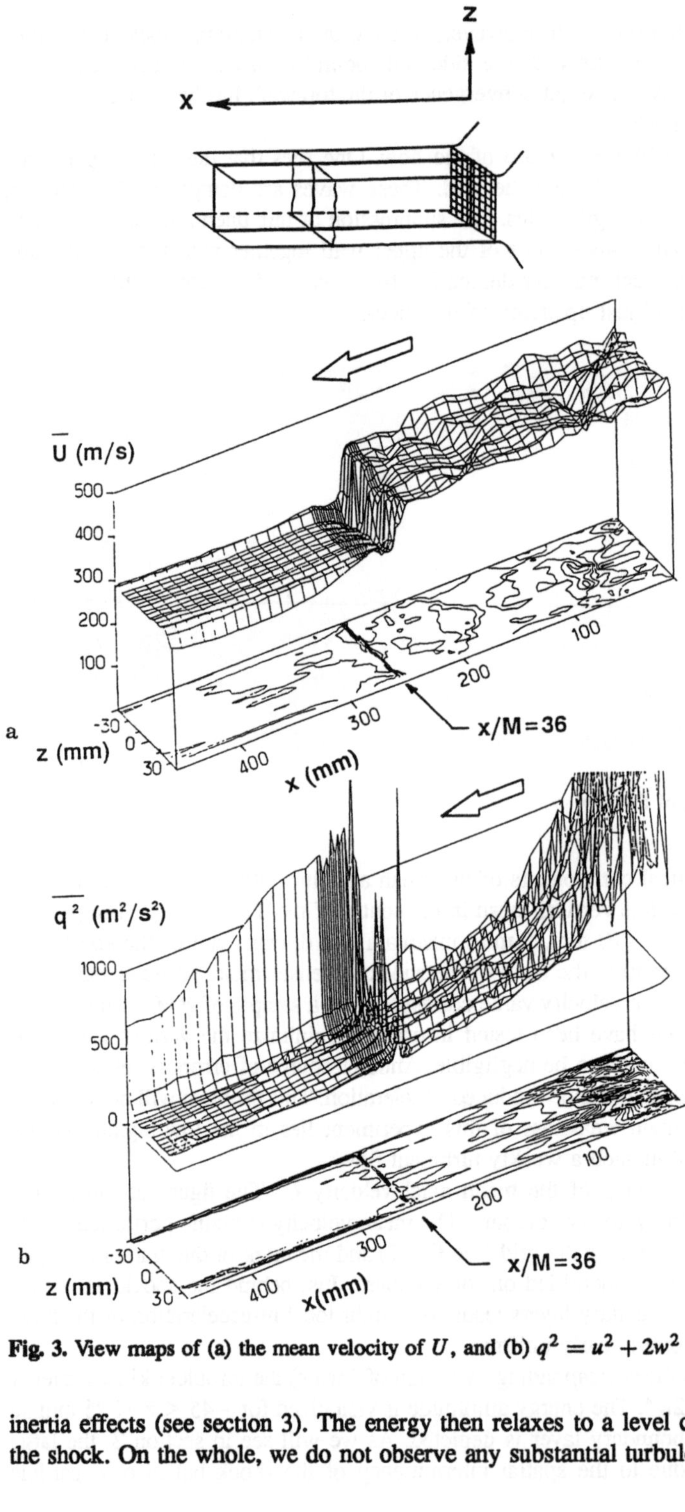

Fig. 3. View maps of (a) the mean velocity of U, and (b) $q^2 = u^2 + 2w^2$

inertia effects (see section 3). The energy then relaxes to a level close to that ahead of the shock. On the whole, we do not observe any substantial turbulence amplification.

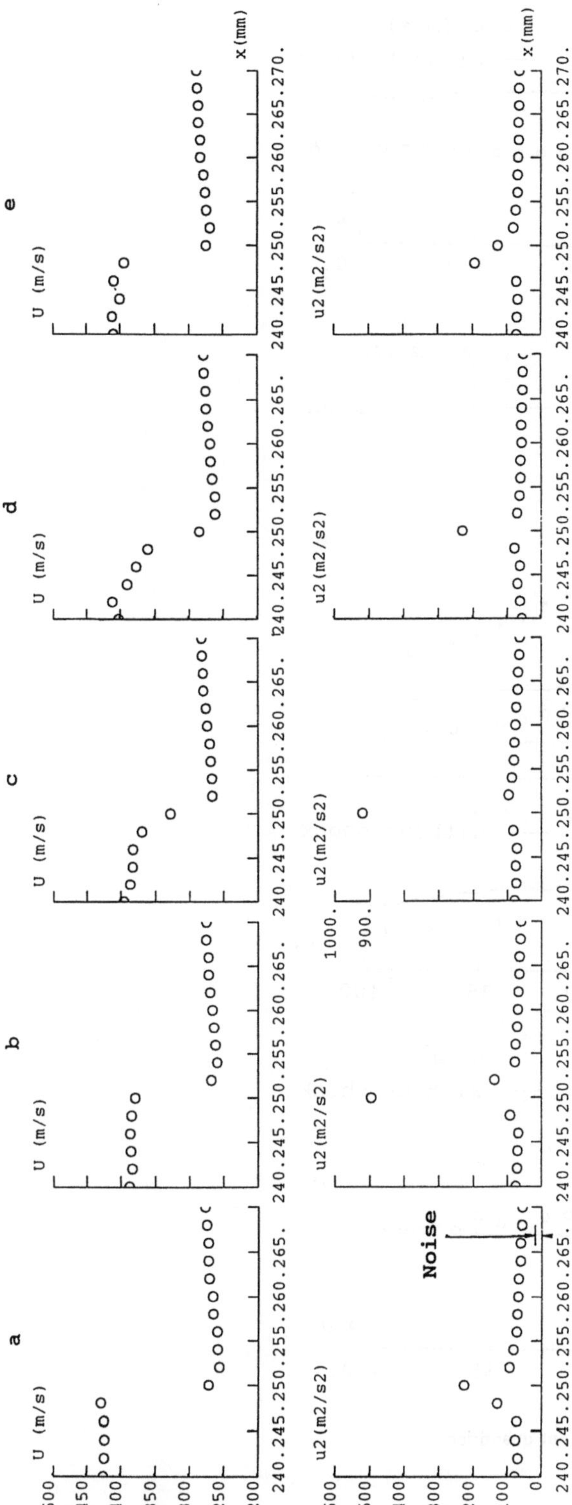

Fig. 4. Mean velocity and longitudinal energy at different locations along the shock of the grid flow experiment

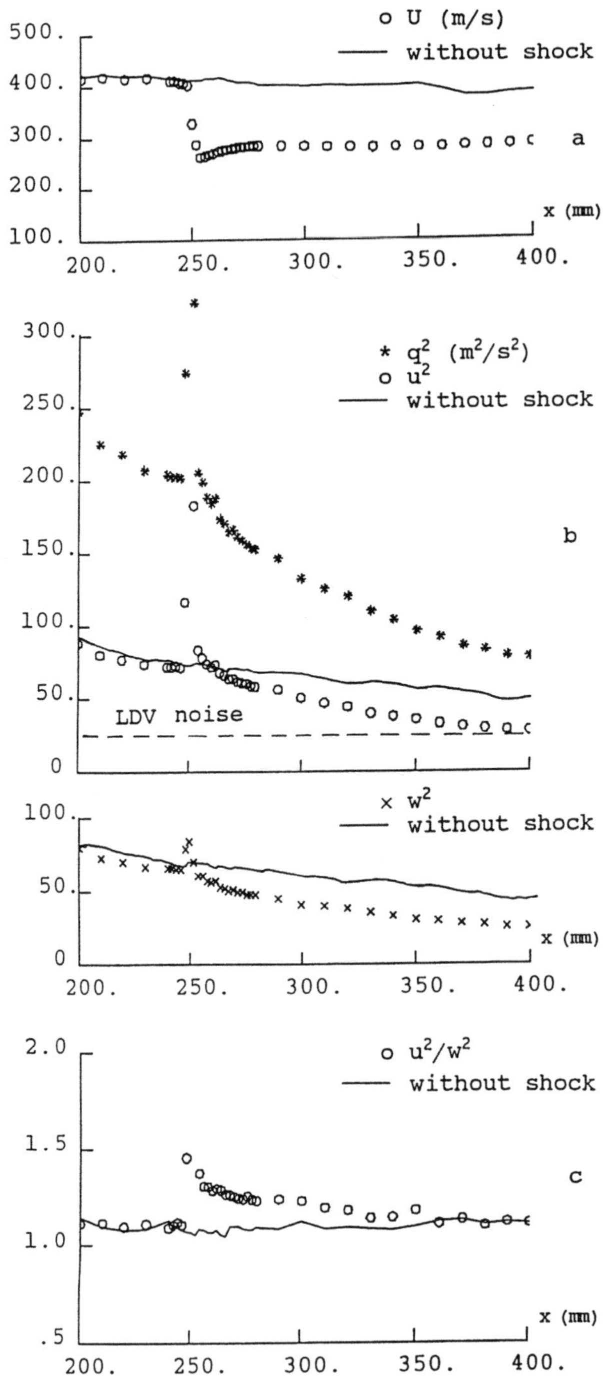

Fig. 5. *Z*-averaged mean and turbulent quantities

Figures 4(a) to (e) show details of some longitudinal evolutions of U and u^2 obtained at five different vertical locations in the nearfield of the normal shock. Note first that the longitudinal probing mesh size (2 mm) is not sufficiently dense to capture the details of the shock region. More refined measurements will be presented in section 3. The normal shock is not perfectly plane and its position varies along the z axis. This contributes to the apparently large thickness of the shock on the schlieren picture of Figure 2. The intensity of the shock also varies: note for instance in Figure 4(c) and (d), the decrease of the mean velocity ahead of the shock due to the effect of the compression waves. The re-acceleration of the mean flow behind the shock (boundary layer thickening effect) is also visible in the figures. The estimated level of the measurement noise is depicted on the figure. As we said, a low signal-to-noise ratio holds for this weakly turbulent flow.

Figure 5 shows the evolution of U, q^2, $\overline{u^2}$, $\overline{w^2}$ and $\overline{u^2}/\overline{w^2}$ calculated by averaging all the available results in the z direction. The integration is limited to values of z comprised between −30 and +20 mm, in order to avoid the influence of the side wall boundary layers. Only the region close to the shock is represented for the upstream part. Measurements performed without the shock have been also reported.

The turbulence rate $(q_0^2/2)^{1/2}/U_0$ ahead of the shock is equal to about 2.5%.

The presence of the peaks precludes a precise evaluation of the levels just behind the shock. Also, the slight local re-acceleration of the mean flow after the shock may contribute to reduce the turbulent energy. However, we can conclude that there is no significant amplification of q^2. Moreover, compared to the measurements without shock, a lower level of energy is obtained downstream. This increase of kinetic energy decay will be characterized in Figure 6.

In Fig. 5(c), the anisotropy ratio $\overline{u^2}/\overline{w^2}$, which is equal to about 1.1 before the shock, increases up to about 1.25 just behind, then it relaxes to its initial value.

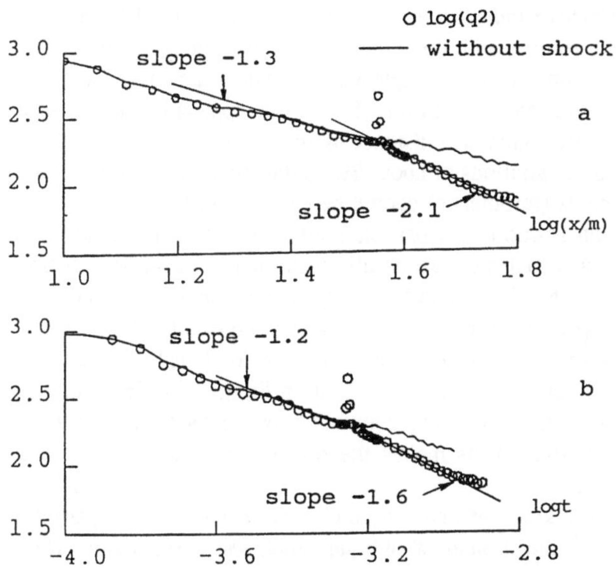

Fig. 6. Turbulent energy decay versus x/m (a) and t (b)

The log-log plot of q^2 versus x/m depicted in Figure 6(a), shows a significant increase of the kinetic energy decay behind the shock. The exponent of the decay law changes from -1.3 before the shock to -2.1 behind. In a uniform flow this change of slope could be directly related to a strong increase in the dissipation rate. However, before concluding, one must take into account the changes in mean velocity, which make for example the residence time of the turbulence downstream of the shock longer than upstream. Figure 6(b) thus shows the corresponding log-log plot of q^2 versus

$$t = \int_0^x \frac{dx}{U(x)}. \tag{1}$$

As expected, the decay is reduced with respect to that in Figure 6(a) but a change of slope is still apparent. One notes the similitude between the exponent for the decay law upstream of the shock with that which holds in incompressible grid generated turbulence. In the last measurement stations downstream, the noise of the measuring system contributes to an increase of energy.

On the whole, one may then conclude that there is a stronger increase of kinetic energy decay behind the shock.

3. On Measurement Accuracy in the Nearfield of the Shock

As mentioned before, LDV measurements are not accurate in the nearfield of the shock.

The first cause is related to laser beam/shock interactions. When the measurement point lies close to the shock, the laser beam which crosses the shock is slightly refracted by changes of density and refractive index along its path. This should modify the characteristics of the measurement volume and, consequently, the measurement itself. This phenomenon primarily depends on the initial angle the incident beam makes with the shock plane. Following the evaluation given in ref. [5], for our apparatus and our laser system configuration, this problem occurs on a distance of about $\Delta x = 2$ mm centered on the shock location, and it leads to variations of about 2% in the mean velocity. However, the effects on the higher moments (turbulence) cannot be easily evaluated.

A more important limitation is due to the particle inertia. Analyses of this problem have been given in several papers, but they are usually restricted to mean flow characteristics (see e.g. [5]). Investigations of the turbulence aspects are much more scarce [6]. The particles in the present experiments are polydisperse and, depending on their size, they react differently when traversing the shock. This leads to a spurious "particle turbulence" which contributes to the peaks of energy appearing in Figure 3(b). Intermittency of the shock also contributes to these peaks. The coupling between shock intermittency, fluid turbulence and particle inertia effects makes the respective contribution of each phenomenon difficult to sort out.

Inertia properties of different kinds of aerosols, including incense smoke particles, have been studied in ref. [5] on the ground of oblique shock test experiments. These experiments have been performed in the Fluid Mechanics Laboratory at ONERA Meudon, in a similar wind tunnel to that used in the present work. The configuration of the LDV

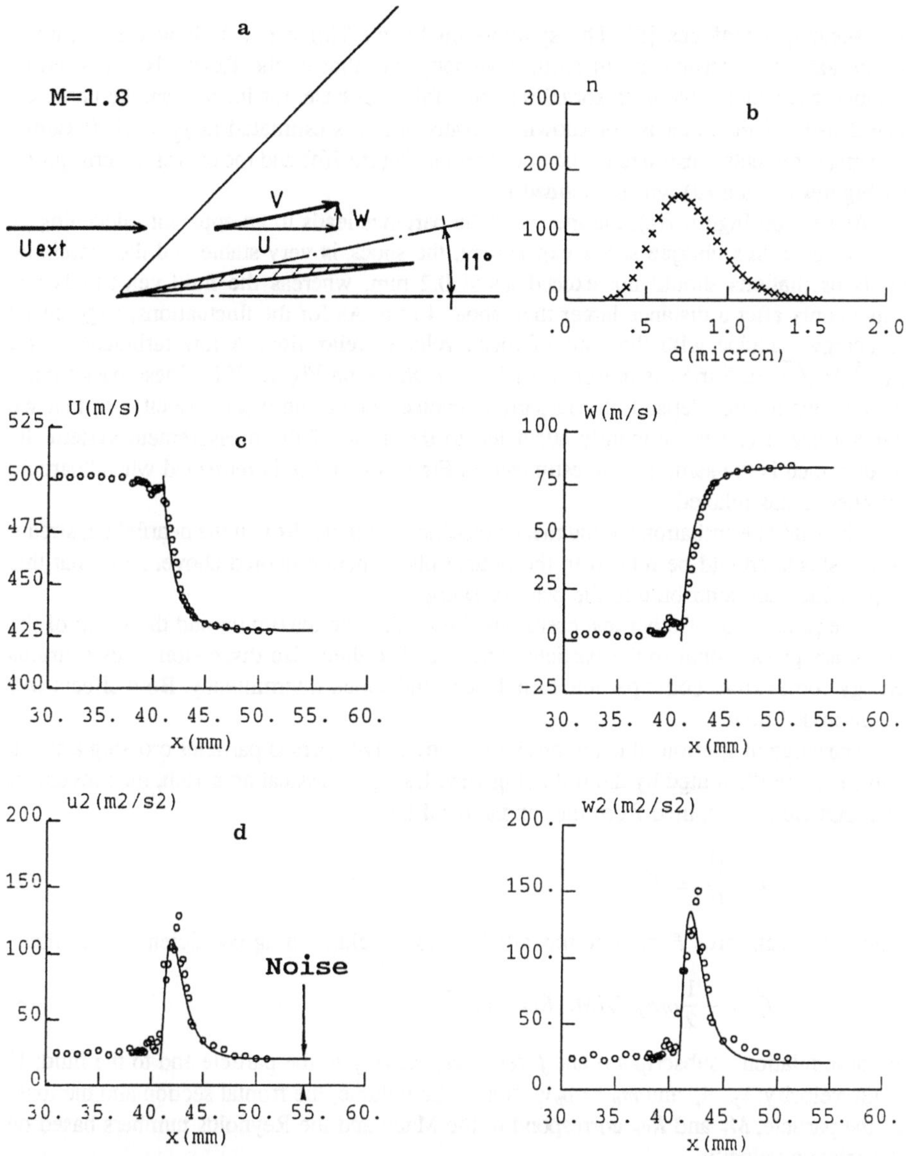

Fig. 7. Behavior of encense smoke particles in the seeding experiment of ref. [5] and theoretical predictions: (a) experimental apparatus; (b) particle size distribution, (c) mean velocity components, (d) spurious energy due to particle inertia

system is also the same. Mean flow characteristics have been presented in ref. [5]. By courtesy of the authors, we have been able to scrutinize the turbulent aspects of their data.

As sketched in Figure 7(a), a steady shock was fixed upon a 11° wedge in a Mach 1.8 flow. Two component LDV measurements have been performed around the shock location, far away from boundary layer influence, and for different kinds of particles and seeding techniques [5]. The symbols on Figure 7(c) and (d) show the evolution of the axial and vertical mean particle velocity and that of the Reynolds stress tensor components which have been sorted out from this data base, for incense smoke particles. The density of the latter is not known precisely and it is estimated as $\rho_p = 1000$ kg/m^3. The measurements mesh size is $\Delta x = 0.2$ mm in Figure 7(b) and the curves superimposed on Figures (c) and (d) will be defined later.

As seen on Figure 7(c), the inertia of the particles leads to an apparent thickening of the shock; in this oblique shock experiment, the shock is very stable and the amplitude of its oscillations should not exceed about 0.2 mm, whereas the fluid velocity is retreaved only after a distance larger than about 4 mm. As for the fluctuations, large peaks of energy coincide with the zone of mean velocity relaxation. A low turbulence level ($\sqrt{\overline{u^2}} \simeq \sqrt{\overline{w^2}} \simeq 5$ m/s) is present ahead of the shock on Figure 7(d). These background fluctuations neither depend on the particle choice nor on the mean velocity, so that we hypothesize it can be primarely attributed to the noise of the measurement system. Its level, which is the same as that estimated in Figures 4 and 5, is retreaved when "particle turbulence"has relaxed.

The slight perturbations of the mean velocities in Figure 7(c), in the nearfield upstream of the shock, should be related to the optical phenomenon quoted above. Note that this region does not contribute to the peak of energy.

The processing of other test cases has shown that the maximum and the width of the peaks are proportional to the particle density and to their size dispersion. This spurious energy comes from both "particle turbulence" and shock intermittency. Both effects will be discussed now.

The surge of spurious fluctuations induced by polydispersed particles crossing a shock wave may be illustrated by the following way. Using a classical approach, the movement of a particle is computed from the fundamental law:

$$m_p \frac{d\vec{V}_p}{dt} = \vec{F} \tag{2}$$

where the drag force F may be modelled by introducing a drag coefficient C_x, so that:

$$\vec{F} = -\frac{1}{2}\rho_f A_p C_x(M_r, Re_r)\,|\vec{V}_r|\vec{V}_r \tag{3}$$

In these relations, subscripts p and f refer respectively to the particle and to the fluid; V is the velocity; v_p, A_p and $m_p = \rho_p v_p$ denote the volume, the frontal section and the mass of the particle, M_r and Re_r correspond to the Mach and the Reynolds numbers based on the relative velocity:

$$\vec{V}_r = \vec{V}_p - \vec{V}_f$$

$$M_r = \frac{V_r}{a}; \quad Re_r = \frac{V_r\sqrt{A_p}}{\nu} \tag{4}$$

For C_x, we have used the formulation given by Henderson [7]. Concerning the dispersion in size, we have supposed that it corresponds to a log-normal distribution. The log-normal law depicted in Figure 7(b) has been fitted by imposing the more probable size, d_0, and the maximal size, d_{max}, of the particle population. These parameters are chosen so as to obtain the "best" agreement with the mean velocity results.

The curves in Figures 7(c) and (d) have been obtained by computing the above relations with $(d_0, d_{max}) = (0.7, 1.6)$ and $\rho_p = 1000$ kg/m^3. The background noise energy has been added to the results of the computation. Note that the slight intermittency of the shock and optical phenomena preclude a precise determination of the theoretical shock location from LDV measurements alone. Thus, the choice of this position (initial point of the computation) must be estimated. One sees in Figure 7(c) that the exact shape of the mean velocity profile is not well described by the theory in the steepen gradient region. In Figure 7(d), the widths of the peaks are well reproduced, but the peak heights are underestimated. However, such a calculation is highly sensitive to the assumed characteristics of the particles, which are misunderstood, and it neglects several important phenomena, e.g. shock intermittency, optical distortions and discrimination by the reception process of the measuring system. Consequently, these results must be taken with caution. They have just been presented in order to illustrate the "particle turbulence" phenomenon in the absence of fluid turbulence.

Concerning the shock intermittency, one notes that, even if shock oscillations occur in a very thin region, particle inertia makes the induced perturbations propagate downstream. Basic intermittency relations show that the movement of a shock on both sides of a fixed measurement point brings an additional energy whose maximum is about $[\Delta U]^2/4$ and $[\Delta W]^2/4$ for $\overline{u^2}$ and $\overline{w^2}$ respectively (see [8]), where ΔU denotes the difference between the fluid velocities ahead and behind the shock. This leads to about 1400 and 2000 m^2/s^2 for this oblique shock experiment. One would have obtained such values with a hot-wire probe, as in ref. [8]. But here, particle inertia dampens this effect and, also, shifts the location of the maximum of the peak energy downstream. On the whole, the movements of the shock surface must contribute to an increase of the "particle turbulence" peak obtained in the case of a standing shock wave.

The characteristics of this peak thus depends on the particle sizes and on the shock unsteadiness. But it also depends on the optical and electronic arrangement of the system. For instance, the smaller particles which will be taken into account are those which scatter a sufficient amount of light. For a particular particle size, and a fixed laser beam power, the scattering will be proportional to the measuring control volume. Modification of the latter can then enable to select different ranges of particle sizes.

Some tests have been conducted in our normal shock experiment with a smaller measuring control volume than that used for the results presented in the previous section. This reduction has been obtained by changing both the focusing lens and the beam expanders. For instance, with this new arrangement, the width of the measuring control volume has been reduced from 200 μm to 120 μm. This allows smaller particles to be taken into account.

Figure 8 shows results obtained with this new configuration, using two different kinds of particles: incense smoke, as previously, and olive oil. The density of the latter (918 kg/m^3) is equivalent to that of incense smoke, but the mean size of oil particles is smaller. The figure shows that for incense smoke, selection of smaller particles has led to a significant reduction if the characteristic length of the particle inertia effects, when compared with Figures 4 and 7. The width of the energy peak has been reduced from about 4 mm to 2 mm. This value decreases down to about 1 mm with olive oil. At the same time the maximum value of the peak is shifted upstream and its maximum increases. One may hypothesize this last effect is mainly related to shock intermittency, whose contribution must increase as the particle inertia decreases. Note that shock unsteadiness is certainly much more important in this turbulent flow-normal shock interaction than in the oblique shock test case.

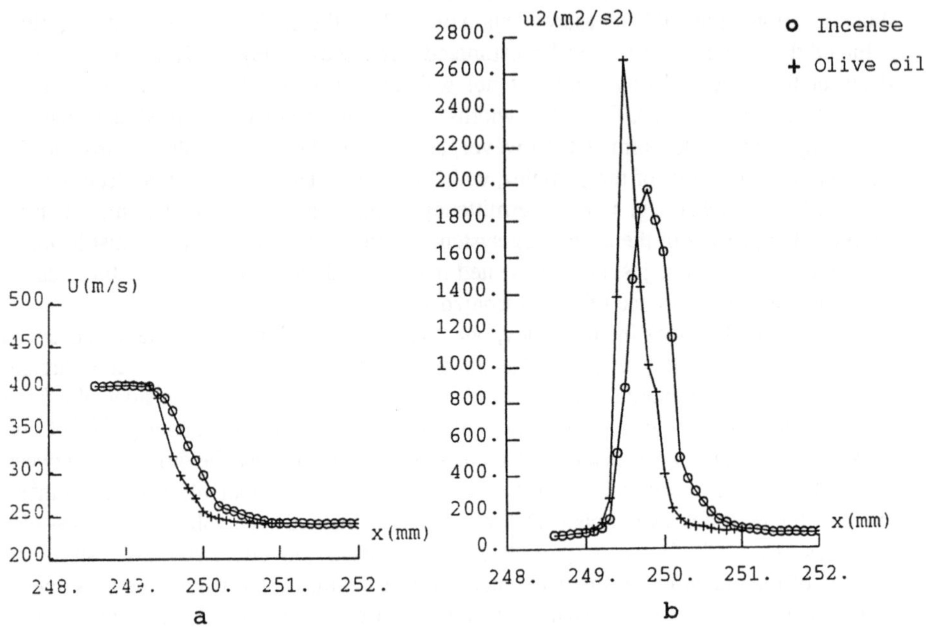

Fig. 8. Refined measurements in the nearfield of the normal shock at $z = -20$ mm, using a reduced probing control volume and two different kinds of seeding particles

At last, one must note that these additional results obtained with a new optical arrangement and different particles, show again no significant amplification of the incoming fluctuations.

Note also, that the spurious phenomenon described above may perturbate the measurements on a larger scale in interactions involving higher velocities, more complex shock structures (as in a boundary layer or in a jet) or heavier particles (usually used in heated or in reacting flows).

Direct numerical simulations, in the same way as those performed in [9] for the mixing layer, could greatly help to scrutinize rigourously this problem.

4. Theoretical Predictions

The generation of thermal and acoustic fluctuations and the amplification of the vorticity fluctuations resulting from the interaction of a turbulent field with a shock wave have been first studied by Ribner [10]. This author has considered the linear response of the flow behind a shock perturbed by an impinging single shear wave. Several authors have then reformulated and completed this linear approach (see e.g. [11] for a review). A second candidate, restricted to mean compression effects, is the linear analysis of homogeneous compressed turbulence, the so-called rapid distortion theory (RDT; e.g. [12]-[14]).

In a recent paper, Lee *et al.* [15], have presented results obtained by using direct numerical simulation (DNS), RDT and the "Linear Interaction Analysis" (LIA) of Ribner. They have applied them to the case of an initially incompressible turbulence passing through a normal shock. As for the linear theories, they conclude that RDT and LIA lead to the same amplification and anisotropy for moderate shock intensities (upstream Mach number $M_0 < 2$). The differences observed for larger shock intensities were attributed to unsteadiness and curvature effects of the shock which are not taken into account by RDT.

These two theories are of very different nature and the similitudes between them are not straigthforward. They will be now discussed briefly.

4.1 Rapid Distortion Theory

In RDT, the complex interaction phenomena between the fluid and the shock discontinuity are ignored. The theory only accounts for the mean compression of the shock, as in a piston engine. The basic equations of RDT are deduced from the linearization of the equations of gasdynamics for a homogeneous medium. One starts from the linearized equation of the fluctuating velocity u:

$$\frac{Du_i}{Dt} + u_j D_{ij} + \frac{p_{,i}}{\bar{\rho}} + \frac{\rho}{\bar{\rho}}\frac{D\bar{U}_i}{Dt} = 0, \tag{5}$$

with p and ρ denoting the fluctuation of pressure and density, D_{ij}, the mean velocity gradient, and D/Dt, the particle derivative:

$$D_{ij} = \bar{U}_{i,j}\ ; \qquad \frac{D}{Dt} = \frac{\partial}{\partial t} + \bar{U}_k \frac{\partial}{\partial x_k}. \tag{6}$$

For the solution of (5) to be homogeneous, one needs:

$$\bar{\rho} = \bar{\rho}(t); \quad D_{ij} = D_{ij}(t); \quad \frac{D\bar{U}_i}{Dt} = 0. \tag{7}$$

In this case, the mean continuity equation, leads to the following compression law:

$$C(t) = \frac{\bar{\rho}(t)}{\bar{\rho}_0(0)} = \exp\left\{\int_0^t -D_{ii}(t)dt\right\} \tag{8}$$

$C(t)$ characterizes the intensity of the compression.

From the last equation (7), it can be shown that the basic flow must fulfill the following condition:

$$\frac{d}{dt}D_{ij} + D_{ik}D_{kj} = 0. \tag{9}$$

The solution of (9) for an axial compression of axis 1 reads:

$$D(t) = \frac{D_0}{1 + D_0 t}; \quad D_{ij}(t) = D(t)\delta_{i1}\delta_{j1} \leq 0; \quad D_0 = D(0) \tag{10}$$

Now, in accordance with (5), condition (9) may be relaxed if one considers solenoidal fluctuations ($\rho = 0$). Using the general methodology given in [13], the solutions of this solenoidal version of RDT can be managed analytically. Considering for example the velocity component u aligned with the compression axis, one can define its initial energy by:

$$\overline{u_0^2} = \iiint U_{11}(\vec{K}) d\vec{K} \tag{11}$$

where $U_{11}(k)$ is the first diagonal component of the second order spectral tensor. Soenoidal RDT leads to the following expression inside the compression:

$$\overline{u^2} = C^2 \iiint U_{11}(\vec{K}) \{1 + P(\mu)\}^2 d\vec{K} \tag{12}$$

with $\mu = K_1/K$, the cosine of the angle between the wave vector K and the compression axis. The directional factor $P(\mu)$,

$$P(\mu) = \frac{\left(1 - C^2\right)\mu^2}{1 - \left(1 - C^2\right)\mu^2} \tag{13}$$

accounts for the pressure effect on the solenoidal field which constrains $U_{ij}(\vec{K})$ to remain in the plane normal to $\vec{K}$.

Applying the theory to the shock, the above relation shows that the result of the interaction strongly depends on the initial angular distribution of the energy. The maximum amplification concerns the wave vectors parallel to the shock ($K_1 = \mu = 0$). Thus, turbulence with energy concentrated in a spectral region almost parallel to the shock will be more amplified than isotropic turbulence. In the most practical applications (e.g. boundary layers), the incident field is anisotropic and the above mentioned directional aspects may be important for the interpretation.

On the other hand, if the damping effect of pressure were ignored ($P(\mu) = 0$), from (12), a simple amplification factor equal to C^2 would be obtained for $\overline{u^2}$, the transversal components remaining unchanged. For the kinetic energy, this leads to the following simple relation:

$$\frac{q^2}{q_0^2} = \frac{\left(2 + C^2\right)}{3} \tag{14}$$

This expression thus gives the amplification of turbulence energy, if pressure was released during the compression. A release of pressure effects in the shock interaction has been adopted in the analysis of Debiève [16] (see also [13]), by arguing that pressure terms do not "have time" to act inside the shock discontinuity.

Now, RDT solutions for an initially anisotropic field cannot be constructed without a precise formulation of the spectra $U_{ij}(\vec{K})$. If considering, for simplicity, an isotropic initial field, the solenoidal RDT solution for the kinetic energy simply amounts to [13]:

$$\frac{q^2}{q_0^2} = \left\{ C^2 \frac{\arctan\sqrt{C^2 - 1}}{\sqrt{C^2 - 1}} + 1 \right\} \tag{15}$$

Taking into account pressure leads through (15) to less amplification than that corresponding to (14).

Figure 9(a) shows the amplification factor of the kinetic energy obtained by applying relations (14) and (15) for moderate values of C.

At least, one can show that compressible RDT leads to amplification factors comprised between the two limits given by (14) and (15) (see [17]). In particular, DNS results concerning initially isotropic turbulence submitted to an axial compression have shown larger amplifications than those of relation (15). These DNS results have been reported in [19].

Note also that, applying the theory to the shock, turbulence intensity is limited by the following criterium:

$$\frac{\left(q_0^2/2\right)^{1/2}}{\Delta U} = \frac{C}{C-1}\frac{\left(q_0^2/2\right)^{1/2}}{U_0} \ll 1; \quad \Delta U = U_0 - U_1, \tag{16}$$

for the rapid distortion hypothesis to be valid. Subscript 1 refers to the flow downstream from the shock.

4.2 Ribner's Theory

LIA, which is based on the Rankine-Hugoniot jump relationships, is much closer to the physics of the interaction than RDT. In particular, Ribner's LIA has shown to be impressively efficient for the analysis of acoustic properties of a shock-vortex interaction [18].

As for the similarities between LIA and RDT, in both cases the amplification factors are fixed by the ratio of the mean densities (see (8)):

$$C = \frac{\bar{\rho}_1}{\bar{\rho}_0} \tag{17}$$

However, inspection of this theory shows very complex behaviors which cannot be easily interpreted. A key point is likely the way the shock surface is supposed to react to the fluid perturbations. When a shear wave (solenoidal field) impacts the shock, it is transmitted and amplified, but entropic and acoustic modes are also generated. The amplification of the impinging shear wave is then largely controlled by the perturbation of the shock surface, which is itself sensitive to the generated acoustic field: as shown by Ribner [10], for particular orientations of the upstream mean flow and of the impinging shear wave vectors, the shock oscillations may be strongly amplified by downstream pressure perturbations. For the sake of simplicity, we can say that the wave vectors of the upstream disturbance field which contribute in this critical regime are those which make an angle with respect to the shock plane smaller than or equal to a "critical" angle θ_c, whose value is smaller than 30°. Wave vectors belonging to this angular domain participate in the emergence of an intense near pressure field behind the shock, and contribute in a large part to the kinetic energy amplification. Wave vectors exactly oriented with an angle θ_c contribute the most to the shock vibration.

Thus, as in RDT, the wave vectors almost parallel to the shock play a particular role in the amplification process.

Another consequence is that the downstream pressure field is composed of two parts: the above mentioned nearfield, which decays rapidly downstream, and a regular acoustic field which propagates away.

Figure 9(b) shows the amplification of kinetic energy calculated from Ribner's theory assuming an isotropic incompressible initial field. We have depicted the contribution of the shear waves and that of the pressure waves. For the latter, we have shown both the nearfield level, just behind the shock, and the asymptotic one, which corresponds to propagating acoustic waves. This result shows that just behind the shock, the nearfield pressure contributes for a large part to the kinetic energy amplification. Away from the shock, the vanishing of the nearfield contribution leads to values which are close to those given by RDT, as quoted in [15].

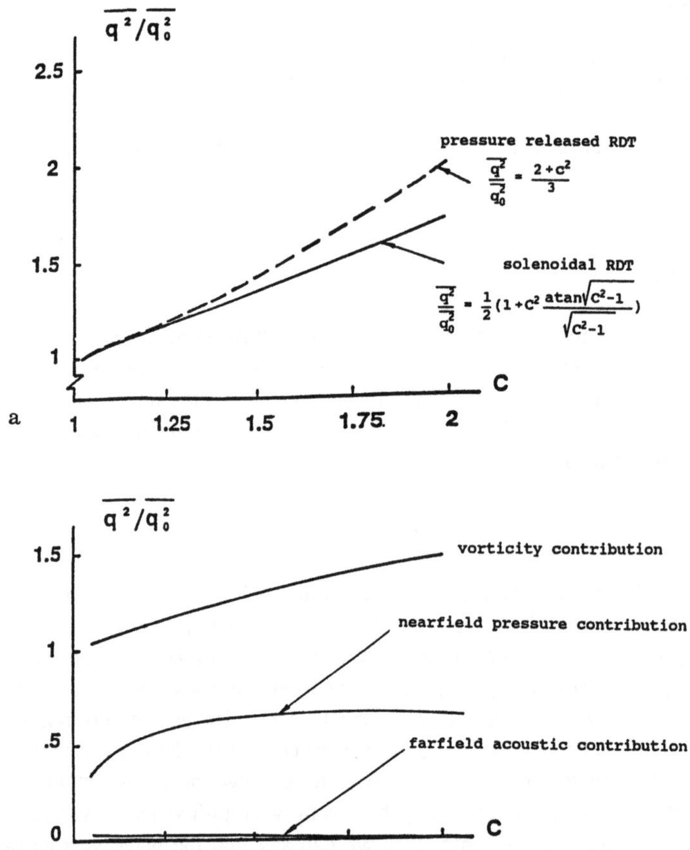

Fig. 9. Kinetic energy amplification across a normal shock given by solenoidalor pressure release RDT (a) and Ribner's LIA theory (b)

4.3 Concluding Remarks

As a conclusion of this review we can say that:

– RDT enables to evaluate the turbulence amplification associated to the shock compression. The lower amplification is obtained if one supposes that the field is solenoidal,

and the release of the pressure effects leads to a maximum. Very simple expressions are obtained for these two limiting cases.
– For moderate shock intensity, and if one considers an initially isotropic and incompressible field, the two theories, LIA and RDT, predict similar amplifications of the turbulent kinetic energy.
– However, the anisotropic aspects of the problem seem to play an important role in the two theories, but for very different physical reasons. In particular the wave vectors almost parallel to the shock play a special role. They are most amplified in RDT, and in LIA, they are responsible for the pressure nearfield and they largely contribute to the amplification phenomena driven by the shock oscillation. Characterization of this anisotropy (from the Reynolds stress tensor, or better, from the lengthscales) may be important for efficient predictions.
– Concerning LIA, the role of the nearfield pressure behind the shock must be emphasized. In particular it may greatly enhance turbulence in the vicinity of the shock. Concerning this point, one can note that this phenomenon could contribute to the peaks of energy which have been scrutinized in section 3.

At last one must evoke a basic similarity between the two theories: both lead to the same transmission laws between the incident and the refracted vorticity wave vectors. The relation reads precisely:

$$k_1 = CK_1, \quad k_2 = K_2, \quad k_3 = K_3. \tag{18}$$

Subscript 1 refers to the direction normal to the shock; K and k denote the wave numbers ahead and behind the shock. Thus the wave distortion is equivalent in the two theories. In particular the compression associated with a shock ($C > 1$) induces a shift of energy towards large wave numbers. An increase in the dissipation rate ϵ may thus be expected. The solenoidal part of ϵ is given by:

$$\epsilon = \nu\overline{\omega^2} \tag{19}$$

where ω denotes the fluctuation of vorticity, and $\overline{\omega^2} \doteq \overline{\omega_i\omega_i}$, the enstrophy. In the case of irrotational mean flow such as a pure compression, a simple Cauchy solution holds for ω (see e.g. [12]). For an axial compression acting on an isotropic initial turbulence, one obtains:

$$\frac{\overline{\omega^2}}{\overline{\omega_0^2}} = \frac{1+2C^2}{3}. \tag{20}$$

5. Discussion

Various aspects of the experiments will be discussed now in the framework of the theories evoked in section 4.

In the experiment, the density ratio is about $C \simeq 1.55$, the turbulence intensity ahead of the shock is about $(q_0/2)^{1/2}/U_0 \simeq 0.025$, and for relation (16), one has $(q_0/2)^{1/2}/\Delta U \simeq 0.07$.

Even if the Reynolds stress tensor is not sufficient to check the (spectral) anisotropy, we may suppose that the latter ahead of the shock, is moderate (see Fig. 5(c)). Thus an isotropy assumption seems nearly justified for this experiment. In the framework of this hypothesis, the kinetic energy amplification predicted by relation (15) is $q^2/q_0^2 \simeq 1.35$. Similar values are obtained from LIA.

Such amplifications are not observed in our results. This constitutes the main paradox of this study.

Note however that the noise level, depicted in Figure 5(b) is not amplified by the shock and must be substracted when considering amplification. But this does not change our conclusion. Another point is that the local re-acceleration just behind the shock may also contribute to a reduction of the energy level. However, if using a RDT evaluation, this reduction is not significative. One must also note that, in view of the highly perturbed incoming flow, a part of the fluctuations measured upstream from the shock may originate form acoustics. Solenoidal RDT does not apply for these acoustic components but LIA does and shows that they are amplified.

In other words, the experimental results in the grid flow baffle the linear theories.

A fact is that this low turbulence flow constitutes a limiting case for LDV measurements. Other limitations than those quoted in this paper may exist. However, absence of net amplification has been also observed in the experiment of ref. [8] where a low intensity free turbulence (emanating from a jet) was probed during its interaction with an oblique shock. The two-dimensional DNS of ref. [15] also shows less amplification than that predicted by RDT. An interpretation of this DNS results, in terms of potential and kinetic energy exchanges, has been proposed in ref. [19].

As for the downstream behavior, the results of Figure 6 confirm an increase in the energy decay as predicted by the analysis given at the end of section 4. An increase in the enstrophy has been also reported in the DNS of ref. [15].

6. Conclusion

On the whole, this experiment has revealed quite unexpected results. The salient observations are an absence of net amplification of the turbulent kinetic energy and an increase of its decay rate downstream of the shock. Such trends go in the same direction than the experimental results of ref. [8] and the numerical results of ref. [15].

The classical linear theories which we have recalled in the paper predict amplifications which have not been observed.

Part of the paper has been devoted to a peculiar problem inherent to LDV application in this flow. We have shown how the polydisperse nature of the particles and the unsteadiness of the shock contribute to the high amplitude peaks of energy which are observed at the shock location. This problem as well as other problems related to LDV application in shock interactions are the subject of ongoing research.

Further efforts will be necessary before sorting out the physics of the interaction. In particular, the experimental results which have been presented must be completed by other explorations with other techniques, such as the hot-wire. The apparatus which has been employed may be also used to scrutinize the influence of different parameters. First, ongoing modifications of the apparatus will enable to reach higher upstream Mach numbers. The possibility of changing the position of the shock inside the decaying grid turbulence is also envisaged. This would enable, in particular, to check the influence of the initial turbulence intensity.

Lastly, the present results may enlighten some of the peculiarities of the complex interactions encountered in aerodynamic applications. From inspection of the properties of the linear solutions (section 4), one may conjecture that the intensity and anisotropy of the incoming turbulence, as well as the behavior of the pressure fluctuations in the nearfield behind the shock, could participate in the amplification of turbulence kinetic energy which has been observed in boundary layers, for instance. First results obtained in the jet flow interaction mentioned in the introduction have shown amplifications analogous to those reported in [1] and [20] for wall bounded flows. As quoted by Ribner, in these complex flows, inhomogeneous mechanisms, as for instance an interaction between the nearfield pressure and the flow boundaries, could also play an important role.

Acknowledgement

We are grateful to the experimental department of ONERA and particulary to C. d'Humières and F. Micheli for their contribution to this work. This research has been supported by DRET under contract no. 89.34.001 BC 210.

References

1. Délery J. and Marvin J.G.: Turbulent shock/boundary layer interaction. AGARdograph 280, 1980.
2. Bushnell D.M: Turbulence modeling in aerodynamic shear flows – Status and problems. AIAA 91-0214.
3. Keller J. & Merzkirch W.: Interaction of a normal shock wave with a compressible turbulent flow. Experiments in Fluids, 8, pp. 241-248, 1990.
4. Honkan A. and Andreopoulos J.: Experiments in a shock wave/Homogeneous turbulence interaction. AIAA paper No 90-1647.
5. Micheli F., d'Humières C. & Papirnyk O.: Aerosol behavior in supersonic flows, 4th International Conference on Laser Anemometry, Cleveland, Ohio (USA), August 1991.
6. Bloomberg J.E., Dutton J.C. & Addy A.L.: An investigation of particle dynamics effects related to LDV measurements in compressible flows, Report UILU ENG 89-4009, 1989.
7. Henderson C.B.: Drag coefficients of spheres in continuum and rarefied flows, AIAA Journal, vol. 14, 1976.
8. Debiève J.F. & Lacharme J.P.: A shock wave/free turbulence interaction. IUTAM Symposium on Turbulent shear layer/shock wave interactions, Editor J. Delery, Springer Verlag, 1985.
9. Samimy M. & Lele S.K.: Motion of particles with inertia in a compressible free shear layer, Phys. Fluids A 3 (8), 1991.
10. Ribner H.S.: Convection of a pattern of vorticity through a shock wave. Lewis Flight Propulsion Laboratory. RP 1164, 1953.

11. Zang T.T., Hussaini M.Y. & Bushnell D.M.: Numerical Computation of turbulence amplification in shock-wave interactions. AIAA Journal, 23, No1, 1984.
12. Lee M.J.: Distortion of homogeneous turbulence by axisymmetric strain and dilatation, Phys. Fluids A, 1,, pp. 1541-1557, 1989.
13. Cambon C.: Linear spectral approach to turbulence subjected to mean velocity gradients or body forces. PEPIT Summer Workshop, Lyon, France july 1989.
14. Sabel'nikov V.A.: Pressure fluctuations generated by uniform distortion of homogeneous turbulence, Fluid Mechanics-Soviet-Research, Vol 4, No 6, 1975.
15. Lee S., Lele S.K. & Moin P.: Direct numerical simulation and analysis of shock turbulence interaction. AIAA paper 91-0523.
16. Debiève J.F.: Etude d'une interaction turbulence onde de choc. Thèse de Doctorat d'Etat. Univ. Aix-Marseille II, 1983.
17. Jacquin L. and Cambon C.: Turbulence amplification by a shock wave and rapid distortion theory. Submitted fo Phys. Fluids. A.
18. Ribner H.S.: Cylindrical sound wave generated by shock-vortex interaction, AIAA Journal, Vol. 23, No 11, 1985.
19. Zeman O.: Compressible turbulence subjected to shear and rapid compression, 8th Symposium on Turbulent Shear Flows, Munich, 1991.
20. Dussauge J.P., Fernholz H.H., Finley P.J. and Smits A.J.: A survey of measurements and measuring techniques in rapidly distorded compressible turbulent boundary layers. AGARDograph NO.315, 1989.

Compressible Homogeneous Shear: Simulation and Modeling

S. Sarkar, G. Erlebacher, and M.Y. Hussaini

Institute for Computer Applications in Science and Engineering, NASA Langley Research Center, Hampton, VA 23665, USA

Abstract

The present study investigates compressibility effects on turbulence by direct numerical simulation of homogeneous shear flow. A primary observation is that the growth of the turbulent kinetic energy decreases with increasing turbulent Mach number. The sinks provided by compressible dissipation and the pressure-dilatation, along with reduced Reynolds shear stress, are shown to contribute to the reduced growth of kinetic energy. Models are proposed for these dilatational terms and verified by direct comparison with the simulations. The differences between the incompressible and compressible fields are brought out by the examination of spectra, statistical moments, and structure of the rate of strain tensor.

1. Introduction

Homogeneous turbulent flow has long been viewed as a simple model of turbulence which is sufficiently realistic to have fundamental physical features that persist in more complex flows. For example, hairpin vortices which have been observed in boundary layer experiments and simulations of channel flow, have been identified through direct numerical simulation (DNS) of homogeneous shear flow by Rogers and Moin (1987). Similarly, the longitudinal streaks observed in wall boundary layers have been shown by the simulations of Lee, Kim and Moin (1990) to exist in homogeneous shear flow when the shear rate is high enough to be comparable to those encountered near the wall in boundary layers. Rogallo (1981), who was the first to perform DNS of this flow, showed by direct comparison with physical experiments that the simulated flow statistics were in good agreement with experimental data.

In the past several years, there has been a strong activity in the area of compressible turbulent simulations. The first fundamental study of compressible isotropic turbulence was conducted by Passot and Pouquet (1987) who demonstrated the presence of shocks in the flow. Their results motivated the study by Erlebacher et al. (1990) to classify the different possible regimes in isotropic turbulence attainable from arbitrary initial conditions. They quantified the conditions for the presence or absence of shocks. Other results on decaying isotropic turbulence can be found in Sarkar et al. (1989), Lee, Lele and Moin (1990), and Zang, Dahlburg, and Dahlburg (1992). Results on forced isotropic turbulence are also available (Kida and Orszag 1990). Simulations have now moved on to the next level of complexity, namely, homogeneous turbulence subjected to a linear mean velocity field. For example, Coleman and Mansour (1991) have considered turbulence subjected to homogeneous compression.

Homogeneous shear turbulence has a production term which continuously feeds energy into the system, which then cascades down to the smaller scales. This flow exhibits some of the properties of incompressible turbulent shear flows, with the added presence of acoustic waves, density fluctuations, and dilatational velocity fields. Direct numerical simulations of such flows can provide complete statistics not available from physical experiments. Compressibility effects on the flow can be obtained from the simulations, and perhaps more importantly, mechanisms responsible for these effects can be identified, and turbulence models for compressibility-related phenomena can be devised. The compressible problem was considered by Feiereisen et al. (1982) who performed relatively low resolution 64^3 simulations and concluded that compressibility effects are small. Recently Blaisdell, Mansour and Reynolds (1991) have also considered compressible shear flow, and identified eddy shocklets in the flow for high enough turbulence Reynolds and Mach number.

Research on the small-scale properties of homogeneous turbulence is primarily focussed on understanding the intermittency of high dissipation regions. It has been numerically established that the rate of strain tensor has a preferred shape in regions of strong dissipation (Ashurst et al. 1987b). Furthermore, the vorticity tends to align with the intermediate eigenvalue (whose most probable sign is positive). While this alignment property of the vorticity is a consequence of a simplification of the Euler equations (Vieillefosse 1984), the shape of the tensor is still unexplained. Alignment of other vectors have been considered by Ashurst et al. (1987a).

In this paper, we use databases from the DNS of homogeneous compressible turbulence to clarify the sources of reduced growth of kinetic energy with increased levels of compressibility. This is then used to propose a new decomposition of pressure into incompressible and compressible components. In turn, this leads to an improved model for the pressure-dilatation term. Finally, we analyze the rate of strain tensor after decomposition into solenoidal and irrotational components. Differences between the two tensors are brought out through the use of one-dimensional probability functions.

2. Governing Equations

The compressible Navier-Stokes equations are written in a frame of reference moving with the mean flow $\overline{u}_1$. This transformation, which was introduced by Rogallo (1981) for incompressible homogeneous shear, removes the explicit dependence on $\overline{u}_1(x_2)$ in the exact equations for the fluctuating velocity, thus allowing periodic boundary conditions in the x_2 direction. The relation between the moving frame x_i^* and the lab frame x_i is

$$x_1^* = x_1 - Stx_2, \quad x_2^* = x_2, \quad x_3^* = x_3.$$

Here S denotes the constant shear rate $\overline{u}_{1,2}$. In the transformed frame x_i^*, the compressible Navier-Stokes equations take the form

$$\partial_t \rho + (\rho u_i')_{,i} - St(\rho u_2')_{,1} = 0, \tag{1}$$

$$\begin{aligned} \partial_t(\rho u_i') + (\rho u_j' u_i')_{,j} &= -p_{,i} + \tau'_{ij,j} - S\,\rho u_2'\delta_{i1} \\ &+ St(\rho u_2' u_i')_{,1} + Stp_{,1}\delta_{i2} - St\tau'_{i2,1}, \end{aligned} \tag{2}$$

$$\begin{aligned} \partial_t p + u_j' p_{,j} + \gamma p u'_{j,j} &= Stu_2' p_{,1} + \gamma Stp u'_{2,1} + \Phi \\ &+ (\gamma - 1)\kappa[T_{,jj} - 2StT_{,12} + S^2 t^2 T_{,11}], \end{aligned} \tag{3}$$

$$p = \rho RT, \tag{4}$$

where $\Phi = \tau_{ij}\partial u_i/\partial x_j$ is the dissipation function, τ_{ij} the viscous stress, u_i' the fluctuating velocity, ρ the instantaneous density, p the pressure, T the temperature, R the gas constant. The molecular viscosity μ, and thermal conductivity κ are taken to be constant. In terms like $u'_{i,j}$ in the above system, the comma denotes a derivative with respect to the transformed coordinates x_i^*.

A Fourier collocation method is used for the spatial discretization of the governing equations. A third order, low storage Runge-Kutta scheme is used for advancing the solution in time. The nonlinear terms are dealiased by discarding modes lying in the highest third of the wave number range resolved by the grid. This procedure fully dealiases the quadratic and not the cubic non-linearities in the Navier-Stokes equations.

3. Statistical Moments and Spectra

We have performed simulations for a variety of initial conditions and obtained turbulent fields with Taylor microscale Reynolds numbers Re_λ up to 45 and turbulent Mach numbers M_t up to 0.7. Note that $Re_\lambda = q\lambda/\nu$ where $q = \sqrt{u_i'u_i'}$ and $\lambda = q/\sqrt{\omega_i'\omega_i'}$, while $M_t = q/\overline{c}$ where $\overline{c}$ is the mean speed of sound. All the simulations have been performed with $\gamma = 1.4$, and Prandtl number $Pr = 0.7$. The computational domain is a cube with side 2π. The results discussed here were obtained with a uniform 128^3 mesh overlaying the computational domain.

In the case of homogeneous shear flow, the mean velocity $\bar{u}_i$ varies linearly in space and remains invariant in time. The mean density $\overline{\rho}$ is uniform initially and does not evolve in time. However, due to the mean viscous dissipation, the mean pressure $\overline{p}$ and temperature $\overline{T}$ increase in time.

3.1 Effect of Compressibility on Kinetic Energy

Figs. 2-5 show results from two selected cases which have identical initial data for the velocity and thermodynamic variables. The initial data is incompressible, that is, the density is constant and the divergence $\nabla.\mathbf{u} = 0$. The initial pressure is calculated from the Poisson equation appropriate for incompressible flows, and the temperature is obtained

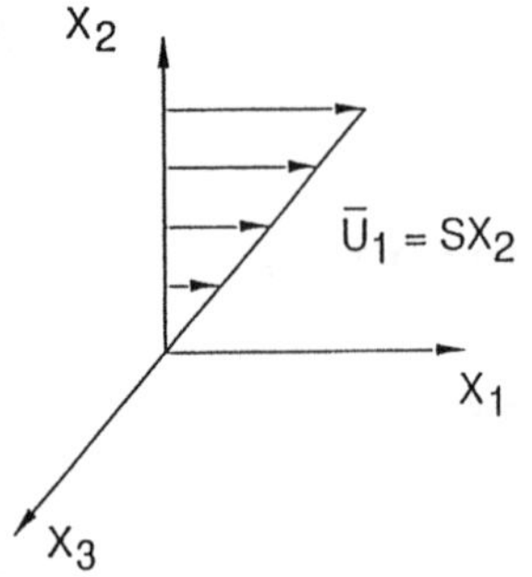

Fig. 1. Schematic of homogeneous shear flow.

Table 1. Parameters for the DNS cases of homogeneous shear flow

Case	S	ν	$M_{t,0}$	$R_{\lambda,0}$	ρ'	d'
1	20	1/150	0.2	14	0	0
2	20	1/150	0.4	14	0	0

from the ideal gas equation of state. The variable parameters for the problem are the shear rate S, viscosity μ, and the speed of sound $\overline{c}$. The two cases have identical values for S and μ, but different values for the speed of sound $\overline{c}$ leading to different initial Mach numbers $M_{t,0}$. Table 1 lists the initial parameters for these cases. The simulation of Case 1 ran up to $St = 18$ when $M_t = 0.46$ and $Re_\lambda = 43$, while Case 2 was continued up to $St = 21$ when $M_t = 0.67$ and $Re_\lambda = 41$.

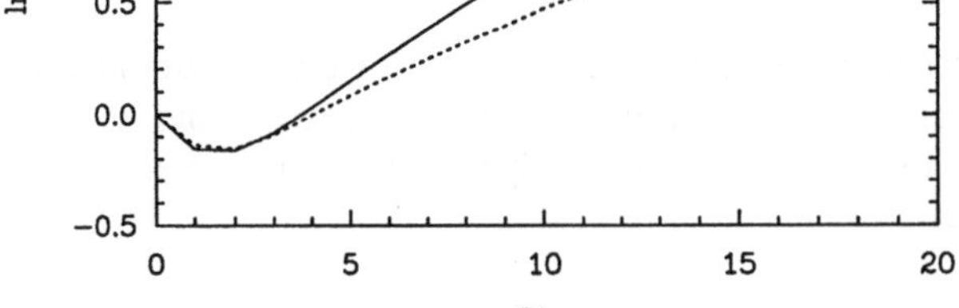

Fig. 2. Time evolution of turbulent kinetic energy.

The Favre-averaged kinetic energy K is defined by $K = \overline{\rho u_i'' u_i''}/2\overline{\rho}$. Note that the overbar over a variable denotes a conventional Reynolds average, while the overtilde denotes a Favre average. A single superscript $'$ represents fluctuations with respect to the

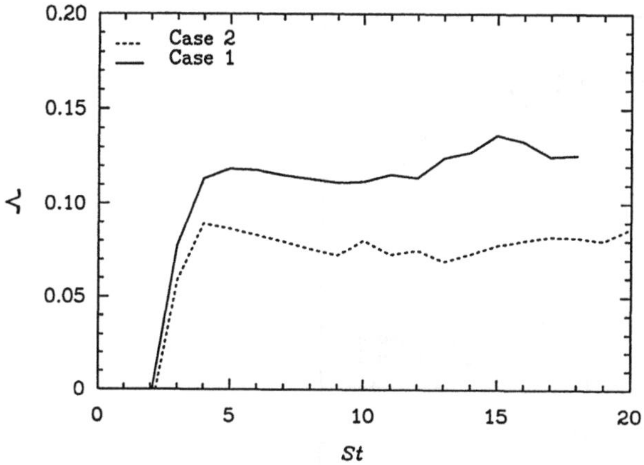

Fig. 3. Growth rate of turbulent kinetic energy.

Reynolds average, while a double superscript $''$ signifies fluctuations with respect to the Favre average. The evolution of the kinetic energy K as a function of non-dimensional time St is shown in Fig. 1. Physical experiments and DNS of the incompressible case indicate that K, after an initial transient, evolves as $K_0 \exp(\Lambda St)$. The feature of exponential growth is carried over to the compressible case, as evidenced by the approximately linear regime in the plot of $\ln(K/K_0)$ in Fig. 2 for $St > 7$. Although the picture of exponential growth survives, the growth rate of K shows a significant decline with a decrease in the speed of sound. Figure 3 shows that the growth rate $\Lambda = d(\ln K/K_0)/d(St)$ decreases by about 30% when M_t increases from 0.2 to 0.4. We note that the reduction of kinetic energy growth rate with increasing compressibility is a consistent trend in all our simulations.

In order to understand the phenomenon of reduced growth rate of kinetic energy, we consider the equation governing the kinetic energy of turbulence in homogeneous shear which is

$$\frac{d}{dt}(\overline{\rho}K) = \overline{\rho}\mathcal{P} - \overline{\rho}\epsilon_s - \overline{\rho}\epsilon_c + \overline{p'd'}, \tag{5}$$

where $\mathcal{P} = -S\widetilde{u_1''u_2''}$ is the production, $\epsilon_s = \overline{\nu\omega_i'\omega_i'}$ the solenoidal dissipation rate, $\epsilon_c = (4/3)\overline{\nu d'^2}$ the compressible dissipation rate and $\overline{p'd'}$ the pressure-dilatation. The last two terms represent the explicit influence of the non-solenoidal nature of the fluctuating velocity field in the kinetic energy budget.

If the notion of exponential growth in homogeneous shear flow is correct, Λ should asymptote to a constant. The relative importance of the various terms on the rhs of (5) can be gauged by their contribution to Λ. We, therefore, normalize (5) by $\overline{\rho}SK$. The terms on the rhs of (5) are plotted in Fig. 4 after being thus normalized. According to Figs. 4a-4b,the dilatational terms are negative and reduce Λ. By $St = 20$, the combined contribution of the dilatational terms is approximately 20% of Λ, implying that they need to be considered in turbulence modeling. The normalized production, $\mathcal{P}/SK$, is plotted in Fig. 4c. The normalized production in Case 2 begins to deviate from that in Case 1 around $St = 5$, and eventually asymptotes to a value noticeably smaller than in Case 1.

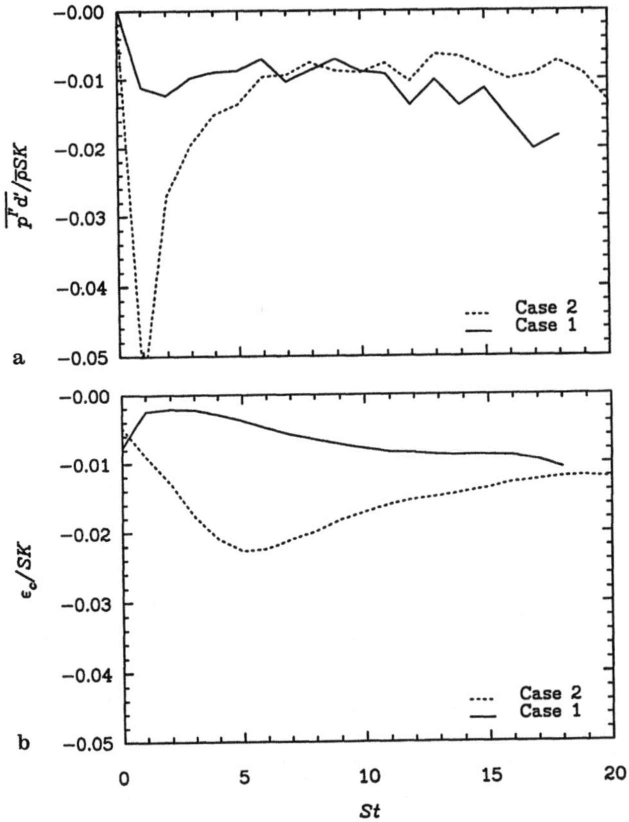

Fig. 4a, b. a: Contribution of pressure-dilatation to the growth rate of K.
b: Contribution of compressible dissipation to the growth rate of K.

Thus, the production is reduced due to compressibility. The normalized dissipation in Fig. 4d also shows a decrease in Case 2 relative to Case 1, but to a smaller degree than the production. The reduction in growth rate of K in Case 2 relative to Case 1 seems to be primarily due to the dilatational terms during the early phase of the evolution and, for the later phase ($St > 8$), is related to a decreased level of production.

3.2 Measures of Compressibility

Apart from the differences in the evolution of turbulence statistics in the case of compressible flow relative to incompressible flow, a question that arises is how compressible is the turbulence? The turbulent Mach number $M_t = \sqrt{2K}/\overline{c}$ and the normalized rms density $\rho_{rms}/\overline{\rho}$ are measures of departure from incompressibility, since both these quantities are zero for strictly incompressible flow. M_t shows a monotone increase in Fig. 5a reaching a maximum of 0.6, which is larger than the upper bound of M_t in free shear layers and wall boundary layers encountered in aerodynamic practice. Figure 5b shows that, after starting from zero density fluctuations, both simulations develop significant rms density levels at $St = 18$, and most of the increase in rms density occurs rapidly within

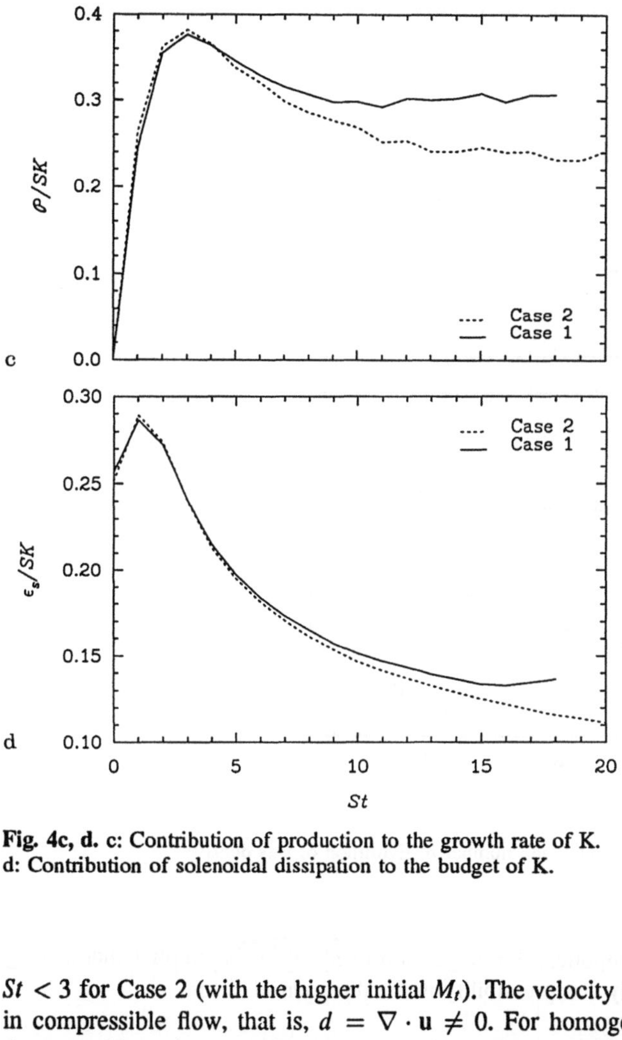

Fig. 4c, d. c: Contribution of production to the growth rate of K.
d: Contribution of solenoidal dissipation to the budget of K.

$St < 3$ for Case 2 (with the higher initial M_t). The velocity field is no longer solenoidal in compressible flow, that is, $d = \nabla \cdot \mathbf{u} \neq 0$. For homogeneous flow, the Helmholtz decomposition gives a unique decomposition of the velocity field into incompressible and compressible components by $\mathbf{u} = \mathbf{u}^I + \mathbf{u}^C$, where $\nabla \cdot \mathbf{u}^I = 0$ and $\nabla \times \mathbf{u}^C = 0$. The departure of the velocity field from solenoidality is another measure of compressibility, and can be represented by the ratio of dilatational variance to enstrophy $\chi_d = \overline{d'^2}/\overline{\omega_i'\omega_i'}$ or by the fraction of kinetic energy which is dilatational $\chi_K = K_C/K$. The former quantity χ_d is more general, because the Helmholtz decomposition required for obtaining K_C is unique only for homogeneous flows. Figure 6 shows that after an initial transient, χ_d increases monotonically for Case 1, but levels off at about 8% for Case 2. The quantity χ_K (not plotted here) is comparable to χ_d, leveling off at approximately 6% for Case 2. Since the mean shear leads to anisotropy of the Reynolds stress tensor, the dilatational fraction of each Reynolds stress component is obtained. Except for the transverse component $\overline{u_2'u_2'}$, the dilatational contribution to the Reynolds stress tensor was a few percent, comparable to the magnitude of χ_K. The ratio of $\overline{u_2^{C'}u_2^{C'}}/\overline{u_2'u_2'}$ is substantially larger, as shown in Fig. 7. Since the dilatational component is a relatively poor mixer of scalar

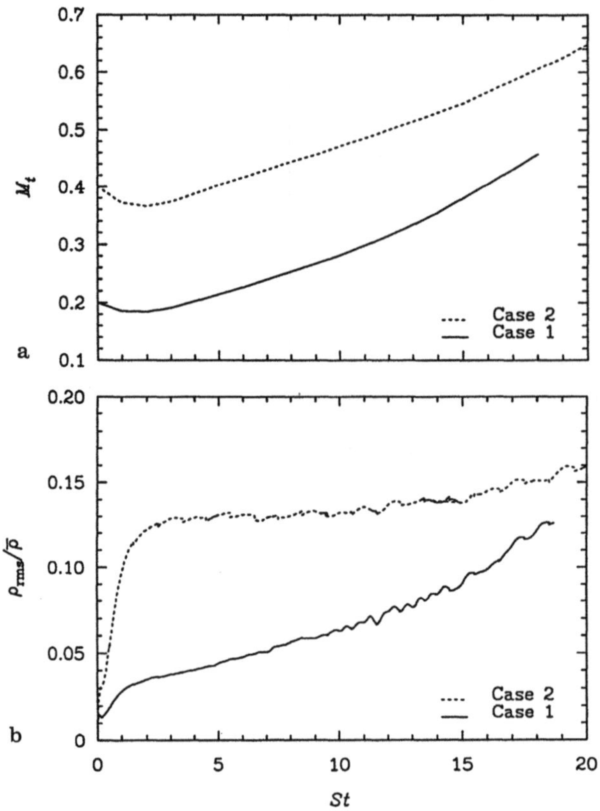

Fig. 5a, b. a: Evolution of turbulent Mach number b: Evolution of rms density

compared to the vortical component, the large ratio of $\overline{u_2^{C'}u_2^{C'}}/\overline{u_2^I u_2^I}$ implies that mixing is preferentially decreased by compressibility in the direction of shear.

3.3 Compressible and Solenoidal Spectra

The Fourier component of the velocity is decomposed into components perpendicular and parallel to the wave number vector from which the solenoidal spectrum $E_s(k)$ and compressible spectrum $E_c(k)$ are calculated. Figure 8 compares the solenoidal and compressible spectra at St=15 (when $M_t = 0.55$) for Case 2. The compressible energy is small relative to the incompressible energy for all modes. The compressible spectrum is flatter at low wave numbers ($k < 16$) relative to the incompressible spectrum. The two spectra have similar slopes in the intermediate wave number range. The solenoidal spectrum is compared in Fig. 9 between an incompressible run ($M_t = 0$) and a compressible run at $St = 17$. The compressible run has an initial M_t of 0.4 and the initial velocity and pressure fields are the same as in the incompressible run. From Fig. 9, it appears that the shape of the solenoidal spectrum is not altered by compressibility, even

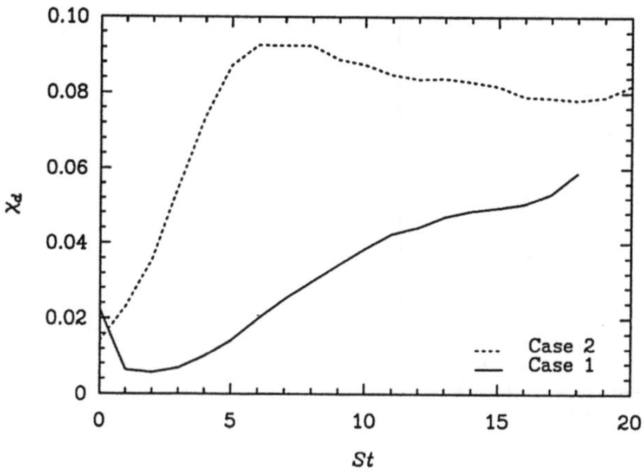

Fig. 6. Evolution of ratio of dilational variance to enstrophy

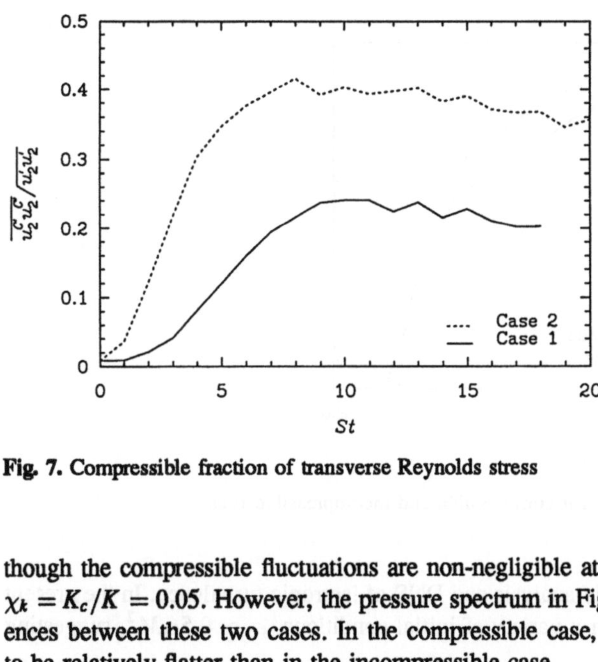

Fig. 7. Compressible fraction of transverse Reynolds stress

though the compressible fluctuations are non-negligible at this time $\rho_{rms}/\overline{\rho} = 0.12$ and $\chi_k = K_c/K = 0.05$. However, the pressure spectrum in Fig. 10 shows significant differences between these two cases. In the compressible case, the pressure spectrum seems to be relatively flatter than in the incompressible case.

4. Modeling the Dilatational Terms

We showed in Fig. 4 that the compressible dissipation ϵ_c and the pressure-dilatation $\overline{p'd'}$ contribute significantly to the kinetic energy budget and therefore require modeling. In Sarkar et al. (1991), we proposed a model for the compressible dissipation $\epsilon_c =$

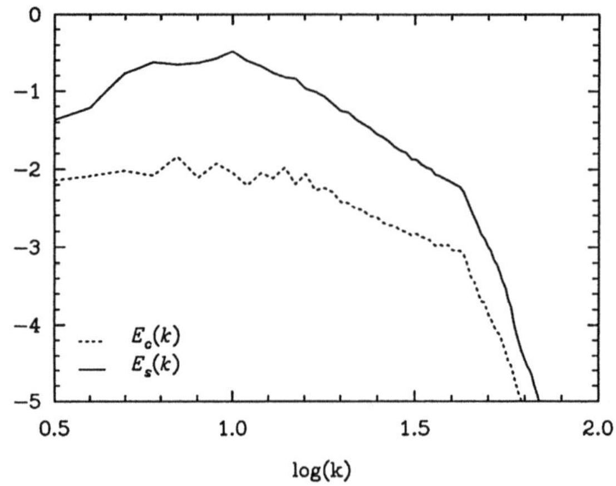

Fig. 8. Energy spectra at $St = 15$ for Case 2

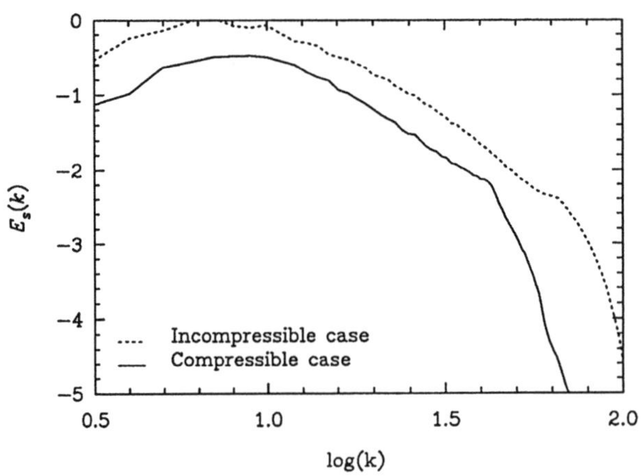

Fig. 9. Solenoidal velocity spectra for compressible and incompressible runs

$\alpha_1 \epsilon_s M_t^2$ based on an asymptotic analysis and DNS of isotropic turbulence. In the present simulations, after starting from a variety of initial conditions, $\epsilon_c \simeq 0.5\epsilon_s M_t^2$, suggesting that $\alpha_1 = 0.5$.

Our direct numerical simulations of isotropic turbulence and homogeneous shear flow provided a data base for the pressure-dilatation and suggested a theoretical approach towards modeling it. The evolution of the pressure-dilatation $\overline{p'd'}$ for Case 1 is depicted by the solid curve in Fig. 11. From numerical experiments, it was found that the nominal time period of the oscillations in $\overline{p'd'}$ decreased approximately linearly with the speed of sound. This suggested that one could isolate the oscillatory part of $\overline{p'd'}$ by decomposing the fluctuating pressure p' into the sum of an incompressible part $p^{I'}$ and a compressible part $p^{C'}$. The incompressible pressure $p^{I'}$ satisfies

$$\nabla^2 p^{I'} = -2\bar{u}_{i,j}\bar{\rho}u'_{j,i} - 2\bar{u}_{i,i}\bar{\rho}u'_{j,j} - \bar{\rho}(u_i'u_j')_{,ij} - 2\bar{u}_{i,ij}(\bar{\rho}u_j') \quad (6)$$

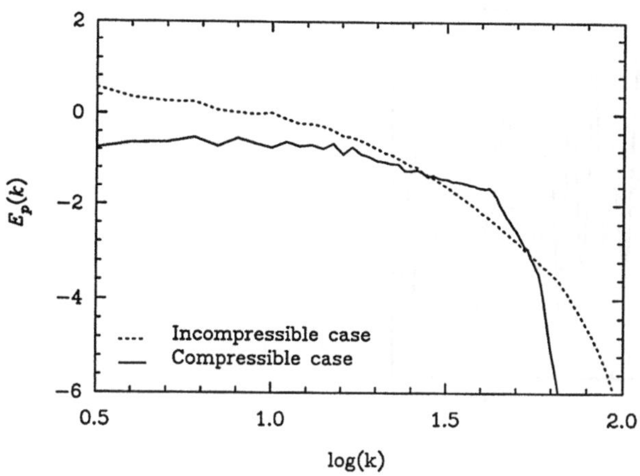

Fig. 10. Pressure spectra for compressible and incompressible runs

while the remainder $p^{C'}$ is the compressible pressure. The rhs of (6) collects all terms that depend on $\overline{\rho}$ in the equation obtained by taking the divergence of the momentum equation. In the simulations, since p' is available from the compressible Navier-Stokes solution and $p^{I'}$ is evaluated from (6), we obtain $p^{C'}$ as the difference $p' - p^{I'}$. Since $p' = p^{I'} + p^{C'}$ we have $\overline{p'd'} = \overline{p^{I'}d'} + \overline{p^{C'}d'}$. The oscillations are substantial only for $\overline{p^{C'}d'}$ (dotted curve in Fig. 11), and furthermore, the peaks and valleys in the evolution of $\overline{p^{C'}d'}$ seem to be much more symmetric around the origin than those in $\overline{p'd'}$ (solid curve in Fig. 11). The component $\overline{p^{I'}d'}$ (dash-dotted curve in Fig. 11) does not have strong temporal oscillations, and shows a systematic decrease with time. In order to gauge the relative importance of the two components $\overline{p^{C'}d'}$ and $\overline{p^{I'}d'}$ of the pressure-dilatation in the evolution of the turbulent kinetic energy, we calculate the time integrals of these components. The integrated contribution of $\overline{p^{I'}d'}$ is about an order of magnitude larger than that of $\overline{p^{C'}d'}$ in Case 2. Examination of other DNS cases indicates that, in general, $\overline{p^{C'}d'}$ has a negligible contribution to the turbulent kinetic energy evolution relative to $\overline{p^{I'}d'}$. Therefore, it seems that only the component $\overline{p^{I'}d'}$ of the pressure-dilatation requires modeling in shear flows.

In order to model $\overline{p^{I'}d'}$, we consider the Poisson equation (6) for the incompressible pressure. After splitting the pressure into a rapid part $p^{R'}$ and a slow part $p^{S'}$, we obtain the following exact expressions for the rapid pressure-dilatation and slow pressure-dilatation:

$$\overline{p^{R'}d'} = 2\overline{\rho}\overline{u}_{m,n} \int \frac{k_m k_j}{k^2} E_{nj} d\mathbf{k}, \tag{7}$$

$$\overline{p^{S'}d'} = \overline{\rho} \int \frac{k_m k_l k_j}{k^2} (i\widehat{u_l u_j}\hat{u}_m^* - i\widehat{u_l u_j}^* \hat{u}_m) \, d\mathbf{k}. \tag{8}$$

Here $\hat{\phi}^*$ denotes the complex conjugate of the Fourier transform $\hat{\phi}$ and E_{nj} represents the spectrum of the Reynolds stress tensor $\overline{u_n' u_j'}$. Using scaling arguments (see Sarkar (1991) for details) to simplify (7)-(8) we find that $\overline{p^{R'}d'}$ depends on the production $\mathcal{P}$ while

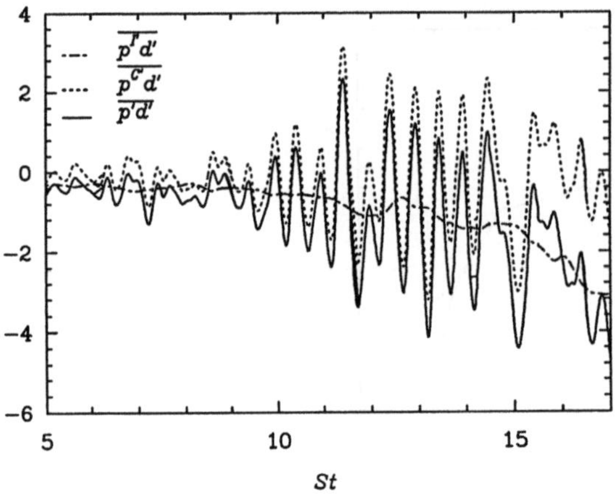

Fig. 11. Evolution of pressure-dilatation in Case 1

$\overline{p^{S'} d'}$ depends on the dissipation ϵ_s. Finally, we propose the following model for the pressure-dilatation:

$$\overline{p'd'} = \alpha_2 \overline{\rho} \tilde{u}_{i,j} b_{ij} q^2 M_t + \alpha_3 \overline{\rho} \epsilon_s M_t^2 + \frac{8}{3} \overline{\rho} \tilde{u}_{i,i} \chi(M_t) q^2, \tag{9}$$

where $b_{ij} = \widetilde{u_i'' u_j''}/q^2 - \delta_{ij}/3$ is the anisotropy tensor, $M_t = \sqrt{2K}/\overline{c}$ the turbulent Mach number, and ϵ_s the solenoidal dissipation. To obtain the functional dependence $\chi(M_t)$ in the last term of Eq. (9), one would require data from a flow with mean homogeneous compression. In this paper, we validate and calibrate the first two terms in the model for $\overline{p'd'}$. In homogeneous shear, the model becomes

$$\overline{p'd'} = -\alpha_2 \overline{\rho} \mathcal{P} M_t + \alpha_3 \overline{\rho} \epsilon_s M_t^2, \tag{10}$$

where $\mathcal{P} = -S\,\widetilde{u_1'' u_2''}$. Because the production $\mathcal{P} = 0$ in decaying isotropic turbulence, the variation of the incompressible pressure-dilatation with ϵ_s can be verified using DNS of isotropic turbulence. The ratio $\overline{p^{I'} d'}/(\overline{\rho} \epsilon_s M_t^2)$ is shown as a function of non-dimensional time in Fig. 12. The decaying isotropic turbulence simulations, D1,D2 and D3 start with $M_{t,0}$ of 0.6,0.5, and 0.4 respectively. $\overline{p^{I'} d'}/(\overline{\rho} \epsilon_s M_t^2)$ reaches an equilibrium value by a time of 0.25, substantiating the validity of the second term in (9). Based on the DNS value of the equilibrium ratio, the model coefficient α_3 in (9) is taken to be 0.2. The remaining part of the model for the pressure-dilatation is calibrated against simulations of homogeneous shear flow. Fig. 13 shows results from three cases S1,S2 and S3 with $M_{t,0}$ of 0.2,0.3, and 0.4 respectively. Cases S2 and S3 have shear S=15, while Case S1 has S=20. After $St = 5$, the ratio $(\overline{p^{I'} d'} - 0.2 \overline{\rho} \epsilon_s M_t^2)/(\overline{\rho} \mathcal{P} M_t)$ evolves in a similar fashion for the three cases and exhibits a slight decrease with time. In the period $5 < St < 15$, the three cases S1-S3 encompass a range of turbulent Mach numbers of $0.21 < M_t < 0.65$. The results of Fig. 13 suggest that, with a coefficient $\alpha_2 = 0.15$, the proposed model (9) is able to parametrize the pressure-dilatation for $0.2 < M_t < 0.6$.

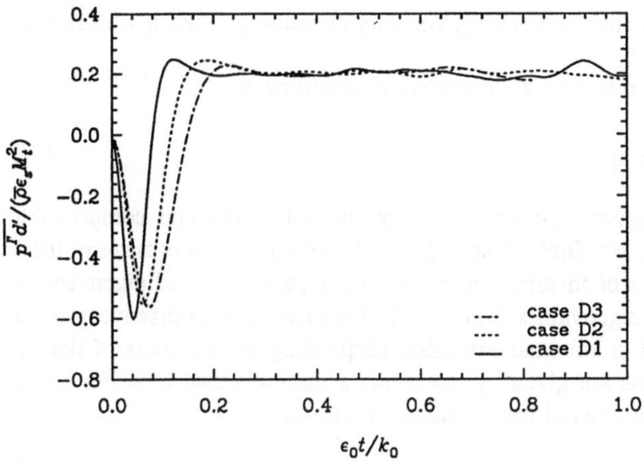

Fig. 12. Incompressible pressure-dilatation in decaying isotropic turbulence

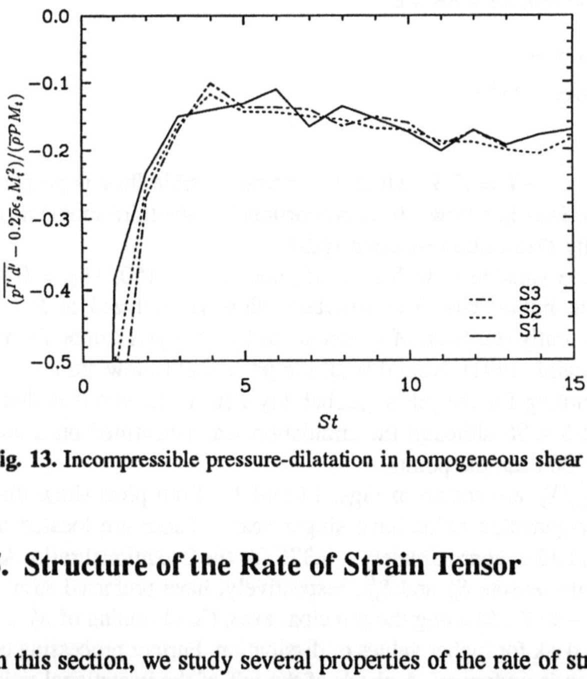

Fig. 13. Incompressible pressure-dilatation in homogeneous shear turbulence

5. Structure of the Rate of Strain Tensor

In this section, we study several properties of the rate of strain tensor S_{ij} to help identify characteristics of the flow which are solely the result of compressibility effects, and how these characteristics differ from corresponding properties in an incompressible turbulent flow. Because the flow is anisotropic, it is useful to consider characteristics of the flow both with respect to the laboratory frame of reference, and with respect to a frame of reference attached to, and rotating with an individual fluid element. Global orientation properties of S_{ij} are established by studying the principal directions of S_{ij} in the laboratory frame of reference. (In isotropic turbulence, this diagnostic does not provide useful information.) On the other hand, local properties can be characterized by considering the relative orientation of vectors such as velocity, vorticity or scalar gradients, with respect to the principal directions of S_{ij}.

We consider the rate of strain tensor, decomposed according to

$$S_{ij} = S_{ij}^C + S_{ij}^I + \frac{1}{3} u_{k,k}\, \delta_{ij} \tag{11}$$

where S_{ij}^I and S_{ij}^C are respectively constructed from the solenoidal and compressible (irrotational) velocities $\mathbf{u}^I$ and $\mathbf{u}^C$. Both S_{ij}^I and S_{ij}^C are deviatoric. The properties of these tensors are expressed in terms of invariant quantities, i.e. eigenvalues and eigenvectors. Because S_{ij} is symmetric, its eigenvalues λ_i, $i = 1,2,3$ are real, and its eigenvectors are in the directions of maximal or minimal extension (depending on the signs of the λ_i) of the tensor. These directions are given by the tensor's three eigenvectors. For future reference, the eigenvalues are ordered from smallest to largest:

$$\lambda_1 \leq \lambda_2 \leq \lambda_3. \tag{12}$$

The first three invariants of the tensor S_{ij}, defined as the coefficients of the polynomial characteristic equation, are related to small scale phenomena, and can be expressed in terms of the eigenvalues. These relationships are:

$$\begin{aligned} I &= -(\lambda_1 + \lambda_2 + \lambda_3) \\ II &= \lambda_1\lambda_2 + \lambda_1\lambda_3 + \lambda_2\lambda_3 \\ III &= -\lambda_1\lambda_2\lambda_3 \end{aligned}$$

If the tensor is deviatoric, $I = 0$, $-II = S_{ij}S_{ji}$ which for incompressible flow is proportional to the dissipation, and in isotropic flow, III is proportional to the third moment of the velocity gradient probability distribution function (pdf).

We present results from one simulation with $S = 15$, and $\mu = 1/150$, $M_{t0} = 0.3$, $R_{\lambda 0} = 20$ and incompressible initial data. The simulated flow is analyzed at $St = 1,3,5,7,9$, which are all sufficiently resolved. More details of this analysis can be found in Erlebacher, Sarkar and Hussaini (1991). Note that all the pdf's that follow are unconditioned and unweighted. Sampling for the pdf's (probability density function) is done on a grid resolution of $48 \times 96 \times 96$ although the simulation was performed on a 96^3 mesh. This provides over 400,000 sample points.

The pdfs of λ_2^I/λ_1^I and λ_2^C/λ_1^C are shown in Figs. 14 and 15. Both plots show that after an initial transient, the eigenvalue ratios have single peaks. These are located at -0.25 for the solenoidal ratio, and at approximately -0.375 for the irrotational ratio. As a consequence, the rate of strain tensors S_{ij}^I and S_{ij}^C, respectively, have preferred strains in the ratios $(-4:1:3)$ and $(-8:3:5)$ along the principal axes. Conditioning of λ_2^I/λ_1^I with respect to ϵ_s sharpens the peak for higher values of dissipation. Further processing of the irrotational ratio distribution is underway. A check of the pdf of the irrotational ratio was also performed from a 128^3 database (Case 2) with 800,000 sample points, and its shape is qualitatively similar with the peak at -0.375. Note that the pdf of the irrotational ratio is more broadband than that of the solenoidal ratio. However, the location of the peak is well defined, and constant in time. This equilibrium structure of $(-4:1:3)$ for S_{ij} is observed in incompressible flow, first by Ashurst, Kerstein, Kerr & Gibson (1987). They considered the statistics of $(\lambda_2^I)^2/(S_{ij}^I S_{ij}^I)$ conditioned on the dissipation and found that the preferred ratio was most dominant in regions of high dissipation. In a later work,

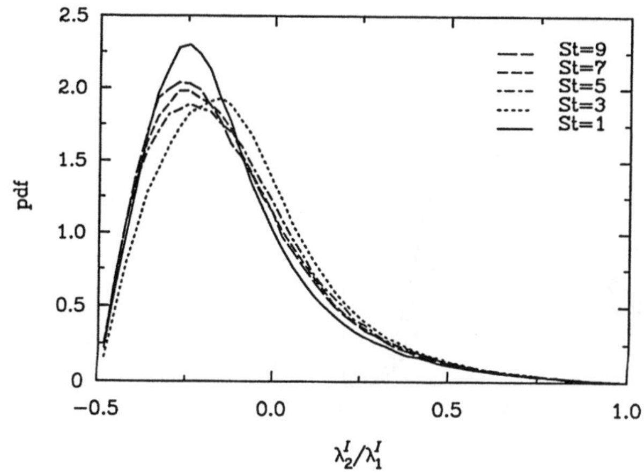

Fig. 14. Pdf of solenoidal eigenvalue ratio

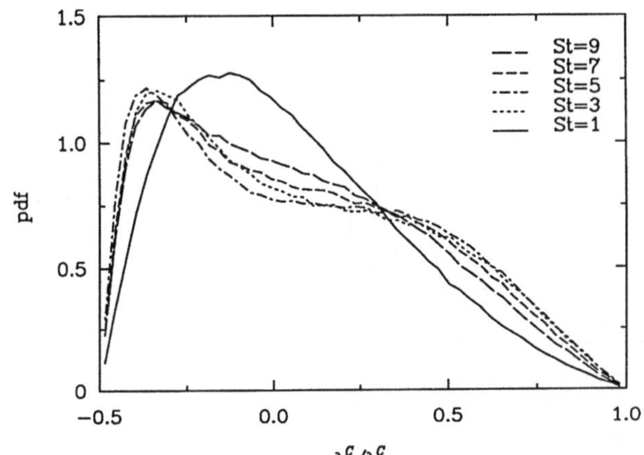

Fig. 15. Pdf of irrotational eigenvalue ratio

Chen et al. (1990) displayed scattergrams of II versus III based on incompressible mixing layer DNS data, which clearly demonstrated that III is proportional to $(-II)^{3/2}$ in the regions of highest dissipation. This is in fact a statement about the preferred shape of the S_{ij}^I principal ellipsoid. Our results indicate that compressibility does not substantially affect this preferred structure of the solenoidal S_{ij}^I.

The distributions of λ_1^I and λ_1^C show some distinctive differences, as illustrated in Figs. 16-17. Both figures show negative skewness of the distributions, and a flattening in time, although the effect is much more severe for λ_1^C. On both plots, the most probable value for the eigenvalue shifts towards more negative values. In the irrotational case, the effect is extremely strong. However, the most probable ratio of λ_2^C to λ_1^C (see Fig. 15) remains invariant in time. Figure 17 also indicates that the amount of change in the

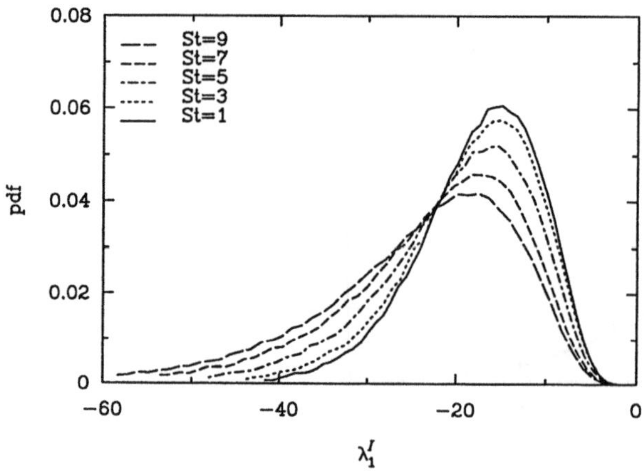

Fig. 16. Pdf of most compressive solenoidal eigenvalue

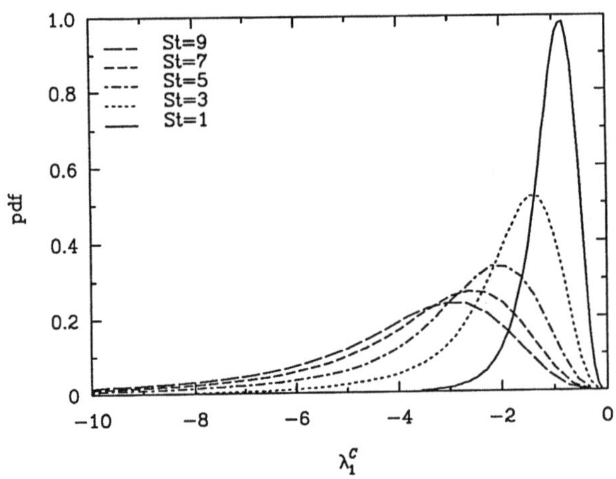

Fig. 17. Pdf of most compressive irrotational eigenvalue

distribution of λ_1^C decreases with time, but there is no strong evidence that the shape of the pdf is evolving towards a steady state.

Information on structural differences in the flow as they relate to preferred directions of straining for both the solenoidal and dilatational components of the flow are presented next. After the eigenvectors of the rate of strain tensors are normalized to unity, we compute the angle θ_{ij} between eigenvector $\mathbf{f}_i$ and the unit vector in the coordinate direction x_j. The pdf's of $|\cos\theta_{ij}^I|$ and $|\cos\theta_{ij}^C|$ are then computed. The cosine of θ is chosen instead of θ so that the probability density function based on a Gaussian distribution for the velocity derivatives is flat. In Figs. 18-19, we respectively plot the pdf of $|\cos\theta_{1i}^I|$ and $|\cos\theta_{1i}^C|$, $(i = 1,2,3)$ at $St = 9$ to illustrate that the solenoidal and irrotational rate of strain tensors have different preferential alignments in the laboratory frame of reference. Note that $R_\lambda = 33$ at this time. The direction of maximum compression of S_{ij}^I

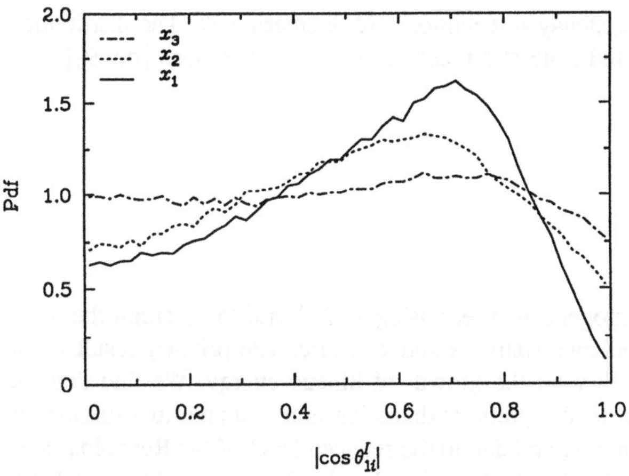

Fig. 18. Pdf of the orientation of the most compressive solenoidal eigenvector with respect to the three coordinate axes

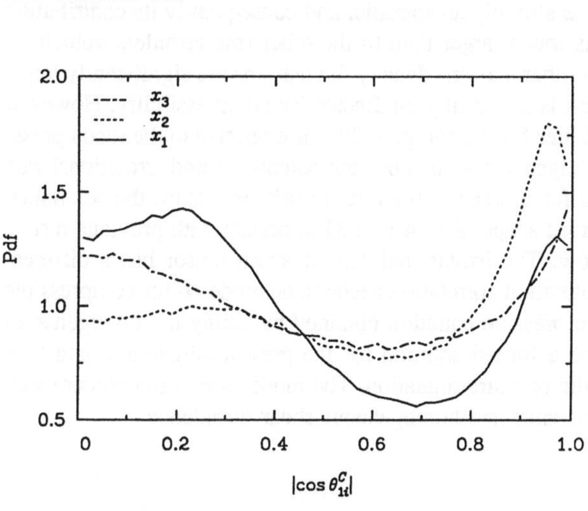

Fig. 19. Pdf of the orientation of the most compressive irrotational eigenvector with respect to the three coordinate axes

($\mathbf{f}^I_1$) tends to be at approximately 45° to both the x and y axes, with no preferred orientation with respect to z. The alignment of $\mathbf{f}^I_3$ is also in the 45° direction, while $\mathbf{f}^I_2$ has no unambiguous preferred direction. The irrotational rate of strain tensor exhibits markedly different properties. Most of the distributions of $\cos\theta_{ij}$ have two strong peaks, thus indicating two preferred directions of alignment. For example, $\mathbf{f}^C_1$ is most often aligned along one of the three coordinate axes. This is true of the other two principal directions, but to a lesser degree. Interestingly enough, all three pdf's of $\mathbf{f}^C_i$ exhibit a valley in the neighborhood of 45°. Statistics based on the database of Case 2 (at a higher shear and higher Reynolds number than the results plotted in the preceding figures) show that the pdf's of $|\cos\theta^I_{1i}|$ and $|\cos\theta^C_{1i}|$ are sharper. In particular, the tendency for $\mathbf{f}^I_1$ to align is strongest in a direction at 30° to the x axis and not 45°. Finally, we notice that all the pdf's of $|\cos\theta_{ij}|$ tend towards a steady distribution with increasing St. The distribution of $|\cos\theta_{ij}^{C}|$ reaches an equilibrated state at a much earlier stage than does $|\cos\theta_{ij}^{I}|$.

6. Conclusions

We have performed DNS of homogeneous shear using a 128^3 grid to ascertain the influence of compressibility on turbulence statistics and structure. The primary result is the stabilizing effect of compressibility on the growth of kinetic energy. We find that the reduction in growth rate is due to the pressure-dilatation and compressible dissipation acting as sinks for the kinetic energy, and due to the reduced level of the Reynolds shear stress. After starting with zero density fluctuations, the simulations develop significant density fluctuations and moderate levels of energy in the dilatational component. The dilatational velocity component is strongly anisotropic, and consequently its contribution to the transverse rms velocity is much larger than to the other rms turbulent velocities. Although, the turbulent Mach number and rms density fluctuations are significantly large, the shape of the energy spectrum is practically unaffected by compressibility. However, the pressure spectrum is much flatter for the compressible case relative to the incompressible case. The eigenvalues and eigenvectors of both the solenoidal and irrotational rate of strain tensors were examined for preferred structure. In principal axes, the solenoidal rate of strain tensor has a preferred shape of $(-4 : 1 : 3)$ in accord with previous results available for incompressible flow. The irrotational rate of strain tensor has a different preferred shape. The pressure-dilatation correlation needs to be modeled for compressible flow. We consider a form of the pressure equation obtained by taking the divergence of the momentum equation, deduce a formal solution for the pressure-dilatation, and then simplify to obtain a model for the pressure-dilatation. The model seems to compare well with DNS data in isotropic turbulence and homogeneous shear turbulence.

References

Ashurst, Wm. T., Chen, J.-Y.,and Rogers, M.M.: Pressure Gradient Alignment with Strain Rate and Scalar Gradient in Simulated Navier-Stokes Turbulence. *Phys. Fluids A*, **30**, (1987a) 3293-3294.

Ashurst, Wm. T., Kerstein, A.R., Kerr, R.M., and Gibson, C.H.: Alignment of Vorticity and Scalar Gradient with Strain Rate in Simulated Navier-Stokes Turbulence. Phys. Fluids **30**, (1987b) 2343-2353.

Blaisdell, G.A., Mansour, N.N, and Reynolds, W.C.: Numerical simulation of Compressible Homogeneous Turbulence, *Stanford University Report No. TF-50* (1991).

Chen, J.H., Chong, M.S., Soria, J., Sondergaard, R., Perry, A.E., Rogers, M., Moser, R., and Cantwell, B.J.: A study of the topology of dissipating motions in direct numerical simulations of time-developing compressible and incompressible mixing layers. *CTR, Proceedings of Summer Program* (1990).

Coleman, G.N., and Mansour, N.N.: Modeling the rapid spherical compression of isotropic turbulence. *Phys. Fluids A*, **3** (1991) 2255-2259.

Erlebacher, G., Hussaini, M.Y., Kreiss, H.O., and Sarkar, S.: The Analysis and Simulation of Compressible Turbulence. *Theoret. Comput. Fluid Dynamics*, **2** (1990) 73-95.

Erlebacher, G., Sarkar, S., Hussaini, M. Y.: Structure of Compressible Turbulence in Homogeneous Shear Flow. *ICASE Report* (in preparation).

Feiereisen, W.J., Shirani, E., Ferziger, J.H., and Reynolds, W.C.: Direct Simulation of Homogeneous Turbulent Shear Flows on the Illiac IV Computer. In *Turbulent Shear Flows 3* (1982) 309-319. Springer-Verlag, Berlin.

Kida, S. and Orszag, S.A.: Energy and Spectral Dynamics in Forced Compressible Turbulence. *J. Sci. Comp.*, **5**, (1990) 85.

Lee, M.J., Kim, and J., Moin, P.: Structure of turbulence at high shear rate. *J. Fluid Mech.*, **216**, (1990) 561-583.

Lee, S., Lele, S.K., and Moin, P.: Eddy-shocklets in Decaying Compressible Turbulence. *CTR Manuscript 117* (1990).

Passot, T.: Simulations numeriques d'ecoulements compressibles homogenes en regime turbulent : application aux nuages moleculaires. *Ph. D. Thesis, University of Paris* (1987).

Rogallo, R.S.: Numerical Experiments in Homogeneous Turbulence. *NASA TM 81315* (1981).

Rogers, M.M., and Moin, P.: The Structure of the Vorticity Field in Homogeneous Turbulent Flows. *J. Fluid Mech.*, **176** (1987) 33-66.

Sarkar, S., Erlebacher, G., Hussaini, M.Y., and Kreiss, H.O.: The Analysis and Modeling of Dilatational Terms in Compressible Turbulence. *J. Fluid Mech*, **227** (1989) 473-493.

Sarkar, S., Erlebacher, G., and Hussaini, M.Y.: Direct Simulation of Compressible Turbulence in a Shear Flow. *Theor. Comput. Fluid Dyn.*, **2** (1991) 291-305.

Sarkar, S.: Modeling the Pressure-dilatation Correlation. *ICASE Report 91-42* (1991).

Vieillefosse, P.: Internal Motion of a Small Element of Fluid in an Inviscid Flow. Physica **125** A,(1984) 150-162.

Zang, T.A., Dahlburg, R.B., and Dahlburg, J.P.: Direct and Large-eddy Simulations of Compressible, Isotropic, Navier-Stokes Turbulence. *Phys. Fluids A* (in press).

Simulation and Modeling of Homogeneous Compressible Turbulence Under Isotropic Mean Compression

G. N. Coleman and N. N. Mansour

NASA Ames Research Center, Moffett Field, CA 94035, USA.

Abstract

Compressible homogeneous turbulence subjected to isotropic mean compression is simulated numerically for high and low turbulent Mach numbers, at various compression rates. We find that at low Mach numbers, the effects of viscosity variations on the development of the turbulent kinetic energy can be significant. A model consistent with the invariance of the Navier-Stokes equations to spherical compression, that takes into account variable viscosity effects is proposed. At high Mach numbers, the distribution of energy between the acoustic and solenoidal fields is found to depend upon initial conditions. The simulation results are also used to evaluate one-point compressible turbulence models.

Introduction

One of the most pressing needs in the field of turbulence modeling today is that of correctly incorporating the effects of compressibility into one-point closures. Because of uncertainties in compressible turbulence models, the ability to adequately predict flows such as those associated with internal combustion engines, hypersonic flight, supersonic combustion and astrophysics is currently rather limited. The goal of this study is to increase our understanding of compressible turbulence by considering the simple limiting case of homogeneous compressible turbulence subjected to isotropic ("spherical") mean compression. Our main objective is to generate direct numerical simulations (DNS) of the spherically compressed case, for various turbulent Mach numbers and compression rates, and use the results to modify turbulence models so that they correctly predict this flow. By "tuning" the models in this manner, it is felt that they will provide better predictions for all flows.

Flow description

The fluid is assumed to be a viscous, compressible ideal gas with constant specific heats. The mean straining is given by the isotropic compression

$$\widetilde{u}_{i,j}(t) = S(t)\delta_{ij}, \tag{1}$$

where the mean dilatation, $\widetilde{u}_{i,i} = 3S(t)$, is negative, and $\widetilde{u}_i$ is the Favre averaged mean velocity. In what follows, Favre averaged quantities are denoted by a tilde, and ensemble averaged variables by angle brackets. Single and double primes represent respectively, deviations from the ensemble and Favre average: eg. $u_i = \langle u_i \rangle + u_i' = \widetilde{u}_i + u_i''$. The list of parameters that govern the flow includes: (1) the initial compression rate, defined here by $(Sq^2/\epsilon)_0$ (a zero subscript is used throughout to denote an initial value), where q^2 is $\langle \rho u_i'' u_i'' \rangle / \langle \rho \rangle$, twice the turbulent kinetic energy per unit mass, and $\epsilon = \epsilon_s + \epsilon_d$ is the total rate of turbulent kinetic energy dissipation due to both solenoidal and dilatational velocity fluctuations: $\epsilon_s = \widetilde{\nu} \langle \omega_i' \omega_i' \rangle$ and $\epsilon_d = \frac{4}{3} \widetilde{\nu} \langle u_{i,i}'' u_{j,j}'' \rangle$ ($\widetilde{\nu} = \widetilde{\mu}/\langle \rho \rangle$, ω_i' is the fluctuation vorticity and μ and ν are the dynamic and kinematic viscosity, respectively); (2) the compression history, $S(t)$; (3) the initial turbulent Reynolds number, $(Re_T)_0 = (q^4/\epsilon\widetilde{\nu})_0$; (4) the initial turbulent Mach number $(M_T)_0 = (q/\widetilde{c})_0$, where $\widetilde{c}$ is the sound speed based on the mean temperature, $\widetilde{T}$; (5) the molecular Prandtl number, Pr; (6) the ratio of specific heats, γ; (7) the temperature dependence of the dynamic viscosity, $\mu(T)$; (8) the initial magnitude and spectra shapes of the density and temperature fields; and (9) the initial partitioning of turbulent kinetic energy between the solenoidal and dilatational fields, and the respective initial spectra shapes.

The DNS code of Blaisdell, Mansour & Reynolds (1991) (henceforth BMR) is used to produce the results. The program computes numerical solutions of the compressible Navier Stokes equations, for various types of mean deformation. All the relevant scales of motion are resolved (so that no turbulence model is required), by utilizing a Fourier spatial discretization and Runge-Kutta time advance algorithm.

We assume

$$\mu(T) = \mu(\widetilde{T}_0) \left(\frac{T}{\widetilde{T}_0} \right)^n , \tag{2}$$

and $\widetilde{\mu} = \mu(\widetilde{T})$. The viscosity exponent, n, is taken to be either 0.67, 0.75 or (for reasons discussed below) $1/(\gamma - 1)$. To maintain the homogeneity of the turbulence, the compression history must follow (BMR)

$$S(t) = S_0 / (1 + S_0 t) . \tag{3}$$

The other parameters used for each run are given in the appendix, as are details regarding generation of the initial conditions.

Two-equation Model

For isotropic turbulence under spherical compression, the off-diagonal Reynolds stresses are negligible. We therefore choose to focus our attention upon two-equation models. A form appropriate for compressible flow is (BMR, Speziale & Sarkar 1991, Reynolds 1980)

$$\frac{1}{2} q^2_{,t} + \widetilde{u}_j \frac{1}{2} q^2_{,j} = \mathcal{P} + \mathcal{D}_{q^2} - \langle u_i'' \rangle \langle p \rangle_{,i} + \Pi_d - \epsilon_s - \epsilon_d , \tag{4}$$

$$\epsilon_{s,t} + \widetilde{u}_j \epsilon_{s,j} = \frac{\epsilon_s}{q^2} (2C_1 \mathcal{P} - 2C_2 \epsilon_s) + 3(1 - C_3) \epsilon_s S + \mathcal{D}_{\epsilon_s} \tag{5}$$

where $\mathcal{P} = -\widetilde{u_i'' u_j''} \widetilde{u}_{i,j}$ is the production of $q^2/2$, $\mathcal{D}_{q^2}$ and $\mathcal{D}_{\epsilon_s}$ represent turbulent and viscous diffusion and $\Pi_d = \langle p' u_{i,i}'' \rangle / \langle \rho \rangle$ is the pressure-dilatation term. The q^2 transport equation is exact (except for the fact that terms involving correlations of fluctuating viscosity – which have been found to be negligible by BMR – have been omitted), while the transport equation for ϵ_s is a model. C_1, C_2 and C_3 are modeling constants; popular values are $C_1 = 1.44$, $C_2 = 1.83$ and $C_3 = (7 - 2C_1)/3$ (Speziale & Sarkar 1991). $\mathcal{P}$ is modeled using an eddy viscosity approach, so that $\mathcal{P} = 0.5C_\mu(q^4/\epsilon_s)(\widetilde{S}_{ij}\widetilde{S}_{ij} - 3S^2) - q^2S$ with $\widetilde{S}_{ij} = \frac{1}{2}(\widetilde{u}_{i,j} + \widetilde{u}_{j,i})$ and $C_\mu = 0.09$ (Patel *et al.* 1985). Given parameterizations for the two compressibility terms, Π_d and ϵ_d, the equation set is closed.

Applied to the special case of isotropic compression, Eqs. (4) and (5) reduce to

$$\frac{dq^2/2}{dt} = \mathcal{P} - \epsilon_s - \epsilon_d + \Pi_d, \tag{6}$$

$$\frac{d\epsilon_s}{dt} = C_1\mathcal{P}\frac{\epsilon_s}{q^2/2} + 3\,(1 - C_3)\,\epsilon_s S - C_2\frac{\epsilon_s^2}{q^2/2} \tag{7}$$

with $\mathcal{P} = -q^2S$. Our objective is to use DNS results to test some proposed closures for Π_d and ϵ_d, and to evaluate the constants in (and form of) the ϵ_s equation. We begin with the later task, by considering the flow in which the velocity fluctuations are small compared to the local sound speed.

Low Mach Number Case

When the turbulent Mach number, M_T, is very small, the fluctuation velocity field is effectively solenoidal, implying that ϵ_d and Π_d are both negligible. During an isotropic compression, they will remain negligible. (This will not necessarily be true for non-isotropic compressions, however, where ϵ_d and Π_d can be important even at low M_T (Zeman & Coleman 1991, Durbin & Zeman 1992).) The low Mach number DNS results presented in this section are thus primarily useful as a means for evaluating the ϵ_s equation. We first turn our attention to compressions at very rapid rates, since this allows us to compare Eq. (7) and DNS results with rapid distortion theory (RDT) (see also Coleman & Mansour 1991).

Rapid Spherical Compression

Dissipation histories from two low Mach number rapid compression runs, with $(Sq^2/\epsilon)_0 = -47$, are given in Fig. 1. The time span shown represents half of the total available, since at $|S_0|t = 1$ the volume collapses to zero (see Eq. (3)). The first of the DNS data (shown by the symbols), is Wu *et al.*'s (1985) Case SQF, which assumed the fluctuating velocity was divergence free, so that M_T was exactly zero. Very good agreement is found between Wu's data and the present simulation results (run c3dh, dotted curve), which uses $(M_T)_0 = 0.04$ (the other run parameters are given in the appendix). Also shown in Fig. 1, as the dashed curve, is the model Eq. (7) prediction, using $C_2 = 1.92$ and $C_3 = (7 - 2C_1)/3$, the standard value quoted above. The large difference between the model and the DNS histories is at first glance puzzling, since this value of C_3 was advocated by Reynolds (1980) and Morel & Mansour (1982) as a means of exactly matching rapid isotropic compression. However, in setting the value for C_3, Reynolds and Morel & Mansour matched RDT for constant kinematic viscosity. Coleman & Mansour (1991) have shown that the effect of variable viscosity is important in the ϵ_s equation (at *any* Reynolds number), and

$$C_3 = \left(7 - \frac{3}{\widetilde{u}_{i,i}\widetilde{\nu}}\frac{d\widetilde{\nu}}{dt} - 2C_1\right) \Big/ 3 \tag{8}$$

will reproduce the RDT and DNS results. In Fig. 1, predictions using Eq. (8) are shown as the chain-dot and solid curves, for C_2 equal to 1.92 and 0, respectively; very good agreement with the simulation data is observed.

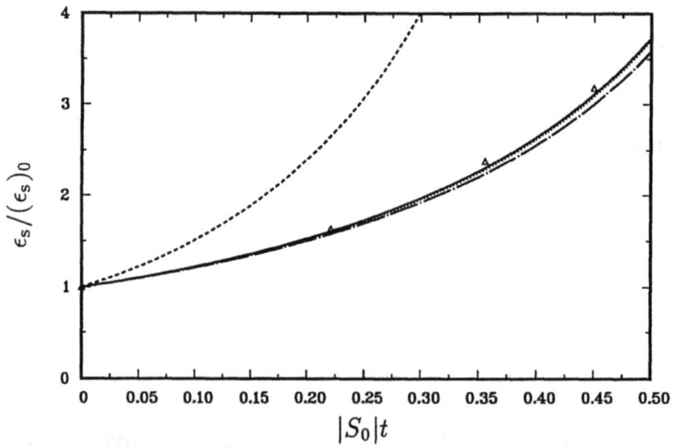

Fig. 1. Evolution of the solenoidal part of ϵ during a rapid spherical compression, $\left(Sq^2/\epsilon\right)_0 = -47$ and $M_T \simeq 0$; $\triangle$, Wu *et al.*'s 1985 DNS Case SQF, $M_T = 0$; ········ , present DNS run c3dh, $(M_T)_0 = 0.04$; ----, Eq. (7) prediction using $C_3 = (7 - 2C_1)/3$, $C_2 = 1.92$; —·—, Eq. (7) prediction using Eq. (8) for C_3, with $C_2 = 1.92$; ——, Eq. (7) prediction using Eq. (8) for C_3, with $C_2 = 0$.

Intermediate and Slow Spherical Compressions

While the correct treatment of terms dominant during a rapid compression is apt to improve the overall behavior of the ϵ_s equation, it provides no guarantee of an accurate prediction for flows experiencing compression at low or intermediate rates. The mean production term is faithfully represented in the ϵ_s equation when Eq. (8) is used for C_3, but the manner in which Eq. (7) models the terms that are important away from the rapid compression regime, namely the vortex-stretching and destruction terms, are crucial in more general situations. This is the subject of the present section.

The key to modeling arbitrary isotropic compressions is found in a conclusion attributed to Spiegel & Frisch in a recent paper by Cambon, Mao & Jeandel (1992): that the form of the Navier-Stokes equations is unchanged by the transformation that maps a decaying isotropic flow to one experiencing spherical compression, provided $M_T = 0$ and the variations of kinematic viscosity are the same for the decaying and compressing flows. This result is demonstrated by the DNS data shown in Fig. 2. The solid and dashed curves trace q^2 histories of an intermediate compression, $(Sq^2/\epsilon)_0 = -2.5$, for cases with constant (run c3dq) and variable viscosity (c3dl), respectively. The later uses in Eq. (2) the exponent $n = 3/4$, and the former $n = 1/(\gamma - 1)$. Setting $n = 1/(\gamma - 1)$ leads to (since for the mean flow the isentropic relations are a very good approximation), $d\widetilde{\nu}/dt = 0$. The dotted curve in Fig. 2 is the result of mapping the isotropic decay evolution (run ixxb), which began from the same state as the compression runs, according to the Spiegel-Frisch transformation:

$$u_{*i}(\mathbf{x}_*, t_*) = \left(\frac{\langle\rho(t)\rangle}{\langle\rho(0)\rangle}\right)^{-1/3} u_i(\mathbf{x}, t)$$

$$t_* = \int_0^t \left(\frac{\langle\rho(\tau)\rangle}{\langle\rho(0)\rangle}\right)^{2/3} d\tau, \tag{9}$$

where the "starred" quantities represent the unmapped isotropic decay variables, and the "unstarred" those of the mapped spherically compressed flow. The fact that the solid and dotted curves correspond shows that the density fluctuations are negligible for these runs, while the divergence of the dashed and dotted curves points to the importance of the viscosities exhibiting the same behavior, for the transformation to be exact.

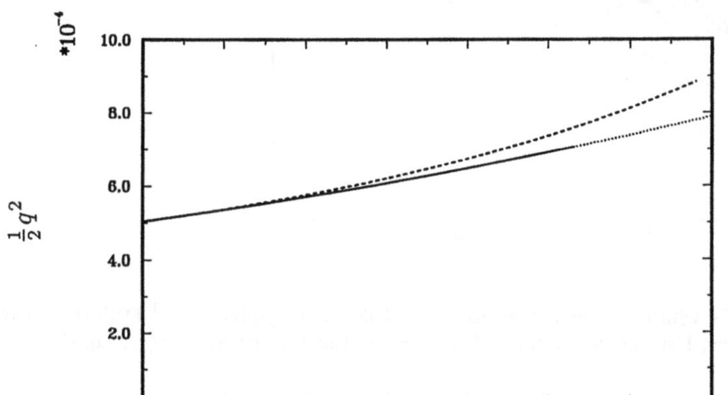

Fig. 2. Evolution of turbulent kinetic energy for intermediate-rate spherical compression, $(Sq^2/\epsilon)_0 = -2.5$ and $(M_T)_0 = 0.03$; ----, run c3dl, $n = 3/4$; ——, c3dq, $n = 1/(\gamma - 1)$ (i.e. $d\widetilde{\nu}/dt = 0$); ········, ixxb, mapped isotropic decay.

The Spiegel-Frisch result implies that to accurately model an isotropic compression of any rate, one need only obtain an accurate model of isotropic decay with time-varying viscosity. We first consider the $\widetilde{\nu}$ = constant case, so that $d\epsilon_s/dt = \widetilde{\nu} d\omega^2/dt$ (where $\omega = \sqrt{\langle\omega_i'\omega_i'\rangle}$). Eq. (7) predicts

$$\frac{d\epsilon_s}{dt} = -2C_2\frac{\epsilon_s^2}{q^2}. \tag{10}$$

To determine C_2, we appeal to the recent direct simulations of incompressible isotropic decay of Mansour & Wray (1991). Fig. 3 shows the development of C_2 with $Re_\lambda = (5Re_T/3)^{1/2}$ predicted by the simulations, and the models of Hanjalić & Launder (1976) and Lumley (1978). The Lumley model consistently underpredicts the DNS data for the range of Re_λ shown, while the Hanjalić & Launder curve reaches the high Re_λ limit too quickly. Although the models and data all appear to agree on the very small Re_λ limit of $C_2 \simeq 1.4$ (a value that is consistent with an energy spectra $\mathcal{E}(k) \sim k^4$ at low wavenumber, for a flow undergoing pure viscous decay), the behavior as $Re_\lambda \to 0$ is very different: no evidence of the $dC_2/dRe_\lambda = 0$ variation assumed by both models is found in the low Re_λ simulation results. Modeling C_2 by a curve fit of the DNS data, we have

$$C_2 = -0.4\exp(-0.13Re_\lambda) + 1.8, \tag{11}$$

which produces the dotted curve shown in Fig. 3.

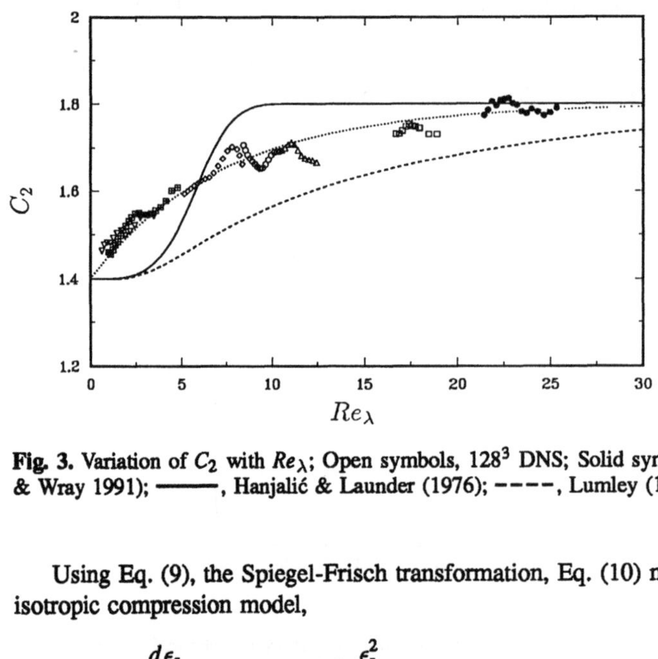

Fig. 3. Variation of C_2 with Re_λ; Open symbols, 128^3 DNS; Solid symbols, 256^3 DNS (Mansour & Wray 1991); ———, Hanjalić & Launder (1976); - - - -, Lumley (1978); ········ , Eq. (11).

Using Eq. (9), the Spiegel-Frisch transformation, Eq. (10) maps to the $\widetilde{\nu}$ = constant isotropic compression model,

$$\frac{d\epsilon_s}{dt} = -4S\,\epsilon_s - 2C_2\frac{\epsilon_s^2}{q^2}. \tag{12}$$

This result was first found by Cambon *et al.* (1992), who only consider the constant viscosity case. It is identical to the result one obtains by applying Eq. (7) directly to spherical compressions when using Eq. (8) with $d\widetilde{\nu}/dt = 0$, and is therefore consistent with Reynolds' and Morel & Mansour's $\widetilde{\nu}$ = constant RDT work.

As we have seen, via Fig. 2, even at intermediate compression rates the variation of $\widetilde{\nu}$ is important during the compression. Therefore, in order to account for the usual situation in which $d\tilde{\nu}/dt \neq 0$, and fully take advantage of the Spiegel-Frisch transformation, a realistic model of variable viscosity isotropic decay must be developed. In an attempt to do so, we consider the transport equation for the enstrophy, ω^2, and assume that the imbalance between the vortex stretching and viscous destruction terms induced by variations of $\tilde{\nu}$ is governed by the parameter $(1/q^2)d\tilde{\nu}/dt$. Since in general, $d\epsilon_s/dt = \tilde{\nu}d\omega^2/dt + \omega^2 d\tilde{\nu}/dt$, this leads to the following model for variable-$\tilde{\nu}$ isotropic decay:

$$\frac{d\epsilon_s}{dt} = \frac{\epsilon_s}{\tilde{\nu}}\frac{d\tilde{\nu}}{dt} - 2C_2\frac{\epsilon_s^2}{q^2} - C_{2_\nu}\frac{1}{q^2}\frac{d\tilde{\nu}}{dt}\left(\frac{\epsilon_s^3}{\tilde{\nu}}\right)^{1/2}. \tag{13}$$

Provided Eq. (13) is valid, the Eq. (9) mapping (of Eq. (13) in "starred" variables) implies that a low Mach number spherical compression of any rate would then be described by

$$\begin{aligned}\frac{d\epsilon_s}{dt} = & -4S\,\epsilon_s + \frac{\epsilon_s}{\tilde{\nu}}\frac{d\tilde{\nu}}{dt} \\ & -2C_2\frac{\epsilon_s^2}{q^2} - C_{2_\nu}\left(\frac{\langle\rho(t)\rangle}{\langle\rho(0)\rangle}\right)^{2/3}\frac{1}{q^2}\frac{d\tilde{\nu}}{dt}\left(\frac{\epsilon_s^3}{\tilde{\nu}}\right)^{1/2}.\end{aligned} \tag{14}$$

The first two terms on the RHS of Eq. (14) follow from the variable viscosity RDT findings of the previous section; C_2 is modeled by Eq. (11). We intend to evaluate the C_{2_ν} term by DNS of variable viscosity isotropic decay. One limitation of Eq. (14) (caused by the use of $(1/q^2)d\tilde{\nu}/dt$ as the relevant parameter) is that it applies only to diatomic gases, such as air. This is because the ratio of the last to the second term, which must not increase during a rapid compression (to avoid violating Eq. (8)), is proportional to $\tilde{\nu}\langle\rho\rangle^{2/3}$ (see Eqs. (8) and (9) of Coleman & Mansour (1991)). Since low Mach number compressions are essentially isentropic, the ratio of the two $d\tilde{\nu}/dt$ terms $\sim \langle\rho\rangle^{n(\gamma-1)-1/3}$ (where n is the viscosity exponent in Eq. 2); in other words, the specific heat ratio must satisfy $\gamma < (1+3n)/3n$ for Eq. (14) to be correct during a rapid compression.

High Mach Number Case

Up to this point, our discussions have dealt with "compressing" flows, those in which compressibility affects only the mean density. There are many situations, however, where density fluctuations are important. In this section, we address the issues related to these flows, in which M_T is not small.

"Synthetic" Initial Conditions

The first issue we consider is the effect of isotropic compression on the acoustic variables. During their study of compressible homogeneous shear flow, BMR found that the mean shear drives the acoustic (dilatational) field to a unique state, regardless of the magnitude of initial conditions – even when no development time was allowed for the turbulence to mature before the shear was imposed. Two parallel runs, c3dr and c3ds, with identical initial Mach numbers ($M_T = 0.3$) and compression rates ($(Sq^2/\epsilon)_0 = -2.5$), but different acoustic initial conditions, were performed to see if spherically compressed turbulence is similar in this regard. One run, c3dr, began with no initial dilatational velocity or fluctuation density, temperature or pressure; the other, c3ds, began with one-fifth of the turbulent kinetic energy in the dilatational component, and fluctuation density and absolute temperature fields both equal to 15 percent of their volume averaged means (with the pressure given by the ideal gas law). All non-zero fields for both runs assumed "top-hat" initial spectra (BMR), and the compression began immediately upon the "synthetic" undeveloped fields (see appendix for other run parameters).

The Mach number histories for the parallel runs are given in Fig. 4a, and are very similar to each other. However, the variations of the dilatational dissipation (Fig. 4b) and pressure-dilatation correlation (not shown) are quite distinct. Since the histories have progressed, without converging, to over half of the total possible compression time, $0 \leq |S_0|t < 1$ (see Eq. (3)), it appears that unlike shear, spherical compression does not drive the acoustic field to a unique state. This behavior is a result of the fact that the only coupling between the solenoidal and acoustic fields is due to nonlinear interactions, which are slow compared to the turbulence time scale at this Reynolds number. It implies that ϵ_d and Π_d models must account for initial conditions if they are to work correctly during isotropic compression. The degree to which one Π_d model succeeds is taken up in the next section, as part of a discussion of high M_T data that are generated using for initial conditions – rather than the synthetic fields of this section – developed isotropic decaying turbulence.

"Natural" Initial Conditions

As with the spherically compressed flow, acoustic variables in decaying isotropic turbulence do not converge to unique values which are independent of their initial conditions (cf. BMR). Because of this, a certain amount of arbitrariness exists in specifying "natural" initial conditions for the high Mach number simulations. For the present computations, the initial acoustic field is the result of beginning with a purely solenoidal field, allowing it to decay and develop dilatational fluctuations, before the compression is imposed, at the point where $M_T = 0.44$, $\rho_{rms} = 0.03\langle\rho\rangle$, $T_{rms} = 0.02\langle T\rangle$ and 9 percent of the turbulent kinetic energy is in the dilatational field. Three separate compression speeds were chosen, defining three cases, runs c3db ($(Sq^2/\epsilon)_0 = -47$), c3dc (-12.5) and c3dd (-1.25).

The initial field was not in exact acoustic equilibrium, since the dilatational energy was not evenly split into the kinetic and potential energy partitions. The ratio of the two is defined by the parameter, $\mathcal{F} = \gamma\langle\rho\rangle\langle p\rangle q_{\text{dil}}^2/\langle p_c' p_c'\rangle$, where p_c is the so-called "compressible pressure" (Sarkar *et al.* 1989). Both high and low M_T isotropic decay simulations have shown that after an initial transient, well developed compressible turbulence reaches acoustic equilibrium, characterized by $\mathcal{F} \simeq 1$ (Sarkar *et al.* 1989, Lee *et al.* 1991, BMR). Instead of being about one, the initial condition used here saw $\mathcal{F} \simeq 0.6$. (During the evaluation of $\mathcal{F}$, we found that for this flow, the compressible pressure fluctuations $\langle p_c' p_c'\rangle$ accounted for approximately 80 percent of the total fluctuations, $\langle p'p'\rangle$.)

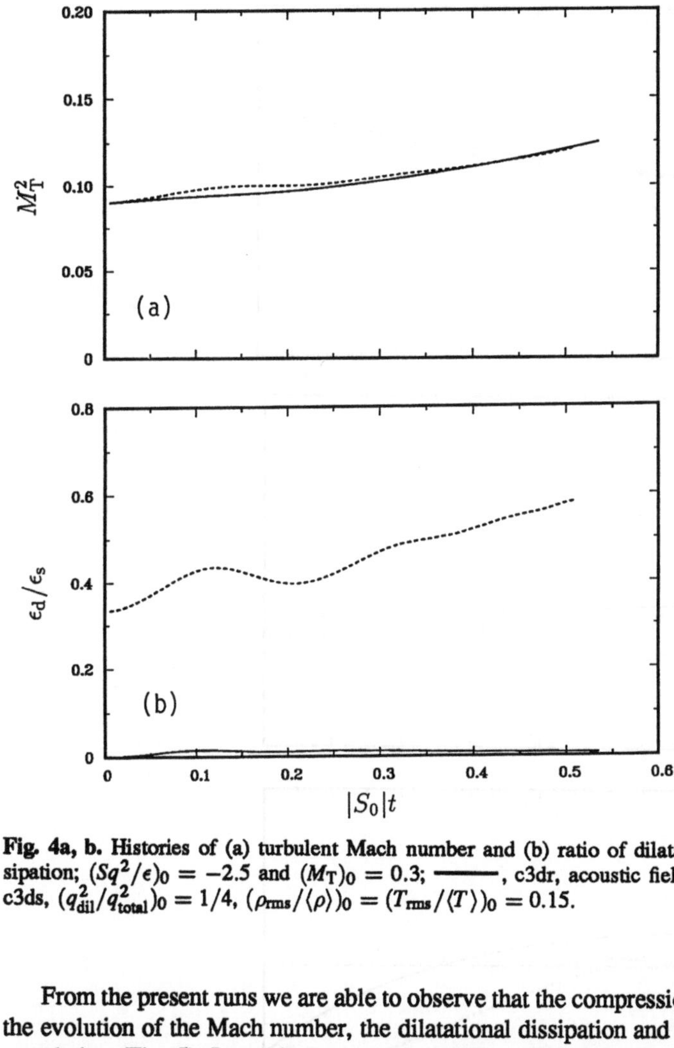

Fig. 4a, b. Histories of (a) turbulent Mach number and (b) ratio of dilatational to solenoidal dissipation; $(Sq^2/\epsilon)_0 = -2.5$ and $(M_T)_0 = 0.3$; ———, c3dr, acoustic fields initially zero; - - - -, c3ds, $(q^2_{dil}/q^2_{total})_0 = 1/4$, $(\rho_{rms}/\langle\rho\rangle)_0 = (T_{rms}/\langle T\rangle)_0 = 0.15$.

From the present runs we are able to observe that the compression rate strongly affects the evolution of the Mach number, the dilatational dissipation and the pressure-dilatation correlation (Fig. 5). In particular, we note in Fig. 5b that for the initial condition used, the level of ϵ_d is very small, never more than 4 percent of the total dissipation. One of the predictions of RDT is that ϵ_d/ϵ_s will remain constant during a rapid spherical compression (Blaisdell 1991); the DNS results support this finding, since the rate of change of ϵ_d/ϵ_s decreases as the compression rate increases.

From a modeling standpoint, the most significant result is the behavior of the pressure-dilatation term shown in Fig. 5c. The magnitude of Π_d is seen to be a significant fraction of the total dissipation, and demonstrates a great dependence on compression rate. Beginning from the initial condition level, where the pressure-dilatation in the unstrained field acts as a source of turbulent kinetic energy (supplied by the potential energy of the pressure field), the compression in all cases causes a reduction of Π_d. At the slow rate, it approaches zero and remains slightly positive; when $|S|$ is large however, the pressure-dilatation is driven to very large negative values, with magnitude many times that of the total dissipation rate (see Fig. 6a), representing an important conversion from kinetic to potential energy.

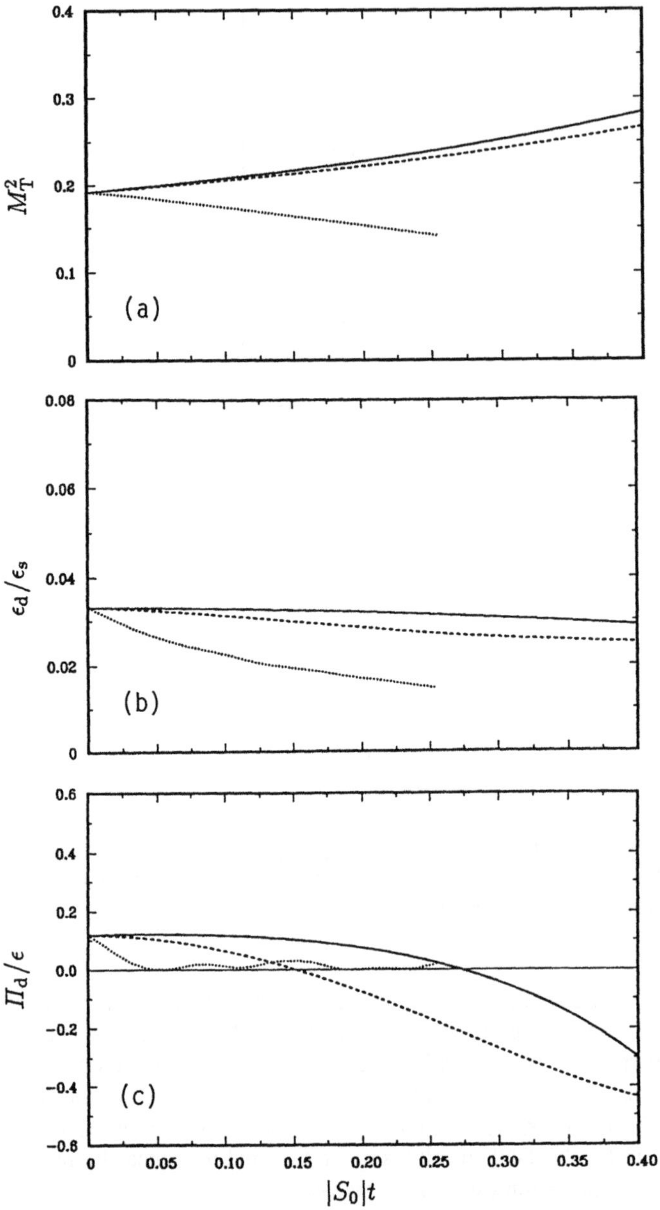

Fig. 5a-c. Evolution of (a) turbulent Mach number, (b) ratio of dilatational to solenoidal dissipation, and (c) pressure-dilatation; $(M_T)_0 = 0.44$; ———, c3db, $(Sq^2/\epsilon)_0 = -47$; ----, c3dc, $(Sq^2/\epsilon)_0 = -12.5$; ········, c3dd, $(Sq^2/\epsilon)_0 = -1.25$.

In Fig. 6, we test the ability of Aupoix *et al.*'s (1990) Π_d model to duplicate the DNS histories for the largest and smallest compression rates (note the expanded vertical and horizontal scales in Fig. 6a). The model assumes

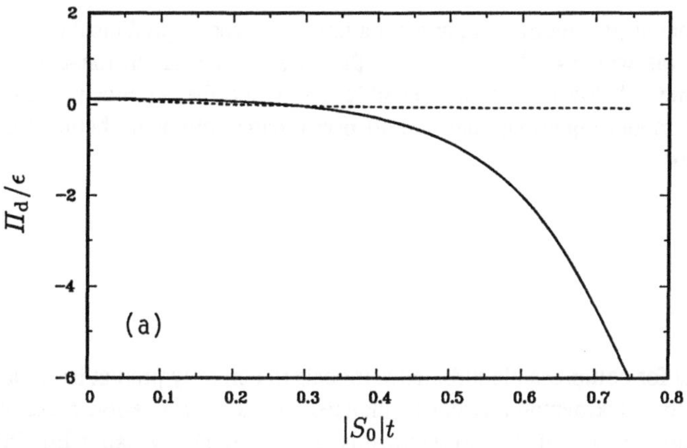

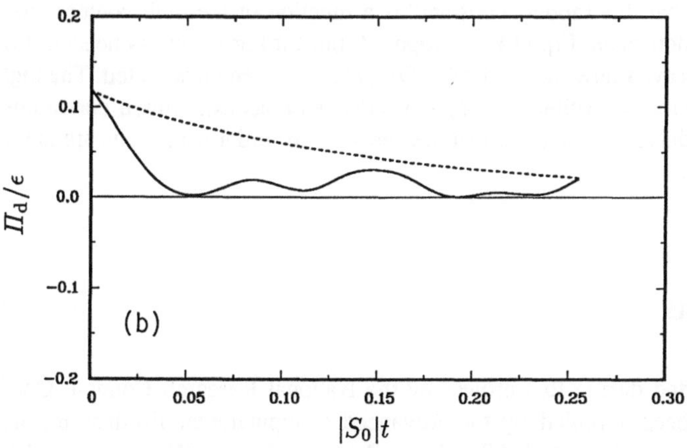

Fig. 6a, b. Pressure-dilatation histories, $(M_T)_0 = 0.44$; (a) Run c3db, $(Sq^2/\epsilon)_0 = -47$; (b) Run c3db, $(Sq^2/\epsilon)_0 = -1.25$. ——, DNS; - - - -, Aupoix *et al.* (1990) Eq. (15).

$$\frac{d\langle\rho\rangle\Pi_d}{dt} = -C_{A1}\langle\rho\rangle\frac{M_T^2}{2\tau_A}\frac{dq^2}{dt} - C_{A2}\frac{\langle\rho\rangle}{\tau_A}\Pi_d, \qquad (15)$$

where $\tau_A = (1/2)^{3/2}(q^2/\epsilon)M_T$, $C_{A1} = 0.25$ and $C_{A2} = 0.20$. While the trend for the $(Sq^2/\epsilon)_0 = -1.25$ case is roughly correct (Fig. 6b), that for the rapid compression (Fig. 6a) is not. Given that the slow compression performance is about as good as that found for isotropic decay (see Figure 5 of Aupoix *et al.* 1990), and the poor response for large $|S|$ found here, it seems as if the major deficiency of the model is the strain rate sensitivity – although part of the poor showing could be due to the fact that it has been tested against a flow in acoustic nonequilibrium. Aupoix *et al.* have acknowledged the model's weakness for flows with strong initial condition dependence and have tentatively proposed model transport equations for both Π_d *and* ϵ_d, which might alleviate some of the shortcomings observed here.

An algebraic model of Π_d, which contains some initial condition dependence, has also recently been proposed (Zeman and Coleman (1991), in these proceedings). When this model is one component of a set of modeling equations, it correctly predicts the DNS histories observed here (as well as DNS results for the case of one-dimensional mean compression; see Appendix below). It is also possible that when the Aupoix proposal is combined with other model equations, the overall performance might be better than implied by the above test.

Conclusion

The DNS data imply that for compressible turbulence models to correctly predict isotropic compression, the variation of kinematic viscosity induced by the compression must be accounted for. For a rapid spherical compression, this may be done by using Eq. (8) for C_3, which implies that this model "constant" is a function of the bulk compression. At non-rapid compression rates, Eq. (14) is proposed, but further testing is needed. For constant-$\widetilde{\nu}$ isotropic decay, a new model for C_2, Eq. (11), has been constructed. The high M_T flows have been found to exhibit a strong dependence on acoustic initial conditions, which implies that models for ϵ_d and Π_d that are sensitive to initial conditions are apt to be the most successful.

Acknowledgements

GNC has been supported during this project by the National Research Council. Computer resources have been supplied by the Advanced Computational Facility and the National Aerodynamic Simulator at NASA Ames Research Center. We are grateful to Prof. G. A. Blaisdell for his contributions to this work.

Appendix

All the DNS runs reported herein use 96^3 grid points, and choose $Pr = 0.7$ and $\gamma = 1.4$ as fixed parameters; variable parameters are listed in Tables 1 and 2. Table 2 gives the acoustic initial conditions: the ratio of dilatational to total turbulent kinetic energy, and the (rms to volume average) density fluctuation and (absolute) temperature fluctuation ratios. The compression rate, turbulent Reynolds and Mach numbers, viscosity exponent (Eq. (2)) and type of initial condition are given in Table 1.

Two methods of generating initial conditions are employed, one termed "synthetic," the other "natural." The runs that use the "natural" approach (those with 'N' entered in the IC column of Table 1), begin with random, isotropic velocity, density and temperature fields with prescribed analytic spectra (either "top-hat" or "exponential" (Blaisdell *et al.* 1991)), and allow the turbulence to decay without imposed strain to a mature state (such that the velocity derivative skewness $\simeq 0.4$), before the compression is imposed

Table 1. Non-acoustic run parameters.

Run	$-(Sq^2/\epsilon)_0$	$(q^4/\epsilon\widetilde{\nu})_0$	$(M_T)_0$	n	IC
c3dh	47.0	21	0.04	0.75	N
c3dl	2.5	52	0.03	0.75	N
c3dq	2.5	52	0.03	$1/(\gamma-1)$	N
ixxb	0	52	0.03	0.75	N
c3dr	2.5	200	0.30	0.67	S
c3ds	2.5	200	0.30	0.67	S
c3db	47.0	500	0.44	0.67	N
c3dc	12.5	500	0.44	0.67	N
c3dd	1.25	500	0.44	0.67	N
c3dk	47.0	21	0.02	0.67	N
c1da	47.0	500	0.44	0.67	N
c1db	47.0	21	0.02	0.67	N

Table 2. Acoustic run parameters.

Run	$(q^2_{\text{dil}}/q^2_{\text{total}})_0$	$(\rho_{\text{rms}}/\langle\rho\rangle)_0$	$(T_{\text{rms}}/\langle T\rangle)_0$
c3dh	0.00	0.00	0.00
c3dl	0.00	0.00	0.00
c3dq	0.00	0.00	0.00
ixxb	0.00	0.00	0.00
c3dr	0	0	0
c3ds	0.20	0.15	0.15
c3db	0.09	0.03	0.02
c3dc	0.09	0.03	0.02
c3dd	0.09	0.03	0.02
c3dk	0.00	0.00	0.00
c1da	0.09	0.03	0.02
c1db	0.00	0.00	0.00

upon the mature field characterized by the Table 1 and 2 values. In the two runs that use "synthetic" initial conditions (denoted in Table 1 by an 'S'), no development time is allowed before the compression begins upon a field defined by the analytic spectra. The last three runs listed, c3dk, c1da, and c1db, are not used in this paper, but along with c3db, are discussed by Zeman & Coleman (1991) in these proceedings. For c1da and c1db, the mean compression is one-dimensional, i.e. $\widetilde{u}_{i,j}(t) = S(t)\delta_{i1}\delta_{1j}$.

References

Aupoix, B., Blaisdell, G.A., Reynolds, W.C. & Zeman, O. (1990): "Modeling the turbulent kinetic energy equation for compressible, homogeneous turbulence", in Proc. of the 1990 NASA/Stanford Center for Turbulence Research Summer Program.

Blaisdell, G.A., Mansour, N.N. & Reynolds, W.C. (1991): "Numerical simulation of compressible homogeneous turbulence", Department of Mechanical Engineering, Stanford Univ., Thermosciences Div. Report TF-50.

Blaisdell, G.A. (1991): Private communication.

Cambon, C., Mao, Y. & Jeandel, D. (1992): "On the application of time dependent scaling to the modeling of turbulence undergoing compression", to appear in *European J. Mech.*

Coleman, G.N. & Mansour, N.N. (1991): "Modeling the rapid spherical compression of isotropic turbulence", *Phys. Fluids A.* 3(9) 2255.

Durbin, P.A. & Zeman, O. (1992): "Rapid distortion theory for homogeneous compressed turbulence with application to modeling", to appear in *J. Fluid Mech.*

Hanjalić,K. & Launder, B.E. (1976): "Contribution towards a Reynolds-stress closure for low-Reynolds-number turbulence", *J. Fluid Mech.* **74**(4) 593.

Lee, S., Lele, S.K. & Moin, P. (1991): "Eddy shocklets in decaying compressible turbulence", *Phys. Fluids A* 3(4) 657.

Lumley, J.L. (1978): "Computational modeling of turbulent flows", *Advances in Applied Mechanics*, **18** 123.

Mansour, N.N. & Wray, A.A. (1991): "Direct numerical simulation data: a tool for turbulence modeling", in Proc. 4th Int. Symp. Comp. Fluid Dynamics, Univ. Calif., Davis.

Morel, T. & Mansour, N.N. (1982): "Modeling of turbulence in internal combustion engines", SAE Paper No. 820040, Warrendale, PA.

Patel, V.C., Rodi, W. & Scheuerer, G. (1985): "Turbulence models for near-wall and low Reynolds number flows: A review", *AIAA J.* **23**(9) 1308.

Reynolds, W.C. (1980): "Modeling of fluid motions in engines: An introductory overview", in *Combustion Modeling in Reciprocating Engines* (ed. J.N. Mattavi & C.A. Amann) Plenum Press.

Sarkar, S., Erlebacher, G., Hussaini, M.Y. & Kreiss, H.O. (1989): "The analysis and modeling of dilatational terms in compressible turbulence", ICASE Report No. 89-79.

Speziale, C.G. & Sarkar, S. (1991): "Second-order closure models for supersonic turbulent flows", AIAA Paper No. 91-0217.

Wu, C.-T., Ferziger, J.H. & Chapman, D.R. (1985): "Simulation and modeling of homogeneous, compressed turbulence", Department of Mechanical Engineering, Stanford Univ., Thermosciences Div. Report TF-21.

Zeman, O. & Coleman, G.N. (1991): "Compressible turbulence subjected to shear and rapid compression", page 283 of this volume.

Compressible Turbulence Subjected to Shear and Rapid Compression

O. Zeman[1] and G. N. Coleman[2]

[1] CTR, Stanford University, Stanford CA 94305, USA
[2] NASA Ames Research Center, Moffett Field CA 94035, USA

Abstract

First, a review is presented of the physics and modeling of compressibility effects in homogeneous decaying, and shear-driven turbulence. Second, new findings are reported concerning the effects of rapid spherical (3D), and one-dimensional (1D) compression on initially isotropic turbulence. On the basis of the direct numerical simulation results of Coleman and Mansour (1991), it is shown that in 1D rapid compression the pressure-dilatation correlation $\overline{pu_{j,j}}$ is responsible for unusually high energy transfer from the kinetic to pressure modes. This transfer mechanism leads to considerable turbulent kinetic energy losses even when turbulence is virtually incompressible. A parametric model for $\overline{pu_{j,j}}$ is developed to capture this anisotropic rapid effect, and the model results are compared with the Coleman-Mansour data. The model is then tested against direct simulations of turbulence/normal shock interactions (Lee, 1991). It is shown that apart from the energy transfer due to $\overline{pu_{j,j}}$, another important mechanism within the shock is associated with the acceleration term (containing the mean pressure gradient).

1. Introduction

The measure of the compressibility effects on turbulence is the so-called r.m.s. Mach number M_t, based on an r.m.s. fluctuating speed and the sonic speed corresponding to the background (mean) temperature. The sonic speed can be considered as an intrinsic velocity scale which limits the propagation of perturbations through the fluid and sets a limit, in a statistical sense, on the velocity fluctuations as well. As the r.m.s. Mach number increases, the velocity fluctuations are accompanied by growing levels of density and pressure fluctuations and a growing degree of coupling between the vortical and dilatational (acoustic) fields. Utilizing Helmholtz' decomposition and Favre averaging, one can identify two explicit terms in the turbulence energy equations which depend on the dilatational field and provide the vortical/acoustic mode coupling. These are: 1) the dilatation dissipation defined as $\epsilon_d = (4/3)\mu\overline{(u_{l,l})^2}$; and 2) the pressure dilatation correlation term $\Pi_d = \overline{pu_{j,j}}/\overline{\rho}$.

Zeman (1990) suggested that turbulence with a sufficiently large M_t may generate shock-like structures or shocklets, which entail high levels of dilatation dissipation (ϵ_d). The existence of turbulent shocklets has been recently confirmed in direct numerical simulations (DNS) of decaying turbulence (Lee, Lele, and Moin 1990) and in the DNS of shear-driven turbulence as well (Blaisdell 1990). In laboratory flows, the most prominent effect of compressibility on turbulence is the inhibition of shear layer growth rate with increasing convective Mach number as observed in experiments of Papamoschou and Roshko (1988), Samimy and Elliott (1989), and others. Zeman (1990) argued that the growth rate inhibition is due to the (anomalous) dilatation dissipation caused by turbulent shocklets and, on the basis of a stochastic shocklet model, inferred a parametric expression to quantify ϵ_d. Inclusion of this anomalous dissipation in a second-order closure model provided the growth rate reduction, agreeing with the experimental data.

The importance of the pressure-dilatation term Π_d for turbulence energetics has, so far, not been fully appreciated. As illustrated in Zeman & Blaisdell (1990) and Zeman (1991), Π_d mediates the exchange between the kinetic and potential energies. The potential energy is meant to be the energy associated with the compression work; it is proportional to the pressure fluctuation variance $\overline{p^2}$. In equilibrium flows such as self-preserving shear layers and flat plate boundary layers, Π_d appears to be negligible. However, in unsteady turbulent flows controlled by initial conditions and in flows with rapid (mean) compression, the contribution of Π_d in the turbulence energy budget can be significant, and its knowledge is indispensable for computation of the variances of pressure and density fluctuations (see also Zeman, 1991a).

In the following a brief review is presented of the current understanding of compressibility effects in two generic (DNS) flows: decaying and shear-driven turbulence. The effects of mean rapid compression on turbulence (with non-zero values of M_t) are the main subjects of this paper. The problems to be addressed are: 1) homogeneous three- and one-dimensional rapid compression (hereafter abbreviations 3D or 1D are used) and 2) response of shear-free turbulence to passage through a normal shock.

2. Compressible Decaying, and Shear-Driven Turbulence

The fundamental energy-governing equations for compressible homogeneous turbulence in the Favre average setting are

$$\frac{1}{2}\frac{\partial q^2}{\partial t} = P_s - (\epsilon_s + \epsilon_d - \Pi_d),$$

$$\frac{\partial \widetilde{T}}{\partial t} = (\epsilon_s + \epsilon_d - \Pi_d)c_v^{-1}, \tag{1}$$

accompanied by the equation of state $\overline{p} = \overline{\rho}R\widetilde{T}$ where tilde and bar denote, respectively, Favre and Reynolds averages. Defined previously, Π_d is the pressure-dilatation term, P_s represents the shear production of turbulence due to the (constant) mean shear (S), and $q^2 = \widetilde{u_j u_j}$ is twice the Favre-averaged kinetic energy. The separation of the total dissipation in the solenoidal (ϵ_s), and dilatational (ϵ_d) parts is feasible due to the Helmholtz' decomposition of the velocity field (Zeman, 1990) and is strictly valid only if the flow is homogeneous in the mean. The definition of the r.m.s. Mach number is $M_t = q/a$ where the sonic speed a is based on the mean temperature $\widetilde{T}$. In the absence of compressibility effects ($M_t << 1$), the Reynolds stress tensor components $\widetilde{u_i u_j}$ can be computed by means of a standard turbulence closure model (for details see Zeman 1991a). Here we discuss merely the physics and modeling associated with the compressible terms.

2.1 Dilatation Dissipation

In high Reynolds number turbulence, eddy shocklets are the only significant source of ϵ_d. On the basis of a stochastic shocklet model, Zeman (1990) inferred a parametric expression

$$\epsilon_d \propto \epsilon_s F(M_t, K). \tag{2}$$

The shocklet dissipation function $F(M_t, K)$ is an integral functional of a pdf $p(m, K)$ of fluctuating Mach number m. The parameter K characterizes a departure from Gaussianity of m and has been defined as the kurtosis i.e. $K = \overline{m^4}/(\overline{m^2})^2$. From the direct simulations of homogeneous turbulence, K was estimated as $K \approx 4$ (Zeman & Blaisdell 1990) and $F(M_t, K)$ approximated as

$$F(M_t) = 1 - \exp\{-((M_t - 0.25)/0.80)^2\}$$

$$F(M_t) = 0, \quad \text{if } M_t \leq 0.25. \tag{3}$$

We note that ϵ_d is negligible (w.r.t. ϵ_s) when $M_t < 0.3$. Physically, this means that shocklets cannot occur below a certain (threshold) value of the r.m.s. Mach number. In the simulations of Lee et al. (1991), the shocklets were found only when M_t exceeded values of about 0.3. In most of the computational examples presented in this paper, ϵ_d is unimportant and the solenoidal dissipation ϵ_s is determined from a standard modeling equation (independent of M_t).

2.2 Pressure-dilatation Correlation

Zeman (1991) proposed that in shocklet-free decaying or shear-driven turbulence, the principal balance in the pressure variance equations is

$$\frac{1}{2}\frac{D\overline{p^2}}{Dt} = -\overline{\rho}^2 a^2 \Pi_d, \tag{4}$$

and proposed a closure for Π_d based on a heuristic argument that in the absence of forcing, the pressure fluctuations will tend to relax to an equilibrium value $p_e(M_t)$ at a rate dictated by an acoustic time scale $\tau_a \propto L/a$, also called sound crossing time (of an eddy of size L). This has led to the closure equation

$$\frac{D\overline{p^2}}{Dt} = -\frac{\overline{p^2} - p_e^2}{\tau_a} = -2\overline{\rho}^2 a^2 \Pi_d. \tag{5}$$

The equilibrium pressure variance p_e^2 was inferred from theory and DNS data as

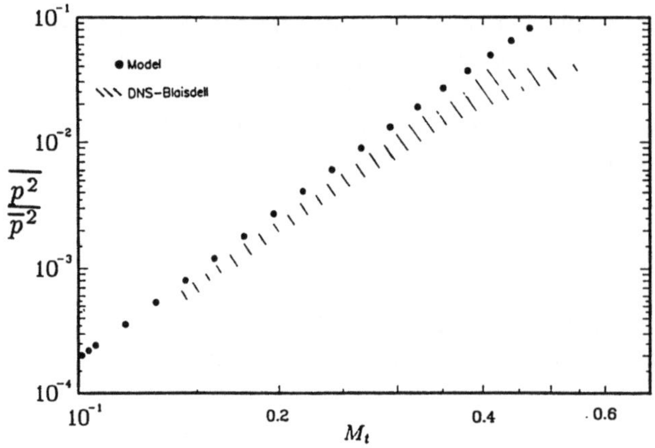

Fig. 1. Pressure variance vs M_t in homogeneous shear turbulence: DNS-model comparison.

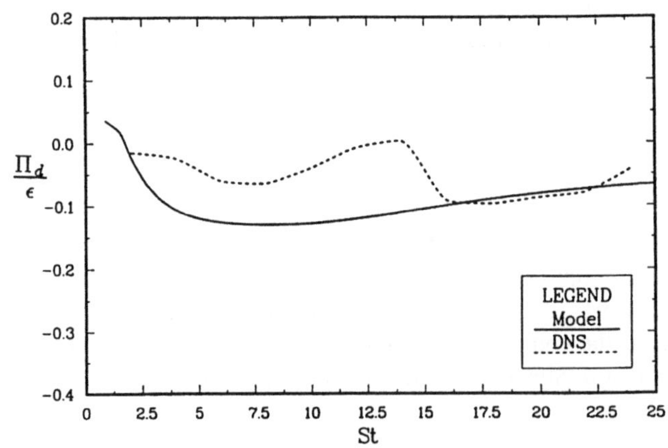

Fig. 2. Evolution of the pressure-dilatation term normalized by total dissipation in homogeneous shear turbulence

$$\frac{p_e^2}{\bar{\rho}^2 q^2 a^2} = P_e(M_t) = \frac{M_t^2 + M_t^4}{1 + M_t^2 + M_t^4}, \tag{6}$$

with the acoustic time scale redefined as $\tau_a = 0.2\tau M_t$. It is noted that in these flows, the turbulence time scale is related to vortical turbulence, i.e., $\tau = q^2/\epsilon_s$.

The DNS data of Blaisdell (1990) confirmed that (4) is a valid assumption for a range of r.m.s. Mach numbers $0.05 \leq M_t \leq 0.7$. The model (5) accounts for the initial data influence on the (transient) behavior of turbulence: when the initial pressure (or density) fluctuation level is much higher (or lower) then the equilibrium value p_e, potential energy $\overline{p^2}$ is transferred to (or extracted from) the kinetic mode. This energy exchange takes place on the acoustic time scale.

In the equilibrium regime of homogeneous shear turbulence when $St \geq 2$, the model represented by Eqs. (4)-(6) can be tested for prediction of two quantities: the pressure dilatation Π_d and the (normalized) pressure variance $\overline{p^2}/\overline{p}^2$. The pressure variance as a function of M_t is shown in Fig. 1, where the cross-hatched area represents the spread of DNS data reproduced from Blaisdell (1990) (plotted only for $t \geq S^{-1}$). The values of $\overline{p^2}$ are very close to, but always smaller than, the equilibrium pressure variance p_e^2 (not shown); the subtle difference between $\overline{p^2}$ and p_e^2 is, according to (5), responsible for the sign and magnitude of Π_d. DNS-model comparison of the ratio Π_d/ϵ_s is shown in Fig. 2. Both figures support the model mechanism of relaxation to equilibrium. We note that another viable model for Π_d in shear turbulence has been proposed by Aupoix et al. (1990). This model, however, is not capable of dealing with the condition of high initial pressure (or density) fluctuation content.

3. Rapid Homogeneous Compression: 3D vs 1D

Recently, Coleman and Mansour (1991) performed direct simulations of compressible turbulence subjected to rapid spherical (3D) and 1D compression for different initial values of the r.m.s. Mach number $M_{to} = 0.02$ and 0.44. These DNS experiments have relevance to the flows of practical interest such as the internal combustion engine flow and turbulence/shock interactions taking place in a compression corner flow. The purpose of this section is to elucidate the physics of rapidly compressed turbulence and describe the recent modeling effort in this area. Perhaps the most important finding is that due to strong interactions between fluctuating pressure, dilatation ($u_{j,j}$), and vorticity in the 1D rapid compression, the pressure-dilatation term Π_d cannot be discarded even for $M_t << 1$. This is contrary to the current believe that turbulence under such conditions can be considered incompressible.

The important parameters of the flows in question are the initial Mach number M_{to}, initial timescale $\tau_o = (q^2/\epsilon_s)_o$, and the initial rate of compression $A(t=0) = A_o$. Denoting the mean velocity field by U_i, we define the spherical compression by $U_{1,1} = U_{2,2} = U_{3,3} = -A(t)$, and 1D compression by $U_{1,1} = \nabla\cdot\mathrm{U} = -A(t)$. The condition of spatial homogeneity demands that

$$A(t) = \frac{A_o}{1 - A_o t}, \tag{7}$$

and the compression is considered rapid if $A_o\tau_o >> 1$.

In the simulations discussed here, $A_o\tau_o = 47$ and therefore, in principle, a rapid theory could be utilized to determine the evolution of q^2 and, if need be, $\overline{p^2}$. The 3D rapid compression of low M_t isotropic turbulence is therefore primarily of interest to us as a reference flow. If, however, M_{to} is high enough to produce initial dilatation and pressure fluctuations, the rapid theory is more involved. Here, we use the results of the modified rapid theory of Sabel'nikov (1975) to determine a free constant in the rapid part of the model for Π_d. The relevant set of equations to describe rapid spherical compression of isotropic turbulence are

$$\frac{1}{2}\frac{\partial q^2}{\partial t} = -\frac{1}{3}q^2\nabla\cdot\mathrm{U} - (\epsilon_s + \epsilon_d - \Pi_d)$$

$$\frac{\partial \widetilde{T}}{\partial t} = -(\gamma - 1)\widetilde{T}\nabla\cdot\mathbf{U} + (\epsilon_s + \epsilon_d - \Pi_d)c_v^{-1}$$

$$\frac{1}{2}\frac{\partial \overline{p^2}}{\partial t} = -\gamma\overline{p^2}\nabla\cdot\mathbf{U} - \gamma\overline{p}\overline{p}\Pi_d, \tag{8}$$

and the usual equation of state. Equations in (8) require a closure for Π_d, and for the total dissipation $\epsilon_t = \epsilon_s + \epsilon_d$. There is virtually no dilatation dissipation in this flow and therefore ϵ_d is neglected. The equation for the solenoidal dissipation ϵ_s requires a special treatment because the molecular viscosity varies during compression. We used the ϵ_s-equation proposed by Coleman and Mansour (1991) which yields a perfect agreement with the DNS data.

Rapid distortion theory suggests that the pressure dilatation term Π_d has a rapid contribution, linear in $\overline{p^2}\nabla\cdot\mathbf{U}$. Hence, a new expression

$$\Pi_d = \{\frac{\overline{p^2} - p_e^2}{2\tau_a} + c_d\overline{p^2}\nabla\cdot\mathbf{U}\}\frac{1}{\overline{\rho}^2 a^2} \tag{9}$$

has been proposed to account for the rapid compression contribution. The free constant c_d has been determined from the rapid theory as $c_d = (5 - 3\gamma)/12$ (note that $c_d = 0$ for monatomic gases, and 0.066 in air). The details of the calculations leading to the determination of c_d are described in Sabel'nikov (1975), and Durbin and Zeman (1992). We note that in analogy with pressure-strain terms, we have in (9) a slow (relaxation) term and a rapid (compression) term. Here, however, the rapid condition requirement is $|\nabla\cdot\mathbf{U}|\tau M_t >> 1$, or in terms of the initial parameters

$$A_o\tau_o M_{to} >> 1.$$

This new rapid condition is pertinent to the dilatational field and is stronger than the condition

$$A_o\tau_o >> 1$$

for the solenoidal field.

Finally, we note that during the rapid compression, the turbulence lengthscale (L) and timescale (τ) no longer depend on ϵ_s, and therefore a relation such as $L \propto q^3/\epsilon_s$ may not be appropriate. We can construct a lengthscale equation which yields the rapid limit satisfying $L^n\overline{\rho} = \text{const.}$, and the equilibrium limit $L \to q^3/\epsilon_s$ (when $\nabla\cdot\mathbf{U} \to 0$):

$$\frac{DL}{Dt} = L\{\frac{0.9}{\tau} + \frac{1}{n}\nabla\cdot\mathbf{U}\}, \tag{10}$$

where n is an integer denoting the dimension of compression (or expansion). Eq. (10) is exactly valid only for rapid spherical compression; in 1D compression ($n = 1$), it approximates the lengthscale variation in the direction of compression. In principle, τ in the modeling equations should be proportional to L/q. We found, however, that L/q decreased too rapidly and had undesirable effects on the model results. We therefore used in the following calculations a timescale based on the geometric average: $\tau = \sqrt{qL/\epsilon_s}$.

DNS computations of rapidly compressed turbulence (in 1D and 3D) of Coleman and Mansour (1991) for two initial Mach numbers $M_{to} = 0.02$ (Case A) and 0.44 (Case B)

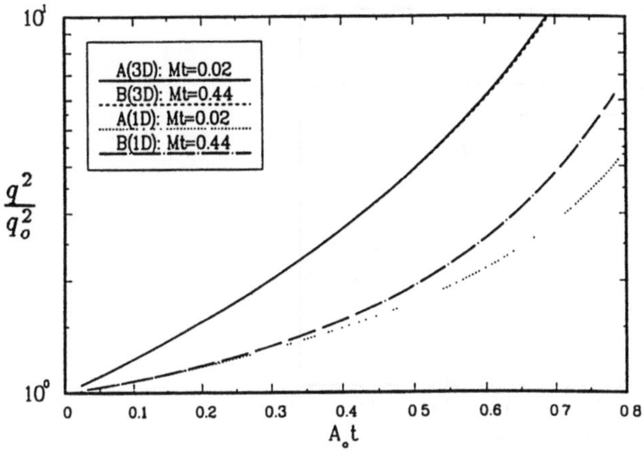

Fig. 3. Evolution of kinetic energy during rapid compression in 1D and 3D. Case A is for $M_{to} = 0.02$; Case B for $M_{to} = 0.44$.

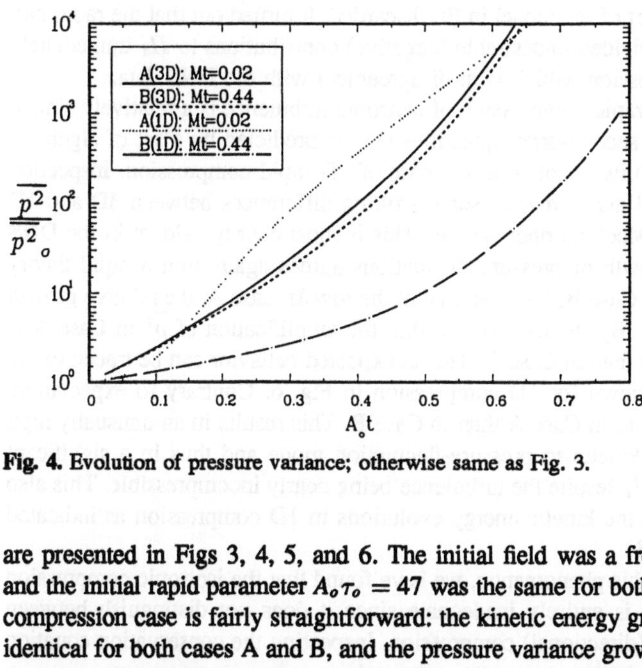

Fig. 4. Evolution of pressure variance; otherwise same as Fig. 3.

are presented in Figs 3, 4, 5, and 6. The initial field was a freely decaying turbulence and the initial rapid parameter $A_o\tau_o = 47$ was the same for both cases A and B. The 3D compression case is fairly straightforward: the kinetic energy growth (Fig. 3) is virtually identical for both cases A and B, and the pressure variance growths (in Fig. 4) are nearly the same. The DNS data agree well (within 4% error) with the rapid compression limit whereby the pressure dilatation Π_d is neglected (also $\epsilon_s = 0$). According to (7) and (8) this limit yields simple expressions:

$$\frac{q^2}{q_o^2} = (\frac{\overline{\rho}}{\rho_o})^{2/3} = (\frac{1}{1-A_o t})^2, \quad \frac{\overline{p^2}}{p^2{}_o} = (\frac{1}{1-A_o t})^{6\gamma}, \tag{11}$$

which are independent of M_t. The model-DNS comparison of the evolution of Π_d/ϵ_t for 3D compression only is shown in Fig. 5. Here, the best match with the DNS data was

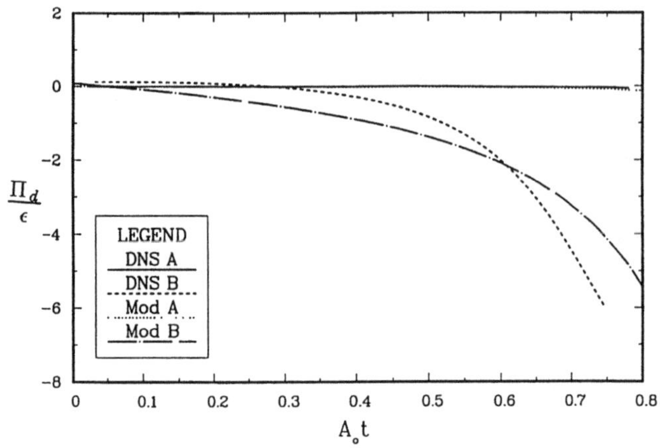

Fig. 5. Evolution of the normalized pressure dilatation for 3D rapid compression: DNS-model comparison.

achieved with the rapid part of the model in (9) discarded. It turned out that the rapid part of (9) (with $c_d = 0.066$) yielded undesirable (negative) contributions to Π_d immediately at the onset of the compression, which is in disagreement with the DNS data.

In conclusion, the 3D rapid compression of isotropic turbulence is a relatively simple flow and the fundamental second-order quantities can be predicted by a set of algebraic equations (11). However, this is not so in the case of 1D rapid compression. Inspection of the DNS data in Figs 3 and 4 reveals some striking differences between 3D and 1D compression for the low Mach number case A. This is most clearly evident in the DNS data of $\overline{p^2}$. While the growth of pressure fluctuations agrees again with a rapid theory estimate (such as (11)) for Case B, it is seen that in the low M_t case A, the relative growth of $\overline{p^2}$ is much more rapid. By the time $A_o t = 0.5$, the amplification of $\overline{p^2}$ in Case A is by an order $O(100)$ larger than in Case B. This unexpected behavior can be traced to the pressure-dilatation term shown for 1D compression in Fig. 6. Contrary to expectation, Π_d/ϵ_t decreases much faster in Case A than in Case B. This results in an unusually high energy transfer from the kinetic to pressure-fluctuation mode and thus in a significant loss of kinetic energy to $\overline{p^2}$, despite the turbulence being nearly incompressible. This also explains the difference in the kinetic energy evolutions in 1D compression as indicated by the DNS data in Fig. 3.

Attempting to model this phenomenon, we have found that the isotropic compression model for Π_d in Eq. (9) is entirely inadequate since it does not distinguish between isotropic and anisotropic (directional) compression. Inspecting the conservation equation for $\overline{pu_{j,j}}$, which can be written as

$$\frac{D\overline{pu_{j,j}}}{Dt} = -(\frac{3}{2} + \gamma)\nabla\cdot\mathbf{U}\overline{pu_{j,j}} - \gamma\overline{p}\overline{(u_{j,j})^2} - 2S^*_{ij}\overline{ps_{ij}} + H.O.T.,$$

one can identify the term $\propto S^*_{ij}\overline{ps_{ij}}$ which is responsible for strong dilatation/vorticity interactions in anisotropic compression. Here, S^*_{ij} is the trace-free deformation tensor defined as $S^*_{ij} = \frac{1}{2}(U_{i,j} + U_{j,i} - \frac{2}{3}\nabla\cdot\mathbf{U}\delta_{ij})$, and $\overline{ps_{ij}}$ is the pressure-strain correlation which, to lowest order, is modeled as $\overline{ps_{ij}} \propto \overline{\rho}\overline{q}^2 S^*_{ij}$. Hence, there exists a negative source term $\propto \overline{\rho}\overline{q}^2 (S^*_{ij})^2$ in the $\overline{pu_{j,j}}$-equation, which is evidently nonzero only for anisotropic

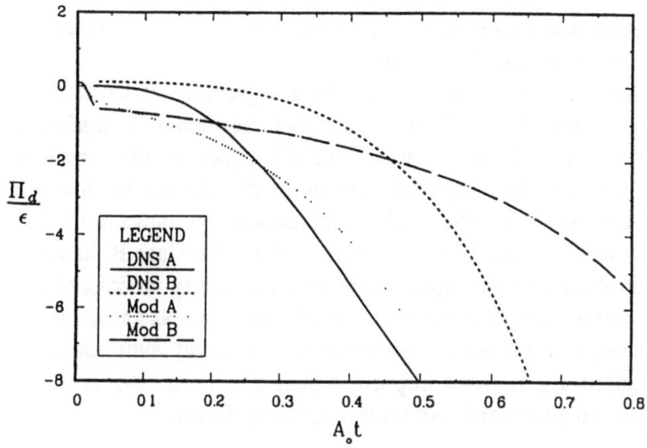

Fig. 6. Evolution of the normalized pressure dilatation for 1D rapid compression: DNS-model comparison.

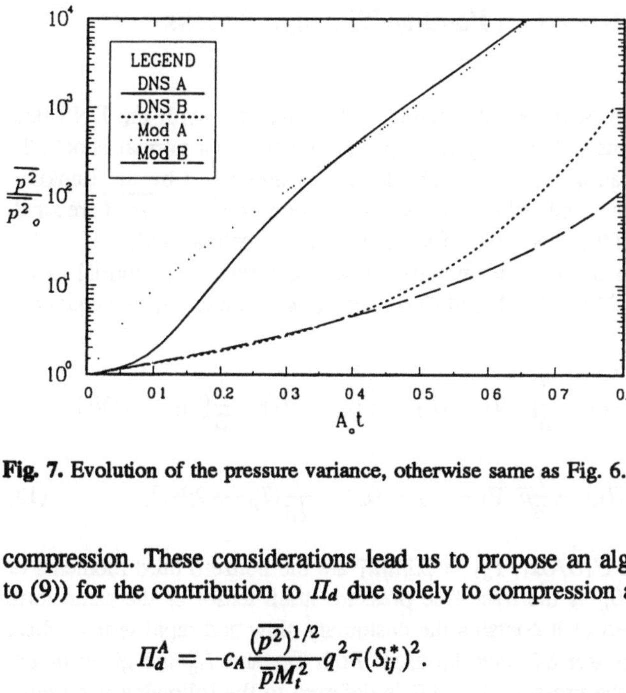

Fig. 7. Evolution of the pressure variance, otherwise same as Fig. 6.

compression. These considerations lead us to propose an algebraic model (to be added to (9)) for the contribution to Π_d due solely to compression anisotropy:

$$\Pi_d^A = -c_A \frac{(\overline{p^2})^{1/2}}{\overline{p} M_t^2} q^2 \tau (S_{ij}^*)^2. \tag{12}$$

with one adjustable constant c_A. The factor $(\overline{p^2})^{1/2}/\overline{p}M_t^2$ expresses the degree of departure from the pressure equilibrium (see Eqs. (5),(6)). This modifying factor was necessary to simulate rapid compression at higher M_t's. A preliminary value of the free constant is $C_A = 0.0004$ (note that for 1D compression $(S_{ij}^*)^2 = \frac{2}{3}(\nabla \cdot \mathbf{U})^2$). It is of interest to note that an analogous term of the form $-\tau q^2 (\nabla \cdot \mathbf{U})^2 (\overline{\rho'^2}/\overline{\rho}^2)/M_t$ has also been suggested by Taulbee and VanOsdol (1991). However, in their final expression, Taulbee and VanOsdol retained only the linear term in $\nabla \cdot \mathbf{U}$, which is similar to the rapid term in Eq. (9).

Similarly, Horstman (1987) proposed a closure for Π_d of the form $-\overline{M}q^2\nabla\cdot\mathbf{U}$ where $\overline{M}$ is the mean local Mach number (in a boundary layer).

The DNS-model comparisons of the pressure dilatation and pressure variance in 1D rapid compression are shown in Figs 6 and 7. It is seen that the model is capable of replicating the important feature of 1D compression: the difference in the kinetic-to pressure energy transfer and the resulting amplification rate of $\overline{p^2}$ between the low and high Mach number cases. It can be shown that without the anisotropic rapid term (12), the predictions for the low M_t case A would be virtually identical with Case B, hence in gross error. The unique effect of anisotropic rapid compression of nearly incompressible turbulence is also evident in shock-produced compression of turbulence as shown in the following section. In conclusion, in 1D rapid compression, the quadratic contribution (in $\nabla\cdot\mathbf{U}$) in the pressure dilatation Π_d is more important than the linear one, and it is expected to be so, for example, in supersonic compression corner flows.

4. Response of Turbulence to a Passage Through Shocks

Here, we present the results of model simulation and comparison with the DNS data of Lee (1991) of the response of initially isotropic turbulence to a normal shock. To simplify the model computations, the mean shock flow is prescribed by the Rankine-Hugoniot relations. This is justified as long as the r.m.s. velocities $u' = \sqrt{q^2/3}$ are small compared to the jump across the shock ΔU; for the present computations $u'/\Delta U \leq 0.1$. The flow is inhomogeneous in and axisymmetric about the direction x_1 normal to the shock; the mean quantities $U(x_1)$, $\overline{\rho}(x_1)$, and $\overline{p}(x_1)$ are known. The turbulence equations to be solved are:

$$\frac{D\widetilde{u_1^2}}{Dt} = -2\widetilde{u_1^2}U_{1,1} - \frac{2}{\overline{\rho}}\overline{p}_{,1}\overline{u}_1 - \Pi_{11}^o - \frac{2}{3}(\epsilon_s - \Pi_d) - \frac{1}{\overline{\rho}}(T_{111} + 2\overline{pu_1})_{,1}$$

$$\frac{1}{2}\frac{Dq^2}{Dt} = -\widetilde{u_1^2}U_{1,1} - \frac{1}{\overline{\rho}}\overline{p}_{,1}\overline{u}_1 - (\epsilon_s - \Pi_d) - \frac{1}{2\overline{\rho}}(T_{jj1} + 2\overline{pu_1})_{,1} \qquad (13)$$

Here, the operator $D/Dt \equiv U_1\partial/\partial x_1$, $T_{ij1} = \overline{\rho u_i u_j u_1}$ are the nonzero third moments or fluxes in the x_1 direction. Π_{ij}^o is the trace-free pressure-strain tensor of the same form as in incompressible turbulence; it contains the customary slow and rapid terms, where the trace-free mean strain tensor S_{ij}^* must be used. Both T_{ij1} and Π_{ij}^o are given in Zeman (1990); discussion of the pressure flux $\overline{pu_1}$ is deferred to the following paragraph. A version of the ϵ_s-equation (in 1D) required to solve (13) is e.g. in Durbin and Zeman (1992), and the L-equation is as in (10) with $n = 1$. A new, inhomogeneous term in Eq. (13) is the so-called acceleration term: the product of the mean pressure gradient $\overline{p}_{,i}$ and the fluctuation velocity average $\overline{u}_i = -\overline{\rho' u_i}/\overline{\rho}$. A customary approach to modeling $\overline{u}_i$ has been utilization of the so-called Strong Reynolds Analogy (SRA) which yields $\overline{u}_1 = -\widetilde{u_1^2}/U_1$. According to the DNS data, this estimate is of the wrong sign and results in a gross error in the energy q^2 (gain instead of loss of kinetic energy); we strongly suggest that the application of SRA be avoided in modeling turbulent boundary layers in the presence of shocks. A more appropriate approach is to use a model rate equation for the mass flux $\overline{\rho' u_i}$:

$$\frac{D\overline{\rho' u_i}}{Dt} \approx -\frac{\overline{\rho' u_i}}{\tau_a} - \overline{u_i u_j}\,\overline{\rho}_{,j} - \overline{\rho' u_j} U_{i,j}. \tag{14}$$

The first term on the RHS of Eq. (14) is a relaxation term, which drives the mass flux to zero (after the shock) on the fast acoustic timescale τ_a as suggested by the DNS data. Within the shock, Eq. (14) yields $\overline{\rho' u_i} < 0$, i.e. the acceleration term in (13) dampens turbulence as expected by the Rayleigh-Taylor analogy. Application of the SRA yields an opposite (destabilizing) effect, which is unphysical.

Modeling the pressure flux $\overline{pu_1}$ is not straightforward. It would appear that in the incompressible limit $M_t \to 0$, $\overline{pu_1}$ should be negligible; however, in the vicinity of the shock, this is not so. It may be more appropriate to utilize the observed fact that the fluctuations of density and temperature are positively correlated (as opposed to SRA) so that

$$\overline{pu_i} = C_p \overline{\rho' u_i} a^2, \tag{15}$$

where the mass flux $\overline{\rho' u_i}$ can be determined from (14). For the strictly adiabatic relation $T'/\widetilde{T} = (\gamma - 1)\rho'/\overline{\rho}$, the constant $C_p = 1$. Note that according to (15), the divergence $(\overline{pu_1})_{,1}$ needed in (13), contains terms that would reduce the magnitude of the acceleration term.

The principal purpose of the preliminary computations presented next is to test the effect of the pressure dilatation and acceleration terms on the overall (modeled) response of turbulence to the shock, and comparison with the DNS results. For simplicity we neglected the total energy flux divergence in (13). This is not entirely justified since the flux divergence is expected to yield energy gain; this gain is, however, of order $O(M_t)$ smaller than the remaining terms in (13).

The simulated flow is decaying isotropic turbulence subjected to compression through a normal shock with the upstream Mach number $M_1 = 1.18$, and with the density (and

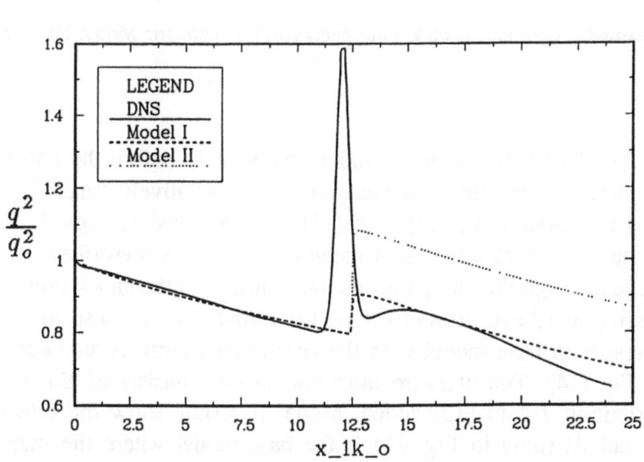

Fig. 8. Response of turbulence kinetic energy to the passage through shock: comparison of model predictions with DNS data of Lee (1991). Model I: with $c_A = 0.0004$ in Eq. (12); Model II: $c_A = 0$.

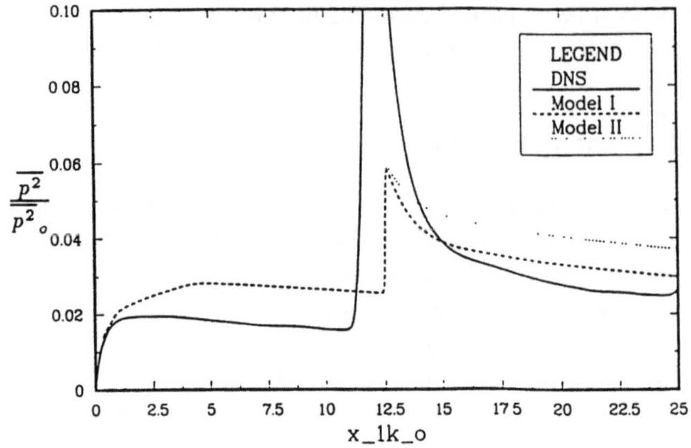

Fig. 9. Response of pressure variance, otherwise same as Fig. 8.

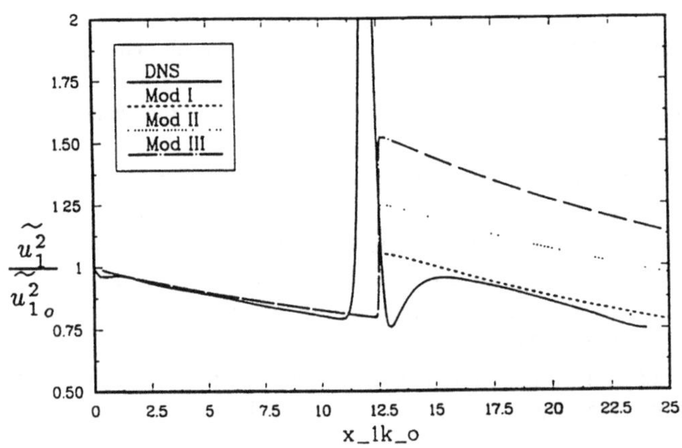

Fig. 10. Response of the streamwise component of kinetic energy($\widetilde{u_1^2}$); except for Model III (see text), same as Fig. 8.

velocity) ratio $\overline{\rho}_2/\overline{\rho}_1 = 1.31$. The r.m.s. Mach number immediately before the shock is $M_t = 0.13$, hence the initial compressibility-induced effects are relatively small. The results, using the modeling equations (13), (14), and (15), are presented in Figs. 8, 9, and 10. We remark that since the shock profiles of mean quantities are prescribed, the model is unable to reproduce the high fluctuation levels associated with the shock flutter. These fluctuations are passive and have little effect on the turbulence response to the shock. Here, Model I represents a base model with the acceleration term included and modeled with the aid of Eq. (14). The pressure dilatation model consists of Eq. (9) with the anisotropic contribution Π_d^A in (12) added. Model II results show the effect of neglecting Π_d^A and Model III (only in Fig. 10) is the base model where the mass flux $\overline{u}_1 = -\overline{\rho' u_1}/\overline{\rho}$ in the acceleration term in (13) is evaluated according to SRA, i.e., $u_1 = \widetilde{T'u_i}/\widetilde{T}$. Since the heat flux $\widetilde{T'u_i}$ (computed from an equation analogous to (14)) is negative, the acceleration term provides, in this case, a large gain in the $\overline{u_1^2}$ budget. According to Fig. 10, Model III overpredicts the $\overline{u_1^2}$ amplification through the shock by about 200%. We emphasize that the model results presented here are only tentative and serve to point out the relative importance of various mechanisms in the shock/turbulence interactions. The detailed work on this problem is in progress.

5. Summary

The subject of the presented work has been investigation of the pressure-dilatation interactions in shear-driven and rapidly compressed turbulence. The principal finding is that turbulence subjected to rapid directional (anisotropic) compression produces high negative values of the pressure-dilatation correlation, which in turn leads to a considerable transfer (loss) of turbulent kinetic energy to pressure fluctuations. This transfer mechanism has been found to be important even in nearly incompressible turbulence and is expected to be of importance in flows of practical interest such as those occuring in the internal combustion engine and turbulence/shock interactions.

A new model for the pressure-dilatation term has been proposed to account for this rapid compression effects. Preliminary modeling results indicate that significant kinetic energy losses to pressure fluctuation energy may occur in the shock/turbulence interactions.

References

Aupoix, B., Blaisdell, G.A., Reynolds, W.C., Zeman, O. (1990): "Modeling the turbulent kinetic energy equation for compressible, homogeneous turbulence", in Proc. of the 8th Symposium on Turbulent Shear Flows, Munich.

Blaisdell, G.A. (1990): "Numerical simulations of compressible homogeneous turbulence", Ph. D. Thesis, Mechanical Engineering Dept., Stanford University.

Coleman, G.N., Mansour N.N. (1991): "Direct simulations of homogeneous compressible turbulence under isotropic mean compression", in Proc. of the 8th Symposium on Turbulent Shear Flows, Munich.

Durbin, P.A., Zeman, O. (1992): "Rapid distortion theory for homogeneous compressed turbulence with application to modeling", J. Fluid Mech., Vol. 242, 349-370.

Horstman, C.C. (1987): "Prediction of Hypersonic Shock-Wave/ Turbulent-Boundary-Layer Interaction", AIAA Paper 87-1367.

Lee, S. (1991): "Direct numerical simulations of compressible turbulence", Ph. D. Thesis, Mechanical Engineering, Stanford University.

Lee, S., Lele, S., Moin, P. (1990): "Eddy-Shocklets in Decaying Compressible Turbulence", *Phys. Fluids* Vol. A 3, 657.

Papamoschou, D., Roshko, A. (1988): "The Compressible Turbulent Shear Layer: an Experimental Study", J. Fluid Mech., Vol. 197, pp.453-457.

Sabel'nikov, V.A. (1975): "Pressure Fluctuations Generated by Uniform Distortion of Homogeneous Turbulence", Fluid Mech. Soviet Res. Vol. 4, pp.46-57.

Samimy, M., Elliott, G.S. (1990): "The Effects of Compressibility on the Characteristics of Free Shear Layers", AIAA Journal, Vol. 28, pp. 429.

Taulbee, D., VanOsdol, J. (1991): "Modeling Turbulent Compressible Flow: the Mass Fluctuating Velocity and Square Density", AIAA Paper 91-0524.

Zeman, 0. (1990): "Dilatation Dissipation: The Concept and Application in Modeling Compressible Turbulence", Phys. Fluids, Vol.A 2, 178-188.

Zeman, O. (1991): "On the decay of compressible isotropic turbulence", Phys. Fluids Vol. A 3, 951-955.

Zeman, O. (1991a): "Towards a constitutive relation in compressible turbulence", in Proc. John Lumley's 60th Birthday Symposium, 1990 (Springer Verlag, New York)

Zeman, 0., Blaisdell, G.A. (1991): "New physics and models for compressible turbulence", *in Advances in Turbulence 3*, eds A.V. Johansson and H.P. Alfredsson (Springer Verlag, New York).

The Effect of Compressibility on Conserved Scalar Entrainment in a Plane Free Shear Layer[1]

Jacqueline H. Chen

Combustion Research Facility, Sandia National Laboratories, Livermore, California 94551-0969, USA

Abstract

This paper describes a numerical investigation of entrainment by organized motions in a compressible plane free shear layer. Insights obtained from the results of recent three-dimensional direct numerical simulations of a temporal compressible mixing layer, at low and moderate convective Mach numbers, are used to explain the effect of the large-scale structure dynamics on the transport of a conserved scalar. At moderate convective Mach number entrainment is found to occur in several intermediate steps producing a marching instantaneous concentration profile. Conserved scalar statistics and fast chemical reaction results at low and moderate convective Mach numbers are compared. Finally, the instantaneous streamline pattern at moderate convective Mach number is presented in a convective frame, and the three-dimensional critical points of the velocity gradient tensor are classified.

1. Introduction and Background

The plane compressible free shear layer has been a topic of renewed interest in recent years due to mixing and combustion issues related to supersonic flight. The persistence of turbulent large-scale structures at high Mach number led Papamoschou and Roshko [1988] to characterize the compressibility of the flow in a convective frame in which the large-scale structures are stationary. In this frame, the intrinsic Mach number of the flow is the convective Mach number (M_c), defined as $M_c = (U_1 - U_2)/(c_1 + c_2)$ where c_1 and c_2 are the respective freestream sound speeds and U_1 and U_2 are the respective freestream velocities. Using the convective Mach number, Papamoschou and Roshko were able to collapse the growth rate data onto a single curve for a wide range of freestream conditions, and further showed that the growth rate decreases as the convective Mach number increases.

More recent results from experiments (Clemens and Mungal [1990], Fourgette *et. al.* [1990]) and direct numerical simulations (Sandham and Reynolds [1991]) of nonreacting plane free shear layers and reacting round jets show that the underlying structure in a free shear layer changes as the convective Mach number increases. Experimental evidence suggests that as the convective Mach number increases the large-scale motions become

[1] Research supported by the United States Department of Energy, Office of Basic Energy Sciences, Division of Chemical Sciences.

significantly more three-dimensional. Flow visualization images taken at $M_c = 0.62$ by Clemens and Mungal [1990] show that the structures are highly three-dimensional, and that the spanwise rollers, which are the dominant structures at $M_c = 0.29$ due to the Kelvin-Helmholtz instability, are no longer present with any degree of regularity. Linear stability predictions by Sandham and Reynolds [1991] show that as the convective Mach number increases above 0.6, the shear layer is more unstable to oblique disturbances than it is to the Kelvin-Helmholtz instability. Therefore, it is not surprising that the structures resulting from the primary instability at moderate convective Mach number are three-dimensional. Based on direct numerical simulation results obtained for a temporal shear layer, Sandham and Reynolds [1991] modeled the large-scale structure resulting from the primary instability at moderate convective Mach number as a pair of lambda vortices, with the downstream vortex inverted. The plan view passive scalar images of Clemens and Mungal [1990] exhibit streamwise streakiness which may represent the legs of the lambda vortices.

For low convective Mach number free shear layers, phenomenological models for mixing and chemical reaction have been developed based on experiments by Konrad [1976] among others that demonstrate the significance of organized large-scale motions in entrainment and mixing processes. In the Broadwell-Breidenthal model (Broadwell and Breidenthal [1982]), free stream fluid is engulfed into the layer by the churning motion of the spanwise rollers over a dimension on the order of the layer thickness. The entrained fluid is subsequently mixed down to the Kolmogorov scale and ultimately exists either in thin laminar diffusion layers or in homogenized cores of the spanwise rollers.

The entrainment mechanism at moderate convective Mach number is poorly understood due to the difficulty in obtaining measurements under extreme flow conditions. Current measurements have focused primarily on passive scalars which have integrated history effects built in. For a complicated three-dimensional flow it is very difficult to deduce from passive scalar images the kinematics responsible for the scalar transport. Direct numerical simulations complement the experiments because they reproduce the low wave number motions very accurately; and, more importantly, because they provide passive scalar as well as pressure, velocity, and vorticity information simultaneously. While high wave number motions at large Reynolds numbers are not currently accessible in simulations, low wave number motions, which are responsible for the entrainment of fluid into the layer, are.

In the present work, simulations were performed of moderate ($M_c = 0.8$) and low ($M_c = 0.4$) convective Mach number free shear layers at similar Reynolds numbers. To compare the different entrainment mechanisms, the following results will be presented: (1) instantaneous structure of the large-scale motions, (2) conserved scalar statistics and fast chemistry results, and (3) three-dimensional topology of the velocity field moving at the convective velocity.

2. Numerical Method

The numerical method used in this work was developed by Sandham and Reynolds [1991]; therefore, only a brief summary is provided here. The full compressible Navier-

Stokes equations are solved for the temporally-evolving mixing layer using a spectral collocation method. The spatial derivatives are obtained using spectral and higher-order finite difference techniques and the solution is advanced in time explicitly using a compact third-order accurate Runge-Kutta method. Periodic boundary conditions are imposed in the streamwise (x) and spanwise (z) directions where the spatial derivatives are obtained spectrally. Characteristic nonreflecting boundary conditions are imposed in the major gradient direction (y) to prevent acoustic waves from affecting the solution. The grid points are uniform in the spectral directions and are stretched algebraically in y, with approximately half of the points inside the shear layer. To minimize aliasing problems, the solution is overresolved and the energy spectrum is monitored during the course of a run. Additional Fourier modes are included as the resolution becomes marginal.

In the present study, all of the initial fields were pertubed with random noise at 2.5% of the freestream velocity. This choice of initial disturbances was selected to mimic an unforced mixing layer. The mean velocity was modeled by an error function, and the mean temperature was derived from the Crocco-Busemann relationship. Random noise was applied to the temperature, density, and all components of velocity field. In addition to the hydrodynamic fields, a conserved scalar with an initial hyperbolic tangent profile was carried along, having values of unity in the high-speed stream and zero in the low-speed stream. The dimensions of the computational box in the streamwise and spanwise directions, $2L_x$ and L_z, were chosen to support the growth of both a fundamental mode and its subharmonic in the streamwise direction, and only a fundamental in the spanwise direction.

The simulations described in the following sections were performed at an initial Reynolds number (Re) of 800 based on the initial vorticity thickness (δ_0) and the velocity difference ($U_1 - U_2$), at convective Mach numbers (M_c) of 0.4 and 0.8, and at Schmidt and Prandtl numbers of unity. The lower Mach number simulation was perfomed to provide a baseline against which the higher Mach number results could be compared.

3. Instantaneous Structure

The developed structure at $M_c = 0.4$ and $M_c = 0.8$ is shown in perspective plots of the conserved scalar and pressure in Figure 1. At $M_c = 0.8$, large values of the vorticity magnitude correspond to the mean vorticity which remains largely in vortex sheets. The rotational regions tend to have lower values of vorticity magnitude; consequently, rotational regions are better delineated by the pressure minima than by the vorticity magnitude. From isocontour plots of the pressure and scalar fields it is evident that the structure at the higher Mach number is much more three-dimensional. The spanwise rollers which dominate the $M_c = 0.4$ simulation are completely absent at $M_c = 0.8$. Instead, they are replaced by oblique vortices which comprise the legs of two lambda vortices, similar to the structure reported in Sandham and Reynolds [1991]. Note that the vortices are tilted with respect to the shear layer centerline, $y = 0$. The head of the upstream lambda is tilted toward the high-speed stream while the head of the downstream lambda is tilted toward the low-speed stream.

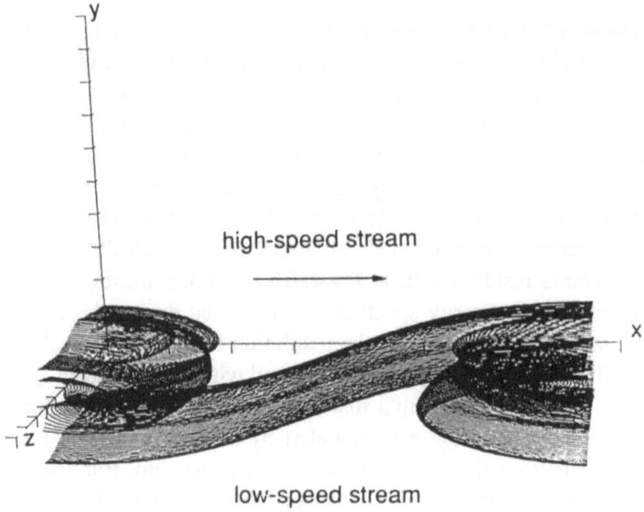

a

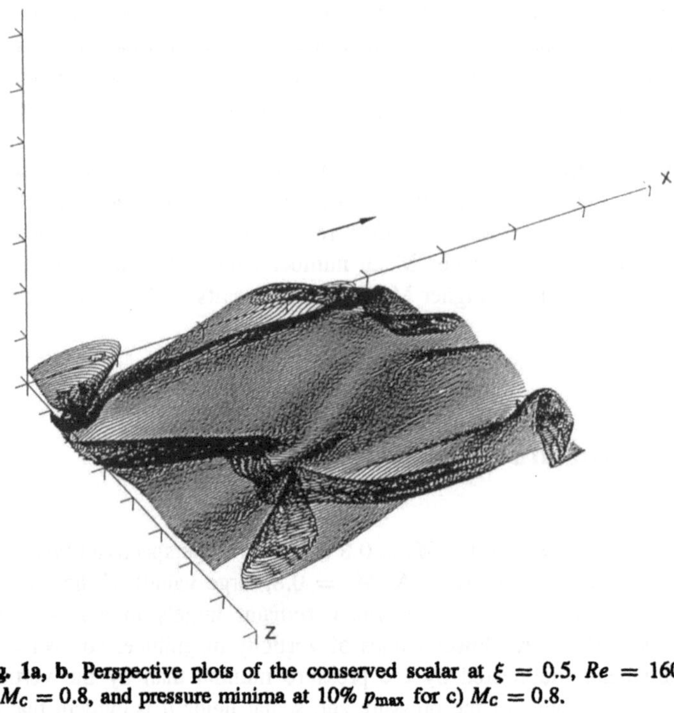

b

Fig. 1a, b. Perspective plots of the conserved scalar at $\xi = 0.5$, $Re = 1600$ for a) $M_C = 0.4$, b) $M_C = 0.8$, and pressure minima at 10% p_{max} for c) $M_C = 0.8$.

Figure 2 shows several side (x-y), plan (x-z), and end (z-y) view slices through the $M_c = 0.8$ conserved scalar data. The plan view slice through the midplane, $y = 0$, (Figure 2a) best reveals the three-dimensionality of the large-scale motions; there is no evidence of any spanwise structure in this plane. Instead, there exist several pockets of partially mixed fluid with the mixture fraction favoring either the low- or the high-

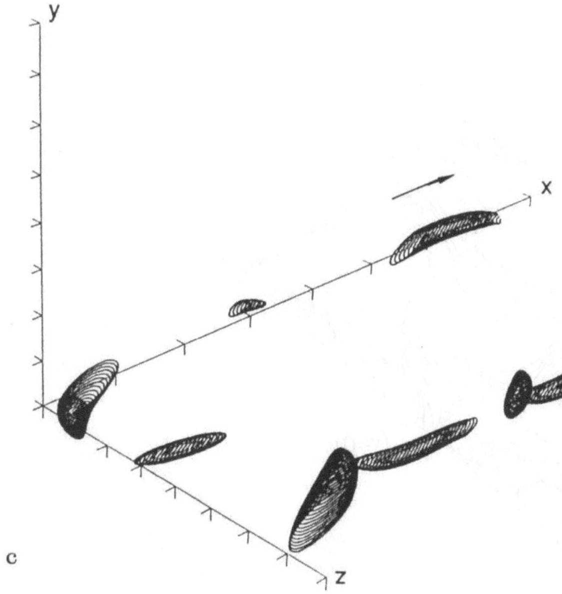

Fig. 1c. For caption see Fig. 1a, b

speed side. These pockets of partially mixed fluid exhibit a streamwise preference, and are adjacent to smaller rotational regions, corresponding to projections of the legs of the lambda vortices in this plane. The end view slices (Figure 2b) taken at several streamwise locations show the manner in which freestream fluid is entrained into the layer. Depending upon the particular streamwise location, irrotational fluid is entrained into the layer in one of three ways: (1) pure low- and high-speed fluid is drawn into the layer in one continuous motion, (2) pure high-speed fluid is entrained and mixed with partially mixed fluid inside the layer, or (3) pure low-speed fluid is entrained and mixed with partially mixed fluid inside the layer. Near the streamwise locations where the tips of adjacent lambda vortices are on top of one another, entrainment occurs mainly by methods (2) and (3) above. In regions in between, entrainment occurs mainly by method (1).

4. Conserved Scalar Statistics and Fast Chemistry Results

Statistics of the conserved scalar field at $M_c = 0.4$ and $M_c = 0.8$ are used to simulate fast chemical reactions in the manner detailed in Mungal and Dimotakis [1984]. The primary objective of this calculation is to obtain a qualitative comparison of the histogram of the conserved scalar and the chemical product profiles at low and moderate convective Mach numbers. In Mungal and Dimotakis [1984], the temperature rise due to a fast reaction is obtained by integrating the product of a weight function and the probability density function (p.d.f) of the conserved scalar. The weight function is adjusted to achieve the desired reactant equivalence ratio It has its peak at the equivalence ratio and decreases

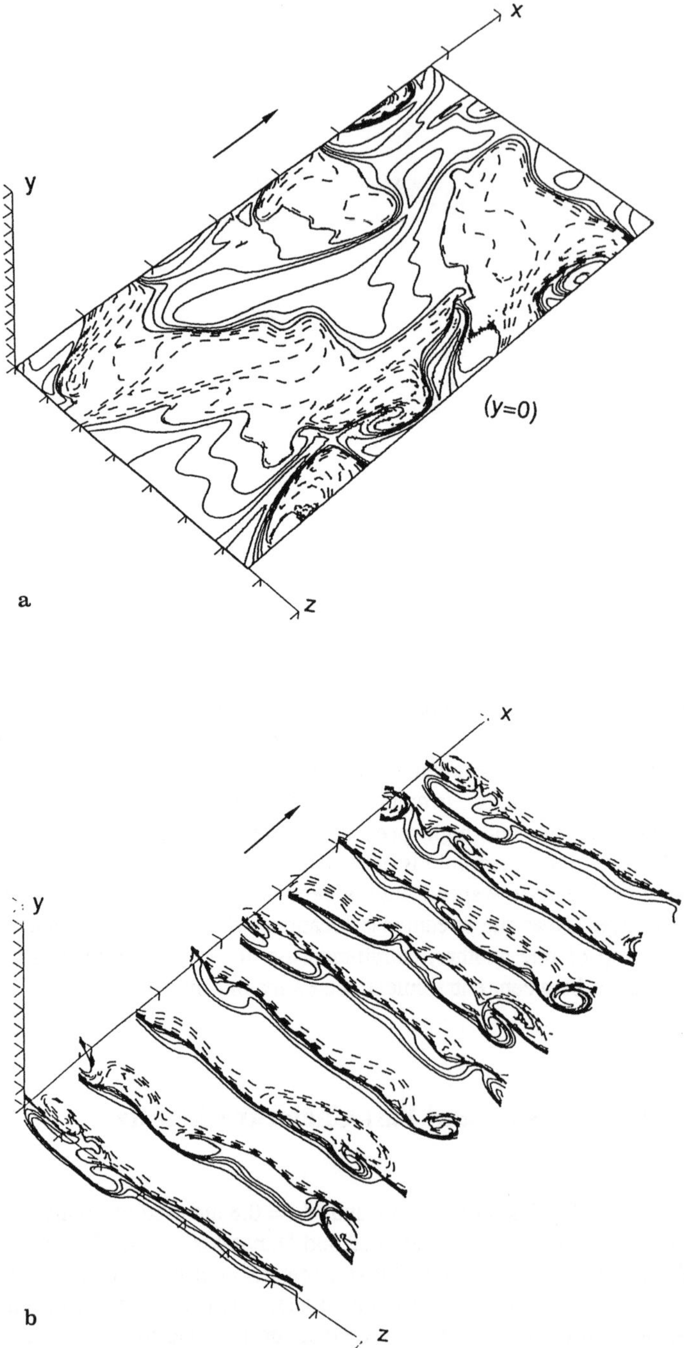

a

b

Fig. 2a, b. Contour plots of the conserved scalar at $Re = 1600$, $M_c = 0.8$ for a) plan, b) end, and c) side view slices. Solid lines correspond to $0 < \xi < 0.5$ and broken lines correspond to $0.5 < \xi < 1.0$.

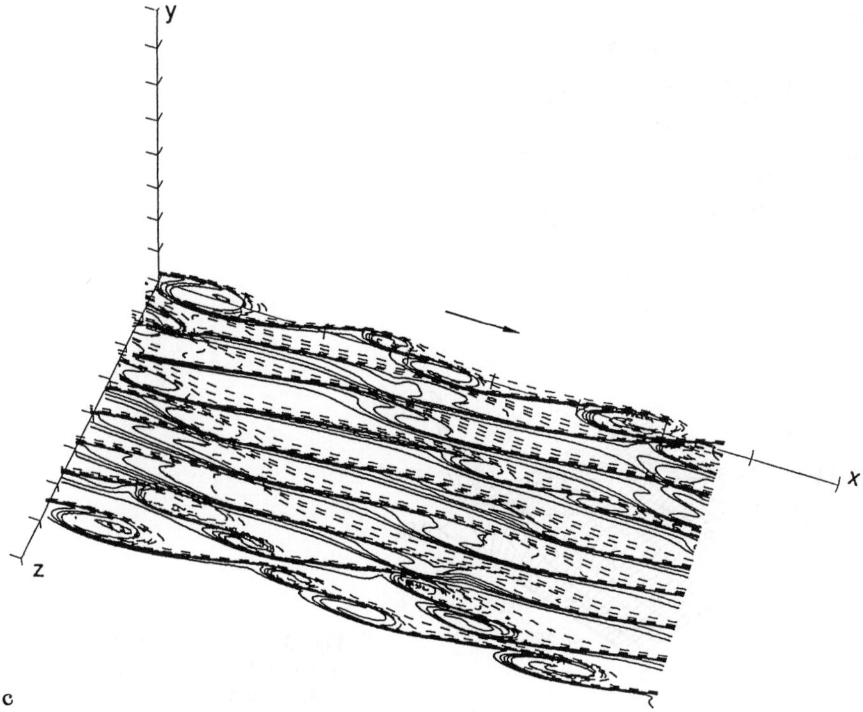

c

Fig. 2c. For caption see Fig. 2a, b

linearly to zero in the two freestreams. The equivalence ratio, ϕ, is defined as the ratio of the reactant concentrations in the freestreams, $\phi = (c_{02}/c_{01})$. Here, a simple one-step reaction, $A + B \rightarrow AB$, is assumed. The product concentration is maximum at the stoichiometric mixture fraction (ξ_s), which for the above reaction, occurs at $\xi_s = \phi/(1+\phi)$. Given the p.d.f. of the conserved scalar, the product profile and thickness can be readily computed. In the present study, the p.d.f. is replaced by a histogram, obtained by averaging the mixture fraction in the homogeneous directions, x and z, from a single realization.

Histograms of the conserved scalar at $Re = 1600$ are presented in Figure 3. Note that there are fundamental differences in the shape of the histogram for the two cases. At $M_c = 0.4$, the histogram exhibits three humps, corresponding to unmixed fluid in the high- and low-speed streams and fluid mixed at the entrainment ratio. The central peak at the entrainment ratio extends across the lateral extent of the layer, in agreement with the p.d.f. obtained in shear layer experiments by Konrad [1976]. In contrast, at $M_c = 0.8$, the histogram marches across the lateral extent of the layer; there is *no* central peak at the entrainment ratio. Moreover, at a given y location there appear to be *two* preferred mixture fractions. The two peaks may originate from 'upwash' and 'downwash' regions in the flow produced by the churning motion of the two counter-rotating legs of the lambda vortex. Low-speed fluid is being pumped up or high-speed fluid is being pushed down. The induced motion by the lambda vortices is illustrated in Figure 4.

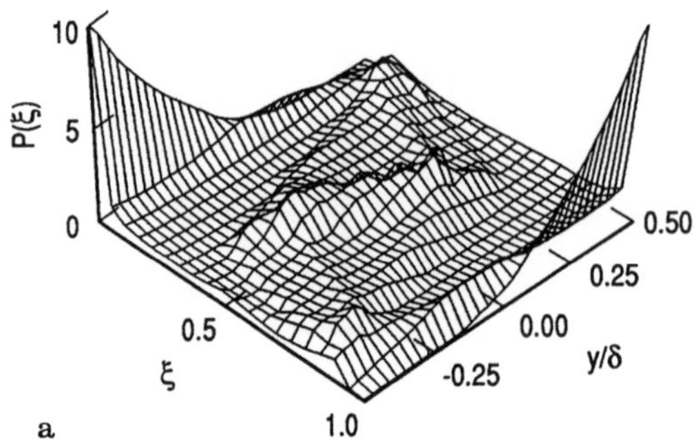

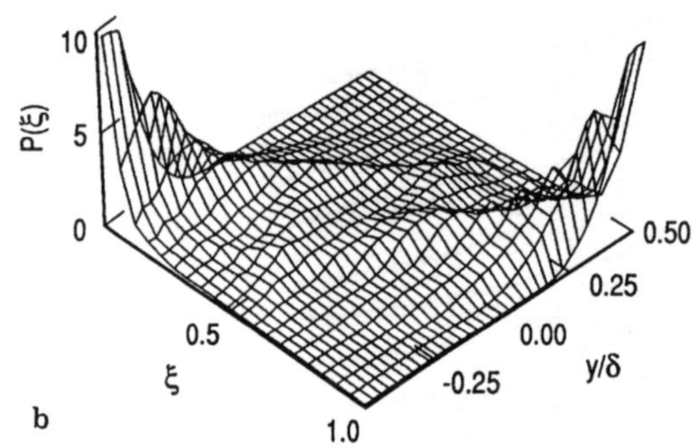

Fig. 3. Histogram of the conserved scalar at $Re = 1600$, a) $M_c = 0.4$ and b) $M_c = 0.8$.

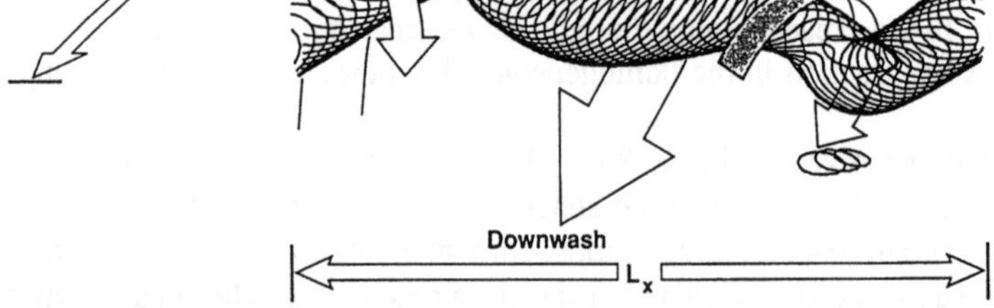

Fig. 4. Conceptual drawing illustrating the generation of 'upwash' and 'downwash' regions by the lambda vortices.

The marching character of the histogram may be attributed to the inclined orientation of the lambda vortices in the lateral direction. The tips of the vortices are located close to one of the freestreams, and therefore, tend to preferentially entrain fluid from that side of the layer. A raised tip close to the high-speed side tends to entrain high-speed fluid which

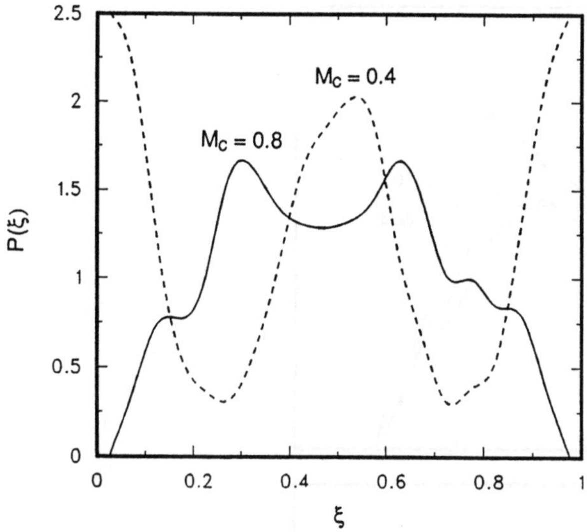

Fig. 5. Midplane ($y = 0$) through the histogram of the conserved scalar at $Re = 1600$.

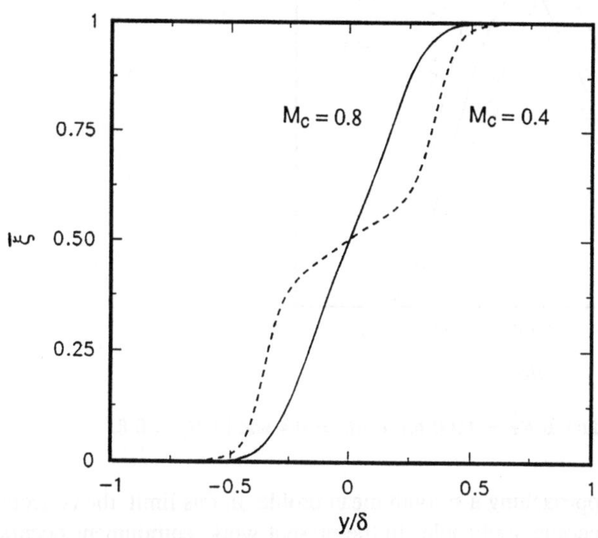

Fig. 6. Mean concentration profile

subsequently mixes with partially mixed fluid inside the layer; conversely, a depressed tip close to the low-speed side tends to entrain pure low speed fluid which then mixes with partially mixed fluid inside the layer. Therefore, rather than having one region of fluid mixed at the entrainment ratio, at moderate Mach numbers there exist two or more regions of fluid at a mixture fraction favoring the high and low-speed sides. In a recent 'sonic eddy' compressible turbulence model by Breidenthal [1990], it is hypothesized that for very large convective Mach numbers, the instantaneous concentration profile would

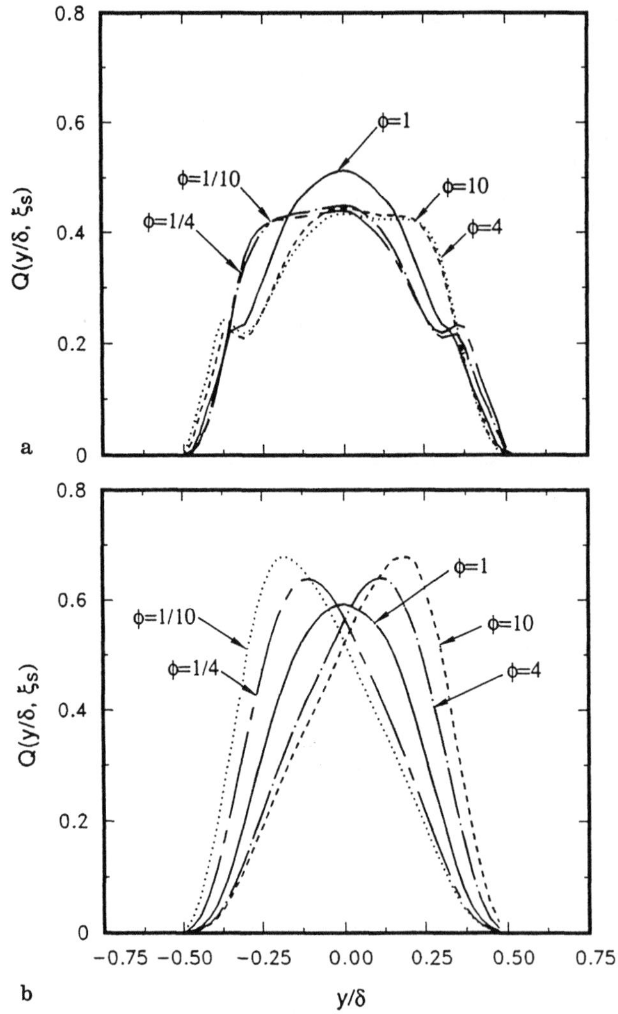

Fig. 7. Normalized chemical product at $Re = 1600$ for a) $M_C = 0.4$ and b) $M_C = 0.8$.

consist of many small steps, approaching a smooth mean profile. In this limit, the concept of gradient diffusion would become applicable. In the present work, entrainment occurs over one step at $M_c = 0.4$, and over two steps at $M_c = 0.8$. It is uncertain, whether, at higher convective Mach numbers, the number of steps would continue to increase, or whether it would remain at two.

Figure 5 shows a slice of the histograms for the two cases corresponding to the midplane, $y = 0$. There are several differences in the shape of the histrogram worth pointing out. First, at $M_c = 0.4$, there are intrusions of pure high- and low-speed fluid deep in the layer as evidenced by the two humps centered at a mixture fraction (ξ) of zero and unity. These intrusions are the tongues of freestream fluid that are engulfed into the layer as the spanwise rollers turn over. On the other hand, at $M_c = 0.8$, there are no tongues of pure fluid present near the center of the layer; therefore, the range of mixture

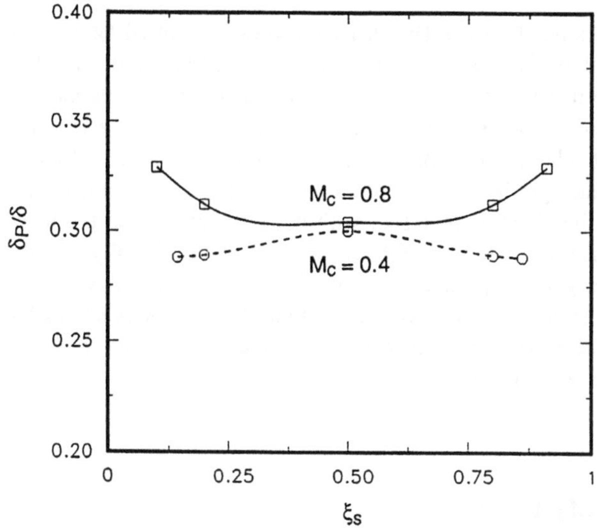

Fig. 8. Normalized product thickness vs. stoichiometric mixture fraction.

fractions that are populated is narrower. Recent compressible shear layer experiments by Clemens *et al.* [1991] also show a narrower range of mixed fluid at $M_c = 0.62$ compared to $M_c = 0.28$. Second, at $M_c = 0.8$, there are two peaks in the mixture fraction, at $\xi = 0.3$ and $\xi = 0.65$, instead of one at the entrainment ratio. As previously discussed, the origin of the two peaks may come from the induced motion of the legs of the vortices.

The conserved scalar mean profiles are shown in Figure 6 for the two cases. A triple inflection point exists in the mean for $M_c = 0.4$ indicating the presence of homogeneous structure cores separated by tongues of pure freestream fluid. On the other hand, at $M_c = 0.8$ the mean varies almost linearly from the low- to the high-speed stream with no evidence of pure fluid penetrating into the layer.

In Figure 7, the histograms of the conserved scalar were used to derive the temperature rise due to a fast reaction at $M_c = 0.4$ and $M_c = 0.8$ for different inverse equivalence ratios corresponding to the so-called 'flip' experiments by Mungal and Dimotakis [1984]. At $M_c = 0.4$ (Figure 7a), the peaks in the normalized product profiles all lie close to the entrainment ratio. This is expected since the histogram is non-marching. Note that there are secondary shoulders present near the edges of the layer indicating that the tongues of nearly pure freestream fluid have penetrated deep into the layer. Similar results were obtained in two-dimensional spatially-evolving mixing layer simulations by Lowery and Reynolds [1986]. However, subsequent three-dimensional simulations showed that streamwise vortices inhibit the lateral extent of the tongues (Lowery and Reynolds [1986], Buell and Mansour [1989]). In the present simulations, the streamwise vortices are relatively weak due to the random initial forcing. In Figure 7b, the normalized product profiles are shown for $M_c = 0.8$. Here there is a larger shift away from the rich reactants as a result of the marching character of the histogram. Second, unlike the lower Mach number case, the maximum product is not attained at an equivalence ratio of unity; instead, the maximum product is formed for the flip experiments, with more product formed on the top and bottom sides of the layer.

The product thickness normalized by the 1% visual thickness is plotted against the stoichiometric mixture fraction in Figure 8 for the two cases. Although the product thickness is overpredicted due to substantial viscous diffusion at the Reynolds number of the present simulation, the qualitative shape of the product thickness as a function of the stoichiometry is correct. Note, that at $M_c = 0.4$, the familiar 'gull wing' shape is reproduced, with the maximum product formed at an entrainment ratio of unity for a temporal layer. At $M_c = 0.8$, the shape is reversed, with more product being produced at a stoichiometric mixture fraction of 0.2 and 0.8, rather than at 0.5. Finally, for the same Reynolds number, slightly more product is formed at higher convective Mach number over a wide range of stoichiometric mixture fraction. This result is somewhat inconclusive, however, since the Reynolds number dependency of the product thickness has not been determined in the present simulations.

5. Topology of the Velocity Field

In this section, the topology of the velocity field at $M_c = 0.8$ is determined from instantaneous streamline patterns, obtained by integrating the instantaneous velocity in a frame of reference moving at the convective velocity. In this reference frame, the flow pattern is plotted in Figure 9 as projected streamlines in the three orthogonal planes. The streamlines were initialized near the critical points, where the velocity magnitude is nearly zero. The location of the critical points relative to the flow structure is determined by comparing Figure 4 with Figure 9. Figure 9a shows the projected streamline pattern in the x-z plane at $y = 0$, while Figures 9b and 9c are y-z and x-y planes near the tips of two adjacent lambda vortices (near A in Figure 9a). The flow pattern clearly reveals two pairs of counter-rotating vortices that are focal in nature separated by three-dimensional saddle points. In between the two lambda vortices, near A in Figure 9a, there is a three-dimensional critical point corresponding to a stable-node/saddle/saddle topology determined from the classification methodology of Chong, Perry, and Cantwell [1990]. A saddle point exists in the y-z and x-y planes, and a stable-node exists in the x-z plane. The saddle point in the y-z plane at A is shown in Figure 9b. This saddle point is between four focal regions corresponding to the legs of the two vortices. The local topology is consistent with the idea of entrainment occurring in two steps across the layer, at least in the vicinity of this critical point. The saddle point in the x-y plane is shown in Figure 9c where the streamline patterns for several adjacent z planes are plotted. Note that the saddle topology changes to two saddle points separated by an unstable node over a very small distance in z. A second critical point exists at B of the unstable-node/saddle/saddle type. This critical point is a saddle in the x-z and y-z planes and an unstable-node in the x-y plane. The unstable node is shown in Figure 9c. A third critical point exists at C which is also an unstable-node/saddle/saddle. However, this critical point is a saddle in the x-z and x-y planes and an unstable-node in the y-z plane. A fourth critical point exists at D of the same type as B. Based on the present flow pattern, it is clear that at $M_c = 0.8$, pure fluid is being entrained from the top and bottom sides in at least two steps, particularly in regions where exists an overlap between the tips of adjacent lambda vortices.

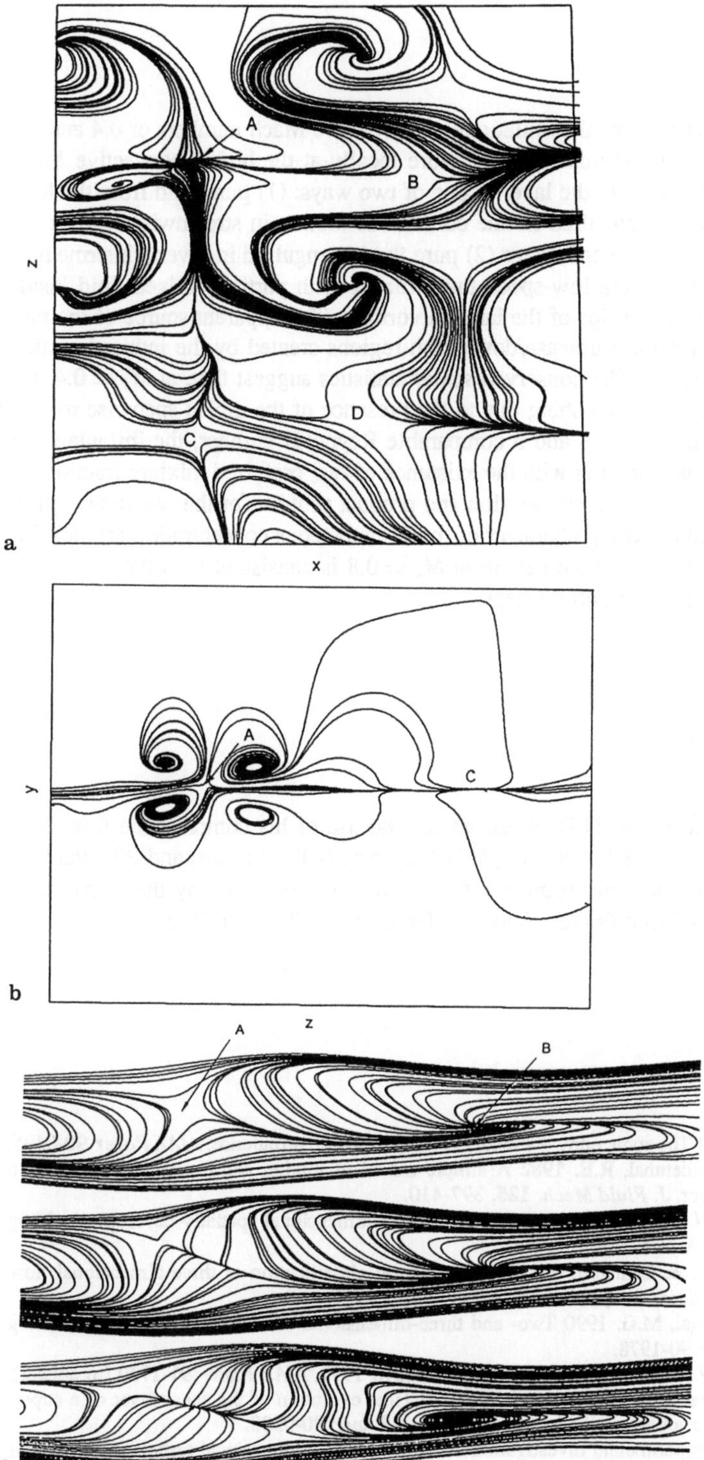

Fig. 9. Projected streamline patterns at $Re = 1600$, $M_c = 0.8$ a) x-z plane at $y = 0$, b) y-z plane at A, and c) x-y plane at A. The symbols A, B, C and D denote the location of critical points.

6. Conclusions

Comparisons of direct numerical simulations at convective Mach numbers of 0.4 and 0.8 suggest that entrainment and mixing occur more locally at the higher convective Mach number. Fluid is entrained into the layer in one of two ways: (1) pure fluid from the low- and high-speed streams is engulfed in one continuous motion in streamwise regions between the tips of the lambda vortices; or (2) pure fluid is engulfed in several intermediate steps, with pure high- or pure low-speed fluid mixing with partially mixed fluid inside the layer in regions near the tips of the lambda vortices. The apparent source of the partially mixed fluid is the local upwash/downwash regions created by the induced motion of the legs of the vortices. The conserved scalar statistics suggest that, at $M_c = 0.4$, the concentration profile is non-marching due to the presence of the global spanwise rollers. On the other hand, at $M_c = 0.8$ and a comparable Reynolds number, the instantaneous concentration profile is marching with the existence of *two* preferred mixture fractions at a given y location. The mean concentration and product profiles for the two cases reflect the non-marching and marching character of the instantaneous concentration profiles. Finally, the instantaneous streamline pattern at $M_c = 0.8$ is consistent with the concept of entrainment occurring over multiple steps.

Acknowledgements

The author is grateful to Dr. N.D. Sandham for the use of his compressible flow code and to Drs. M.G. Mungal, B.J. Cantwell, N.N Mansour, N.T. Clemens, and P.H. Paul for many enlightening discussions. Support for this work was provided by the Department of Energy's Office of Basic Energy Sciences, Division of Chemical Sciences.

References

Breidenthal, R.E., 1990 The sonic eddy – a model for compressible turbulence. *AIAA Paper* 90-0495.

Broadwell, J.E. & Breidenthal, R.E. 1982 A simple model of mixing and chemical reaction in a turbulent shear layer. *J. Fluid Mech.* **125**, 397-410.

Buell, J. & Mansour, N. 1989 Asymmetric effects in three-dimensional spatially-developing mixing layers. *Turbulent Shear Flows* **7**.

Chong, M., Perry, A.E. & Cantwell, B.J. 1990 A general classification of three-dimensional flow fields. *Physics of Fluids A* **2**(5), 765-777.

Clemens, N.T. & Mungal, M.G. 1990 Two- and three-dimensional effects in the supersonic mixing layer. *AIAA Paper* 90-1978.

Clemens, N.T., Paul, P.H., Mungal, M.G., & Hanson, R.K. 1991, *AIAA Paper* 91-1720 (to appear).

Fourgette, D.C., Mungal, M.G. & Dibble, R.W. 1990 Time evolution of the shear layer of a supersonic axisymmetric jet at matched conditions. *AIAA Paper* 90-0508.

Konrad, J.H 1976 An experimental investigation of mixing in two-dimensional turbulent shear flows with application to diffusion-limited chemical reactions. *Ph.D. Dissertation, Cal. Tech.*

Lowery, P & Reynolds, W.C. 1986 Numerical simulation of a spatially-developing forced, plane mixing layer. Mechanical Engineering Report TF-26.

Mungal, M.G. & Dimotakis, P.E. 1984 Mixing and combustion with low heat release in a turbulent shear layer. *J. Fluid Mech.* **148**, 349-382.

Papamoschou, D. & Roshko, A. 1988 The compressible turbulent shear layer: an experimental study. *J. Fluid Mech.* **197**, 453-477.

Sandham, N.D. & Reynolds, W.C. 1991 Three dimensional simulations of large eddies in the compressible mixing layer. *J. Fluid Mech.* **224**, 133-159.

IV

Buoyancy, Rotation, and Curvature Effects

Introductory Remarks

J.C. Wyngaard

Department of Meteorology, Pennsylvania State University, University Park, Pa. 16802, USA

Buoyancy, rotation, and curvature effects are important in many of the turbulent flows found in engineering and in geophysics. Earlier in this century, as the need developed for better turbulence prediction tools in engineering design, researchers turned to geometrically simple shear flows to study turbulence mechanics in its barest form. Today, while the goal of understanding the physics of even these simplest flows is still unrealized, our ability to cope with turbulence in applications has markedly improved. These six papers signal that clearly in their direct approaches to the complicating effects of buoyancy, rotation, and curvature.

Three thoughts surfaced as I read this collection. First, these papers build on some of the oldest traditions in turbulence, which I find comforting in these times of rapid change. Second, the papers cause me to marvel at the prescience of some of our early turbulence researchers. Even if the influence of these pioneers is not explicitly acknowledged by the authors, I believe I can see its effects here. Finally, a measure of the progress we have made in turbulence, thanks to those who did not stop at traditional boundaries between engineering and geophysics, is that we now use essentially one language to discuss turbulence in these disparate fields.

A bit more than twenty years ago, the first serious attempts were made to use second-order closure in boundary-layer meteorology. One of its appeals is that buoyancy and rotation effects, frequently controlling in geophysical flows, appear explicitly in the second-moment equations. An early leader here was Coleman du P. Donaldson, who used second-order closure to develop a useful and workable model of atmospheric turbulence and transport (Donaldson, 1972). Donaldson's approach, if approximate, was straightforward – a simple, systematic closure of the second-moment equations – and its results seemed intriguingly physical. Before long, second-order closure had caught the attention of a number of geophysical flow researchers. Fed in part by parallel efforts within the engineering community, the field mushroomed. Models for shear flows were soon modified for use in geophysical flows; they seemed to perform well in the atmospheric surface layer, which had a well-documented structure.

By the end of the decade, second-order closure was being applied to a wide variety of geophysical problems. John Lumley was particularly influential in developing guiding principles for its closures (Lumley, 1978), particularly for the convective boundary layer where the influences of buoyancy are pervasive. Some of this decade of experience in second-order modeling of the buoyancy-influenced atmospheric boundary layer is embodied in the first paper in this chapter, "Some further exploration of turbulence models for buoyancy driven flows" by K. Hanjalic and S. Vasic.

By the 1980s the second-order modeling of planetary boundary layers was outstripping our data base. It became clear that making further progress required higher-quality, more detailed structural data on turbulence. Such data are the staples of engineering turbulence research, but are generally much more difficult and expensive to obtain in geophysical flows. A decade earlier, however, J. Deardorff of the National Center for Atmospheric Research (NCAR) in Boulder, Colorado had developed one of the first "large-eddy simulation" (as it is now called) codes. His first calculation (Deardorff, 1970), of a turbulent channel flow, used only 6720 grid points on a CDC 6600 machine (on today's machines we can use over 10^6) but captured much of the essential dynamics of the large eddies. He soon applied the code to the atmospheric boundary layer and over the next dozen years revealed many of its structural secrets – including the inner workings of many of its second-moment balances. Deardorff's early success with LES stimulated its further development and application in geophysics and in engineering, in flows ranging from boundary layers to severe convective storms.

In reflecting on this recent history I am reminded of the punch line of a remark by Stan Corrsin. Following his demonstration of the impossibility of obtaining data on high Reynolds number flows through numerical solution of the Navier-Stokes equation on a digital computer, Corrsin (1961) wrote: "The foregoing estimate is enough to suggest the use of analog instead of digital computation; in particular, how about a analog consisting of a tank of water?" Deardorff and colleagues did indeed use such an analog – in the form of a laboratory convection tank 1 meter on a side – to confirm many of his LES findings and to generate new data as well, particularly on turbulent diffusion.

Corrsin's estimate concerned what we now know as direct numerical simulation, or DNS – the "brute-force" numerical solution of the Navier-Stokes equations, which as he showed is possible only at small Reynolds numbers. LES differs from DNS in that it concentrates the available computer power on the energy-containing range and models the rest. Our second paper, "Large-eddy simulation of the convective boundary layer: A comparison of four computer codes" is written by four contemporary practioners of geophysical LES – F.T.M. Nieuwstadt, P.J. Mason, C.H. Moeng, and U. Schumann. In this overview of the codes and their capabilities it is reassuring to see that their results agree quite well.

LES is particularly useful for exporing new phenomena in cases where measurements are difficult. In atmospheric boundary-layer applications, for example, LES has shown that buoyancy has effects far beyond those suggested by the explicit buoyancy terms in the equations – e.g., beyond those predicted by adding buoyancy to a shear-flow model. The third paper in this chapter illustrates this principle. In "Effects of solid body rotation on the transport of turbulence", L. Shao, M. Michard and J.P. Bertoglio use LES to study the effect of solid-body rotation on turbulence. They find that the effects are large and manifested through the pressure-velocity correlations; this would have been very difficult, if not impossible, to infer experimentally. Very importantly, the authors also demonstrate the utility of such results in improving turbulence models.

The real thrust of the Corrsin quote, of course, is the power of direct measurements in turbulence research. The next two papers, "Scaling and structure of turbulent eddies in curved shear flows" and "Experimental investigation of the three dimensional boundary layer on a rotating disk", build on this strongest of turbulence traditions. In the first, A.B.L. Holloway and S. Tavoularis document curvature effects on energy-containing turbulence structure in shear flows, which are analogous to buoyancy effects in the atmosphere. In the second, H. Littel and J.K. Eaton present measurements of the boundary layer on a rotating disk, which bears a similarity to the geophysical Ekman layer. The data sets should be good test cases for numerical simulation.

In the thirty years since Corrsin's comment, computer size has grown vastly, making DNS of modest Reynolds number flows routine today. The relevance of such results to engineering or geophysical flows at much larger Reynolds number rests on the validity of Reynolds number similarity. The results are increasingly seen as of interest in their own right, however, particularly as a testbed for the examination of theoretical notions. These DNS "data sets" are far more complete (and can also be more reliable) than those from direct measurements. The last paper, "Reorganization of coherent vortices in shear layers under the action of solid-body rotation" by O. Métais, S. Yanase, C. Flores, P. Bartello, and M. Lesieur illustrates this well.

References

Corrsin, S., 1961: Turbulent flow. *Am. Scientist*, **49**, 300-325.

Deardorff, J.W., 1970: A numerical study of three-dimensional turbulent channel flow at large Reynolds numbers. *J. Fluid Mech.*, **41**, 453-480.

Donaldson, C. du P., 1973: Construction of a dynamic model of the production of atmospheric turbulence and the dispersal of atmospheric pollutants. *Workshop on Micrometeorology*, D. Haugen, Ed., American Meteorological Society, Boston, 313-392.

Lumley, J.L., 1978: Computational modeling of turbulent flow. *Adv. Appl. Mech.*, **18** 123-176.

Some Further Exploration of Turbulence Models for Buoyancy Driven Flows

Kemal Hanjalić and Slavko Vasić

Mašinski fakultet, University of Sarajevo, Omladinsko šetalište bb, 71000 Sarajevo, Bosnia Hercegovina

Abstract

Several levels of truncation of the second-moment closure method have been explored in search for a simpler form of model which would enable satisfactory predictions of buoyancy-dominated turbulent flows, but still retain a form suitable for incorporation into the Navier-Stokes computer solver for complex geometries. The models have been tested by numerical computation of several external and internal buoyant flows including penetrative convection of an unstable mixed layer, thermal boundary layer on a heated vertical plate and natural convection in two-dimensional enclosures with mixed boundary conditions. Although higher order models produce generally better agreement with experiments, a computationally more convenient explicit algebraic expression for the turbulent heat flux in conjunction with the low-Re-number k and ε equations gave generally satisfactory predictions of the considered cases.

1. Introduction

Turbulent flows dominated by thermal buoyancy fall into the category of turbulence phenomena which are regarded as hardly tractable by the Reynolds averaging description. The main reason for this mistrust is the evidence of the existence of well organized coherent structures dominated by distinct, usually large scale patterns such as thermals, plumes, or large eddies. These are conspicuous in the power spectrum, and dominate the turbulent transport, mixing and entrainment of fluid into an environment with different mean temperatures. Some recent, more detailed experimental exploration of the turbulence structure in buoyancy driven flows seems to give further support to this preconception.

An illustrative example is the case of a Rayleigh-Bénard cell between two horizontal plates heated from below, known for long time to exhibit several patterns depending on the Rayleigh number (e.g. Chu and Goldstein, 1973), ranging from steady Bénard rolls to fully turbulent motion with unstable roll break down in between. Recent reconsideration of the case and a more detailed exploration by several researchers (e.g. Heslot et al. 1987, Tanaka and Miyata 1980) brought new evidence of several sub-patterns in the transition region, which Heslot et al. called Oscillation, Chaos and Transition. Moreover, two different types of turbulence (Heslot et al. called them "soft" and "hard" turbulence) separated at $Ra = 4 \times 10^7$, were distinguished in what was previously regarded simply as turbulent free convection. The external manifestation of the difference between the "soft" and "hard" turbulence may be regarded as marginal and is reflected in the change of the slope of the Nusselt- vs Rayleigh number in log-log coordinates, what was also noticed earlier by some other authors. Heslot et al. found that the slope retains the classic value of 1/3 only in "soft" turbulence, while in "hard" turbulence it decreases to about 2/7. A major difference, however, was detected in the statistics of the temperature fluctuations in the central region of the cell outside the thin boundary layers on each plate. In this region Tanaka and Miyata recognized large irregular and non-stationary convective motions having a size comparable to the distance between the plates. They also identified another conspicuous structure of small eddies of the size of the Taylor microscale which mix up the warm or cold fluid elements discharged from the lower or upper boundary layer with ambient fluid. According to the authors these large scale motions are responsible for transport of warmer or cooler fluid lumps together with small eddies away from regions close to the boundary layers towards the opposite wall.

Another example is the penetrative convection and entrainment of well mixed turbulent fluid into stably stratified layer across the interface between the turbulent and non-turbulent state. Observation and measurements in nature as well as laboratory experiments have substantiated the notion of large scale coherent motions which – like in the previously discussed case – transport the warmer fluid elements and the small eddy structure through the mixed layer almost to the interface with stable fluid, and control the process of ejections and entrainment of one fluid into another (Deardorff et al. 1969).

Of course, none of these phenomena, nor many others associated with coherent structures, are describable by Reynolds averaged equations. Consequently they are rightly regarded as resistant to modelling by Single Point Closure Methods (SPCM) and their mathematical description calls for other types of models. But the important question is whether, and to what extent, the specific organization of turbulence structure, preferential scales, and other structural peculiarities affect statistically averaged quantities, and, if they do, can those effects be adequately accounted for by the mathematical tools employed by SPCM? Clear evidence of the existence of well organized structures of distinct scales and of different flow patterns within the turbulent regime in the same configuration, according to the values of some external parameters, leads to serious doubts about using this modelling practice even for predicting averaged quantities in free convection. However, some success has already been claimed in dealing with non-standard behaviour by invoking higher order modelling schemes instead of $k - \varepsilon$ or similar eddy-viscosity-type models. So, for example, some early recognized features like a strong interaction – and mutual feed-back between the velocity – and density (temperature, species concentration) fluctuating fields, or counter-gradient turbulent diffusion (known to be associated with large scale motion) have been reasonably well represented by the employment of second-moment closure methods, or their simpler derivatives.

Other routes are, of course, possible. Dimensional arguments proved in the past to be very powerful tools in establishing general dimensionless correlations between overall properties. At the other extreme is the technique of direct numerical simulation of the whole turbulence spectrum or only of large eddies. In spite of some success in simulating large structures ("billows") developed locally in a stratified shear layer of the thermocline, the method is still too complex and expensive for computations of the complete flow domain of a real environmental or engineering situation.

Practical needs to predict natural convection in many engineering problems with complex geometries and boundary conditions, such as encountered in building structures, in coolants of nuclear reactors or electronic equipment, crystal growth, as well as in environmental flows in atmosphere and water accumulations, cannot, however, be resolved either by dimensional argument or by direct numerical simulation. The first method is intractable because too many parameters are involved. Direct numerical simulation, even if confined to only large eddies, requires both further exploration and development of models of the smaller scale motion, and, more importantly, the use of computers with much higher performances then presently available. Hence we may state that Single-Point Closure Methods or similar techniques still remain the only tool to predict main features of complex flows.

If buoyancy driven flows are expected to be successfully described by SPCM it seems that the minimum level of modelling should be the Second-Moments Closure (SMC); this implies the solution of differential transport equations for all second order single-point correlations. A consistent approach would require a parallel treatment of all variables in question. So in cases when other contaminants (pollutants, salinity, water vapor) are present, in addition to temperature, the number of equations to be solved increases substantially. For a complex three-dimensional geometry with e.g. irregular boundaries this would pose a formidable problem. However, it is not the mere number of differential equations which is the major problem, but more the uncertainty in modelling various terms which represent mutual interactions between the velocity, temperature and concentration fields. In addition, in free convective flows in engineering equipment, regardless of how large is the bulk Ra number, there exist regions in generally turbulent flows where the molecular effects are significant. This feature requires the application of low-Re-number turbulence models which in turn require a much finer numerical mesh and are known to pose additional computational difficulties. Besides, a general SMC with low-Re-number modification, which would be applicable to buoyancy driven flows, has not been designed yet and it will take some time until such models reach the level of verification comparable to those employed for the computation of shear flows.

The application of a full SMC for solving problems of buoyancy driven flows in more complex geometries seems at present, therefore, to be still premature. Instead, it seems plausible to adopt a compromise which would retain the basic eddy diffusivity formulation, suitable for incorporation into the currently popular Navier-Stokes codes, but modified to account for important buoyancy effects.

The present paper deals with the problem of modelling and computation of buoyancy effects on turbulence with a specific look at the computation of buoyancy driven flows in enclosures. A typical flow in an enclosure with mixed boundary conditions has been decomposed first into simpler processes; these have been analyzed from the standpoint of currently available single-point prediction methods and their performances compared with available experimental data. The paper focuses, therefore, first on the analysis of relatively simple test cases in which "clean" buoyancy effects can be isolated and various modelling ideas explored in detail. Two groups of external flows have been examined, one in which the density stratification vector is collinear with the gravitational vector and the other with the two vectors being perpendicular. Several cases of 2-dimensional internal flows in single- or multi-zone enclosures with mixed boundary conditions were also studied by employing the hybrid model which emerged as optimum from the previous study of simpler cases.

2. Governing Equations and Turbulence Models

Studies of natural convection employ in most cases the Boussinesq approximation by which the local density is expressed in terms of temperature ensuring a direct coupling between the mean momentum and energy equations. In order to avoid possible errors when dealing with higher temperature variations or with liquids for which the density-temperature variation departs from linearity, we have applied the Boussinesq approximation only to eliminate the density fluctuation ρ' in terms of temperature fluctuations θ in the equations for turbulence properties, i.e. $\rho' = -\beta\rho\theta$, where β is the thermal expansion coefficient. Apart from that we have retained the general forms of the mean momentum and energy equations.

We start the discussion of closure of the mean flow equation set by focusing first on the turbulent heat flux $\rho\overline{\theta u_i}$. Although for complex geometries we eventually employed a form of modified eddy diffusivity model, the differential transport equation for $\rho\overline{\theta u_i}$ served as a source of information from which the modification of the simple model was derived. Application of the simplest level of modelling (e.g. Launder 1988) of the differential equations for second moments yields the following equation for the turbulent heat flux:

$$\frac{\mathrm{D}(\rho\overline{\theta u_i})}{\mathrm{D}t} = \frac{\partial}{\partial x_j}\left[C_{\theta}\rho\frac{k}{\varepsilon}\left(\overline{u_i u_k}\frac{\partial\overline{\theta u_j}}{\partial x_k} + \overline{u_j u_k}\frac{\partial\overline{\theta u_i}}{\partial x_k}\right) + \mu\frac{\partial\overline{\theta u_i}}{\partial x_j}\right] -$$
$$-\rho\left[\overline{u_i u_j}\frac{\partial T}{\partial x_j} + (1 - C_{\theta 2})\overline{\theta u_j}\frac{\partial U_i}{\partial x_j} + (1 - C_{\theta 3})\beta g_i\overline{\theta^2} - C_{\theta 1}\overline{\theta u_i}\frac{\varepsilon}{k} + \varepsilon_{\theta i}\right] \quad (1)$$

where $C_{\theta}, C_{\theta 1}, C_{\theta 2}$ and $C_{\theta 3}$ are empirical coefficients.

In comparison with the usual practice, equation (1) contains two additional terms which account for molecular effects and which are expected to be of importance in flow regions where the turbulent Peclet number is small. The molecular diffusion term takes the above form only in case of Pr close to unity and can serve as a reasonable approximation also for $Pr > 1$, while for $Pr < 1$ the same form of the term can be retained, but with μ replaced by thermal diffusivity $\rho\alpha = (\mu/Pr)$. The last term in the equation represents the cross correlation $\varepsilon_{\theta i} = (\nu + \alpha)(\overline{\frac{\partial\theta}{\partial x_j}\frac{\partial u_i}{\partial x_j}})$ which is usually neglected for high Pe numbers.

The neglect of $\varepsilon_{\theta i}$ is not justifiable in flow regions where the turbulence is low and we have explored a model for this term analogous to that of ε_{ij} for low Re-numbers in the form

$$\varepsilon_{\theta i} = f_{\varepsilon\theta i}\sqrt{\varepsilon\varepsilon_{\theta}}\frac{\overline{\theta u_i}}{\sqrt{\overline{\theta^2}k}} \quad (2)$$

where $\varepsilon_{\theta} = \alpha(\overline{\frac{\partial\theta}{\partial x_j}\frac{\partial\theta}{\partial x_j}})$ is the dissipation of the temperature variance $\overline{\theta^2}$. Here $f_{\varepsilon\theta i}$ is a function of the turbulent Peclet number $Pe_t = \frac{k^2}{\alpha\varepsilon}$, which goes to zero at sufficiently high Pe_t and takes the value of unity if Pe_t approaches to zero. Several simple forms of the function in terms of $Re_t = \frac{k^2}{\nu\varepsilon}$, (since the tests were carried out for air with Pr closed to unity) similar to the analogous function for ε_{ij} produced only marginal (not always desired) effects and we have subsequently omitted the term from our computation. For the sake of generality this term will be retained temporarily in the analysis that follows.

The closure of the $\rho\overline{\theta u_i}$-equation requires, in addition to the turbulent stress tensor, the temperature variance $\overline{\theta^2}$, the dissipation rates of turbulence energy ε and of the temperature variance ε_θ – and appropriate low-Re/Pe number modifications. The level of uncertainty involved in modelling the full transport equations for all these variables exceeds at present the potential benefits. We have, therefore, used the full model (adopting the basic form of the stress equation, e.g. Launder 1988) only to obtain the solution of the cases of penetrative convection. Almost identical results were obtained by solving only the kinetic energy equation instead of those for individual normal stresses and this practice was adopted for all other considered cases. In fact, it seems that an optimum level of modelling of more complex buoyant flows at the present stage of development would be to solve the four transport equations for scalar variables k, $\overline{\theta^2}$, ε and ε_θ with an algebraic expression for the turbulent heat flux vector. The approximate form of the equation set considered in the present case with the low-Re-number modifications in lines with those of Jones and Launder (1972) and Hanjalić and Launder (1976) reads as follows:

$$\frac{\mathrm{D}(\rho\overline{\theta^2})}{\mathrm{D}t} = D_\theta + 2\rho P_\theta - 2\rho\varepsilon_\theta \tag{3}$$

$$\frac{\mathrm{D}(\rho k)}{\mathrm{D}t} = D_k + \rho P + \rho G - \rho\varepsilon \tag{4}$$

$$\frac{\mathrm{D}(\rho\varepsilon)}{\mathrm{D}t} = D_\varepsilon + C_{\varepsilon1}\rho P\frac{\varepsilon}{k} + C_{\varepsilon3}\rho G\frac{\varepsilon}{k} - C_{\varepsilon2}f_\varepsilon\rho\frac{\varepsilon\tilde{\varepsilon}}{k} + E \tag{5}$$

$$\frac{\mathrm{D}(\rho\varepsilon_\theta)}{\mathrm{D}t} = D_{\varepsilon\theta} + C^\theta_{\varepsilon1}\rho P_\theta\frac{\varepsilon_\theta}{\overline{\theta^2}} - C^\theta_{\varepsilon2}f_{\varepsilon\theta}\rho\frac{\varepsilon_\theta\tilde{\varepsilon_\theta}}{\overline{\theta^2}} - C^\theta_{\varepsilon4}f_{\varepsilon\theta}\rho\frac{\varepsilon_\theta\tilde{\varepsilon}}{k} + E_\theta \tag{6}$$

where

$$P = -\overline{u_iu_j}\frac{\partial U_i}{\partial x_j}; \qquad G = -\beta g_i\overline{\theta u_i}; \qquad E = 2\mu\frac{\mu_t}{\rho}\left(\frac{\partial^2 U_i}{\partial x_j\partial x_k}\right)^2 \tag{7}$$

$$P_\theta = -\overline{\theta u_j}\frac{\partial T}{\partial x_j}; \qquad E_\theta = 2\rho\alpha\alpha_t\left(\frac{\partial^2 T}{\partial x_j\partial x_k}\right)^2 \tag{8}$$

$$\tilde{\varepsilon} = \varepsilon - \nu\left(\frac{\partial(k)^{\frac{1}{2}}}{\partial x_n}\right)^2 \qquad \tilde{\varepsilon_\theta} = \varepsilon_\theta - \alpha\left(\frac{\partial(\overline{\theta^2})^{\frac{1}{2}}}{\partial x_n}\right)^2 \tag{9}$$

In the last expressions x_n denotes the coordinate normal to the nearest wall. P and G stand for the turbulence energy production by strain and buoyancy respectively, P_θ is the production of temperature variance, while E and E_θ are the terms which account for molecular effects. The function $f_{\varepsilon\theta}(Pe_t)$ should ensure a transition from the initial to the final stage of decay of temperature variance when Pe_t falls bellow the critical value in the same manner as the function f_ε does for the velocity fluctuations.

D- denotes the total diffusion term in which the turbulent part was modelled by the general gradient hypothesis:

$$D_\phi = \frac{\partial}{\partial x_j}\left(C_\phi\rho\frac{k}{\varepsilon}\overline{u_ju_k}\frac{\partial\phi}{\partial x_k} + \mu\frac{\partial\phi}{\partial x_j}\right) \tag{10}$$

3. A Simpler Model for Complex Flows

Following Gibson and Launder (1976) the differential transport equation for $\overline{\theta u_i}$ can be truncated to the algebraic form:

$$\overline{\theta u_i} = -\phi_\theta \frac{k}{\varepsilon}\left[\overline{u_i u_j}\frac{\partial T}{\partial x_j} + (1 - C_{\theta 2})\overline{\theta u_j}\frac{\partial U_i}{\partial x_j} + (1 - C_{\theta 3})\beta g_i \overline{\theta^2} + \varepsilon_{\theta i}\right] \tag{11}$$

where

$$\phi_\theta = \left[C_{\theta 1} + \frac{1}{2}\left(\frac{P+G}{\varepsilon} - 1\right) + \frac{1}{2}\frac{\varepsilon_\theta k}{\overline{\theta^2}\varepsilon}\left(\frac{P_\theta}{\varepsilon_\theta} - 1\right)\right]^{-1} \tag{12}$$

Solution of the penetrative convection with the above set of equations with standard coefficients (Table 1 and those for ε_θ-equation of Nagano and Kim, 1988) did not bring any advantage; in fact it produced a deterioration of the prediction of all properties of the penetrative convection. Some attempts to adjust the coefficients in the ε_θ-equation did not yield much success and considering other uncertainties inherent in the model of that equation, it was subsequently discarded. Instead, the usual practice of relating ε_θ to ε by means of the ratio of thermal-to-mechanical turbulence scale $R = \frac{\overline{\theta^2}}{2\varepsilon_\theta}\frac{\varepsilon}{k}$ was adopted. As it will be shown later, the constant value $R = 0.8$, gave satisfactory results in all computations which were compared with experiments. Some improvements were, however, noticed by using the expression $R^{-1} = 1.5(1 + A_{\theta 2})$, where $A_{\theta 2} = \frac{\overline{\theta u_i}\,\overline{\theta u_i}}{\overline{\theta^2} k}$ as suggested by Launder (1990). In line with this practice we also decided to reduce the $\overline{\theta^2}$-equation to the algebraic form. This may not be justified because this transport equation is simple and easy to solve, but its retaining does not seem to give much advantage if R is assumed constant or a simple function of $A_{\theta 2}$. The truncation of the $\overline{\theta^2}$-equation yields:

$$\overline{\theta^2} = -C_\theta^\theta \frac{k}{\varepsilon}\overline{\theta u_j}\frac{\partial T}{\partial x_j} \tag{13}$$

where

$$C_\theta^\theta = 2\left(\frac{P+G}{\varepsilon} - 1 + R^{-1}\right)^{-1} \tag{14}$$

Now $\overline{\theta^2}$ can be eliminated from the algebraic expression (11). We introduce a generalized "buoyancy parameter" which can be interpreted as a form of gradient Richardson number (a ratio of buoyant production of stresses and the turbulence energy dissipation)

$$B_{ik} = \beta g_i \left(\frac{k}{\varepsilon}\right)^2 \frac{\partial T}{\partial x_k} \tag{15}$$

Note that B_{ik} is an asymmetric second rank tensor and that the first index denotes the direction of the gravitation vector. This yields a modified form of the algebraic expression (11) for $\overline{\theta u_i}$ (with neglect of $\varepsilon_{\theta i}$):

$$\overline{\theta u_i} = -\phi_\theta \left[\frac{k}{\varepsilon}\left(\overline{u_i u_j}\frac{\partial T}{\partial x_j} + \xi \overline{\theta u_j}\frac{\partial U_i}{\partial x_j}\right) - \eta B_{ij}\overline{\theta u_j}\right] \tag{16}$$

where

$$\xi = (1 - C_{\theta 2}); \qquad \eta = C_\theta^\theta (1 - C_{\theta 3}) \tag{17}$$

If the heat flux vector $\overline{\theta u_j}$ in the equation (16) is expanded into components, it is possible to express the individual flux component in a computationally more convenient explicit form in terms of standard isotropic eddy viscosity. (Note: Indices in Kronecker delta are identifying ones, hence no summation applies; for convenience when a summation is not intended, the indices will be written in parentheses):

$$\overline{\theta u_i} = -\frac{\nu_t}{\sigma_{ij}^T}\frac{\partial T}{\partial x_j} - \left(\frac{\nu_t}{\sigma_j^{U_i}}\frac{\partial U_i}{\partial x_j} - \eta B_{ij}\overline{\theta u_j}\right)(1 - \delta_{(ij)}\delta_{(ij)}) \tag{18}$$

$\nu_t = C_\mu f_\mu \frac{k^2}{\varepsilon}$ is the conventional kinematic eddy viscosity and the anisotropy of the field is accounted for by tensorial coefficients which can be interpreted as turbulent Prandtl-Schmidt numbers for two terms containing the mean temperature and mean velocity gradients:

$$\sigma_{ij}^T = \frac{C_\mu[1 + \phi_\theta(\xi\frac{\partial U_{(i)}}{\partial x_{(i)}} - \eta B_{(ii)})]}{\phi_\theta \frac{\overline{u_i u_j}}{k}} \qquad \sigma_j^{U_i} = \frac{C_\mu[1 + \phi_\theta(\xi\frac{\partial U_{(i)}}{\partial x_{(i)}} - \eta B_{(ii)})]}{\phi_\theta \xi \frac{\overline{\theta u_j}}{k}} \tag{19}$$

It seems reasonable to assume that in buoyancy dominated flows an accurate prediction of individual stress components may be of less importance as compared with the turbulent heat flux vector. Hence, in all computations with the eddy diffusivity model we have used the standard stress-strain relationship $\overline{u_i u_j} = \frac{2}{3}k\delta_{ij} - \nu_t(\frac{\partial U_i}{\partial x_j} + \frac{\partial U_j}{\partial x_i})$. In some cases the ratio $\frac{\overline{u_i u_i}}{k}$ was assumed constant for each normal stress component, but this did not yield significant effects. Hence for more complex cavity cases we employed the latter approximation.

In a two-dimensional flow of rectangular geometry when the gravitational vector is aligned with one of the coordinates, for example $g_i(0, -g, 0)$, $B_{11} = B_{12} = 0$ and the equation (16) yields the explicit algebraic expressions for the horizontal and vertical heat flux (neglecting $\frac{\partial U_1}{\partial x_1}$ and $\frac{\partial U_2}{\partial x_2}$):

$$\overline{\theta u_1} = -\phi_\theta \frac{k}{\varepsilon}\left[\overline{u_1^2}\frac{\partial T}{\partial x_1} + \overline{u_1 u_2}\frac{\partial T}{\partial x_2} + \xi\overline{\theta u_2}\frac{\partial U_1}{\partial x_2}\right] \tag{20}$$

$$\overline{\theta u_2} = -\frac{\phi_\theta}{1 - \phi_\theta \eta B_{22}}\left[\frac{k}{\varepsilon}\left(\overline{u_1 u_2}\frac{\partial T}{\partial x_1} + \overline{u_2^2}\frac{\partial T}{\partial x_2} + \xi\overline{\theta u_1}\frac{\partial U_2}{\partial x_1}\right) - \eta B_{21}\overline{\theta u_1}\right] \tag{21}$$

The general expression (18) is, however, expected to be useful in more complex geometries when the gravitation does not align with any coordinate of the system.

We have used the standard values of those empirical coefficients which have been generally adopted in literature, as seen in the Table 1. A difference appears in specifying $C_{\varepsilon 3}$, ϕ_θ and C_θ^θ. So e.g. we found that for a pure buoyancy driven turbulence when the mean rate of strain is absent, the buoyant production governs the ε-equation and the solutions are very sensitive to the choice of $C_{\varepsilon 3}$. In fact this case may be employed for tuning this coefficient and we found that in conjunction with the employed model $C_{\varepsilon 3}$

Table 1. Specification of coefficients and functions

C_k	C_θ	$C_{\theta 1}$	$C_{\theta 2}$	$C_{\theta 3}$	C_ε	$C_{\varepsilon 1}$	$C_{\varepsilon 2}$	$C_{\varepsilon 3}$
0.20	0.11	3.0	0.4	0.4	0.15	1.44	1.92	0.8/1.44

should be about 0.8. This is in agreement with earlier findings of Svenson (1978) and others for a horizontal buoyant plume. For vertical heated walls the results seem to be less sensitive to the choice of $C_{\varepsilon 3}$. In line with practice of other authors (e.g. Ince and Launder 1989) we adopted for vertical walls $C_{\varepsilon 3} = C_{\varepsilon 1}$.

The model implies variable coefficients ϕ_θ and C_θ^θ computed from equations (12) and (14) respectively. For more complex geometries this may not be computationally convenient. A plot of ϕ_θ across the boundary layers on a heated vertical plate or on a cavity wall showed that over most of the cross-section $(P+G)/\varepsilon \approx 2$ yielding $\phi_\theta \approx 0.28$. The same arguments lead to $C_\theta^\theta \approx 0.9$. These values have been employed in the present computations with the model described here.

$$f_\mu = exp[\frac{-3.4}{(1+\frac{Re_t}{50})^2}] \qquad f_\varepsilon = 1 - 0.3exp(-Re_t^2) \qquad (22)$$

4 Results and Discussion

4.1 Penetrative Convection of Unstable Mixed Layer

We consider first the well known experiment of Deardorff et al. (1969) (also Willis and Deardorff, 1974) of the development of a turbulent mixed layer in the lower region of initially stably stratified quiescent fluid heated from below, and its upward penetrative convection into the stable layer above it. A characteristic feature of a mixed layer is an intensive transport of heat in the vertical direction, while the temperature profiles remain basically uniform, demonstrating clearly the failure of the simple local gradient representation of turbulent flux. Another feature is the entrainment of adjacent overlaying (or underlaying, in case of cooling from above) non-turbulent fluid into the mixed layer and a consequent development of a sharp density jump (thermocline) at the interface of turbulent and non-turbulent fluid. The penetration of the mixed layer causes a reversal of the sign of the heat flux in the region close to the interface (compensating cooling of the stable fluid in the first case and heating in the second case).

The problems of penetrative convection have been used in the past as test cases for the verification of turbulence models of various types. Zeman and Lumley (1976) and André et al. (1976) employed the differential models at the second-moment level while Mellor and Yamada (1982) or Svenson (1978), used the algebraic expressions for the turbulent heat flux. André et al. solved also the transport equations for the third-order moments but applied a clipping of the third moments if they exceeded limits specified by the realizability conditions. A major difference between the models appears in treating the length scale equation.

We started the computations by employing the full Second-Moment Closure scheme which implies the solution of all differential equations for non-zero components of turbulent stress tensor $\overline{u_i u_j}$ and turbulent heat flux vector $\overline{\theta u_j}$. In the present one-dimensional geometry only the vertical heat flux component is different from zero, while the two normal stresses in the horizontal plane are equal. The closure was achieved by solving also the equation for the temperature variance $\overline{\theta^2}$ and turbulence energy dissipation rate ε. As mentioned in the Introduction, the initial computations with the equation for ε_θ was not very successful. Due to uncertainties in modelling the dynamics of this property we introduced the ratio of the thermal-to-mechanical time scale R and carried out computations with both a constant value of 0.8 and with a variable R in terms of $A_{\theta 2}$ as specified earlier. The use of the latter variant brought actually a slight deterioration of the agreement of the heat flux profiles with the experiments – although it improved to some extent the predicted peak of the temperature variance – and has been subsequently discarded in favour of constant R. In such a way the model solves the differential equations for the following variables (in addition to the mean temperature equation): $\overline{u^2}, \overline{w^2}, \overline{\theta w}, \overline{\theta^2}$ and ε[1]. As it appeared, this "full" second order model does not produce in the considered one-dimensional situation any advantage in comparison with a simpler model in which the equations for the individual normal stresses were replaced by that for the kinetic energy k. In fact, since all buoyant production goes into the vertical stress component $\overline{w^2}$, this component will have a dominant contribution to the kinetic energy of turbulence and k will have a similar profile as $\overline{w^2}$. Figure 1 shows a comparison of predicted and measured development of the temperature field in time in a water tank with initially stable temperature profile after the onset of bottom heating, obtained by three different models and two numerical schemes. In the first two figures (1a and 1b) a simplified version of full flux model $k - \varepsilon - \overline{\theta w} - \overline{\theta^2}$ was used in which the modelled differential transport equations for turbulent heat flux $\overline{\theta w}$ and temperature variance $\overline{\theta^2}$ were solved and in this way the need to employ an eddy diffusivity formula has been avoided. The difference between the models employed in Figures 1a and 1b lies in the treatment of the molecular effects and wall boundary conditions. In the first case (Figure 1a) the high Re-number version was used with a wall function for the specification of kinetic energy at the first grid point next to the wall, which was determined from the assumption of local production-dissipation equilibrium, as proposed by Svenson (1978). Figure 1b was obtained by employing the model of low-Re-number effects of Jones and Launder (1972).

It is expected that the molecular damping of the turbulence could be influential at the interface between the mixed layer and stable region as indicated by some experiments (e.g. Turner 1979, p.295). Hence, for the considered experiment of Deardorff et al. it seems appropriate to use a low-Re-number model which will account for molecular effects. Although both versions of the model produce results which are in a relatively good agreement with experiments, (particularly when compared with other simpler models), Figures 1a and 1b show some differences in predictions in the near-wall region and at the upper edge of the mixed layer. It could be concluded that the low-Re-number version performs better, in particular in the middle of the considered time period, where the predicted temperature profiles are almost coincident with the measured ones. However, at later times this model seems to predict a more pronounced "counter-gradient slope" at the

[1] For the cases of penetrative convection the vertical coordinate is denoted by z and the velocity component by w, to comply with the usual practice in environmental fluid mechanics; in all other cases we used y and v respectively.

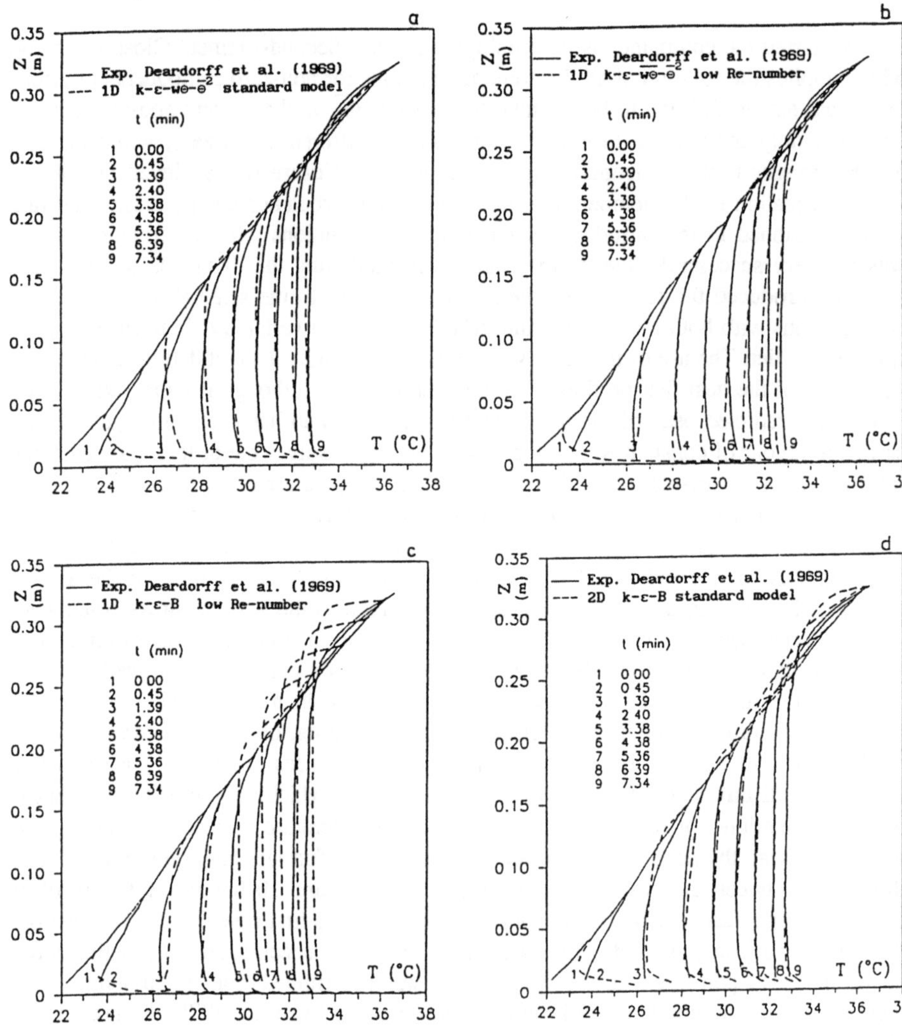

Fig. 1. Time development of temperature profiles in a mixed layer heated from below

interface (slightly higher positive temperature gradients then recorded by experiments), and, contrary to our expectations, slightly underpredicts the cooling and the entrainment at this stage of the process. This behaviour is probably a consequence of inadequacy of the applied low-Re-number modifications which were adopted simply from the practice in modelling viscous effects in shear flows. The fact that we introduced molecular effects only in k- and ε equation but not in $\overline{\theta^2}$- and $\overline{\theta w}$ equations should not matter much. The experimental evidence suggests that the turbulence intensity and, in particular, the intensity of temperature fluctuations are strong at the interface. The temperature variance exhibits a sharp peak at the interface and retains a substantial value well above it, only to decrease sharply with height at the edge of the stable layer. What seems to be needed is the model of the molecular effects which comes sharply into function, but only after the turbulence Re number falls below a certain small value, while the low-Re-number functions of Jones-Launder (applied here) and similar models seem to provide a more gradual influence, aimed at accounting not only for the molecular effects but also for wall damping - (here completely absent). Indeed, our results for both versions of the model give, at later time instants, substantial values of both the kinetic energy and temperature variance (and also of the turbulent Re number) at the interface, so that the predicted effect of low-Re-number modification in the interface region is too strong.

Since our goal was to find a simpler model which retains basically an eddy diffusivity formulation of turbulent stresses and fluxes, we also tested several variants of $k - \varepsilon$ models. The version which accounts partially for the departure from the local gradient formulation of the turbulent heat flux is the so-called $k - \varepsilon - B$ model (where B stands for B_{22} – the only non-zero component of the buoyancy parameter, B_{ij}). Figure 1c shows the results obtained by this model with the same one-dimensional numerical treatment of the considered equations. The results show much poorer agreement with the experiments in comparison with the two previous cases. However, the same model incorporated into the two-dimensional elliptic numerical code, by which momentum equations were solved for both the horizontal and vertical direction, generated some fluid circulation, which produced better agreement, as shown in Figure 1d. Apparently, a slight convective movement of the fluid, which could not be predicted by the one-dimensional code, ensured a better prediction of the heat balance and, consequently, of the entrainment in the stable region of the mixed layer. Although the quality of the predictions shown in Fig. 1d is not so good as in case 1a or 1b, it could be regarded as satisfactory and the $k - \varepsilon - B_{ij}$ model was subsequently tested in computations of recirculating flows in enclosed domains.

As mentioned earlier, almost identical results were obtained by employing the differential or algebraic stress models. However, the application of the standard $k - \varepsilon$ model yielded markedly poorer agreement with the experiments. Although the heating rate was predicted reasonably well, a major shortcoming of the standard $k - \varepsilon$ model is manifested in its inability to generate almost uniform temperature profiles in the mixed layer, which should be attributed to a basic deficiency in the simple gradient transport hypothesis.

In order to further illustrate the findings discussed above we present, in Figure 2, the profiles of the total heat flux obtained by the three models and the one-dimensional numerical procedure corresponding to Figures 1a, 1b and 1c. Each figure shows three selected profiles which correspond to time instants in the later period when the convective mixed layer was already well developed and in which the uncertainties of the initial conditions have already faded away. This parameter is more sensitive to the turbulence model than the temperature profile. All three graphs show almost linear variation of the total heat flux in the mixed layer. The molecular contribution is restricted only to the very thin region close to the bottom and at the interface. Apart from these thin regions, the profiles can be interpreted as actually representing the turbulent heat flux. It is important to point out that all three models predict the change of sign of the heat flux at the upper edge of the mixed layer and its entrainment region in which the heat flux is oriented downwards. However, the agreement of predictions in this region varies from one model to another and the best results were obtained by the most complex model applied. The curvature of the profiles in this zone can be interpreted mathematically after differentiating the equation for thermal energy with respect to the vertical coordinate. This shows that the curvature of the flux profile is caused by the time change of the temperature gradient at that position. These changes are largest in the interfacial zone which separates the

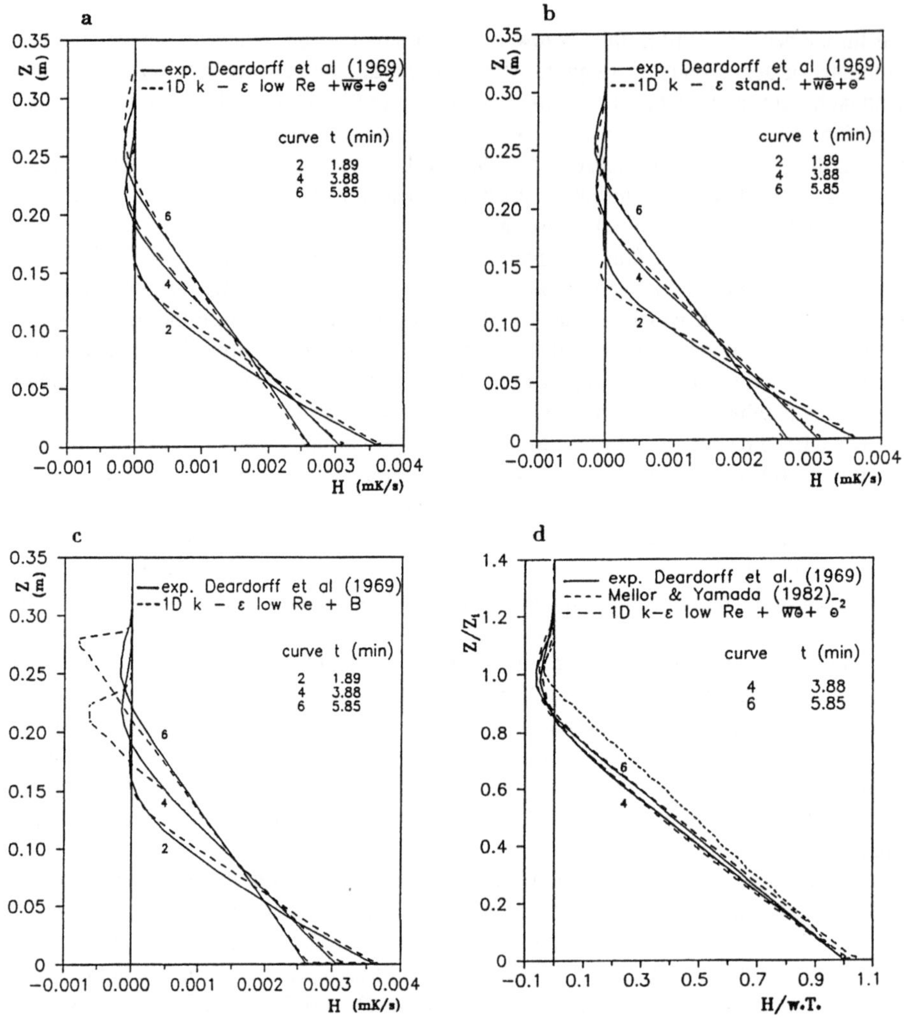

Fig. 2. Heat flux profiles at selected time instants. Figs. 2a to 2c correspond to the same models as Figs. 1a to 1c. Fig. 2d shows non-dimensionalized results from Fig. 1b.

convective layer from the stable zone. Here we find the largest curvature of the flux profile. It is in this region that the simple model, Figure 2c, shows the largest departure from the experimental data. Figure 2d shows the same profiles plotted in dimensionless form, scaled with the interface height z_i and convective temperature- and velocity scales, T_* and w_*, respectively. As could be seen, both the computed and measured data for different time instants collapse almost to a single curve and agree well in between. For comparison, the computed curve of Mellor and Yamada (1982) is also included.

The most important parameter of practical relevance is the height of the mixed layer and its development in time. In order to avoid any ambiguity in defining the position of the interface, we have compared the computations and measured data of two characteristic positions. The first one is the position at which the heat flux changes sign. The second one

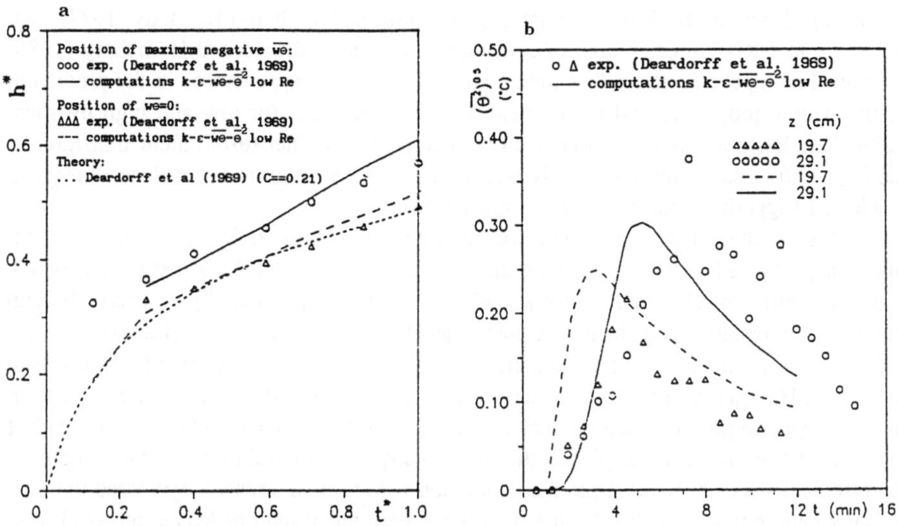

Fig. 3. (a) Non-dimensional height of the interface in function of time; (b)Time development of the temperature variance at two heights

corresponds to the maximum negative value of the heat flux. Both of these are presented in Figure 3a together with the experimental data and the theoretical curve of Deardoff et al. Excluding the initial period where the computations are not realistic because of the uncertainty in specifying the initial turbulence properties, the agreement between the computations and experiments is very good. This applies not only to the higher order models, but also to the $k - \varepsilon - B$ model.

As discussed in the Introduction, the temperature variance plays an important role in free convection because it provides a major source of turbulent heat flux in the absence of a mean temperature- or of a mean velocity gradient. The comparison of our computations, plotted in non-dimensional form with measurements in a similar mixed layer by Willis and Deardorff (1974), agreed well over the larger part of the mixed layer height, except that we were not able to reproduce a high peak in $\overline{\theta^2}$ at the interface . However, the comparison of the time-development of the temperature variance at two vertical positions close to the interface for the experiment considered presently, shown in Figure 3b, indicates that the computed maximum value of $\overline{\theta^2}$ at the interface (z=29.1cm) agrees well with the measurements. At both positions, however, the predicted results show a faster time response then found by measurements.

4.2 Free Convection Along a Heated Vertical Plate

A development of a thermal boundary layer along a heated, infinite, vertical flat plate in an isothermal environment is the simplest representative of a class of free convective flows in which the heat flux vector is perpendicular to the gravitational vector. A collection of experimental data can be found in the literature for the heat transfer coefficient as well as for the mean flow- and turbulence properties (e.g. Cheesewright (1968), Miyamoto et al. (1982). Earlier numerical computations like those of Plumb and Kennedy (1977) used the $k - \varepsilon$ low-Re-number model of Jones and Launder, while To and Humphrey (1986) employed an algebraic stress/flux model and achieved an improvement in the prediction of mean flow properties and of the heat transfer coefficient for low and high overheat ratios. The latter authors also managed to predict the laminar-to-turbulent transition in the $Nu_x - Gr_x$ correlation reasonably well by feeding an amount of turbulence energy at a selected triggering position to the laminar flow .

Following the adopted approach, we carried out the computation of the free convection along a heated vertical plate with the semi-algebraic model described earlier, which solves the differential equations for k and ε in the forms (4) and (5), respectively, and the algebraic equations for heat flux components (20) and (21). A small amount of turbulence was specified at the plate leading edge. The present model differs from that used by To and Humphrey essentially in the treatment of the turbulent normal stresses and in the employed empirical functions that account for the low Re-number effects. Instead of algebraic expressions, we employed the simple eddy viscosity relationship for computing the turbulent stresses. A subsequent computation with normal stresses, expressed in terms of the kinetic energy roughly in line with the experimental data of Miyamoto et al. (i.e. streamwise component equals to k and the lateral one equals to $\frac{2}{3}k$), did not produce much of an effect. This is understandable, since the shear production of the kinetic energy is larger than the buoyancy production in most part of the boundary layer (e.g. Miyamoto et al). Besides, the latter one also depends more upon the shear stress than on normal stresses, as follows from equation (21).

As reported by To and Humphrey and by other authors, even the simple $k - \varepsilon$ low-Re-number model predicts well the Nu_x vs Gr_x relationship in the fully turbulent regime, but it is the position of the transition and the behaviour of the flow properties in the transition region that are sensitive to the chosen model. First we present the Nu_x vs Gr_x correlation in Figure 4. As could be seen the computations reproduce well the $Nu_x - Gr_x$ development along the plate over the large span of Gr numbers, including both the laminar and turbulent regions as well as the position of the transition. The prediction of

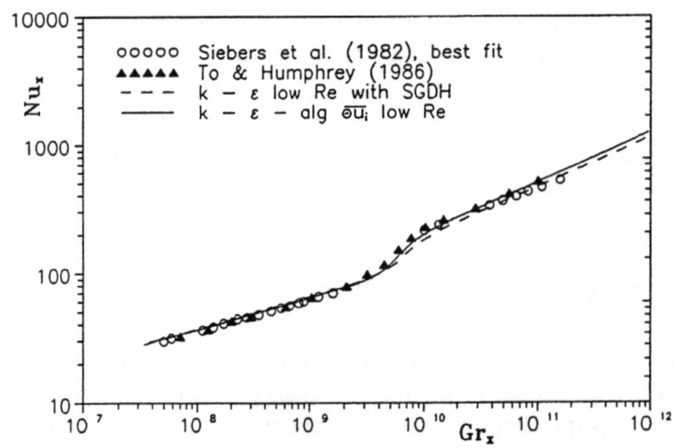

Fig. 4. Local Nusselt number vs Grashoff number along a heated vertical plate; dotted line shows a best fit through the experimental data of Siebers et al. (taken from To and Humphrey, 1986)

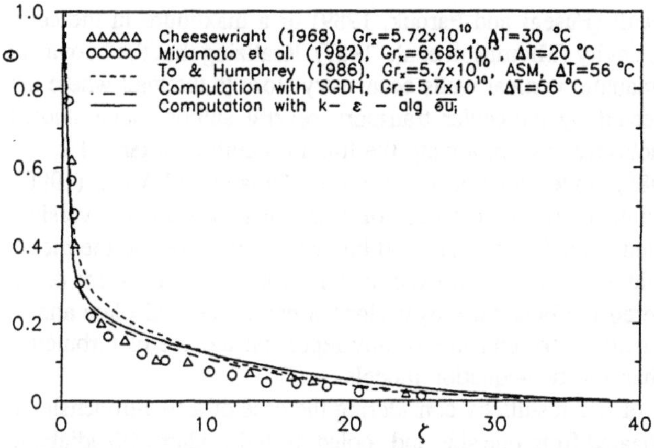

Fig. 5. **Mean temperature profile across the boundary layer at a heated vertical plate**

the latter involves a degree of arbitrariness. It was found that the initial level of turbulence does not influence the location of transition, provided it is kept within reasonable limits and in an adequate relation to the initial level of dissipation rate.

Figure 5 shows the mean temperature profile for $Gr_x = 5.7 \times 10^{10}$ which corresponds to the position at the plate just above the transition. For comparison, the experimental data of Cheesewright (1968) and Miyamoto et al. (1982) are plotted as well as the computations obtained by the ASM model (To and Humphrey) and by standard low-Re-number $k - \varepsilon$ model. The agreement with the data may be regarded as reasonable although the computations yield slightly higher temperature levels in the inner region of the boundary layer. This feature seems to be exhibited by most models and may reflect a possible imprecision in predictions of the transition position. It is interesting to note that the simple $k - \varepsilon$ model (denoted by SGDH, which stands for simple gradient diffusion hypothesis), with inclusion of the buoyant production (which was neglected by To and Humphrey in their $k - \varepsilon$ model) produced even better agreement with both sets of measured temperatures, although the predictions of the Nu-number are poorer (Fig. 4).

4.3 Natural Convection in Enclosures with Side Heating and Cooling

The two cases considered previously fall into the category of external flows, (sometimes referred to as free convection). Buoyancy driven flows confined in enclosures, (natural convection) are regarded to be more complex because of the mutual interaction between the boundary layer-like flows along the enclosure walls and the core region. Besides, experimental evidence suggests that effects of three-dimensionality are often present, even in strictly two-dimensional geometries, causing a sensitivity of the confined buoyancy driven flows to the enclosure configuration and frequently reported non-repeatability of the measured results.

Computations of the turbulent natural convection in two-dimensional cavities have been reported by several authors. The application of the high-Re-number $k - \varepsilon$ model with wall functions produced several anomalies such as a double peak in the kinetic energy profile close to the wall (Fusegi and Farouk, 1989) or a maximum in the eddy viscosity in the middle of the cavity (Thompson et al. 1987). Realizing that the buoyancy effects in flows along non-adiabatic vertical walls occur very close to the wall where the flow is also very much affected by molecular transport, several authors have studied the natural convection in enclosures by employing the low-Re-number variant of $k - \varepsilon$ model, e.g. Betts et al. (1985), Ince and Launder (1989), Hanjalić and Vasić (1990). The latter models seem to be generally more successful than the standard $k - \varepsilon$ version, although most authors reported a need to introduce additional modifications to the model in order to achieve acceptable agreement with experiments. Apart from the work of To and Humphrey, who considered the boundary layer along a heated vertical plate and in an open cavity, the present authors are unaware of any report on the use of turbulence models of an order higher than the two-equation models.

We begin the discussion of our results by considering the case of a two-dimensional cavity with 5:1 aspect ratio, heated from one side and cooled from the other with adiabatic bottom and top walls. Experimental investigations of this geometry by Cheesewright, King and Ziai (1986), Cheesewright and King (1990), Bowles and Cheeswright (1989) constitute a good experimental basis for validation of the models. Computational codes and these data have been used in the present work. The same case has been computed by Ince and Launder, (1989), with the Low-Re-number $k - \varepsilon$ model of Jones and Launder (1972) with the coefficients as given in Table 1 and the functions specified by equation (22) in the present paper. However, in order to achieve the agreement with the experimental data of Cheesewright et al. (1986), they found it necessary to apply what they called the generalized gradient diffusion hypothesis (GGDH) to express the vertical turbulent heat flux (essentially the first term in equation (16) but with ϕ_θ replaced by a constant ≈ 0.16, which is twice smaller than ϕ_θ for an equilibrium situation). In such a way they introduced the lateral mean temperature gradient (and also the lateral mean velocity gradient through the eddy viscosity expression for the shear stress) into the term representing the buoyant generation of turbulence energy, but seem to have neglected the streamwise gradients. Besides, Ince and Launder added a new near-wall source term in the ε-equation, proposed by Yap 1987, (quoted by Launder 1988) which increases the level of ε, and decreases the length scale slope in the wall region to the level found in wall flows in local equilibrium.

We repeated the computations by using the same model employed by Ince and Launder and essentially a similar finite volume numerical procedure (Perić 1987) with a non-staggered, nonuniform grid of 60×60 control volumes (CV), Figure 6a, as well as a finer grid with 80×100 CV. The profile obtained for the mean vertical velocity across the cavity midheight cross-section ($y/H = 0.5$), the dotted line in Figure 7a, agrees very well with the measurements (a slightly better agreement was obtained with $C_{\varepsilon 3} = 0.8$ used for the computation of the penetrative convection than with 1.44, as used by Ince and Launder).

The computed turbulent shear stress and the heat flux components did not show, however, the same quality of agreement. As seen in Figure 8a, the predicted $\overline{uv}$ has a similar shape and a same order of magnitude as obtained by the measurements. The experimental data show an appreciable asymmetry reflecting the deficiency mentioned earlier in securing full control of the boundary conditions in the experiment. Since both the measured and computed integral of the shear stress across the cavity seem to be very close, one might expect that fully symmetric boundary conditions in the experiment

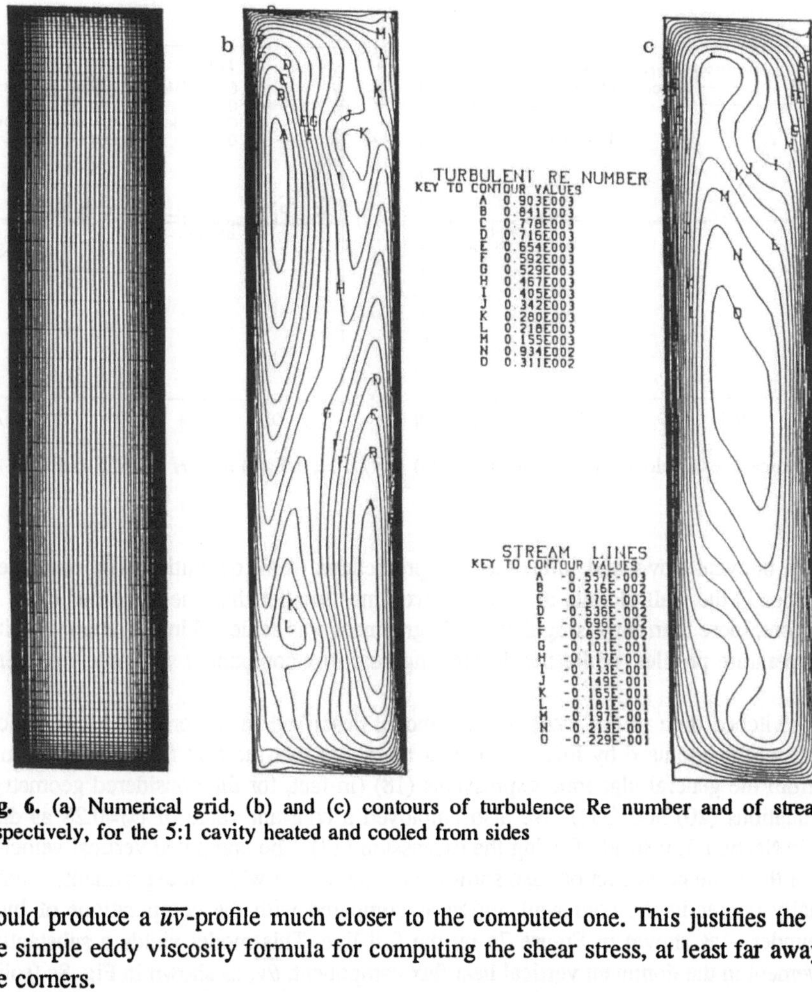

Fig. 6. (a) Numerical grid, (b) and (c) contours of turbulence Re number and of streamlines, respectively, for the 5:1 cavity heated and cooled from sides

would produce a $\overline{uv}$-profile much closer to the computed one. This justifies the use of the simple eddy viscosity formula for computing the shear stress, at least far away from the corners.

The agreement between the computed and measured heat flux components is considerably poorer. Figure 8b shows a comparison of computations of the horizontal heat flux component, $\overline{\theta u}$, obtained by the model of Ince and Launder (dotted line), with data of Bowles and Cheesewright, measured directly by a LDA and a thermocouple. The results correspond to the position $y/H = 0.63$, since experimental data for the mid-plane were not available. The predictions close to the hot wall agree reasonably well with the measurements, while those at the cold wall show a higher discrepancy. Bowles and Cheesewright admitted that the measured data are not free from error because the laser beam had to be angled toward the wall, but their calculation from an integral balance applied to the measured data of Cheesewright and King produced a similar profile, suggesting that the data have a realistic shape and order of magnitude. In light of the mentioned asymmetry in the experiment, the satisfactory agreement of the computed and measured data at the hot wall may be regarded more as a coincidence.

Most surprising, however, is a striking disagreement between the computed and measured profiles of the vertical heat flux component, $\overline{\theta v}$, shown in Figure 8c. As discussed earlier, this flux component plays a major role in the dynamics of the turbulence kinetic

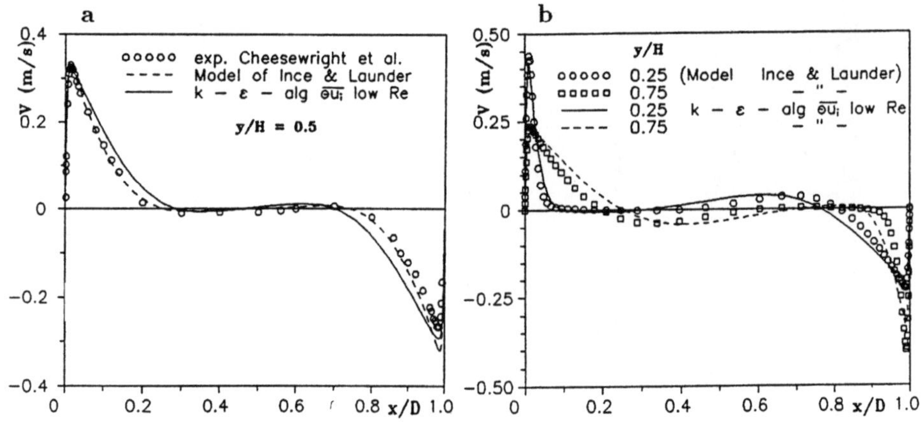

Fig. 7. Vertical mean velocity across the cavity: (a) at $y/H = 0.5$, (b) at $y/H = 0.25$ and 0.75

energy in buoyant flows and influences the predictions of all quantities. The computed peak values in the wall vicinity are four to five times smaller than the measured ones. It is, therefore, more than surprising that good agreement was achieved in the mean velocity and temperature profiles, while the dominating heat flux component was predicted very poorly.

We switched then to the more general model described in the present paper which differs from the one used by Ince in Launder mainly in that the heat flux vector is computed from the general algebraic expressions (18) (in fact, for the considered geometry, from equations (20) and (21)). We also employed a constant value of ϕ_θ=0.28 as discussed in Section 3, instead of using the expression (12). The computed vertical velocity profile at the same cross-section also shows good agreement with the experiments – only marginally poorer in the cold wall region as compared with the computations of Ince and Launder – as shown in Figure 7a by the full line. This model yields a substantial improvement in the dominant vertical heat flux component, $\overline{\theta v}$, as shown in Fig. 8c (solid line).

In order to analyze the influence of various terms in the $\overline{\theta u_i}$-equation, we perfomed the computations with several variants of the model by first neglecting the mean velocity gradients, then with replacement of ϕ_θ by the smaller value used by Ince and Launder, and finally, with the introduction of the additional term of Yap in the ε-equation. None of the variants delivered results in full agreement with the experimental data nor with the previous solutions. The latter version, which differed from the model used by Ince and Launder only in the inclusion of the streamwise temperature gradient, produced at the midplane a more laminar-like solution with much higher velocity peaks, what leads to a conclusion that this gradient, albeit smaller than the cross-stream one, is still influential. Hence, it seems that a consistent application of the full GGDH even with the Yap correction of the ε-equation does not itself yield the desired quality of the predictions for the considered case.

In fact, an inspection of the turbulence field, illustrated in Figure 6b by contours of turbulence Re numbers (note: $Re_t \approx 11\nu_t/\nu$), reveals a concentration of turbulent viscosity with ν_t/ν reaching a value of about 100 in the upper corner close to the hot

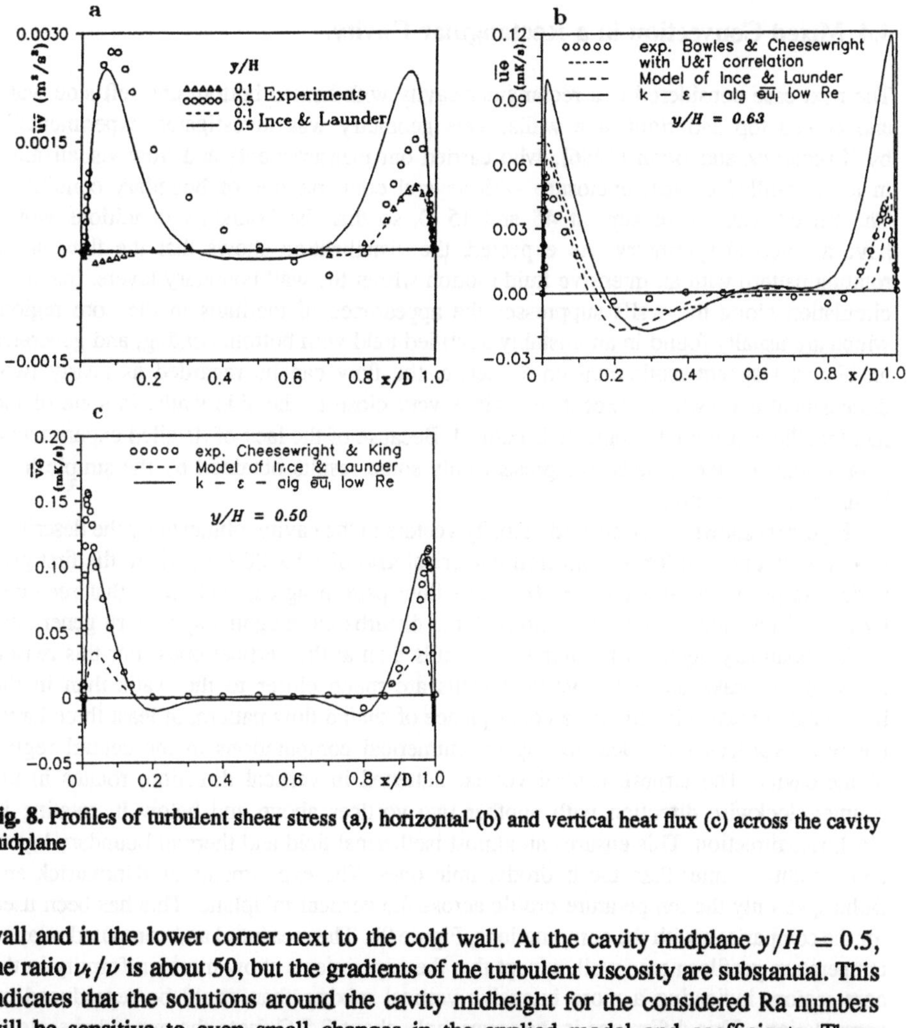

Fig. 8. Profiles of turbulent shear stress (a), horizontal-(b) and vertical heat flux (c) across the cavity midplane

wall and in the lower corner next to the cold wall. At the cavity midplane $y/H = 0.5$, the ratio ν_t/ν is about 50, but the gradients of the turbulent viscosity are substantial. This indicates that the solutions around the cavity midheight for the considered Ra numbers will be sensitive to even small changes in the applied model and coefficients. These changes will cause a shift of the eddy viscosity gradients up or down, without affecting much the maximum turbulence levels concentrated in the upper left and lower right cavity corners.

An argument in favour of this conclusion is the plot of the vertical mean velocity at $y/H = 0.25$ and 0.75 in Figure 7b, obtained by both the model of Ince and Launder and the present one. The two solutions agree very well. However, it is interesting to note that similar solutions at these cross-sections, as well as at $y/H = 0.1$ and 0.90, have been obtained by other variants of the model mentioned earlier, which gave erroneous predictions at the midplane.

Of course, the above discussion is burdened by an uncertainty in regard to possible effects of three-dimensionality and a consequent unsteadiness of the flow pattern which have been detected particularly in the low turbulence regions (upper right- and lower left corner) in similar experiments, but which can not be reproduced by the two-dimensional computations.

4.4 Mixed Convection in a Rectangular Cavity.

The next case considered is a rectangular cavity with heated bottom and left side walls and cooled top and right side walls. This geometry was investigated experimentally by Kirkpatrick and Bonh (1986), who carried out measurements and flow visualization in a water-filled cubical enclosure with several combinations of boundary conditions. The heated walls were kept at 45 and 15 C, so that the boundary conditions would have a diagonal symmetry. As expected, the visualization shows that the flow has a rotating pattern with an intensive fluid motion within the wall boundary layers. Intensive circulation along the walls suppresses the appearance of thermals in the core region, which are usually found in an unstably stratified field with bottom heating, and generates an almost stagnant isothermal core. Hence, the flow can be regarded as nearly two-dimensional everywhere, except in regions very close to the side walls, in spite of the fact that the considered geometry is cubical. Because of the lack of detailed experimental data to validate the models, we present only some results obtained by the simple low-Re-number $k - \varepsilon$ model.

Figure 9a shows the computed velocity vectors in the cavity, obtained by the described model with a non-uniform, staggered numerical grid of 80×80 CV, where the first grid node was placed at about $y^+ = 0.5$. The flow pattern agrees well with that recorded by flow visualization. The gravitational force disturbs the diagonal symmetry generating thicker boundary layers at the horizontal walls than at the vertical ones. For this reason the velocity peaks along the vertical walls are much closer to the walls than in the horizontal boundary layers. As a consequence of such a flow pattern, at least three large, but weak vortices were detected by the numerical computations in the central region of the cavity. The largest, central vortex, flattened in vertical direction, rotates in the counter-clockwise direction with another two vortices above and below it, rotating in clockwise direction. This ensures an almost isothermal field and thermal boundary layers considerably thinner than the hydrodynamic ones. The experiments of Kirkpatrick and Bohn give only the temperature profile across the vertical midplane. This has been used for the comparison with the computations, Figure 9b. The measured and computed relative temperature profiles agree well, except that the experiments show more uniformity in the core region, indicating a more intensive vertical mixing than could be reproduced by computations. This difference is, however, only about 0.5 C (note the expanded scale of $T - T_{bulk}$ on the abscissa) and could be attributed to side effects in the experiments not accounted for by computations such as radiation, three-dimensionality, non-uniformity of the wall temperatures, etc. The same effects are responsible for the disagreement noticeable very close to the bottom wall. The measurements show here a higher sub-cooling of the fluid below the bulk temperature by 1.2 C while the maximum computed sub-cooling is about 0.8 C. The measurements could not be performed very close to the wall where the fluid temperature rapidly approaches the wall value with an overheat of 15 C. Hence, no detailed comparison of the computations was possible in this region.

In order to verify the temperature predictions in the near wall areas, we have compared the measured and computed values for the Nusselt number. For $Ra = 2.15 \times 10^{10}$ (based on the temperature difference $T_w - T_{bulk}$), the computations yielded a Nusselt number at the horizontal walls of 254 which is about 16% smaller than the experimentally obtained value of 302. Better agreement was achieved for the vertical walls for which the computed Nusselt number was 224, which is 7% smaller than the measured value of 242. This

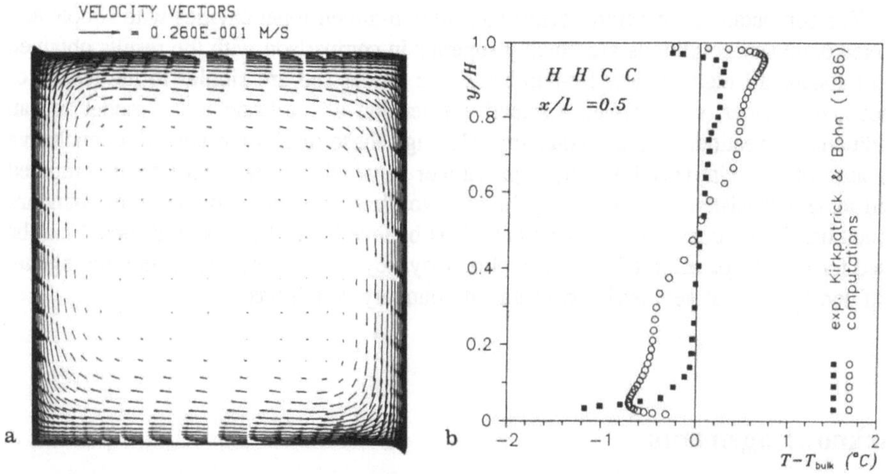

Fig. 9. Mean velocity vectors (a) and vertical temperature profile at the horizontal midplane (b) in a square cavity; left side wall and bottom wall at 45 C, right side wall and top wall at 15 C.

discrepancy can also be partially attributed to various imperfections and a lack of control of the experiments. However, it is symptomatic that the computations yielded a lower level of turbulence (less intensive core mixing, smaller wall heat transfer coefficients) which is caused by the already discussed inadequacy of the applied simple model.

5. Conclusions

Because the application of turbulence models at a Second Moment Closure level for the computation of complex buoyancy-driven turbulent flows seems presently premature, considering the uncertainties inherent in the models as well as numerical inconveniences, a compromise has been sought in a model which will retain a basic eddy viscosity form, but account in a consistent manner for buoyancy effects that dominate turbulence dynamics. An algebraic expression for the turbulent heat flux has been derived in a computationally more convenient, explicit form which can be incorporated into the existing Navier-Stokes solvers, with turbulent stresses still expressed by the conventional eddy viscosity formula.

For an unstable, stratified field heated from below or cooled from above, in which a mixed layer is generated with an almost uniform temperature profile, (and, therefore, intractable by simple eddy viscosity models), it was shown that the derived expression produces results in agreement with measurements almost as good as that obtained with the differential, second-moment closure model. The results for a thermal boundary layer on a heated vertical plate in an infinite surrounding have also been improved and have reached the quality of that obtained by other authors by using the full algebraic stress/flux model.

The computations of natural convection in two-dimensional cavities with simple and mixed boundary conditions are notably superior in comparison with the results obtained with a standard high- or low-Re-number $k-\varepsilon$ model and are similar to the results of Ince and Launder, who applied a selective extension of the basic $k-\varepsilon$ model and an additional corrections to the ε-equation. Although these results are not fully conclusive because of the relatively low Rayleigh number at which the data have been compared and some inconsistencies in ensuring the full two-dimensionality and control of boundary conditions in the considered experiments, it is believed that the proposed model can be used for the computation of more complex buoyancy driven flows in various orthogonal and non-orthogonal geometries with mixed boundary conditions.

Acknowledgements

This work was supported by the Science Fund of Bosnia and Hercegovina. A part of the computations was carried at in the Computer Centre of the Friedrich-Alexander Universität Erlangen-Nürnberg during the stay of one of the author (S. Vasić) at the Lehrstuhl für Strömungsmechanik. The authors acknowledge the valuable help on numerical matters from Dr. M. Perić. Graduate students S. Kenjereš and Z. Perošević assisted in performing some additional computations and completing the results in graphical form.

References

André J. C., De Moor G., Lacarrere P., Therry G., du Vachat R. (1977): The clipping approximation and inhomogeneous turbulence simulation, Turbulent Shear Flows I, Springer, 307

Betts, P. L. Dafa'Alla A.A. (1986): Turbulent buoyant air flow in a tall rectangular cavity, Proc. ASME Meeting HTD **60** 83

Bowles A., Cheesewright R. (1989): Direct measurements of the turbulence heat flux in a large rectangular air cavity, Exp. Heat Transfer **2**, 59

Cheesewright R. (1968): Turbulent natural convection from a vertical plane surface, J. Heat Transfer, **90**, 1

Cheesewright R., King K. J., Ziai S. (1986): Experimental data for the validation of computer codes for the prediction of two-dimensional buoyant cavity flows, Proc. ASME Meeting HTD, **60**, 75

Cheesewright R., King K. J. (1990): Stress distribution in natural convection in a rectangular air cavity, Proc. 9th Int. Heat Transfer Conf., **2**, 161

Chu T. Y., Goldstein R. J. (1973): Turbulent convection in a horizontal layer of water, J. Fluid Mech. **60** 141

Deardorff J. W., Willis G., E., Lilly D. K. (1969): Laboratory investigation of non-steady penetrative convection, J. Fluid Mech. **35**, 7

Fusegi T., Farouk B. (1989): Laminar and turbulent natural convection, Numerical Heat Transfer **15**, 303

Gibson M. M., Launder B.E. (1976): On the calculation of horizontal, turbulent, free shear flows under gravitational influence, J. Heat Transfer **98** 81

Hanjalić K., Launder B.E. (1976): Contribution towards a Reynolds-stress closure for low-Reynolds-number turbulence, J. Fluid Mech. **74/4**, 593

Hanjalić K., Vasić S. (1990): Numerical simulation of free convection in single- and multiple-zone rectangular cavities, Proc. 9th Int. Heat Transfer Conf. **2**, 579

Heslot F., Castaing B., Libchaber A. (1987): Transition to turbulence in helium gas, Physical Review A, **36**/12, 5871

Ince N. Z., Launder, B. E. (1989): On the computation of buoyancy-driven turbulent flows in rectangular enclosures, Int. J. Heat and Fluid Flow, **10**, 110

Jones W.P., Launder B. E. (1972): Prediction of laminarization with a two-equation model of turbulence, Int. J. Heat Mass Transfer **15**, 301

Kirkpatrick A.T., Bohn M. (1986): An experimental investigation of mixed cavity natural convection in the high Rayleigh number regime, Int. J. Heat Mass Transfer **29**

Launder B. E. (1988): On the computation of convective heat transfer in complex turbulent flows, J. Heat Transfer **110** 1112

Mellor G.L., Yamada T. (1982): Development of a turbulence closure model for geophysical fluid problems, Rev. Geophysics and Space Physics **20**, 851

Miyamoto M., Kajino H., Kurima J., Takanami I.(1982): Development of turbulence characteristics in a vertical free convection boundary layer, Proc. 7th Int. Heat Transfer Conf. **2**, 323

Nagano Y., Kim C. (1988): A two-equation model for heat transport in wall shear flows, J. Heat Transfer, **110**, 583

Perić M.(1987): Efficient semi-implicit solving algorithm for nine-diagonal coefficient matrix, Num. Heat Transfer **11** 251

Plumb O. A., Kennedy L.A. (1977): Application of a $k - \varepsilon$ model to natural convection from a vertical isothermal surface, J. Heat Transfer **99**, 79

Svenson U.(1978): A Mathematical model of the seasonal thermocline, Univ. Lund Rept. No. 1002, Sveden

Tanaka H., Miyata H. (1980): Turbulent natural convection in a horizontal layer heated from below, Int. J. Heat Mass Transfer **23** 1273

Thompson C. P., Wilkes N. S., Jones I. P. (1987): Numerical studies of buoyancy driven turbulent flow in a rectangular cavity, Int. J. Num. Meth. Engng. **24**, 89

To W. M., Humphrey J. A. C. (1986): Numerical simulation of buoyant turbulent flow, I and II, Int. J. Heat Mass Transfer, **29**, 573

Turner J. S. (1979): Buoyancy Effects in Fluids, Cambridge Univ. Press

Willis G.E., Deardorff J.W. (1974): A laboratory model of the unstable boundary layer. J. Atm. Sci. **31**, 1297

Zeman O., Lumley J.L. (1976): Modelling buoyancy driven mixed layers, J. Atmos. Sci., **33**, 1974

Large-Eddy Simulation of the Convective Boundary Layer: A Comparison of Four Computer Codes

F.T.M. Nieuwstadt[1], P.J. Mason[2], C-H. Moeng[3] and U. Schumann[4]

[1] Laboratory of Aero and Hydrodynamics, Delft University of Technology, 2628 AL Delft, The Netherlands
[2] Meteorological Office, Bracknell, Berkshire RG12 2SZ, England
[3] National Center for Atmospheric Research, Boulder, CO 80307, USA
[4] DLR, Institute of Atmospheric Physics, W-8031 Oberpfaffenhofen, Germany

Abstract

To test the consistency of large-eddy simulation we have run four existing large-eddy codes for the same case of the convective atmospheric boundary layer. The four models differ in various details, such as: the subgrid model, numerics and boundary conditions.

The agreement between the four models is excellent. In general model results lie within the scatter of available observations. Most of the disagreement between the model results can be attributed to the parameter C_S of the subgrid model. The effect of other differences between the four models, such e.g. due to numerics, is found to be small.

1. Introduction

In large-eddy simulation we seek to compute the large scales of turbulence. This is accomplished by performing on the Navier-Stokes equations a filter operation, designed to remove the small scales from these equations. The resulting set of equations is then solved numerically. Large-eddy simulation is considered to be a tool to study the details of turbulence and it is enjoying an increasing popularity.

The most successful application of large-eddy modelling has been to the atmospheric boundary layer and in particular the convective boundary layer. The reason is of course that the convective boundary layer is dominated by large structures. These are known as thermals and they are responsible for most of the vertical transport processes. They make the convective boundary layer an ideal subject for large-eddy simulation.

The study of the convective boundary layer with large-eddy models has started with the pioneering work of Deardorff in the seventies (e.g. Deardorff 1974). Since then extensive and detailed simulations have been reported by Moeng (1984), Mason (1989) and Schmidt & Schumann (1989). A more general review of large-eddy simulation of convection is presented by Nieuwstadt (1990)

Here, we will not add another simulation to this list. Instead this study has a different objective.

The philosophy of large-eddy modelling is to isolate the dynamics of the large scales. The small scales are treated by a more or less simple parameterization: the so-called subgrid model. This approach hinges on the hypothesis that the actual details of the subgrid model have only minor effects on the large scales. As this may be considered the fundamental premise of large-eddy modelling there are also other uncertainties. Any large-eddy model requires a large number of assumptions about details of the model, ranging from the choice of a numerical scheme to the selection of boundary conditions. It would be very undesirable if a large-eddy simulation turns out to be sensitive to these modelling details.

In this paper we aim to address these questions and investigate whether present large-eddy models are in line with the philosophy mentioned above. Do they lead to consistent simulations for the case of the convective atmospheric boundary layer? Such consistency would present a further argument that large-eddy simulations can be trusted to give a realistic representation of atmospheric turbulence.

To this end four research groups represented by the authors of this paper have decided to test their large-eddy models for a simulation of the convective boundary layer. Besides using different subgrid models, these codes also differ in many other details. We run the models starting from the same initial conditions and using the same forcing. By comparing the results we expect to learn the sensitivity of these models to the various parameters.

First we shall discuss in some detail the differences between the models. Next we shall specify the initial conditions and the calculation procedure. Then we turn to a discussion of the results. In particular we will consider in some detail the effect of the subgrid model.

2. Description of Large-Eddy Codes

As already mentioned above we shall use in this study existing large-eddy codes which have already been described elsewhere (Mason, 1989; Moeng, 1984; Nieuwstadt & Brost, 1986 and Schmidt & Schumann, 1989). Therefore, we shall restrict ourselves in this section to highlighting the various differences between the four models.

2.1 Subgrid Model

The main purpose of a subgrid model is to "dissipate" the energy which is passed down the cascade from the large eddies. Views at present differ on the complexity needed to model the subgrid terms.

The most simple parameterization is the well-known Smagorinsky model, which relates the subgrid stresses to the deformation of the resolved velocity field by means of an eddy viscosity, K_M. This eddy viscosity is expressed as

$$K_M = \ell^2 S \tag{1}$$

where S is the absolute value of the strain rate and ℓ is the so-called mixing length. This mixing length is proportional to the filter length.

The Smagorinsky model is used by Mason and for a more detailed discussion we refer to Mason & Callen (1986) and Mason (1989). In particular Mason (1989) introduces an extension of (1) to convective turbulence. Here, it suffices to mention that in the context of the Smagorinsky model a fundamental role is played by the coefficient C_s. It is defined as the ratio between the mixing length and a characteristic grid size Δ

$$C_s = \ell/\Delta. \tag{2}$$

Mason (1989) has considered the effect of several values for C_s and recommends $C_s = 0.32$.

Alongside the Smagorinsky model for subgrid stresses, an analogous model is used to parameterize subgrid heat fluxes. It relates the temperature flux to the resolved temperature gradient by means of an eddy diffusivity K_H. The K_H is connected to the K_M by means of a turbulent Prandtl number

$$Pr = \frac{K_M}{K_H}. \tag{3}$$

A slightly more complicated subgrid closure model is used by Moeng and Nieuwstadt. It is also based on a relation between the subgrid stress and the resolved deformation rate but in this case the eddy viscosity is defined as

$$K_M = 0.12\Delta e^{1/2} \tag{4}$$

where e is the subgrid energy. A separate equation for the subgrid energy is formulated, which is then solved along with the other equations of the large-eddy model (Moeng, 1984). If we simplify the subgrid energy equation to a balance between shear production and dissipation, we may derive a representative value for C_s. In case of the subgrid model of both Moeng and Nieuwstadt, we find $C_s = 0.18$.

Schumann adopts the most complicated subgrid model. It is a second order closure model, in which the following simplifications are made. Only for the subgrid energy a full equation, including time derivatives and transport terms, is solved. In the other equations of the subgrid model these terms are neglected. Moreover, the production terms in these equations are limited to the isotropic part. As a result the subgrid model can be expressed as a decoupled algebraic system of equations, which can be explicitly solved. For further details we refer to Schmidt and Schumann (1989), where it is also derived that their representative value for C_s is: $C_s = 0.165$.

Another detail of the subgrid models which varies between the four codes is the formulation of the mixing length. In the interior of the boundary layer all four authors take the mixing length ℓ proportional to a characteristic grid size Δ. Moeng and Nieuwstadt use the geometric mean $\Delta = (\Delta x\, \Delta y\, \Delta z)^{1/3}$, whereas Schumann takes the arithmetic average: $\Delta = (\Delta x + \Delta y + \Delta z)/3$. Mason uses a non-uniform grid and he provides no explicit expression of Δ in terms of the grid spacing.

Furthermore, Mason and Schumann allow ℓ to become proportional to height when the surface is approached. Nieuwstadt takes this effect into account through a different parameterization of the dissipation term in the equation for e near the surface.

Finally, Mason, Moeng and Nieuwstadt limit the mixing length to a smaller value in stable conditions. This plays for instance a role in the stable layer above the boundary layer (Schumann, 1991).

Nevertheless, we shall see that the most important parameter influencing the effect of the subgrid model is C_s and perhaps to a lesser extent the Prandtl number Pr. For convenience we have collected these parameters in table 1.

Table 1. Representative values of the coefficient C_s and the turbulent Prandtl number Pr for the subgrid model of the four large-eddy codes.

	Mason	Moeng	Nieuwstadt	Schumann
C_s	0.32	0.18	0.18	0.165
Pr	0.46	0.33	0.33	0.42

2.2 Numerics

The models of Mason, Nieuwstadt and Schumann use a full staggered grid. This means that the velocity components are defined on the sides of the rectangular grid volume with the pressure in the centre. Moeng uses staggering only in the vertical direction, so that in her model all variables are defined on the same x, y-position.

The subgrid energy e, used in the subgrid models of Nieuwstadt and Schumann is also defined in the centre of the grid volume. This means that in these models the exchange coefficients are located on the same position. Mason and Moeng adopt a different view and collocate the exchange coefficients with the w-point to avoid vertical averaging of z-derivatives.

The resolved temperature is governed by an advection-diffusion equation. Therefore, the most obvious location for the temperature is the centre of the grid volume. This view is taken by all participants except Nieuwstadt, who stores the temperature at the w-point. His argument is to avoid vertical averaging of temperature in the w-equation, where buoyancy is the dominant forcing term.

Next we consider the numerical schemes. With respect to time advancement Moeng and Schumann use the Adams-Bashford scheme, whereas Mason and Nieuwstadt employ the leap-frog method. The leap-frog scheme is unstable for the diffusion terms. Therefore, the latter authors resort for these terms to the simple forward Euler scheme.

Mason and Nieuwstadt treat the advection terms by scheme proposed by Piascek and Williams (1970). This is a second-order method designed to conserve variances even if the continuity equation is satisfied only approximately. Schumann uses standard second-order differencing for the momentum equations which conserves kinetic energy if the continuity equation is satisfied. For the temperature, he uses a second-order upwind scheme which preserves positivity. All three authors take second-order, central differencing to solve the diffusion terms. Moeng adopts a somewhat different approach. She uses a pseudo-spectral method to calculate all horizontal derivatives and a finite difference technique to calculate the vertical derivative. To avoid aliasing errors in the pseudo-spectral technique all Fourier expansions are truncated beyond the wave number $2/3\, k_{max}$.

All four authors use a mixed fast-Fourier and finite difference technique to solve the Poisson equation for the pressure.

2.3 Boundary Conditions

In all four models horizontal boundary conditions satisfy periodicity. However, differences are found in the specification of the vertical boundary conditions.

At the lower boundary the vertical velocity w is set equal to zero. For the horizontal velocities one resorts to Monin-Obukhov similarity. This means that a relation is specified between the surface stress and the horizontal velocity in the first grid cell. Furthermore, it is assumed that the velocity and stress are parallel.

The models of Mason, Moeng and Schumann specify the temperature flux at the surface. Surface similarity is then used to obtain the surface temperature T_s (i.e. the temperature at $z = z_0$) from the temperature calculated in the centre of the first grid volume. Nieuwstadt has to adopt a different procedure, because he defines the temperature at the location of the w-variable. He then uses surface-layer similarity to specify a relationship between the surface-temperature flux and the temperature difference across the first grid cell. Note that in convective conditions the temperature profile near the surface is strongly curved. This curvature cannot be resolved within the first grid cell. As a result the subgrid exchange coefficients in Nieuwstadt's model are revised in relation to the standard surface-layer expressions.

With respect to the upper boundary conditions all four models agree on stress free conditions: i.e. $\partial u/\partial z = \partial v/\partial z = 0$. Furthermore the temperature gradient at the top of the calculation domain is set equal to the gradient Γ, which is prescribed in the initial conditions as the temperature gradient above the boundary layer.

There are however, some important differences in the treatment of the vertical velocity at the upper boundary. The goal is to avoid reflecting gravity waves from the top of the model. Moeng and Schumann prescribe a radiation boundary condition which allows gravity waves to propagate out of the calculation domain. Mason uses an expanding grid near the top of the calculation domain in conjunction with a damping layer. Nieuwstadt also uses a damping layer but no expanding grid. The role of the damping layer is to dissipate gravity waves before they can reflect back into the boundary layer. This is accomplished by adding to the equations of motion in the upper part of the calculation domain a relaxation term. This relaxation term then damps fluctuations at a prescribed relaxation time scale. Nieuwstadt takes the relaxation time scale equal to $50\,s$ at the highest calculation level. The relaxation time scale in his model is then increased by a factor of 5 at each computation level for the next 9 levels down.

3. Calculation

The calculations are performed in a rectangular domain. Its horizontal dimensions are $6400 \times 6400\,\mathrm{m}^2$. The vertical domain size is 2400 m for the computations of Moeng, Nieuwstadt and Schumann and 8000 m for the computation of Mason (remember that the latter author uses a non-uniform, expanding grid in the upper part of the calculation domain).

To limit the computational effort for this comparison study we took only a modest number of grid points. All authors use 40 grid points for both horizontal directions.

This implies a horizontal grid spacing of $\Delta x = \Delta y = 160\,\mathrm{m}$. In the vertical direction Nieuwstadt and Schumann use also 40 grid points so in their case $\Delta z = 60\,\mathrm{m}$. Moeng takes 48 grid points which implies $\Delta z = 50\,\mathrm{m}$. As mentioned before Mason specifies a non-uniform grid with 68 grid points. Near the surface his grid spacing is $\Delta z = 20\,\mathrm{m}$ and within the boundary layer a typical grid spacing amounts to $\Delta z = 80\,\mathrm{m}$. Near the top of the boundary layer the grid spacing is again reduced to about $\Delta z = 30\,\mathrm{m}$. The grid then expands to $\Delta z = 250\,\mathrm{m}$ towards the top of the calculation domain.

Convection is driven by a constant temperature flux at the surface of $Q_s = \langle w'T' \rangle_0 = 0.06\,\mathrm{K\,m\,s^{-1}}$. This flux is directly prescribed at the surface in the models of Mason, Moeng and Schumann. The model of Nieuwstadt requires a different procedure because the heat flux is not defined at the surface due to the w-location of the temperature variable. In his case a constant temperature increase was prescribed at the surface with a magnitude of $(dT/dt)_0 = 0.16K\,\mathrm{hour}^{-1}$. With this value of $(dT/dt)_0$ the temperature flux varied slightly during the calculation, say between $0.061\,\mathrm{K\,m\,s^{-1}}$ and $0.058\,\mathrm{K\,m\,s^{-1}}$. However, averaged over the whole calculation period the temperature flux was found to be equal to $0.06\,\mathrm{K\,m\,s^{-1}}$.

The calculations were set up in such a way that boundary-layer height would reach a value in the neighbourhood of 1600 m. Therefore, we introduce as a scaling height $z_{i0} \equiv 1600\,\mathrm{m}$. We can now define a convective velocity scale w_{*0} as: $w_{*0} = [(g/T_0)Q_s z_{i0}]^{1/3}$. The gravitational acceleration g is taken here to be: $g = 9.81\,\mathrm{m\,s^{-2}}$ and the reference temperature $T_0 = 300\,\mathrm{K}$. A temperature scale can now be defined as $T_{*0} = Q_s/w_{*0}$ and a time scale as $t_* = z_{i0}/w_{*0}$. With the values of z_{i0} and Q_s mentioned above, w_{*0} becomes equal to $1.46\,\mathrm{m\,s^{-1}}$, T_{*0} equal to 0.041 K and $t_* = 1096\,\mathrm{s}$.

Note that in most presentations, to be discussed in the following section, we shall use instead w_* and T_* to scale the results. These variables are called the convective scales defined by

$$
\begin{aligned}
w_* &= \left(\frac{g}{T_0} Q_s z_i\right)^{1/3}, \\
T_* &= \frac{Q_s}{w_*},
\end{aligned}
\tag{5}
$$

where z_i is an actual boundary-layer height. That means a boundary-layer height representative for the time period for which the results are evaluated. This actual boundary layer height has in principle a different value for each calculation.

Other parameters of interest are the constant temperature gradient Γ above the boundary layer. This was specified to be $\Gamma = 0.003\,\mathrm{K\,m^{-1}}$. Furthermore, the roughness length, which enters the surface similarity relationships, is taken to be $z_0 = 0.16\,\mathrm{m}$. The density ρ has been set to $1\,\mathrm{kg\,m^{-3}}$ because its value is irrelevant in the normalized results

As initial conditions we prescribe the following profiles for the temperature T and the three velocity components u, v, w.
For $z < z_{i1} = 0.844 z_{i0}$

$$
\begin{aligned}
T &= T_0 + 0.1r\left(1 - \frac{z}{z_{i1}}\right)T_{*0} \\
w &= 0.1r\left(1 - \frac{z}{z_{i1}}\right)w_{*0} \\
u &= v = 0
\end{aligned}
$$

and for $z > z_{i1}$

$$T = T_0 + (z - z_{i1})\Gamma$$
$$u = v = w = 0.$$

In these equations r stands for a random number uniformly distributed between -0.5 and 0.5.

In individual calculations these initial profiles are sometimes slightly modified. For instance, Nieuwstadt prescribes for the horizontal velocity components, u and v, at the lowest grid cell the value $0.01rw_{*0}$ to avoid problems with the surface-layer similarity relationships. Furthermore, he increases the temperature at the surface by 2 K with respect to the value specified by the profile above. Finally the subgrid energy is initialized in the models of Moeng, Nieuwstadt and Schumann with the profile proposed by Schmidt and Schumann (1989),

$$e = 0.1w_{*0}^2(1 - \frac{z}{z_{i0}}). \quad (6)$$

Starting from these initial profiles we ran our models with a constant forcing Q_s for a period of $10t_*$. It was expected that after this period turbulence would reach a quasi-stationary state. Next, calculations were continued for $10 < t/t_* < 11$. Within this time interval various quantities were calculated as averages over N_t timesteps. These are the quantities which we shall discuss in the following section. It also means that our results have been averaged over only one time-scale. As we shall see, this is too short to obtain stable statistics especially for the higher order moments.

Table 2. Several statistics of the computations; N_x, N_y, N_z: number of grid points in the two horizontal and the vertical directions; L_x, L_y, L_z: horizontal and vertical size of the computation domain; Δt: timestep; T_{CPU}: CPU time per time step; N_T: total number of time steps to calculate the time period $0 < t/t_* < 11$; N_t number of timesteps used in the calculation of averages.

	Mason	Moeng	Nieuwstadt	Schumann
N_x, N_y	40	40	40	40
N_z	68	48	40	40
L_x, L_y (m)	6400	6400	6400	6400
L_z (m)	8000	2400	2400	2400
Δt (s)	0.65	3	4.48	10.96
T_{CPU} (s)	23.55	1.15	0.86	1.33
Storage (M words)	0.64	0.53	1.75	2.3
Computer	IBM 3084	Cray-YMP	Cray-XMP	Cray-YMP
N_T	20000	4000	2750	1100
N_t	40	10	4	10

For further information we have summarized in table 2 various statistics of the four computations. Mason uses the smallest time step as required for stability in view of his small vertical grid spacing near the surface. Nieuwstadt's code performs most efficiently but the total computation time is smallest in Schumann's case because of the larger time step. The larger integration time per time step in Schumann's model is a consequence of the second-order upwind scheme for the temperature variable and the more complex

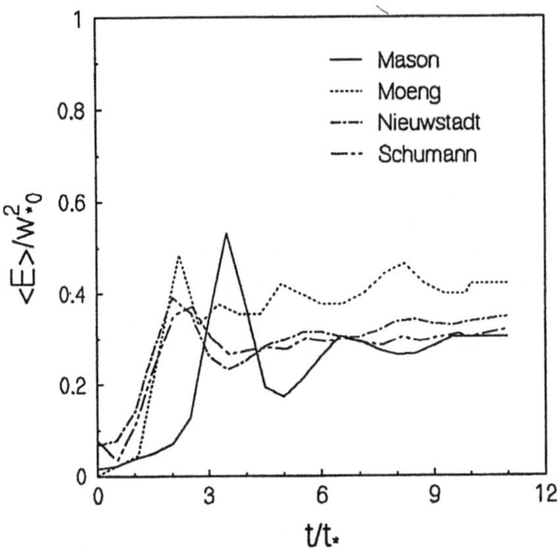

Fig. 1. The total kinetic energy vertically averaged across the boundary-layer height as a function of dimensionless time

subgrid model. His rather large storage requirements come from analysis parts which are executed during the integration. It is also interesting to note that the pseudo-spectral scheme of Moeng performs quite efficiently.

4. Results

First, we consider whether our simulations have indeed reached a state of quasi-stationary turbulence. To that end we have plotted in Fig. 1 the kinetic energy averaged across the boundary layer as a function of the dimensionless time t/t_*. Note that we have plotted the total energy $< E >$ where $E = (u'^2 + v'^2 + w'^2)/2$, i.e the resolved plus subgrid contribution $< e >$. The energy quickly grows in the initial phase of the computation. After an overshoot it settles towards a constant value, which can be interpreted as reaching a state of quasi-stationary turbulence. The overshoot is presumably caused by initial development of large convective rolls, which at later time turn into more random motions. In the case of Mason's computation the overshoot occurs at a later time and it is also larger than for the other models. We believe this to be due to the larger value of C_s in Mason's subgrid model, which leads to a rapid initial loss of energy from the random perturbations. The fact that Moeng's results are somewhat larger than the other calculations seems to agree with her data for the variances to be presented later.

Another proof of quasi-stationarity is given in Fig. 2, where we show the profile of the temperature flux averaged over the time period $10 < t/t_* < 11$ as a function of dimensionless height. Quasi-stationarity should lead to a linear temperature flux profile in the boundary layer. This is very well satisfied by all models.

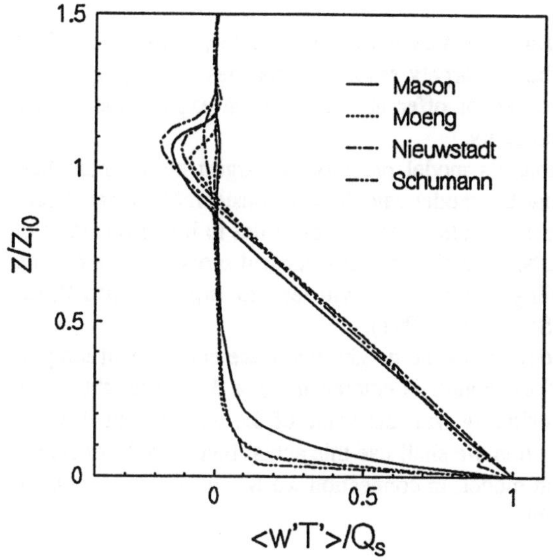

Fig. 2. The temperature flux profile as a function of dimensionless height z/z_{i0}

Tabel 3. The actual boundary-layer height z_i, the convective velocity scale w_* and the entrainment temperature flux $- < w'\theta' >_i$ representative for the time period $10 < t/t_* < 11$; the surface temperature T_s; the average boundary layer temperature above the surface layer $\overline{T}$ (both T_s and $\overline{T}$ are calculated at $t/t_* = 11$).

	Mason	Moeng	Nieuwstadt	Schumann
z_i/z_{i0}	1.0563	1.0312	1.0688	1.0900
w_*/w_{*0}	1.018	1.010	1.022	1.029
$- < w'\theta' >_i /Q_s$	0.147	0.106	0.118	0.188
T_s (K)	302.17	301.53	302.54	302.5
$\overline{T}$(K)	300.55	300.57	300.58	300.57

Another well-known property of the convective boundary layer is that the mean temperature is approximately constant over most of the boundary layer. This is corroborated by the calculated temperature profiles (not shown). As additional information on the temperature profile we give in table 3 the value of the surface temperature T_s and the average boundary-layer temperature above the surface layer $\overline{T}$ calculated at $t/t_* = 11$.

The contribution by the subgrid heat flux is small except near the surface, where subgrid effects should dominate by definition. The anomalous value of the temperature flux near the surface in Schumann's model should be disregarded because it is the result of using interpolations for statistics but upwind fluxes in the simulations.

Near the top of the turbulent boundary layer we observe a negative temperature flux. It results from the fact that the boundary layer grows into the stable layer above. This process is called entrainment. The minimum value of the temperature flux is defined as the entrainment flux, $- < w'\theta' >_i$. Its value for the four models is given in table 3.

Note that the magnitude of this entrainment flux varies considerably between the four computations. As the entrainment flux is largely resolved, this must be a property of the individual models. However, we cannot offer at this stage an explanation of this behaviour in terms of the model parameters.

Furthermore we note that Schumann's model predicts the largest entrainment heat flux. This is partly due to the fact that his model calculates a considerable subgrid contribution to the entrainment flux. The explanation for this result lies in his formulation of the subgrid mixing length ℓ. Remember that Schumann's subgrid model uses a mixing length which is not limited by stability. Therefore it will lead to larger subgrid fluxes near the top of the boundary layer (Schumann, 1991).

The boundary-layer height z_i is defined as the height where the entrainment temperature flux occurs. Its value for the four models is entered in table 3. The height of the boundary-layer seems to correlate positively with the value of the entrainment flux. As already mentioned in the previous section we shall use this actual boundary-layer height to scale the calculation results of each model. In connection we will also use w_*, defined by (5). Its value is also entered in table .

4.1 Variances

In Fig. 3 we present the variance of the vertical velocity fluctuations as a function of z/z_i. We have plotted in this figure both the total variance, i.e. resolved plus subgrid, and the subgrid contribution.

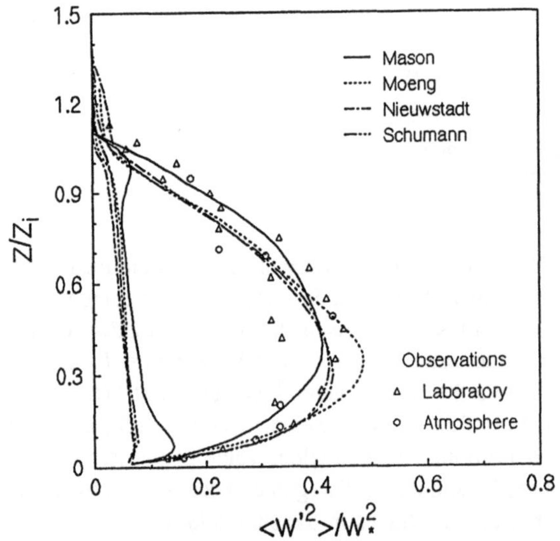

Fig. 3. The variance of vertical velocity fluctuation as a function of dimensionless height z/z_i; plotted is the total variance (resolved plus subgrid) and the subgrid contribution; laboratory data are from Willis & Deardorff (1974) and Deardorff & Willis (1985); atmospheric data from Lenschow et al. (1980).

The subgrid model of Schumann allows a direct estimate of the subgrid contribution to $< w'^2 >$. Moeng and Nieuwstadt calculate the subgrid energy e. They then assume isotropy so that for the subgrid contribution to $< w'^2 >$ results $2/3e$. Mason estimates the subgrid energy from the following diagnostic relationship

$$e = a^{-2}\ell^2 S^2(1 - Ri_f)$$

where Ri_f is the flux Richardson number equal to $Ri_f = g/(T_0 Pr)\partial T/\partial z/S^2$. The constant a is taken as 0.3. The contribution to $< w'^2 >$ is then calculated by the same method as used by Moeng and Nieuwstadt.

The agreement between the four models is clearly excellent. Only Mason's model exhibits a slightly different profile. The other models only differ in their prediction of the maximum value of $< w'^2 >$.

We have plotted in Fig. 3 also some observations obtained from laboratory experiments (Willis & Deardorff, 1974 and Deardorff & Willis, 1985) and from atmospheric field tests (Lenschow et al., 1980). Moreover, the present results agree closely with those of Schmidt & Schumann (1989) who used a much finer grid (160 × 160 × 48). It seems that the differences between the four models are smaller than the scatter in the experimental data. It is therefore not possible to conclude from this figure which model gives the best simulation. We will see that the same statement applies to other comparisons with observational data.

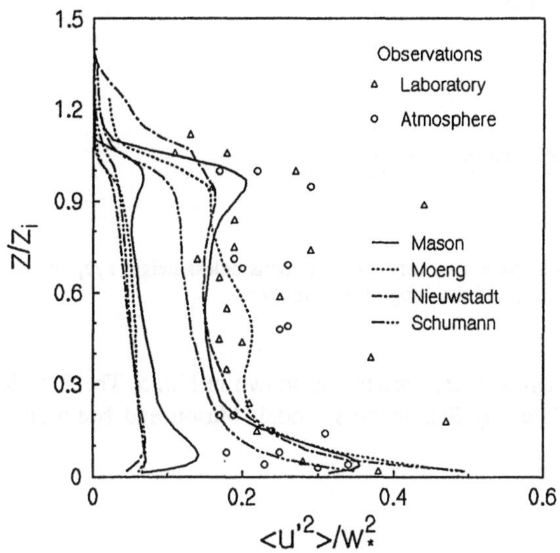

Fig. 4. The variance of horizontal velocity fluctuation as a function of dimensionless height z/z_i; plotted is the total variance (resolved plus subgrid) and the subgrid contribution.

In Fig. 4 we show a dimensionless plot of the total horizontal velocity variance and its subgrid contribution. The subgrid contributions are calculated by the same method as discussed above for Fig. 3.

In this case differences between the four models are somewhat larger but they lie still within the scatter of the observations, also shown in this figure.

Points to be noted are: Mason's model shows a strong increase towards the top of the boundary layer. His results indicate dominance of large-scale thermals or rolls which span over the whole boundary-layer depth with strong horizontal components at the surface and near the inversion. In the next section we will find that this effect is due to the subgrid model. Moeng's profile exhibits a mid-layer maximum, which we believe to be unrealistic. It explains her somewhat larger values of the boundary-layer averaged energy found in Fig. 1. Finally, the results of Nieuwstadt show a large value of the variance near and even above the boundary-layer top. This is caused by his upper boundary condition, which uses a damping layer. It seems that the damping layer fails to remove fluctuations sufficiently.

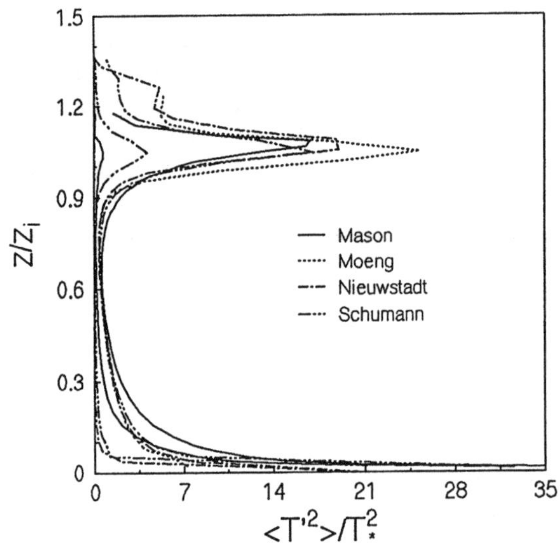

Fig. 5. The variance of temperature fluctuations as a function of dimensionless height z/z_i; plotted is the total variance (resolved plus subgrid) and the subgrid contribution.

A dimensionless profile of the temperature variance is shown in Fig. 5. The subgrid contribution can be directly calculated by Schumann's model. Mason and Nieuwstadt estimate this contribution from

$$< T'^2 > = a_T^{-4} \frac{H^2}{e}$$

where H is the subgrid temperature flux. The value of the constant a_T is taken as 0.67.

The largest disagreement between the four models is found near the top of the boundary layer. This can be partly explained by differences in the temperature gradient. This gradient enters in the production term of the temperature variance and therefore determines the magnitude of $< T'^2 >$. It turns out that Moeng's model, which has the largest variance, has indeed also the largest temperature gradient near the top of the boundary-layer. However, we should also mention here that she uses a somewhat different procedure to calculate her statistics compared with the other participants. Moeng scales her results at each calculation step with scaling variables appropriate for that time step. After that she performs the average over the N_t time steps. The other authors average first their calculation results over the N_t time steps and then scale them with z_i and w_* representative for the whole averaging period. It is expected that the procedure of Moeng leads to somewhat higher variances.

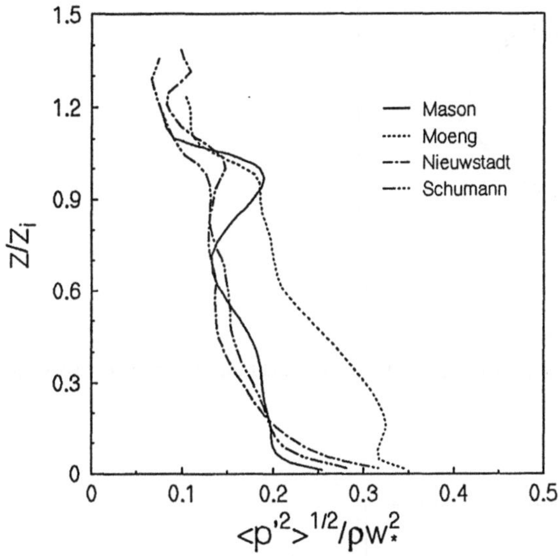

Fig. 6. The variance of pressure fluctuations as a function of dimensionless height z/z_i.

Finally, we give in Fig. 6 a dimensionless profile of the pressure variance. Except the data of Moeng all other models agree very well. An explanation for the deviating behaviour of Moeng's model cannot be offered.

Note that the pressure variance of Mason's model increases strongly towards the top of the boundary layer. The same behaviour was already found in Fig. 4 and we shall come back to it in the next section.

4.2 Higher Order Moments

In Fig. 7 we show the dimensionless profile of the third moment of the vertical velocity fluctuations. This variable is strongly influenced by the thermals in the convective boundary layer. A boundary layer consisting of isolated thermals with inside these a strong vertical upward motion, necessarily leads to a positive third moment as found in Fig. 7.

The agreement between the four models is quite acceptable in view of the observations which are also shown in the same figure.

The models of Moeng, Nieuwstadt and Schumann lead to a negative value of $< w'^3 >$ near the surface. This is considered unrealistic because it would imply that downdrafts become stronger than updrafts when they approach the surface. Mason (1989) and Schmidt & Schumann (1989) have investigated the background of negative values of $< w^3 >$ in

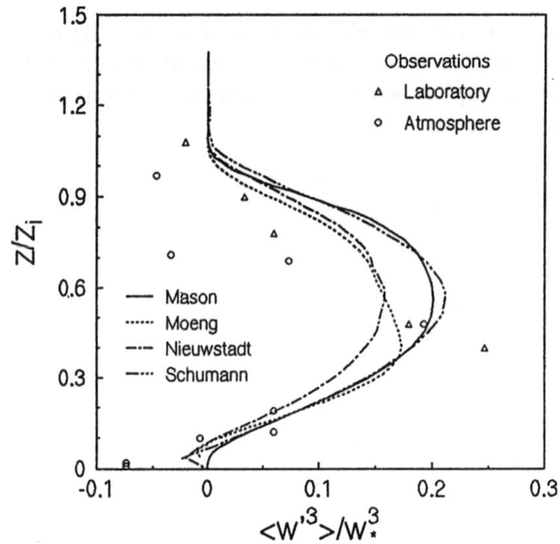

Fig. 7. The third moment of the vertical velocity fluctuations as a function of dimensionless height z/z_i.

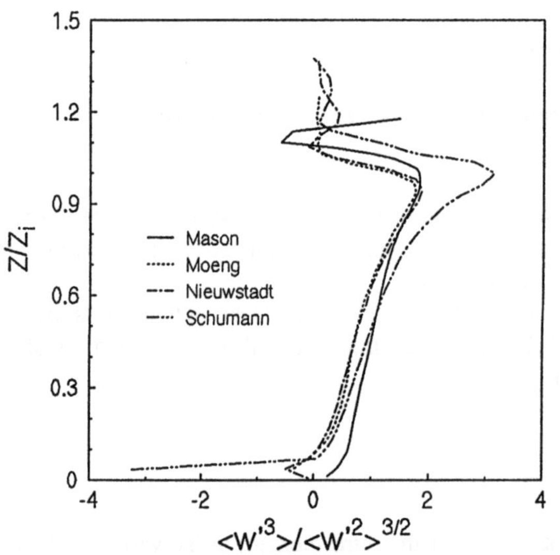

Fig. 8. The skewness *Sk* of the vertical velocity fluctuations as a function of dimensionless height z/z_i.

more detail. They conclude that it is a consequence of the subgrid model. Our calculations confirm this as we shall see in the next section.

The third moment can be combined with the variance to form a dimensionless parameter called skewness: $Sk = < w'^3 > / < w'^2 >^{3/2}$. Note that in this case both $< w'^2 >$ and $< w'^3 >$ are calculated from resolved motions only. The profile of *Sk* is shown in

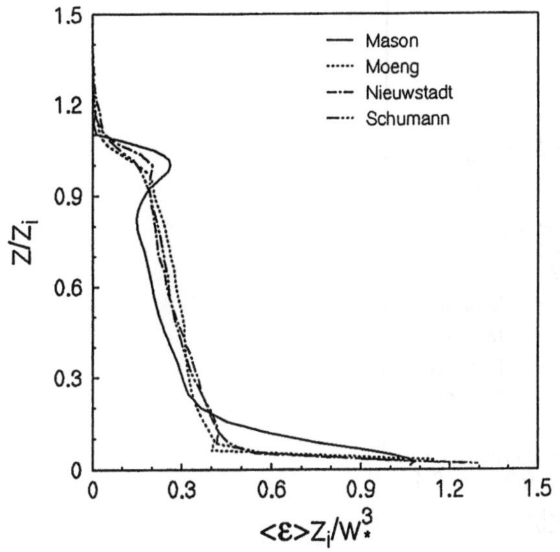

Fig. 9. The dissipation of turbulent kinetic energy as a function of dimensionless height z/z_i.

Fig. 8. The results of all models are again quite close. Schumann's data show a rather high value of Sk near the top of the boundary layer. This might be attributed to the fact that according to Fig. 7 his value of $< w'^3 >$ is somewhat larger in this region compared with the other models. Nevertheless Fig. 7 also shows that this difference is rather small but its effect on Sk is magnified by the small values of $< w'^2 >$ near the top of the boundary layer.

For a more detailed discussion of skewness in the convective boundary layer we refer to Moeng and Rotunno (1990). The comparison of the simulation results for Sk with observations is discussed by Lemone (1990).

4.3 Energy budget

The kinetic energy budget of turbulence in a quasi-stationary convective boundary layer reads

$$0 = \frac{g}{T_0} < w'\theta' > - \frac{\partial}{\partial z}(< w'E' > + \frac{1}{\rho} < p'w' >) - \epsilon \tag{7}$$

where we have omitted the shear production term, which is zero in this case because $< u >= 0$.

The first term in (7) denotes the production of kinetic energy by buoyancy forces. It is proportional to the temperature flux which we already have shown in Fig. 2. This production is balanced by the viscous dissipation ϵ which is computed from the subgrid model. Its vertical profile is given in Fig. 9. The agreement between the four models is excellent. Mason's calculation shows again a maximum towards the top of the boundary layer, the background of which will be disclosed in the next section. Moreover, we see

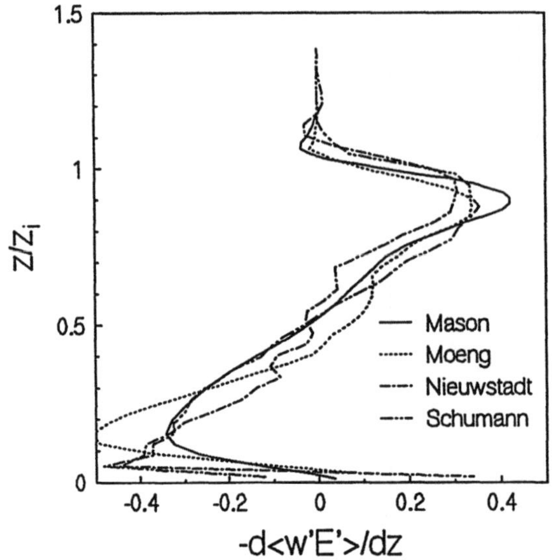

Fig. 10. The divergence of the vertical transport of kinetic energy (with as definition $E' = (u'^2 + v'^2 + w'^2)/2$) by velocity fluctuations, normalized by w_*^3/z_i, as a function of dimensionless height z/z_i.

that near the surface the ϵ calculated by the models of Moeng, Nieuwstadt and Schumann increase very rapidly. The model of Mason exhibits a more gradual profile. Schmidt & Schumann (1989) have pointed out that such rapid increase of ϵ usually goes together with a negative third moment near the surface as a dissipation excess requires downward transport of kinetic energy. This is indeed confirmed by our calculations and we remind the reader to the negative values of $< w^3 >$ found in Fig. 7.

As the vertical profiles of the buoyancy production and dissipation are different we need the transport terms to satisfy the budget. The transport of kinetic energy by vertical velocity fluctuations, $< w'E' >$, turns out to be the most important. Its profile is shown in Fig. 10. Despite some scatter, probably due to the inadequacy of averaging over only one time scale, the agreement between the four models seems again quite acceptable. However, this can no longer come as a surprise because we have already found good agreement for the production and the dissipation terms. The energy budget then forces agreement of the transport terms.

The profile of the pressure transport term $< p'w' >$ is shown in Fig. 11. As suggested above its magnitude is indeed small. In view of this fact it is difficult to draw any conclusions from the rather large differences between the four models which we observe in Fig. 11. Moreover we note again that some of the scatter in Fig. 11 is probably again caused by the insufficient averaging period.

4.4 Conditional averages

In this subsection we discuss the properties of thermals by performing a conditional averaging procedure. It means that we calculate the average of a certain quantity only

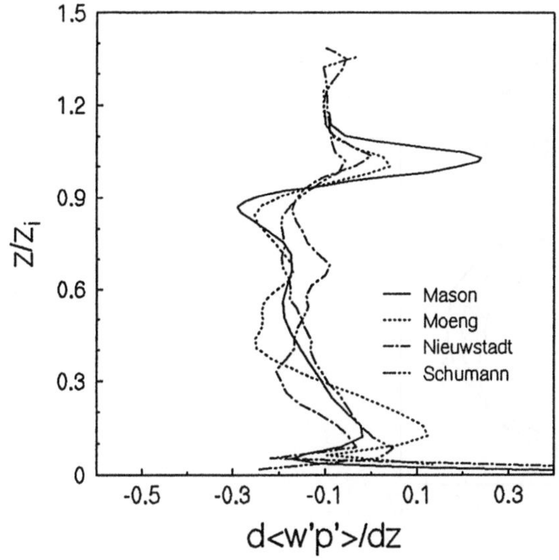

Fig. 11. The divergence of the vertical transport of kinetic energy by pressure fluctuations, normalized by $\rho w_*^3/z_i$, as a function of dimensionless height z/z_i.

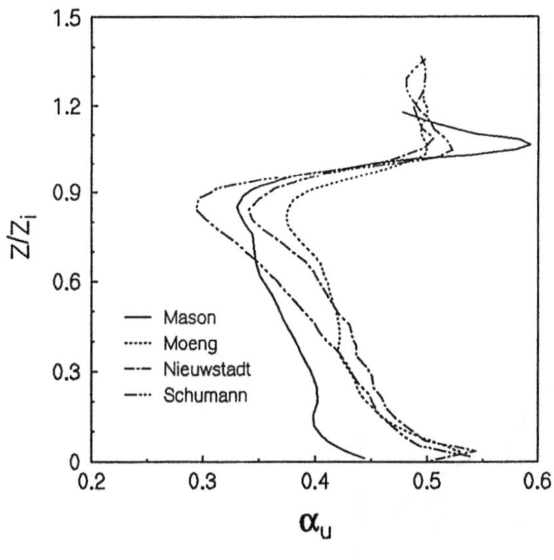

Fig. 12. The area fraction occupied by updrafts as a function of dimensionless height z/z_i.

over the area which is occupied by thermals, i.e. where $w > 0$. Let us first consider the area fraction α_u occupied by the thermals. It is shown in Fig. 12. It is interesting to note that this figure seems to be the mirror image of the skewness shown in Fig. 8.

In Figs. 13 and 14 we show the profiles of vertical velocity w_u and the mean temperature excess $T_u - < T >$ in thermal areas characterized by $w > 0$. The agreement between the four models is surprisingly good.

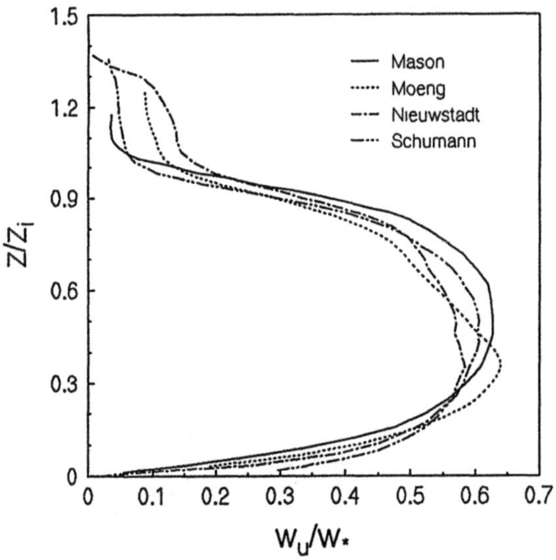

Fig. 13. The conditionally averaged vertical velocity in areas with $w > 0$ as a function of dimensionless height z/z_i.

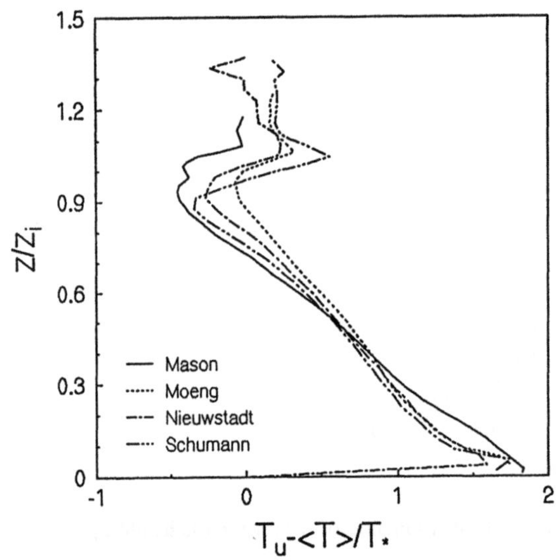

Fig. 14. The conditionally averaged temperature excess in areas with $w > 0$ as a function of dimensionless height z/z_i.

A point to note is the large value of w_u in Nieuwstadt's results above the boundary layer. This is again the consequence of his upper boundary condition.

For a more extensive discussion of conditional averages in the convective boundary layer we refer to Schumann and Moeng (1991a).

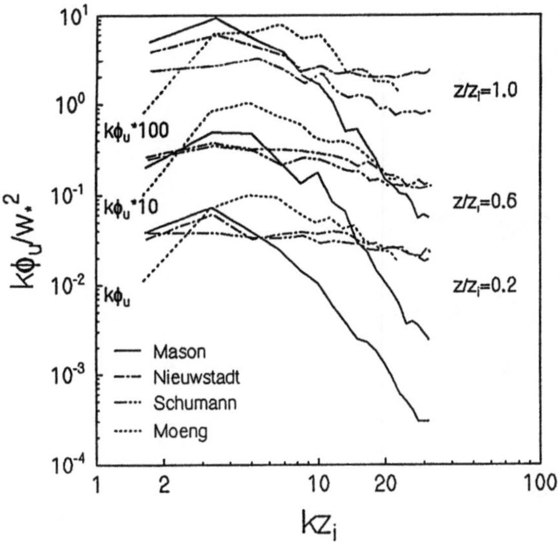

Fig. 15. The spectra of the resolved horizontal velocity fluctuations at three levels in the boundary layer.

4.5 Spectra

In Figs. 15 and 16 we show the spectra of the resolved horizontal and vertical velocity fluctuations at three levels in the boundary layer. The spectra of the resolved temperature fluctuations are given in Fig. 17. Remember that Moeng uses a spectral cut-off to avoid aliasing in her pseudo-spectral method. Therefore, her spectra stop beyond wave number $kz_{i0} = 22$ whereas the other spectra continue until $kz_{i0} = 29.84$ (in the figures the spectral values at the last wave number $kz_{i0} = 31.41$ have been omitted).

One fact which becomes immediately clear when we look at these figures is that the spectra of Mason have a much steeper slope at high wave numbers than the spectra calculated by the other models. This must clearly be an effect of the subgrid model and we shall come back to it in the next section.

However, we point to the fact that in all three figures the spectra of all models are again very close at low wave numbers, i.e $kz_i \leq 10$. This should not come as a total surprise because it is consistent with the good agreement for the variances between all four models. Nevertheless, it supports the fundamental premise of large-eddy modelling, which we have stressed in our introduction: the large eddies do not depend strongly on the details of the subgrid model.

Finally, we should note that none of the spectra seems to follow the well-known $-2/3$ slope of the inertial range. However, this may be due to the rather poor resolution of our calculations, which is only 40^2 in the horizontal. There is some evidence (see e.g. Moeng and Wyngaard, 1988 and Schmidt and Schumann, 1989) that the spectra become more realistic at higher resolution.

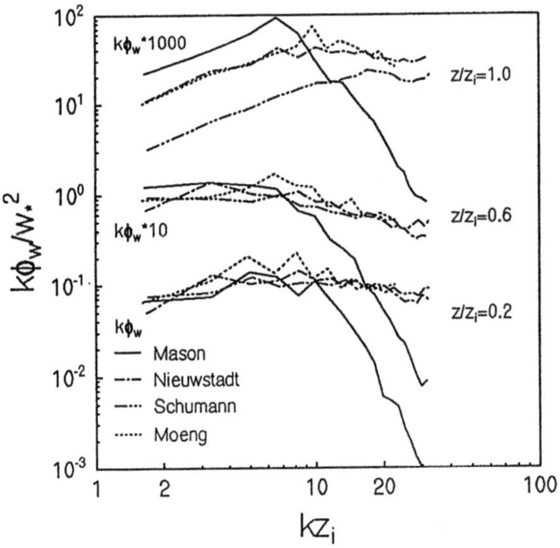

Fig. 16. The spectra of the resolved vertical velocity fluctuations at three levels in the boundary layer.

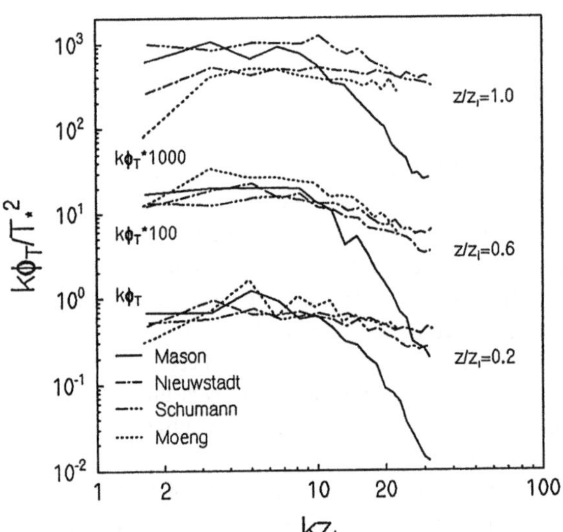

Fig. 17. The spectra of the resolved temperature fluctuations at three levels in the boundary layer.

5. Discussion

A main conclusion which we can draw from the results shown in the previous section is that the four models seem to fall into two categories with Mason in the one and Moeng,

Nieuwstadt and Schumann in the other. (However, in the latter group the models of Nieuwstadt and Schumann seem to agree slightly better). The most obvious support for this distinction into two groups is given by the results for the spectra shown in Figs. 15-17. In addition, Mason's model leads also frequently to a different shape of the vertical profile in comparison with results of the other models. Examples are Figs. 3, 4, 6 and 9.

The main difference in Mason's model with respect to the others is the values of the constant C_s. We have seen in table that for Mason's subgrid model $C_s = 0.32$, whereas the value of the other models vary around 0.18. According to the definition of C_s given in (2) this means that the mixing or filter length in Mason's model is larger with respect to the grid spacing. In other words, much of the variance at higher wave numbers is filtered out in comparison with the other models with smaller values of C_s. This immediately explains the behaviour of the spectra in Figs. 15-17.

To test this effect of C_s, Mason has rerun his model with a value of $C_s = 0.23$. At the same time he changed to an equidistant grid within the boundary layer with a representative vertical grid spacing of ~ 45 m. However, we expected that the effect of the equidistant versus non-uniform grid will be minimal. Therefore, the new results will primarily reflect the influence of C_s in Mason's model.

Let us consider a few of the results obtained from this additional calculation. They are plotted in the same format as used in the previous section. This means that in the following figures the results of Moeng, Nieuwstadt and Schumann have remained exactly the same.

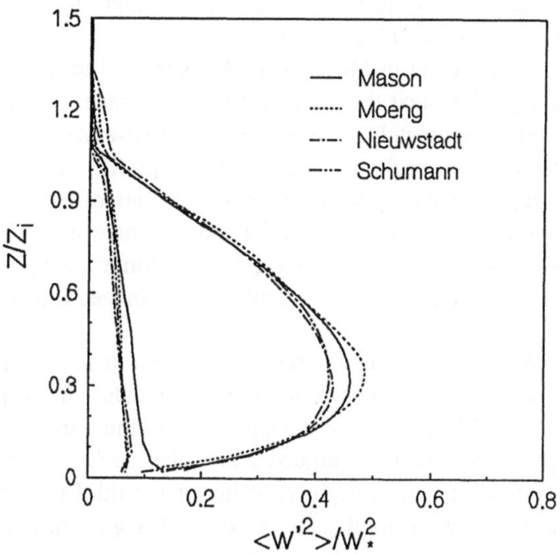

Fig. 18. The variance of vertical velocity fluctuation as a function of dimensionless height z/z_i; plotted is the total variance (resolved plus subgrid) and the subgrid contribution. In comparison with Fig. 3 we have rerun Mason's model with $C_S = 0.23$ instead of $C_S = 0.32$. The results of Moeng, Nieuwstadt and Schumann have not been changed.

First we show the profile of the vertical variance in Fig. 18. If we compare this with Fig. 3 it is clear that the shape of Mason's profile has become much closer to the result

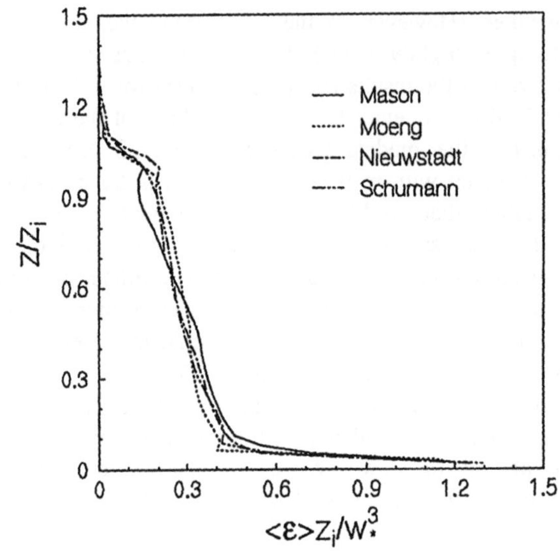

Fig. 19. The dissipation of turbulent kinetic energy as a function of dimensionless height z/z_i. In comparison with Fig. 9 we have rerun Mason's model with $C_s = 0.23$ instead of $C_s = 0.32$. The results of Moeng, Nieuwstadt and Schumann have not been changed.

of the other models. As a matter of fact the models are now almost indistinguishable and differ only in their prediction of the maximum value of $< w'^2 >$.

Another effect which we consistently found in Mason's results presented in the previous section was a maximum in the profiles near the top of the boundary layer (see e.g. Figs. 4, 6, and 9). This phenomenon has completely disappeared for the new calculation. As an example we show in Fig. 19 the new results for the dissipation profile which we should compare with Fig. 9. The fact that the upper level maximum has disappeared may suggest that the large scale thermals in the new calculation have become somewhat weaker resulting in stronger small-scale motions. This suggestion is borne out by the results of the third moment $< w'^3 >$, which for the new calculation are indeed found to be smaller.

Comparison of the Figs. 9 and 19 shows also that the behaviour of ϵ near the surface for the new results of Mason is now much closer to that of the other models. The rather gradual increase in ϵ, which we observed in Fig. 9, has disappeared and the ϵ increases now fast when the surface is approached. We have argued above that such a sudden increase of ϵ near the surface is consistent with a negative value of the third moment $< w'^3 >$. Indeed, we find for Mason's new results that the $< w'^3 >$ becomes negative at the lowest grid point.

Finally, we show in Fig. 20 the new spectra of the vertical velocity fluctuations. The agreement between the four models has much improved in comparison with Fig. 16, especially at the higher wave numbers. This result clearly confirms the effect of C_s. However, the slope of Mason's spectra in Fig. 20 is still somewhat steeper than that of the other models. This is consistent with the fact that $C_s = 0.23$ is still slightly larger than the representative C_s values of the other models.

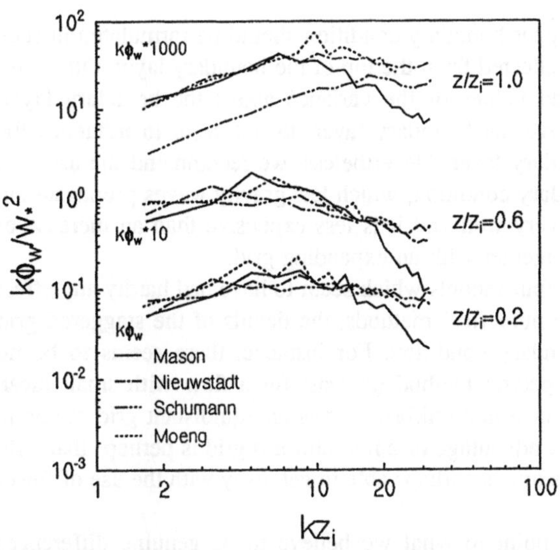

Fig. 20. The spectra of the resolved vertical velocity fluctuations at three levels in the boundary layer. In comparison with Fig. 16 we have rerun Mason's model with $C_S = 0.23$ instead of $C_S = 0.32$. The results of Moeng, Nieuwstadt and Schumann have not been changed.

6. Conclusion

Our comparison of four large-eddy codes has showed clearly that large-eddy simulation leads to a generally consistent picture of convective turbulence despite differences in many details of the models. This fact should strengthen our confidence in the realism of the simulated turbulence.

However, at this stage such statement can only be made for large-eddy simulation of convective turbulence. In other words, convective turbulence seems to be a rather easy case for large-eddy simulations. A relatively coarse resolution as used in this calculation is able to represent much of the salient characteristics of convective turbulence. In a separate calculation, Schumann (1991) found that an even coarser resolution, such as only 15 grid points distributed over the vertical domain of 2400 m, leads to about the same results. Shear turbulence seems to be more challenging test case for large-eddy modelling. Therefore, we are contemplating to perform a similar comparison study for the neutral boundary layer.

Based on this study we may draw several conclusions about details of our large-eddy models. The most important one concerns the subgrid model and in particular the value of the parameter C_s, which can be interpreted as the ratio between the mixing or filter length and the grid spacing. A large value of C_s leads to reduced values of the variance at high wave numbers in accordance with the notion of a larger filter length. Moreover, a large value of C_s results in somewhat stronger thermals at larger scales which influence the vertical profiles of some variables. Nevertheless, the effect of C_s on the large scales seems to be relatively minor, which supports the fundamental hypothesis of large-eddy modelling: the large eddies are little sensitive to the details of the subgrid model.

We have also found that the upper boundary condition should be formulated in such a way that gravity waves are not reflected from the top of the boundary layer. Otherwise we should expect to see increased values of the variance above the boundary layer. However, increased variances above the boundary layer do not seem to influence the simulation results inside the boundary layer. Nevertheless, we recommend the use of a radiation condition as upper boundary condition, which lets gravity waves propagate out of the domain. It seems to work very well and it is less expensive than an increase of the vertical computation domain together with an expanding grid.

Other differences between our four models which seem to have had hardly any effect on the calculation results, are: the numerical methods, the details of the staggered grid and the details of the lower boundary condition. For instance, there seems to be no advantage in applying a pseudo-spectral method (at least for a flow with small mean horizontal velocity). Also the use of a non-uniform versus an equidistant grid seems to have negligible influence. The only advantage of a non-uniform grid is perhaps that with a careful distribution of grid points one is perhaps able to get away with the use of fewer grid points.

Nevertheless, we should also point to what we believe to be genuine differences between the four models. These are differences, which can be attributed to the resolved scales. Examples are the entrainment temperature flux, the horizontal velocity fluctuations and perhaps also differences in the predicted maximum values of $< w'^2 >$ and $< w'^3 >$. The explanation for these admittedly rather small differences is not clear.

Acknowledgement

U. Schumann is grateful do Dr. Jutta Graf, who cooperated with him in this project. F.T.M. Nieuwstadt acknowledges discussions with J. Eggels and H. Cuypers and support from NWO-MFO project 752-365-022.

References

Deardorff, J.W. 1974 Three dimensional numerical study of turbulence in an entraining mixed layer. Boundary-Layer Met. **7**, 199-226

Deardorff, J.W. & Willis, G.E. 1985 Further results from a laboratory model of the convective planetary boundary layer. Boundary-Layer Met. **32**, 205-236

Lemone, M. 1990 Some observations of vertical velocity skewness in the convective planetary boundary layer. J. Atmos. Sci. **47**, 1163-1169

Lenschow, D.H., Wyngaard, J.C. & Pennel, W.T. 1980 Mean-field and second-moment budgets in a baroclinic, convective boundary layer. J. Atmos. Sci. **37**, 1313-1326

Mason, P.J. 1989 Large-eddy simulation of the convective atmospheric boundary layer. J. Atmos. Sci. **46**,

Mason, P.J. & Callen N.S., 1986 On the magnitude of the subgrid-scale eddy coefficient in large eddy simulation of turbulent channel flow. J. Fluid Mech. **162**, 439-462

Moeng, C.-H. 1984 A large-eddy simulation for the study of planetary boundary layer turbulence. J. Atmos. Sci. **41**, 2052-2062

Moeng, C.-H. & Rotunno, R. 1990 Vertical velocity skewness in the buoyancy driven boundary layer. J. Atmos. Sci. **47**, 1149-1162

Moeng, C.-H. & Wyngaard, J.C. 1988 Spectral analysis of large-eddy simulations of the convective boundary layer. J. Atmos. Sci. **45**, 3574-3587.

Nieuwstadt, F.T.M. 1990 Direct and large-eddy simulation of free convection. Proc. 9th Internat. Heat Transfer Conference, Jerusalem 19-24 August 1990, Amer. Soc. Mech. Engrg., New York, Vol I, 37-47

Nieuwstadt, F.T.M. & Brost, R.A. 1986 Decay of convective turbulence. J. Atmos. Sci. **43**, 532-546

Piascek S.A. & Williams, G.P. 1970 Conservation properties of convection difference schemes. J. Comput. Phys., **6**, 392-405

Schmidt, H. & Schumann, U. 1989 Coherent structures of the convective boundary layer derived from large-eddy simulations. J. Fluid Mech. **200**, 511-562

Schumann, U. 1991 Subgrid length scales for large-eddy simulation of stratified turbulence. Theor. Comput. Fluid Dynamics, **2**, 279-290

Schumann, U. & Moeng C.-H. 1991 Plume fluxes in clear and cloudy convective boundary layers. J. Atmos. Sci., **48**, 1746-1757

Willis, G.E. & Deardorff, J.W. 1974 A laboratory model of the unstable planetary boundary layer. J. Atmos. Sci. **31**, 1297-1307

Effects of Solid Body Rotation on the Transport of Turbulence

L. Shao, M. Michard and J.P. Bertoglio

Laboratoire de Mécanique des Fluides et d'Acoustique,URA CNRS 263, Ecole Centrale de Lyon, 36 av. Guy de Collongue, 69130 Ecully, France

Abstract

The mechanism of turbulence transport is studied using a Large Eddy Simulation of a shearless turbulence mixing layer, subjected to solid body rotation. The behaviour of both transport by triple velocity correlations and transport by pressure-velocity correlations is found to be completely different from what was found without rotation. A modified one-point model, that accounts for the effects of rotation, is proposed for the transport term by triple correlations.

1. Introduction

In a previous study, the mechanisms responsible for transport of turbulent kinetic energy and transport of the components of the Reynolds stress tensor were investigated in a simple case of inhomogeneous turbulence: the case of a shearless turbulence mixing layer, in which two nearly homogeneous fields, having different energy and length scales, were put side by side and were interacting (see the sketches in figures 1 and 2). The study was made using a Large Eddy Simulation (Shao, Bertoglio and Michard, 1990). Results were found to be in qualitative agreement with the experimental data of Veeravalli and Warhaft (1989). The main conclusion of the study was that triple velocity correlations were found to be responsible for the energy transport and that pressure-velocity correlations were found to tend to counterbalance the triple correlation transport, their effects remaining small.

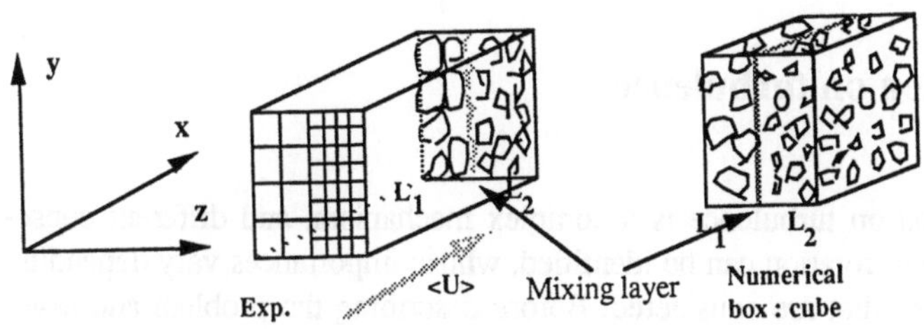

Fig. 1, 2.

Comparisons with results of the classical one-point closure models, like $k-\varepsilon$ and Reynolds stress models, were also presented in another paper (Shao, Le Penven and Bertoglio, 1990)

Concerning triple velocity correlations, the Launder, Reece and Rodi (1975) second order model:

$$<u_i u_j u_k> \;=\; -C_s \frac{k}{\varepsilon}\left(<u_k u_m> \frac{\partial <u_i u_j>}{\partial x_m} + <u_j u_m> \frac{\partial <u_i u_k>}{\partial x_m} + <u_i u_m> \frac{\partial <u_j u_k>}{\partial x_m} \right) \tag{1}$$

in which $k = 1/2<q^2> = 1/2<u_i u_i>$ is the turbulent kinetic energy and ε its dissipation rate, was found to be qualitatively suitable. However, the value $C_s = 0.11$ usually retained in the expression for the triple correlation, leads to underestimated values of $<u_i u_j u_k>$, in particular on the small scale side of the mixing layer when strong effects of intermittency are present. A correction to account for intermittency effects was proposed (Shao and Bertoglio, 1991).

Similarly, in the case of the $k-\varepsilon$ model, the expression

$$\frac{1}{2}<u_i u_i u_k> = -\frac{C_\mu}{\sigma_k}\frac{k^2}{\varepsilon}\frac{\partial k}{\partial x_k} \tag{2}$$

was also found to give qualitatively suitable results but to underestimate triple correlations on the small scale side when the classical value $\sigma_k = 1$ was adopted.

A possible explanation for the underestimation of the triple correlation levels by classical models is the following. The shearless mixing layer can be suspected to be the case in which transport mechanisms, acting alone without interacting with other effects, are the most efficient. One can imagine that when other effects, like shear, strain or rotation, are present the build up of triple correlations can be strongly inhibited. The present study is aimed at investigating how transport mechanisms are affected when rotation is present. The effect of solid body rotation is considered. The reason for the choice of rotation as the first external effect to be studied is not only that it is the simplest effect that can be introduced in a Direct Simulation, but also that it has important applications for inhomogeneous turbulent fields of practical interest, like swirling flows, and flows in turbomachinery.

2. Effects of rotation on turbulence

The influence of rotation on turbulence is a complex mechanism, and different consequences of the presence of rotation can be identified, whose importances vary depending on the different situations that are considered. Before describing the problem addressed here, one can briefly discuss the various effects of rotation, and summarize the various modifications that are usually introduced in simple models to account for these effects.

The effect of solid body rotation on homogeneous grid-generated turbulence is now well identified. If turbulence is initially isotropic, the action of rotation typically takes place at the level of the third-order velocity correlations in wave-space. Due to the fact that Fourier modes are rotating about the directions of the wave-vectors, a scrambling effect occurs in the triadic interaction between modes, resulting in an inhibition of the build up of triple correlations. Energy transfer from the large scales to the dissipative eddies is reduced. In the framework of one-point models, heuristic modifications of the e equation were proposed to account for this cascade inhibition mechanism (Bardina, Ferziger and Reynolds, 1983, Aupoix, Cousteix and Liandrat, 1983). It has to be pointed out that, in this case, another consequence of the presence of rotation can be detected: rotation is also creating anisotropy. The obtained anisotropy is essentially a length scale anisotropy, and therefore the phenomenon is difficult to take into account in existing one-point models. Only models introducing information on the anisotropy of length scales or on energy distribution in Fourier space could permit a description of this effect. Efforts are presently being made along these lines (Cambon, Jacquin and Lubrano,1991).

In the case of initially anisotropic turbulence subjected to solid body rotation, the anisotropy of the Reynolds stress tensor is directly affected by the presence of rotation. The principal axes of the Reynolds stress tensor are rotating under the effect of the Coriolis term. Furthermore there is a more subtle effect due to the rapid part of the pressure strain correlation. This last effect is related to mechanisms that are easy to describe in wave-space, but difficult to model at the level of one-point models. Attempts to take these effects into account are currently being made (Lubrano and Cambon, 1991).

In the case of the simultaneous action of rotation and mean velocity gradients, the most important effect is a coupled effect between production by mean gradients and re-orientation of the Reynolds stress tensor by rotation. This occurs, for example, when turbulence is subjected to a mean shear in a rotating frame. This case is particularly important because of applications to the behaviour of flows inside the blade to blade channels of centrifugal impellers. Either stabilizing or destabilizing effects can occur when rotation is in the plane of the shear, depending on the sign of rotation compared to the shear. These effects are taken into account in Reynolds stress models at the level of the production and the Coriolis terms. The major problem is then whether models for the pressure strain correlation properly account for rotation. Even if existing models are not suitable for every value of the rotation rate compared with the shear rate, they lead to satisfactory predictions, at least qualitatively (see Bertoglio 1982 and Speziale, Gatski and Mhuiris, 1990). When other types of mean gradients are considered, more complex effects can be detected (plane strain: "elliptic flows", or axial compression, see Cambon et al. 1985 or Malkus 1990). It is beyond the scope of the present paper to discuss these effects that are difficult to include in simple models.

The effects of rotation summarized in the previous paragraphs are effects that can be described in the framework of homogeneous turbulence. Even if simple models are not yet describing these effects completely, it is clear that the mechanisms are, at least, well identified. As stated above, efforts are currently being made to devise improved models. The purpose of the present study in to investigate a different consequence of the action of rotation on turbulence, one which has not yet been studied. We address the problem of an effect typically taking place in inhomogeneous turbulence: the influence of rotation on the transport terms of turbulent kinetic energy and of Reynolds stress components. The physical mechanism is, in this case, essentially the inhibition of the build up of triple velocity correlations at one-point. It is a mechanism taking place in physical space, and not in Fourier space as is the case when the influence of rotation on homogeneous turbulence is considered.

3. Study of the influence of rotation on turbulence transport terms

The study is aimed at understanding the influence of rotation on the inhomogeneous transport terms at the level of the equation for the double velocity correlation. The transport terms are the triple correlation term: $\partial/\partial x_k \langle u_i u_j u_k \rangle$, and the pressure term: $\partial/\partial x_k \langle p/\rho(\delta_{jk} u_i + \delta_{ik} u_j) \rangle$. In order to investigate the consequences of rotation in the case of a very simple inhomogeneous field, a Large Eddy Simulation of a shearless turbulence mixing layer is performed in a rotating frame. The numerical technique is similar to the one described in Shao, Bertoglio and Michard (1990). Simulations are run on a cubic box with periodic boundary conditions. The initial field consists of two different homogeneous velocity fields put side by side. z denotes the inhomogeneity direction, and Y is the axis of rotation (Ω parallel to oy). Subgrid scales are parameterized by using the model proposed by Chollet and Lesieur (1981). Due to the fact that periodicity of the boundary conditions is assumed, two mixing layers are simulated instead of one: one in the middle of the numerical box, the other one at the edges. No mean flow is present in the computation (see the sketch in figure 3). Computations are run on regular grids with 64^3 nodes, using a pseudo-spectral numerical scheme.

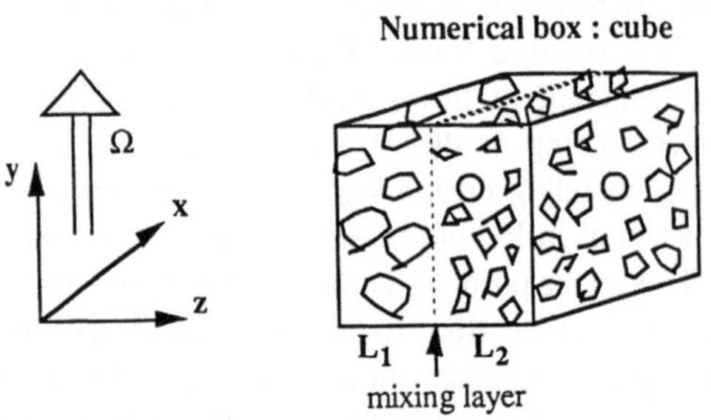

Fig. 3.

At the beginning of the computations, on the left side of the box, a turbulent field with strong intensity and large length scales is introduced. On the right side, a turbulent field with lower intensity and smaller length scales is prescribed. The ratio of turbulent kinetic energies between the two sides is nearly equal to two, and the ratio of length scales is approximatively two. This case corresponds to test case 64-02 in the steady frame study proposed in Shao, Bertoglio and Michard (1990). Results are averaged in planes of homogeneity, that is to say in (x, y) planes. They are also averaged over different realizations in order to improve statistical sampling.

The equations for the double velocity correlations at one point, or Reynolds stress tensor, can be easily written for the present flow. It is convenient to write them in the rotating frame. Then, the important terms clearly appear in the equations for the diagonal components. Taking into account the symmetries of the problem and homogeneity in the

x and y directions, one obtains:

$$\begin{aligned}\frac{D<u^2>}{Dt} &= -\frac{\partial<wu^2>}{\partial z} - 0 + \frac{2}{\rho}<p\frac{\partial u}{\partial x}> + \text{viscous term} + \text{Coriolis term}\\ \frac{D<v^2>}{Dt} &= -\frac{\partial<wv^2>}{\partial z} - 0 + \frac{2}{\rho}<p\frac{\partial v}{\partial y}> + \text{viscous term} \qquad (3)\\ \frac{D<w^2>}{Dt} &= -\frac{\partial<w^3>}{\partial z} - \frac{2}{\rho}\frac{\partial<pw>}{\partial z}\\ &\quad + \frac{2}{\rho}<p\frac{\partial w}{\partial z}> + \text{viscous term} + \text{Coriolis term}\end{aligned}$$

The first two terms on the Right Hand Sides are respectively, transport terms by triple velocity correlations and by pressure-velocity correlations. The third terms are for pressure redistribution, whose effect is the interchange of energy between components of the Reynolds stress tensor, and which do not affect energy transport. The viscous terms essentially account for molecular dissipation. Viscous effects also contribute to turbulence transport, but their contribution is not discussed here since it vanishes for high Reynolds number flows. The last terms are Coriolis terms, whose trace is zero, since Coriolis forces are not producing energy.

The equation for twice the turbulent kinetic energy is:

$$\frac{D<q^2>}{Dt} = -\frac{\partial<wq^2>}{\partial z} - \frac{2}{\rho}\frac{\partial<pw>}{\partial z} + \text{viscousterm} \qquad (4)$$

in which $<wq^2> = <w^3> + <wv^2> + <wu^2>$.

The important terms for the transport of turbulence are therefore $<w^3>$, $<wu^2>$, $<wv^2>$ and $<pw>$.

The results of the L.E.S. clearly show that rotation does reduce the triple correlation term $<u_i u_j u_k>$, as shown in figure 4b where the component responsible for the transport of $<w^2>$ in the z direction, $<w^3>$, is plotted as a function of z. In the present computations, turbulence first evolves without being subjected to rotation between $t = 0$ and $t = 0.04$ s. Then rotation is suddenly switched on at $t = 0.04$ s. The value of Ω is 50 rd/s, corresponding to an initial value of the Rossby number approximatively equal to 0.7 on the high turbulence side of the mixing layer. In figure 4b the normalized quantity, $<w^3>/<w^2>^{3/2}$, is found to first increase up to a value of about 0.4, reached at the end of the non-rotating phase ($t = 0.04$ s). For larger values of time, this quantity decreases drastically: at time $t = 0.08$ s, it is nearly equal to the statistical noise (in this case, results are averaged on 3 realizations only). In the same case without rotation, larger values of $<w^3>/<w^2>^{3/2}$ were obtained (case 64-02 in Shao, Bertoglio and Michard, 1990) as shown in figure 4a where this quantity continues to increase between $t = 0.04$ s and $t = 0.08$ s.

In figure 5, the evolution with time of the turbulent kinetic energy profiles is plotted. The cases with (figure 5b), and without (figure 5a) rotation are compared. It immediately appears that higher levels of turbulent kinetic energy are obtained at the end of the simulation on both sides of the box when rotation is present. This is due to the inhibition of the energy transfer between eddies of various sizes, an effect which appears to be consistent with what is found in homogeneous turbulence. A more interesting remark can be made concerning energy distribution along z: despite the fact that triple correlations are

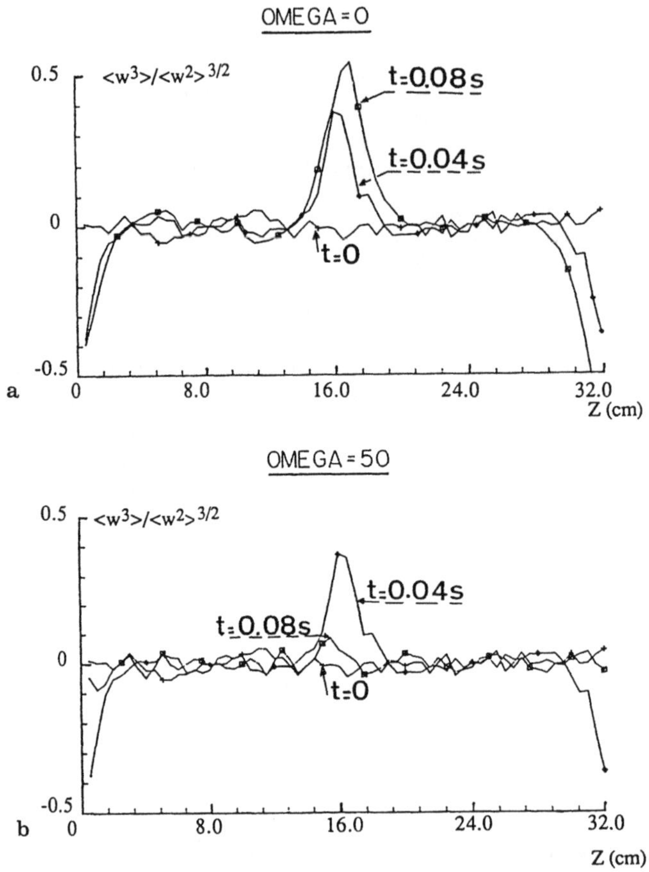

Fig. 4. Profiles of the normalized triple correlation $<w^3>/<w^2>^{3/2}$ as functions of the inhomogeneity direction Z for different times. Results averaged over 2 realizations.
Fig. 4a: without rotation.
Fig. 4b: with rotation $\Omega = 50$ rd/s (rotation switched on at time $t = 0.04$ s).

reduced by rotation, the diffusion of turbulent kinetic energy towards the low turbulence side of the box is still comparable to what was found without rotation. The only possible explanation is therefore that, in the rotating case, a large part of the turbulence transport must be due to the pressure term. When pressure velocity contributions are evaluated, this conjecture is corroborated by the fact that they are much larger than in the case without rotation. In the steady frame case, pressure-velocity correlations tend to counterbalance transport by triple correlations, while they are found to contribute positively to the diffusion mechanism in the rotating case (see figures 6a and 6b). In the non-rotating case, turbulence transport is due to triple velocity correlations, the effect of pressure being to slightly reduce transport. In the rotating case, transport is mainly due to pressure velocity correlations.

In a recent study, Riley (1990) has found very similar results in the case of turbulence subjected to stable stratification: when a slab of stratified turbulence is immersed in a non-turbulent field, diffusion of turbulence appears to be comparable to the diffusion

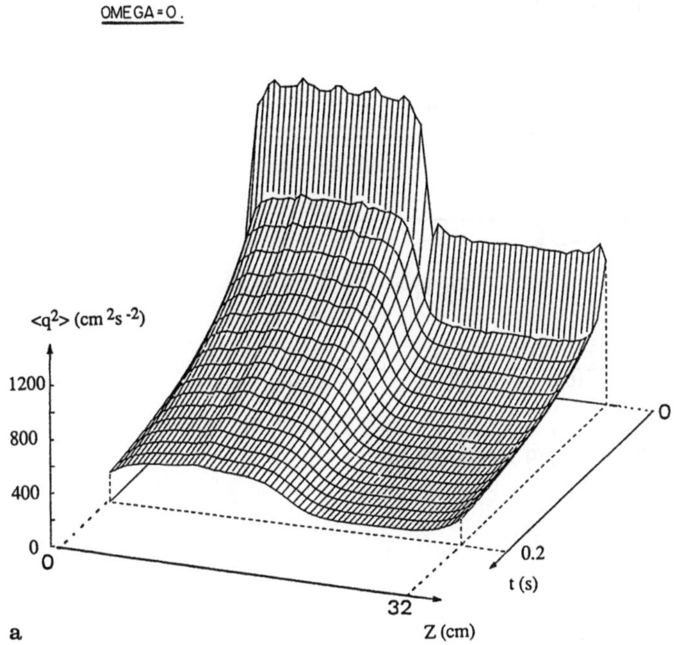

a

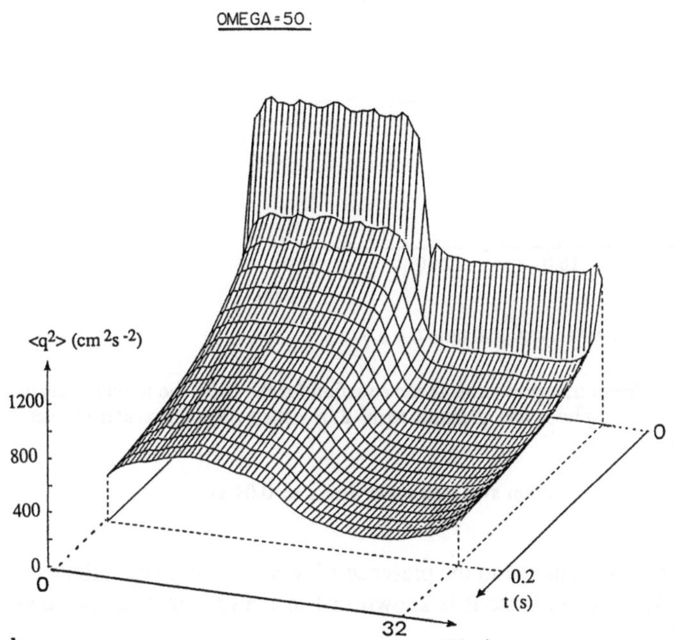

b

Fig. 5. Evolution with time of the turbulent kinetic energy profiles. Results averaged over 3 realizations. Fig. 5a: without rotation. Fig. 5b: with rotation $\Omega = 50$ rd/s (rotation switched on at time $t = 0.04$ s).

obtained in the non-stratified case, even though triple correlations are strongly reduced.

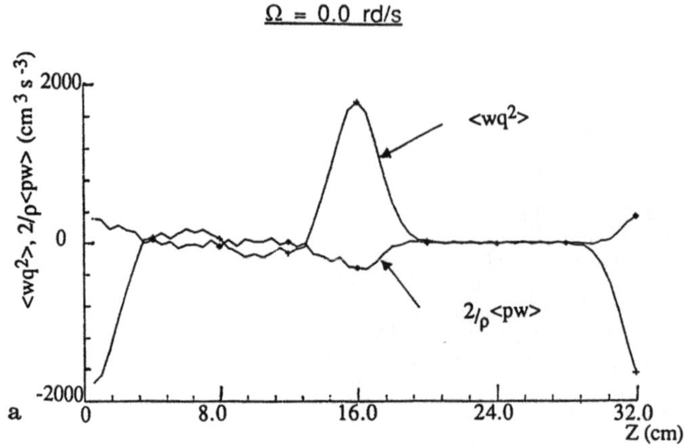

fig 6a : $\frac{2}{\rho}\langle pw\rangle$ compared with $\langle wq^2\rangle$ at time 0.1 s

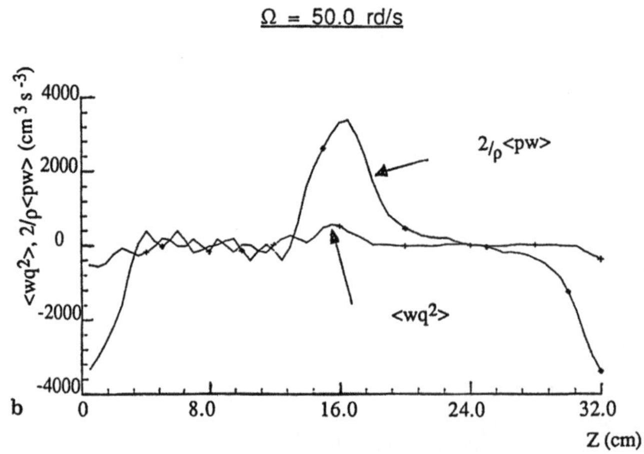

fig 6b : $\frac{2}{\rho}\langle pw\rangle$ compared with $\langle wq^2\rangle$ at time 0.1 s

Fig. 6. Comparison of the different transport terms in the equation for the twice the turbulent kinetic energy: triple correlation term: $<q^2w>$ and pressure term: $2/\rho$ $<pw>$. Results averaged over 3 realizations; time = 0.1 s.
Fig. 6a: without rotation.
Fig. 6b: with rotation $\Omega = 50$ rd/s (rotation switched on at time $t = 0.04$ s).

In Riley's work, this effect is attributed to the presence of waves in stratified turbulence. An interesting analogy is possible, since it is known that in the case of rotating flows, waves also exist: inertial waves.

If results not averaged over different realizations are plotted as functions of time, irregular profiles are obtained when rotation is present, while, in the non rotating case, smoother behaviours are found. This is, in particular, clear in figures 7a and 7b where profiles of $<w^2>$ are compared. This behaviour can not be considered as evidence for the existence of waves, but it is sufficiently important to be noticed.

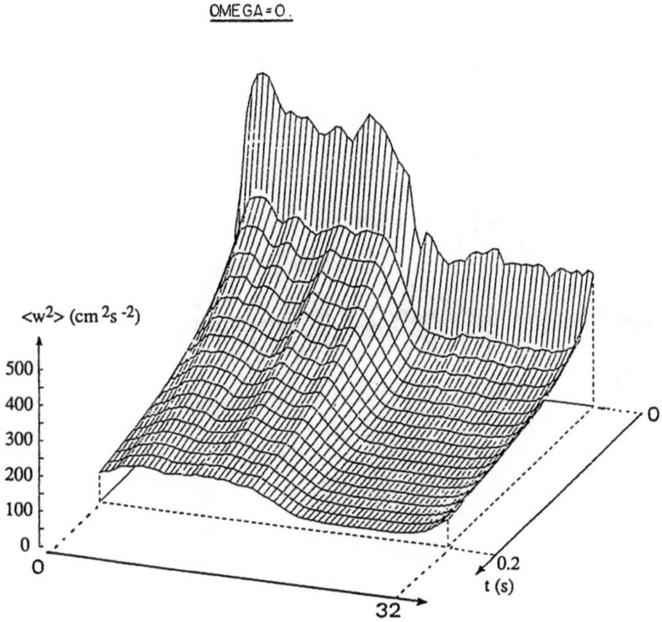

a

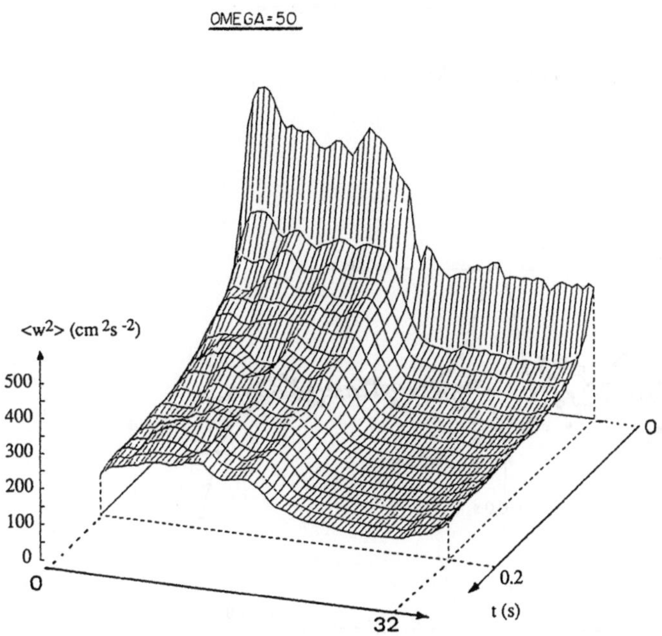

b

Fig. 7. Evolution with time of $<w^2>$ profiles. Results corresponding to one realization only.
Fig. 7a: without rotation.
Fig. 7b: with rotation $\Omega = 50$ rd/s (rotation switched on at time $t = 0.04$ s).

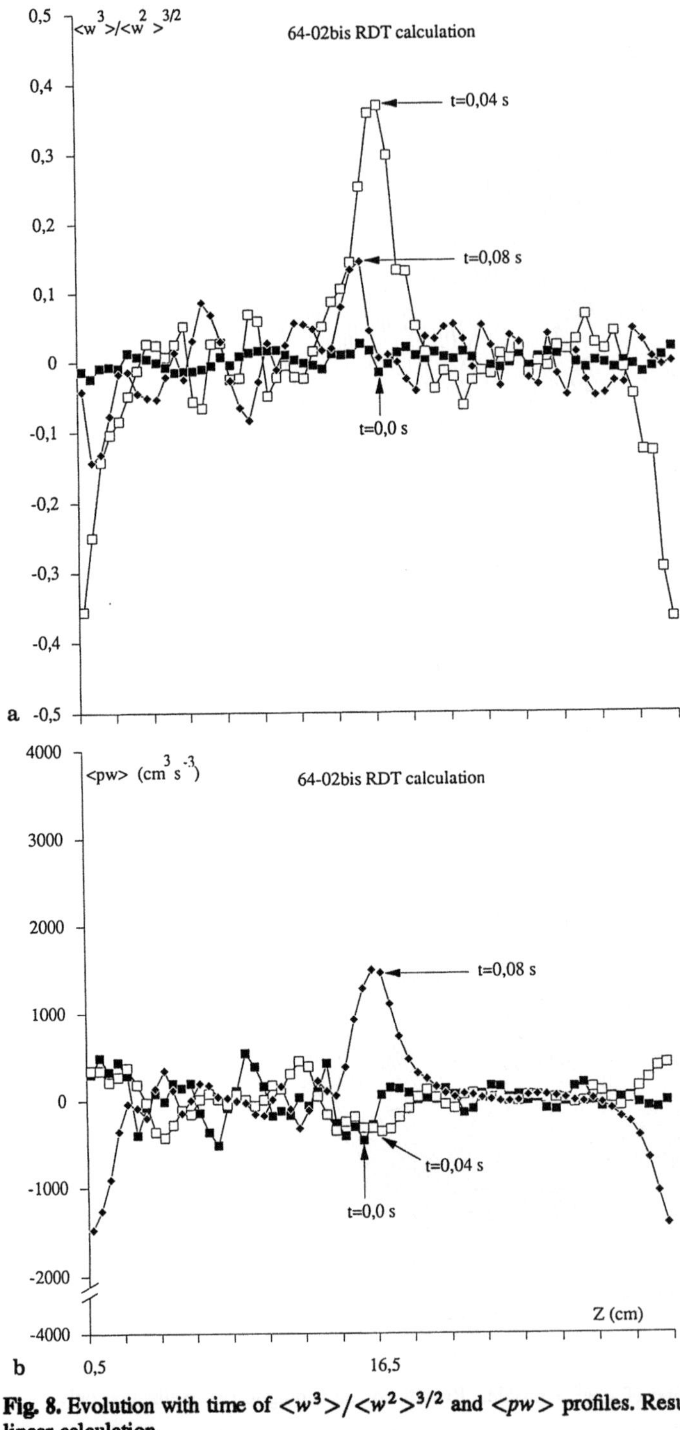

Fig. 8. Evolution with time of $<w^3>/<w^2>^{3/2}$ and $<pw>$ profiles. Results corresponding to the linear calculation.
Fig. 8a: profiles of normalised triple velocity correlations $<w^3>/<w^2>^{3/2}$.
Fig. 8b: profiles of pressure-velocity correlation $<pw>$

In order to determine whether the pressure effect is due to the rapid part of pressure or to non-linear effects, a Rapid Distortion study was made, by simply discarding the non-linear terms in the numerical code. The results clearly show that the same behaviours are found in the linear case: the strong inhibition of the triple velocity correlations (fig. 8a) and the large increase of the pressure-velocity correlation (fig. 8a). It is therefore the rapid part of pressure which is responsible for the observed effects. The hypothesis that turbulence transport is due to the presence of inertial waves in the rotating case becomes plausible.

4. One-Point Models

The aim of the present section is to include in one-point models the reduction of the triple correlation induced by rotation. A model is devised by taking into account the linear term due to Coriolis forces in the equation for the triple velocity correlation. In a second step, a simplified model is proposed.

In the one-point model approach, the equation for the triple correlations is written (see Hanjalic and Launder, 1972), and the fourth order moments are expressed by introducing Quasi-Normal assumptions. The approach is here extended to the case of rotating turbulence, in which we do not neglect the effects introduced by the presence of rotation. Thus, the equation for triple correlation is written as:

$$\frac{\partial<u_i u_j u_k>}{\partial t} = -\Bigg(\frac{\varepsilon}{C_s k}<u_i u_j u_k> + <u_k u_m>\frac{\partial<u_i u_j>}{\partial x_m} + <u_j u_m>\frac{\partial<u_i u_k>}{\partial x_m} + <u_i u_m>\frac{\partial<u_j u_k>}{\partial x_m} + 2\varepsilon_{iln}\Omega_l<u_j u_k u_n> + 2\varepsilon_{kln}\Omega_l<u_i u_j u_n> + 2\varepsilon_{jln}\Omega_l<u_k u_i u_n>\Bigg) \qquad (5)$$

In equation (5), ε_{ijk} is the permutation tensor.

The quantity is a damping term introduced in order to model the effect of pressure. It can also be regarded as taking into account the effects of the fourth-order cumulants.

When the asymptotic values for $<u_i u_j u_k>$ deduced from equation (5) are used in Reynolds stress computations, results show that triple correlation levels are reduced by the presence of rotation – as expected. Consequently, the inhomogeneous transport of turbulence appears to be reduced when compared to the non-rotating case (see figure 9).

A simpler model is also proposed, in which triple correlations are expressed using the classical formulation corresponding to the Hanjalic and Launder model (equation (1)), with C_s replaced by $C_s' = C_s/(1 + R_o^{-1})$. R_o is here the Rossby number defined as $R_o = \varepsilon/k\Omega$ Comparisons between results obtained with the simple model and with the complete form are very satisfactory, as appears in figure 9. However, it has to be pointed out that more work is needed to model the pressure-velocity correlation in this case, since one of the conclusions of the Large Eddy Simulation study is that pressure appears to play a very significant role when rotation is present. The conclusion of the RDT calculation is that the pressure effect is a linear effect, which is consistent with the fact that transport

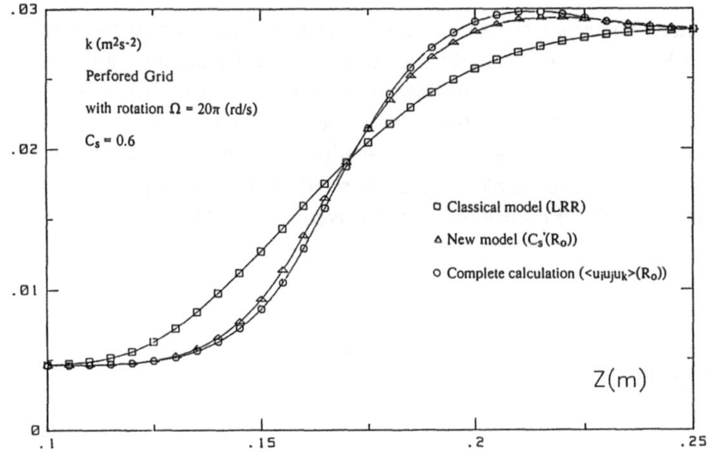

Fig. 9. Results obtained using the improved one-point models: comparison between results obtained with the complete and simplified models, $\Omega = 20\pi$ rd/s, initial profiles corresponding the V.W. experiment (perforated grid).

is due to inertial waves. At the present time, it seems difficult to model this effect at the level of one-point closure. The problem involves quantities like group velocities that are difficult to take into account in classical models. The case of the axis of rotation parallel to the inhomogeneity direction oz could also be investigated. In this situation, effects more similar to the mechanisms observed in the experimental study of Hopfinger et al. (1982) are likely to be found.

References

Aupoix B., Cousteix J. and Liandrat J., 1983; Effects of Rotation on Isotropic Turbulence, Fourth Int. Symp. Turb. Shear Flows, Karlsruhe, Sept. 83.

Bardina J., Ferziger J.H. and Reynolds W.C., 1983; Improved Turbulence Models based on L. E. S. of Homogeneous, Incompressible Turbulent Flows, Report T.F. 19, Stanford Univ., May 83.

Bertoglio J.P., 1982; Homogeneous Turbulent Field within a Rotating Frame, A.I.A.A. Journ., Vol. 20, no 9.

Cambon C., Teissedre C. and Jeandel D., 1985; Etude d'effets couplés de Déformation et de Rotation sur une Turbulence Homogène, Journ. Méc. Théor. Appl., 4.

Cambon C., Jacquin L. and Lubrano J.L., 1991; Towards a New Reynolds Stress Model for Rotating Turbulent Flows , submitted to Physics of Fluids.

Chollet J.P. and Lesieur M., 1981; Parameterization of Small Scales of Three-Dimensional Isotropic Turbulence utilizing Spectral Closures, Journ. of Atm. Sci., Vol. 38.

Hanjalic and Launder B.E., 1972 A Reynolds stress model of turbulence and its application to thin shear flows , J.F.M. 52 pp 609-638.

Hopfinger E.J., Browand F.K. and Gagne Y, 1982; Turbulence and Waves in a Rotating Tank. J. F. M.125.

Jacquin L., Leuchter O., Cambon C. & Mathieu J., Homogeneous Turbulence in the Presence of Rotation, J.F.M. 220, pp 1-52, 1990

Launder, Reece and Rodi , 1975 Progress in the developpement of Reynolds stress closure J.F.M. 68.

Lubrano J.L. and Cambon, C. 1991, Personal communication.
Lumley J., 1978; Computational Modeling of Turbulent Flow, Adv. Appl. Mech., 18, pp. 123-176.
Malkus, 1990 ; Europ. Turb. Conf., Stockholm, July 90.
Riley, 1990, Seminar on Stratified Turbulence, Ecole Centrale de Lyon, Sept. 90.
Shao L., Bertoglio J.P. and Michard M., 1990; Large Eddy Simulation of the Interaction between Two Distinct Turbulent Velocity Scales, Europ. Turb. Conf., Stockholm, July 90, to appear in Springer-Verlag.
Shao L., Le Penven L. and Bertoglio J.P., 1990; Study of the Modelling of the Transport Terms in One-point Closures using L. E. S. of Turbulence, Int. Symp. on Eng. Turb. Model. and Measur., Dubrovnik, Sept. 90.
Shao L. et Bertoglio J.P., Effet de l'Intermittence sur les Termes de Transport Turbulent 10e Cong. Méca. Fr. Sept.1991
Speziale C.G., Gatski T.B. and Mhuiris N.M.G., 1990; A Critical Comparison of Turbulence Models for Homogeneous Shear Flows in a Rotating Frame, Phys. Fluids, A2, 9, Sept. 90, pp. 1678-1684.
Veeravalli and Warhaft , 1989 The shearless turbulence mixing layer J.F.M. 207, 191-229

Lumley J.L. and Cambon, C. 1991: Personal communication.
Lumley J. 1978; Computational Modeling of Turbulent Flows. Adv. Appl. Mech., 18, pp. 123-176.
MalKes, 1990 ; Europ. Turb. Conf., Stockholm July 90.
Riley, 1990, Seminar on Stratified Turbulence. Ecole Centrale de Lyon, Sept. 90.
Shao L., Bertoglio J.P. and Michard M., 1990: Large Eddy Simulation of the Interaction between Two Distinct Turbulent Velocity Scales, Europ. Turb. Conf., Stockholm, July 90, in [illegible] Springer-Verlag.
Shao L., Le Penven L. and Bertoglio J.P., 1990, Study of the Modelling of the Transport Terms in One-point Closures using L.E.S. of Turbulence, Int. Symp. on Eng. Turb. Mod. and Meas., Dubrovnik, Sept. 90.
Shao L. et Bertoglio J.P., Effet de [illegible] 1991
Speziale C.G., Gatski T.B. and Mac Giolla Mhuiris N., 1990: A Critical Comparison of Turbulence Models for Homogeneous Shear Flows in a Rotating Frame. Phys. Fluids A 2, 9, Sept. 90, pp. 1678-1684.
[illegible]

Scaling and Structure of Turbulent Eddies in Curved Sheared Flows

A.G.L. Holloway[1] and S. Tavoularis[2]

[1] Department of Mechanical Engineering, University of Brunswick, Fredericton, New Brunswick, Canada, E3B 5A3
[2] Department of Mechanical Engineering, University of Ottawa, Ottawa, Ontario, Canada, K1N 6N5

Abstract

One-dimensional spectra in uniformly sheared, nearly homogeneous, turbulent flows with various magnitudes of mean shear and mean streamline curvature are presented and analyzed. These spectra confirm our previous observations (Holloway and Tavoularis, 1992) that curvature affects the partition of the kinetic energy among its three components and, especially, the covariance of the streamwise and transverse velocity fluctuations. It is also evident from these spectra, particularly form the coherence, that the energy containing eddies are most strongly affected by curvature, while the fine structure is affected indirectly, probably through changes in the input to the energy cascade. For relatively mild curvature, the coherence of large-scale motions diminished in cases where the mean velocity increased away from the center of curvature and increased in cases where the mean velocity decreased away from the center of curvature. For "strong" curvature in the same direction as the mean shear, the sign of the coherence of large eddies was reversed and momentum was transported up the gradient of mean velocity and away from the center of curvature. For certain combinations of mean shear and curvature, the coherence of large scales was opposite to that of the small scales.

Nomenclature

$C_{uv}(k_1)$ the coherence function
D_{sn} mean strain rate
$F_{uu}(k_1)$, $F_{vv}(k_1), F_{ww}(k_1), F_{uv}(k_1)$ streamwise, one-dimensional spectra of the velocity fluctuations and the shear stress
k_1 streamwise component of the wavenumber vector
L_{uu} the integral length scale of the streamwise fluctuations
P the production rate of turbulence kinetic energy
R_c radius of curvature of the tunnel centerline
Re real part of a complex function
R_L turbulence Reynolds number based on the integral length scale
R_λ turbulence Reynolds number based on the Taylor microscale
s, n, z streamwise, transverse and spanwise coordinates

S	curvature paramter ($= (U_c/R_c)/(dU/dn)$)
U, V, W	streamwise, transverse and spanwise mean velocity components
u, v, w	streamwise, transverse and spanwise velocity fluctuations
U_c	streamwise mean velocity component on the tunnel centerline
ϵ	the rate of viscous dissipation of the turbulence kinetic energy
η	Kolmogorov microscale
θ	turning angle
λ_u	streamwise Taylor microscale
ν	kinematic viscosity
τ	dimensionless time
τ_0	dimensionless time at the entrance to the curved section
v	Kolmogorov velocity
Ω_z	mean rotation rate

Introduction

It has been well documented that curvature has a strong effect on the structure of sheared turbulence (for a recent review of the literature see Holloway and Tavoularis, 1992). It has also been asserted that the fina structure of turbulence is affected by curvature to a lesser degree than eddies in the energy containing spectral range. This conclusion could be based, for example, upon the boundary layer measurements of Ramaprian and Shivaprasad (1978), which show that both convex and concave surface curvature affect the magnitudes of integral length scales more than those of the Taylor microscales. However, the action of curvature in boundary layers is obscured by the presence of wall effects, the entrainment of irrotational flow, mean shear nonuniformity and inhomogeneity of the turbulence. For the present investigation, an effort was made to evince the effects of flow curvature on the different scales of sheared turbulence in relative isolation from wall, entrainment and inhomogeneity effects. An ideal flow for studying this interaction would be one where turbulence statistics, curvature and shear would be uniform. Such a flow can be conceived by extending the notion of rectilinear, homogeneous, shear flow to a flow with mean streamlines which form concentric rings and are subjected to a uniform shear in the radial direction. It is easy to see that inhomogeneity of the turbulence would inevitably develop under such conditions, as a result of the transverse nonuniformities of curvature and strain. Nevertheless, it seems worthwhile to explore the plausibility of an approximately homogeneous curved shear flow by introducing additional constraints.

Consider a flow field bounded, on average, by two coaxial circular cylinders with a difference in radius that is large in comparison with the scales of the turbulence but small compared to the mean radius. The mean streamline curvature can be considered as nearly constant within the volume and the turbulence structure in the core of this volume can be assumed as independent of wall effects, at least over a sufficiently small time interval. If the turbulence within this volume were homogeneous initially, it would likely remain homogeneous for some time.

An approximation to that flow was materialized by modifying the apparatus used in previous experimental studies on nearly homogeneous shear flows (Tavoularis and Karnik, 1989). Uniformly sheared turbulent flows produced with the use of a shear generator were allowed to reach a quasi-self-preserving structure in a straight rectangular duct and then were passed tangentially into a curved duct, also of rectangular cross-section. The near homogeneity of the turbulence and the near uniformity of the shear were preserved, while the structure of the turbulence was modified by the flow curvature.

The present type of flows can best be described in a curvilinear orthogonal system of coordinates, similar to those commonly used in studies of curved boundary layers. These will be denoted by s, n and z, as shown in Fig. 1. The s coordinate is the length of a circular arc, with radius R_c, coincident with the flow volume centerline. The n coordinate is normal to s, has its origin on the volume centerline and is positive when directed away from the centre of curvature. The radial extent of the flow volume is assumed to be small compared to R_c. The spanwise coordinate, z, is normal to both s and n and has a direction following the right-hand rule.

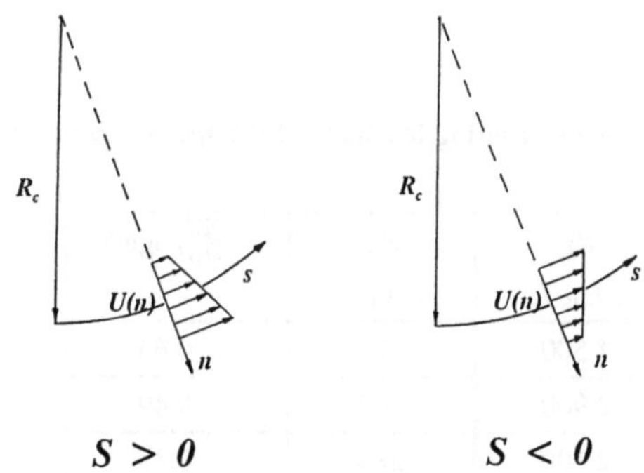

Fig. 1. The flow configuration

The local, instantaneous, velocity vector can be decomposed into means, U, V and W, and fluctuations, u, v and w, parallel, respectively, to the s, n and z directions. Assuming that the mean shear is uniform and in the n direction alone, the mean velocity can be represented as

$$(U,V,W) = (U_c + (dU/dn)n, 0, 0) \tag{1}$$

Details of the experimental apparatus and measurements of the development of the turbulent stresses and length scales for the present flow have been reported by Holloway and Tavoularis (1992). A primary conclusion of that work was that the effects of curvature appeared to scale with the parameter[3] $S = (U_c/(R_c)/(dU/dn)$, where U_c was the radius of curvature at the centerline of the duct. Flows with different values for S were generated by adjusting both the mean shear and R_c. Flows with different values of shear, each developing towards a self-preserving structure, were produced using a shear generator and various combinations of screens and grids. Shears of opposite direction were produced by inverting the shear generator. In addition, two curved ducts were built,

[3] This parameter arises naturally in the dimensionless Reynolds stress equations and can be viewed as the ratio of time scales associated with shear and curvature

with corresponding $R_c = 2$ m and 5 m, using different combinations of radii of curvature and shears. 20 distinct flows were examined, with S ranging from −0.5 to more than 1.

Measurements of the turbulent stresses indicated that the growth of the turbulent kinetic energy was enhanced for negative values of S and suppressed for positive values of S. Whether the kinetic energy of turbulence grew or decayed, it seemed to do so exponentially and the components of the Reynolds stress tensor approached constant ratios, at least for flows with relatively small magnitudes of S, which attained self-preserving structures well within the bounds of the facility. It should be noted, however, that the structures of flows with large magnitudes of S were still developing at the exits of the curved tunnels, although exhibiting trends consistent with the other cases.

Several of the flows described above have been chosen as representative cases; their main characteristics are shown in Table 1. These flows have been ranked by their value of S, under the assumption that this was the determining factor, but possible effects of differences in the values of shears, R_L and R_λ for these flows should not be overlooked. The two flows with the largest positive S were of particular interest, because they exhibited a turbulent momentum flux in the same direction as the mean shear, opposite to the implications of gradient transport.

Table 1. Summary of some flow characteristics at the measuring locations of the spectra shown in Fig. 2 to 5 and 12; in all cases, $U_C \approx 10$ m/s

Symbol	S	$\lvert dU/dn \rvert$ (s^{-1})	R_C (m)	R_L $(u'L_{uu}/\nu)$	R_λ $(u'\lambda_u/\nu)$	$-\frac{\overline{uv}}{u'v'}\,\mathrm{sgn}\,S$
□	−0.25	19	2	4,500	370	0.63
○	−0.10	52	2	2,900	340	0.49
△	0.00	64	∞	2,100	280	0.45
+	0.03	61	5	2,900	340	0.38
×	0.10	53	2	880	170	0.32
◇	0.65	8	2	1,430	200	−0.20
⊞	> 5	< 0.1	2	880	125	−0.18

Instrumentation and Measuring Procedures

The velocity was measured using a custom-made, cross-wire array (TSI 1248bJT15) and conventional hot wire anemometry procedures. The sensing elements were made of tungsten and had a diameter of 5 μm and a length of 1.2 mm. They were separated by 0.5 mm and had nominal inclinations of $\pm 45^{\circ}$ with respect to the axis of the probe body. In the present flows the Kolmogorov microscale ranged from 0.14 to 0.30 mm, which was less than a quarter of the length of each hot wire. The analysis of Wyngaard

(1968) suggests that measurements of fluid motions with sizes less than the wire length would require corrections.

The anemometer signals were low-pass filtered to a cut-off frequency of 10 kHz, converted to velocities using a modified form of King's law, digitized at two different rates, 2 kHz and 25 kHz, and divided into statistically independent records, each of 1024 points. Frequency spectra of the fluctuating velocity components and the shear stress were computed from the discrete time series using an FFT algorithm and then converted to the corresponding streamwise, wave number spectra, $F_{uu}(k_1)$, $F_{vv}(k_1)$, $F_{ww}(k_1)$ and $F_{uv}(k_1)$, with the use of Taylor's frozen flow approximation. While $F_{uu}(k_1)$, $F_{vv}(k_1)$ and $F_{ww}(k_1)$ are real and even functions of k_1, $F_{uv}(k_1)$ is complex with an even real part and an odd imaginary part. Because the imaginary part of $F_{uv}(k_1)$ is odd, only the real part of $F_{uv}(k_1)$ contributes to the shear stress. The corresponding spectra computed from the two sets of time series collapsed in the overlapping ranges. However, only the high frequency ranges of the spectra computed from the records sampled at 25 kHz and the low frequency ranges of the spectra computed from those at 2 kHz were retained. The Nyquist criterion was not strictly observed for the 2 kHz sample but the portion of the spectrum which was significantly distorted was removed. The spectral estimates for the longest wavelengths were subjected to higher uncertainty.

The streamwise integral length scale, L_{uu}, was determined by extrapolating the corresponding spectra to zero wavenumber, as

$$L_{uu} = \frac{\pi}{\overline{u^2}} \lim_{k_1 \to 0} F_{uu}(k_1) \tag{2}$$

Results

Typical measurements of the spectral density $F_{uu}(k_1)$, normalized with the Kolmogorov scales, have been plotted in Fig. 2. The corresponding flows spanned a range of turbulent Reynolds number, $R_\lambda = u'\lambda/\nu$, between 130 and 370. The Pao-Corrsin universal equilibrium spectrum (Hinze, 1975) has also been plotted for comparison. At large wavenumbers (e.g for $k_1\eta > 0.005$), the shape of $F_{uu}(k_1)$ appeared to be unaffected by flow curvature or shear, as the data from all flows essentially collapsed onto a curve, close to the theoretical estimate.

At the lower wavenumbers, $F_{uu}(k_1)$ varied widely from flow to flow, due to the different velocity and length scales. These differences may not be entirely attributed to flow curvature, but also to differences in the history of each flow. For example, the flows with $S = -0.10$ and $S = 0.10$ had comparable initial kinetic energies but curvatures of opposing sense, which caused their kinetic energies to deviate exponentially and to approach a ratio of about two at the streamwise position where these spectra were measured. Their length scales were affected in a similar way.

For an evaluation of the effects of mean curvature and shear on the spectra, the present results will be compared to the spectra of isotropic turbulence. Among the several approximate forms which have been proposed for the three-dimensional spectrum of isotropic turbulence (Hinze, 1975), one leading to a simple expression for $F_{uu}(k_1)$

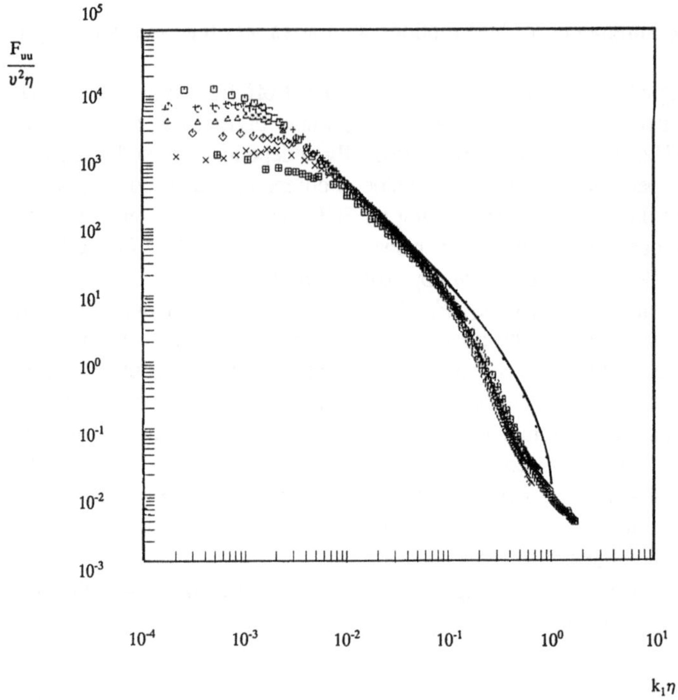

Fig. 2. F_{uu} normalized with the Kolmogorov scales; symbols as in Table 1. The solid line represents the Pao-Corrsin theoretical estimate.

and having a good agreement with measurements in grid-generated turbulence is von Kàrmàn's interpolation formula,

$$\frac{F_{uu}(k_1)}{\overline{u^2}L_{uu}} = 0.63\left[1 + (1.33k_1L_{uu})^2\right]^{-5/6} \tag{3}$$

The spectral densities $F_{uu}(k_1)$, $F_{vv}(k_1)$ and $F_{ww}(k_1)$ have been plotted in Fig. 3, 4 and 5, respectively, normalized with the streamwise component of the kinetic energy, $\overline{u^2}$, and the steamwise integral scale, L_{uu}. The von Kàrmàn interpolation formula has also been plotted for comparison. The corresponding spectra for different flows differed significantly for $k_1L_{uu} > 10$, but, for $k_1L_{uu} < 10$, the curves for different flows were close to each other and only modestly more energetic than predicted by equation (3). One should note, however, that the selected normalization has forced these spectra to collapse at the lowest wavenumbers and to cover the same area.

In each flow, $F_{vv}(k_1)$ and $F_{ww}(k_1)$ were different for $k_1L_{uu} < 5$ and generally less energetic than in isotropic turbulence of equal energy. These spectra approached one another at $k_1L_{uu} \approx 10$ and then they diverged again in the viscous range. The proportion of the turbulent kinetic energy contained in $\overline{v^2}$ was greatest for the flow with $S = -0.25$ and lowest for the flow with $S = 0.65$, while no discernible trend was shown by $F_{ww}(k_1)$. It would also appear from these spectra that shear and curvature tended to increase the anisotropy of the partition of kinetic energy among its three components at the lower wavenumbers.

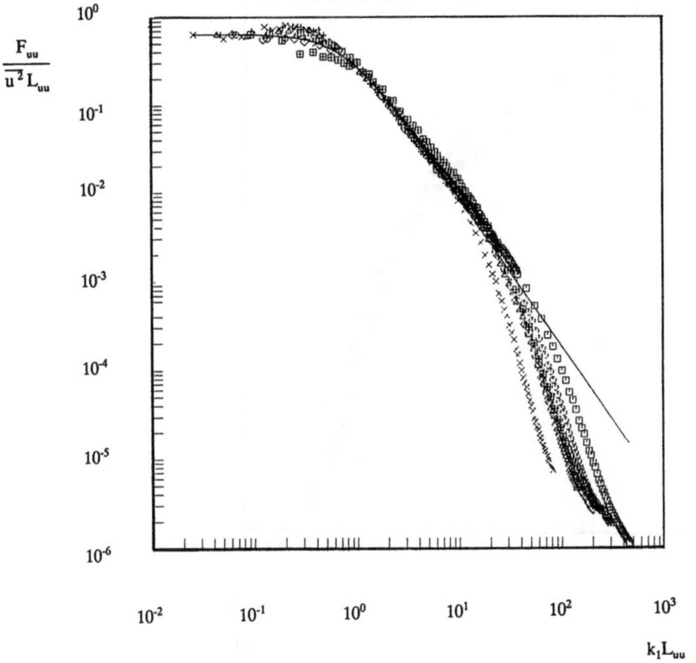

Fig. 3. F_{uu} **normalized with the energy scales; symbols as in Table 1.**

Although one could probably distinguish an inertial subrange in $F_{uu}(k_1)$, $F_{vv}(k_1)$ and $F_{ww}(k_1)$ showed little evidence of inertial subranges. This, however, could be partly attributed to the lower Reynolds number of these components, due to their lower energy levels and length scales.

The one-dimensional cross-spectral density, $F_{uv}(k_1)$, has been normalized with $F_{uu}(k_1)$ and $F_{vv}(k_1)$ to form the coherence function,

$$C_{uv}(k_1) = \frac{F_{uv}(k_1)}{\sqrt{(F_{uu}(k_1)F_{vv}(k_1))}} \tag{4}$$

In isotropic turbulence, $C_{uv}(k_1)$ is zero at all wavenumbers, while, in general, $-1 < C_{uv}(k_1) < 1$. For shear-dominated flows, $C_{uv}(k_1)$ should have a sign opposite to that of the shear, because momentum would, on average, be transferred down the gradient of mean velocity.

It has been suggested by Holloway and Tavoularis (1992) that the general tendency of mean flow curvature is to produce a turbulence structure that transports momentum away from the center of curvature, which, in the present circumstances, would result in the production of a positive coherence. If, at the start of curvature, the mean shear were negative, with mean velocity decreasing away from the center of curvature, then the initial coherence of the flow entering the curved section would be positive and it would be further enhanced by flow curvature, although this enhancement would be subjected to an upper bound. In contrast, if the shear were initially positive, with mean velocity increasing away from the center of curvature, then the initial coherence of the flow

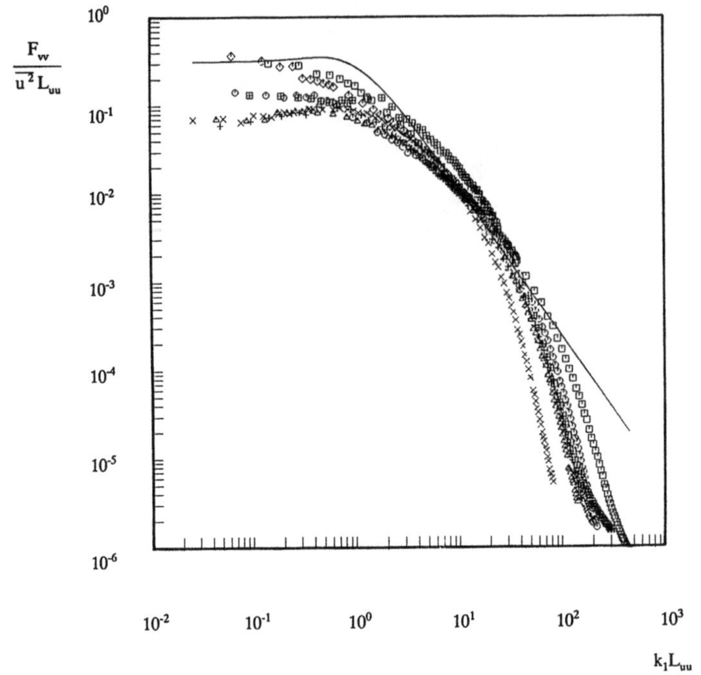

Fig. 4. F_{vv} normalized with the energy scales; symbols as in Table 1.

entering the curved section would be negative and centrifugal actions would force it to attain lower magnitudes, even to become positive, were curvature effects to dominate production by mean shear. Therefore, positively sheared flows are likely to experience the greatest changes in their structure, and, thus, in their coherence, under the influence of streamline curvature.

As in our previous publication (Holloway and Tavoularis, 1992), the streamwise position will be denoted, in dimensionless form, as

$$\tau = \tau_0 + \frac{s}{U_c}\left|\frac{dU}{dn}\right| \tag{5}$$

where τ_0 is the total strain at the entrance to the curved section. This parameter describes adequately the development of "high shear" flows, while for "low shear" flows, the "turning angle"

$$\theta = \frac{U_c}{R_c}\frac{s}{U_c} = (\tau - \tau_0)\, S \tag{6}$$

would be more appropriate.

The development of $Re\{C_{uv}(k_1)\}$ under the influence of flow curvature has been illustrated in Fig. 6 to 11. At the start of curvature ($\tau = \tau_0$), all flows had similarly shaped coherences, although, under the present notation, their signs were positive when $S < 0$ and negative when $S > 0$. In all cases, $|Re\{C_{uv}(k_1)\}|$ decreased monotonically to zero at large wavenumbers ($k_1 L_{uu} > 10$), which is another indication that local isotropy prevailed

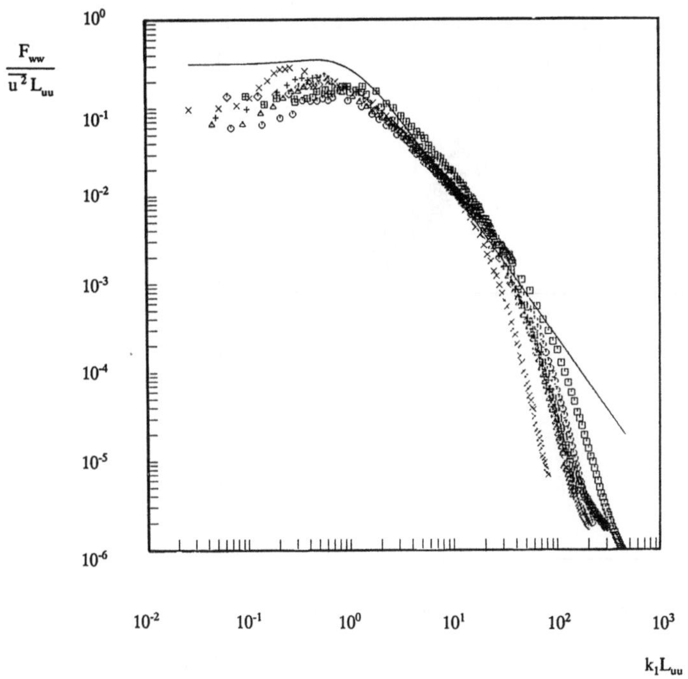

Fig. 5. F_{WW} normalized with the energy scales; symbols as in Table 1.

in the fine structure. At the low wavenumber end ($k_1L_{uu} < 0.3$), the coherence was nearly constant, approaching a maximum magnitude of about 0.7. This is considerably greater than the value of the correlation coefficient, $|\overline{uv}/u'v'| = 0.45$, in straight shear flows, indicating that the large-scale motions were those contributing mostly to the shear stress.

In the curved sections, the coherence followed one of two clearly identifiable evolution patterns, depending on the sign of the parameter S. In flows with $S < 0$ (Fig. 6 and 7), the coherence gradually increased at all wavenumbers, although maintaining a shape similar to that in straight flows, and, after a development region, it approached an asymptotic curve. In the case with the lowest S ($= -0.25$), the maximum of $|Re\{C_{uv}(k_1)\}|$ reached values as high as 0.9, signifying an almost perfect correlation between large-scale streamwise and transverse motions. In sharp contrast, in curved flows with positive S (Fig. 8 to 11), the initially negative coherence tended to decrease in magnitude at all wavenumbers. In the two cases with substantially large S (Fig. 10 and 11) the effects were strong enough to completely reverse the sign of the coherence over a range of low wavenumbers, the upper bound of which seemed to shift upwards with increasing S. A clear observation, at least for the positive-S flows, is that it was the large-scale motions ($k_1L_{uu} < 1$) that were most sensitive to curvature, although the maximum sensitivity did not seem to be at the largest measured wavenumbers but in the energy containing eddy range. Another general observation is that, in the curved flows which attained self-preservation, the asymptotic shape of the coherence strongly depended upon S. One may further speculate (although data to fully document this statement are not available), by extrapolating

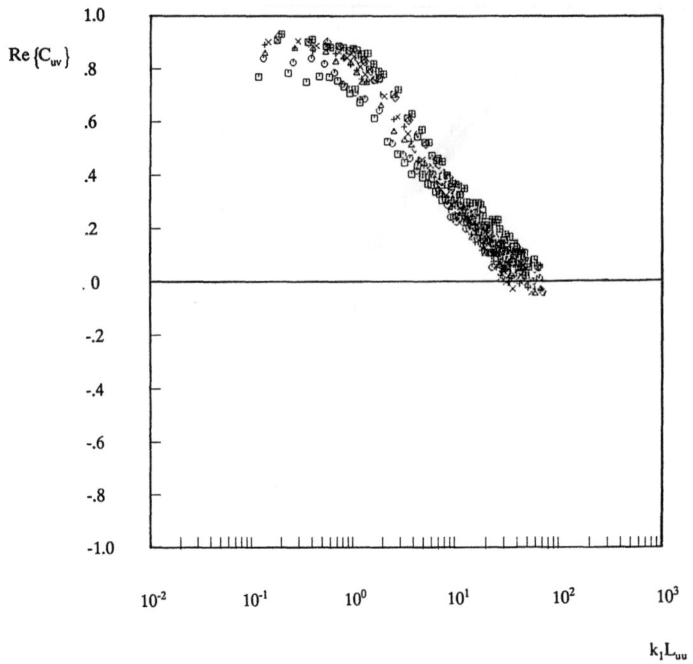

Fig. 6. Development of $Re\{C_{uv}\}$ along the centerline of the curved wind tunnel; $S = -0.25$, $\tau - \tau_0 = 0$ (□), 0.2 (○), 0.6 (△), 1.0 (+), 1.4 (×), 1.8 (◇), 2.2 (⊞) and 2.6 (⊠).

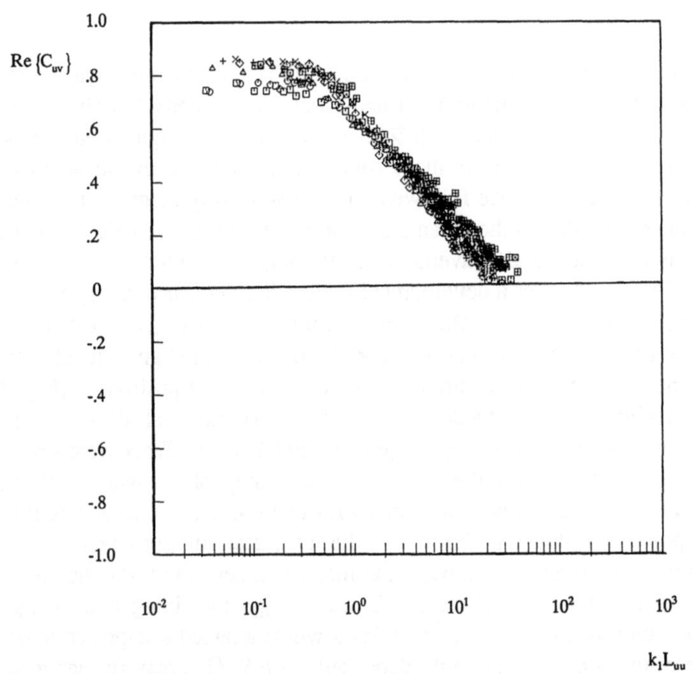

Fig. 7. Development of $Re\{C_{uv}\}$ along the centerline of the curved wind tunnel; $S = -0.10$, $\tau - \tau_0 = 0$ (□), 0.6 (○), 1.7 (△), 2.9 (+), 4.0 (×), 5.2 (◇), 6.3 (⊞) and 7.3 (⊠).

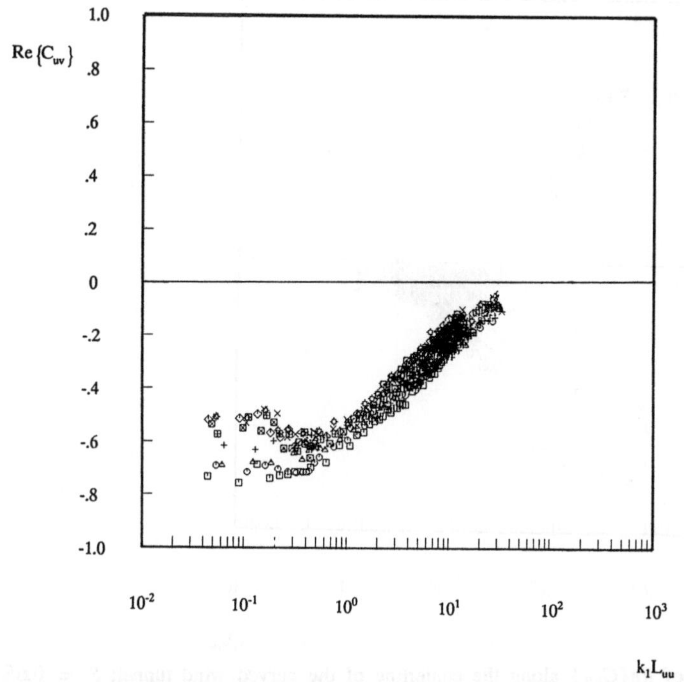

Fig. 8. Development of $Re\{C_{uv}\}$ along the centerline of the curved wind tunnel; $S = 0.03$, $\tau - \tau_0 = 0$ (□), 0.6 (○), 2.0 (△), 3.3 (+), 4.7 (×), 6.0 (◇), 7.3 (⊞) and 8.7 (⊠) and 10 (⊕).

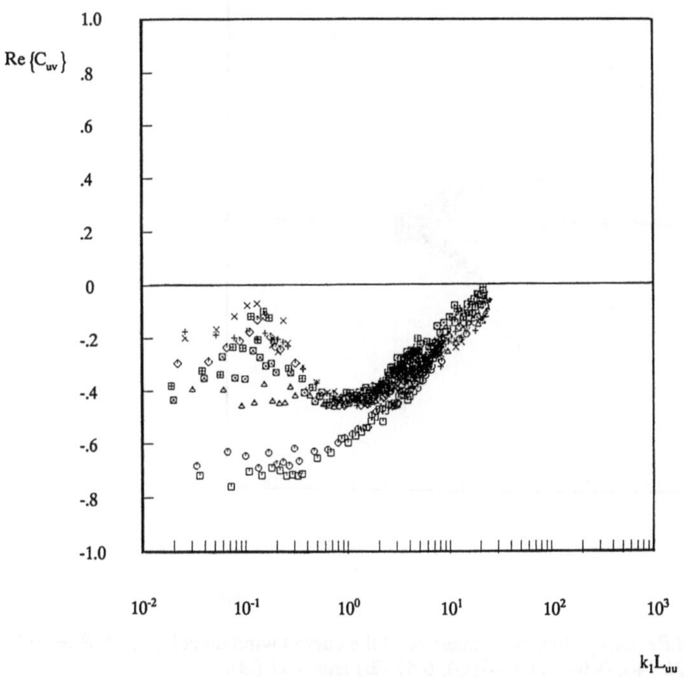

Fig. 9. Development of $Re\{C_{uv}\}$ along the centerline of the curved wind tunnel; $S = 0.10$, $\tau - \tau_0 = 0$ (□), 0.5 (○), 1.5 (△), 2.6 (+), 3.6 (×), 4.6 (◇), 5.7 (⊞) and 6.7 (⊠).

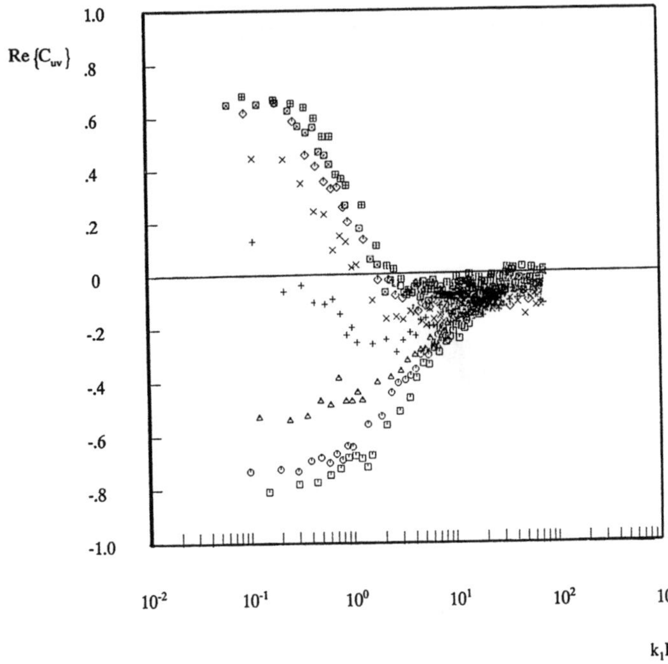

Fig. 10. Development of $Re\{C_{uv}\}$ along the centerline of the curved wind tunnel; $S = 0.65$, $\tau - \tau_0 = 0$ (□), 0.08 (○), 0.25 (△), 0.41 (+), 0.58 (×), 0.75 (◇), 0.91 (⊞) and 1.1 (⊠).

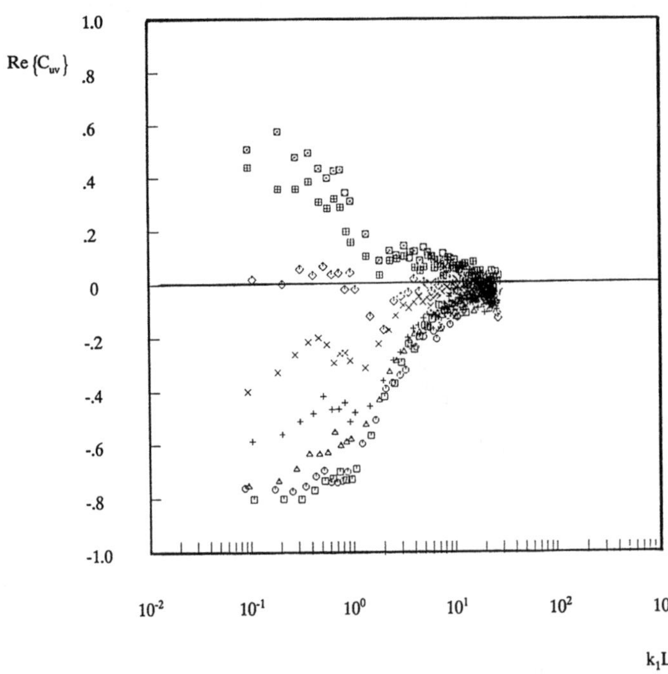

Fig. 11. Development of $Re\{C_{uv}\}$ along the centerline of the curved wind tunnel; $S > 5, \theta = 0$ (□), 0.05 (○), 0.15 (△), 0.26 (+), 0.36 (×), 0.47(◇), 0.57 (⊞) and 0.67 (⊠).

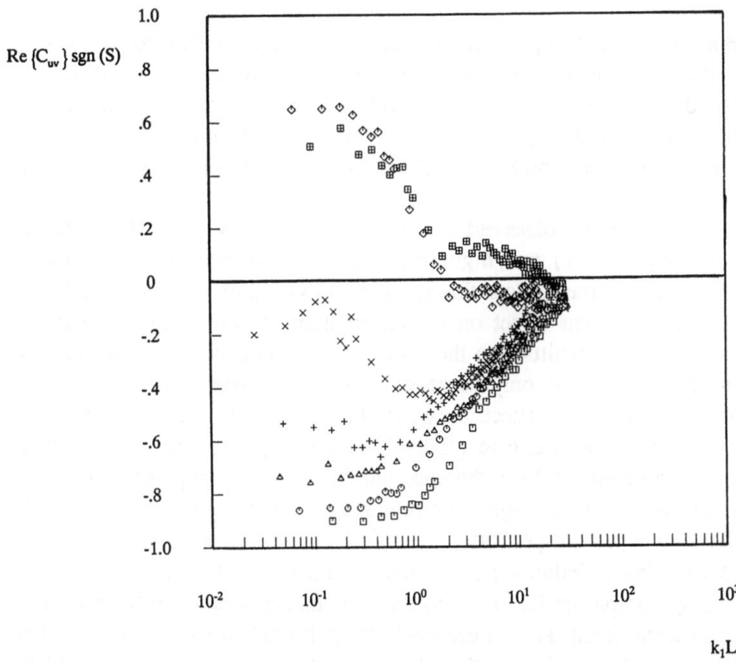

Fig. 12. Comparison of far donwstream shapes of $Re\{C_{uv}\}$ sgn S in flows with different S; symbols as in Table 1.

present large-S measurements, that the coherences of very strongly curved flows would approach the same shape, irrespective of the sign of S.

To further compare the asymptotic shapes (for large τ or θ) of the coherence at different values of S, we have plotted, in Fig. 12, the quantity $Re\{C_{uv}\}$ sgn S (sgn $S = 1$, if $S > 0$, and -1, otherwise) measured at locations near the downstream end of the curved sections. It is evident that, as $|S|$ increased, the low-wavenumber values of $Re\{C_{uv}(k_1)\}$ approached the bounds of ± 1.

The imaginary part of the coherence was found to be relatively weak at all wavenumbers and highly scattered. this part does not, on the average, influence the turbulent shear stress but does make a contribution to the spatial covariance, as it contributes to its asymmetry. This part was largest in the flows with positive S in the wavenumber range where the real part of the coherence changed sign. Because of the unclear significance of this quantity and in view of its small, non-systematically varying values, the corresponding measurements will not be included in the present report.

Discussion

The experimental data seem to indicate that, although the shear and the curvature act uniformly on all turbulent motions, it is the large-scale motions that are affected the most.

In the high wavenumber range ($k_1 L_{uu} \geq 10$) the partition of energy differed little from that in isotropic turbulence and the coherence was small. At lower wavenumbers, the spectral densities of the spanwise and transverse velocity fluctuations were, in general, an order of magnitude lower than that of the streamwise component and the coherence of the streamwise and transverse motions varied widely from values as high as 0.9 to as low as −0.7.

In an attempt to analyze the observed effects of curvature, we will first separate the spectra into low ($k_1 L_{uu} < 1$) and high ($k_1 L_{uu} > 1$) wavenumber ranges. Hinze (1975) suggested that, in rectilinear shear flows, the mean vorticity, dU/dn, could only be expected to have a significant effect on the distribution of energy in wavenumber space, if it is of comparable magnitude to the root mean squared vorticity fluctuations, $(\epsilon/\nu)^{1/2}$. Weak mean shear would only serve as input to the energy cascade toward higher wavenumbers and the energy spectrum would be nearly isotropic. This prediction would, presumably, be well applicable to relatively small-scale motions. On the other hand, the distortion of large eddies by turbulence would be weak compared to that due to the mean shear, so that, as far as large-scale motions are concerned, turbulence could perhaps be considered as a passive process.

By analogy to the above deductions, one may evaluate the importance of mean streamline curvature by comparing its radius to the instantaneous radius of curvature of a turbulent motion of a particular size, presumably proportional to its wavelength. The present experiments were designed such that the mean streamline curvature would be much smaller than the curvature of the energy containing eddies (in all cases studied, $L_{uu} < 0.05\,R_c$). However, rough spectral estimates have been presented for much larger motions, with wavelengths up to $0.5\,R_c$. In view of wall effects, the breakdown of Taylor's frozen flow approximation and other experimental limitations, the appropriateness of the latter measurements is questionable.

Attempts have been made to explain the observed development of turbulence in terms of mean flow distortion. In rapid distortion theory, the development of the turbulence is assumed to be completely determined by the mean distortion but, in other approaches, some model of the effects of turbulence on its own evolution is used. The mean distortion can be considered as the sum of a mean strain rate, D_{sn}, and a mean rotation rate, Ω_z, equal, respectively, to

$$D_{sn} = \frac{1}{2}\frac{dU}{dn}(1 - S) \tag{7}$$

and

$$\Omega_z = \frac{1}{2}\frac{dU}{dn}(1 + S) \tag{8}$$

Equations (7) and (8) show that flow curvature explicitly affects the rates of both the mean flow rotation and the mean strain. Rectilinear shear flow ($S = 0$) and unsheared curved flow ($S \to \pm\infty$) can be considered as the limiting cases for which the mean rotation and the mean strain are equal. Two other special cases are worth mentioning, irrotational flow ($S = -1$) and rigid body rotation ($S = 1$). With the flow viewed from a frame rotating at a steady rate of U_c/R_c, the mean rotation rate would become

$$\Omega_z - \frac{U_c}{R_c} = \frac{1}{2}\frac{dU}{dn}(1 - S) \tag{9}$$

while the mean strain would, of course, be unaffected. It then follows that, in this frame, a turbulent flow appears to be sheared at the rate $(dU/dn)(1-S)$. Therefore, rotating and curved shear flows have some similarities.

Rapid distortion analyses have been performed by other investigators for three of the above special cases, some results of which will be summarized here, as relevant to the present objectives.

Townsend (1976) calculated the repartition of the kinetic energy of initially isotropic turbulence subject to an irrotational strain, which corresponds, in the present context, to a flow with $S=-1$. He found a general trend of a decrease in the streamwise component and increases in the transverse and spanwise compoents, especially the transverse one, and predicted that the shear stress correlation coefficient would reach a maximum magnitude of 0.45 after the flow was subjected to a small amount of strain. None of the present experimental flows were irrotational on the mean. In the flow with the least amount of rotation ($S=-0.25$), $F_{vv}(k_1)$ grew stronger compared to $F_{uu}(k_1)$, while the relative strength of $F_{ww}(k_1)$ showed little change from its level in rectilinear shear flow (Fig. 4 and 5). The coherence for this flow (Fig. 6) achieved values as high as 0.9 in its, presumably, self-preserving state.

Analyses of rectilinear, uniformly sheared flows ($S=0$) performed by several authors (Hinze, 1975) predicted a relative increase in the streamwise kinetic energy component and reductions in the transverse and spanwise components. The shear stress correlation coefficient was found to reach maximum magnitudes of about 0.6. In all present shear flows in the straight section, $F_{uu}(k_1)$ was nearly an order of magnitude greater than $F_{vv}(k_1)$ and the coherence reached an symptotic maximum of about 0.7, although the shear stress correlation coefficient was about 0.45, lower than the rapid distortion prediction.

When $S=1$, there is no straining and energy cannot be extracted from the mean flow by the turbulence. Furthermore, Jacquin *et. al.* (1990) have shown that rotation would have no effect on initially isotropic turbulence. Analysis of the effects of rotation on initially non-isotropic turbulence by Mansour *et al.* (1991) suggests that the components of kinetic energy would oscillate about their initial values at a rate which would depend on the mean rotation rate. Among the present flows, the one which was the closest to rigid body rotation, on the mean, had $S=0.65$. The kinetic energy of this flow was decaying because the viscous dissipation exceeded the turbulence production, however, Fig. 4, 5 and 10 show clearly that $F_{vv}(k_1)$ and $F_{ww}(k_1)$ were declining faster that $F_{uu}(k_1)$ and that the coherence attained a sign opposite to that in rectilinear shear flow. Measurements of this flow were not made for sufficiently large rotation angles, so that it is uncertain whether the structure was experiencing oscillations of the type suggested by Mansour *et al.* (1991).

In conclusion, rapid distortion analyses account solely for some effects of mean strain and mean rotation. Although they predict correctly some aspects of the turbulence development, they cannot describe the observed patterns of self-preserving growth or decay of the Reynolds stresses nor do they provide insight into the relative sensitivity of different size motions to mean shear, curvature and turbulent straining.

The present results provide a basis for attempting to separate the effects of mean streamline curvature from those of mean shear. As explained in our previous paper (Holloway and Tavoularis, 1992) using physical and mathematical arguments, both of these distortions are capable of producing turbulence. The production of turbulent kinetic energy in transversely homogeneous flows of the present type contains both effects, as

$$P = -\overline{uv}\frac{dU}{dn} + \overline{uv}\frac{U_c}{R_c} \tag{10}$$

In this relationship, the mean shear and the rotation rate appear in separate terms but both terms contain the shear stress, which is affected by both shear and curvature, thus coupling the two production mechanisms. It is also clear that, even in the absence of mean shear, curvature can produce turbulence, if acting on an anisotropic flow. It has generally been assumed by previous investigators that curvature simply distorts and reorients the turbulent eddies produced by mean shear. This may be so, however, one ought to also investigate the alternative possibility of two distinct, although interdependent, production mechanisms, each producing a different range of dominant turbulent structures, whose orientations and sizes are imposed by the mother mechanisms. In doing so, one should be aware that inspection of macroscopic turbulent features, such as the shear stress correlation coefficient, would not necessarily reveal the details of the two production mechansims. For example, the shear stress is related to the cross-spectral density by the integral relationship

$$\overline{uv} = \int_0^\infty Re\,\{F_{uv}(k_1)\}\,dk_1 \tag{11}$$

This creates the possibility that a certain turbulent flow, albeit far from isotropic, could appear as having a vanishing shear stress, if it consisted of the right proportions of two coexisting types of turbulent eddies with opposite orientations, i.e. with opposite signs of their coherences, thus totally concealing the subtleties of two active production mechanisms. Although inspection of measured spectra should be capable of exposing some more structural details, this approach is not free of ambiguity either. Experimental limitations, such as the resolution of hot-wires and the inapplicability of crucial assumptions to the largest measured eddies, are always points of concern. Another limitng factor is the fact that the measured one-dimensional, streamwise spectra cannot describe a three-dimensional structure, because they contain the energy of all eddies whose wavenumber projection on the s-axis is equal to k_1.

Having said the above, let us examine whatever evidence there is, concerning the two production mechanisms. Testing of the "two distinct structures" hypothesis can best be made using results from the extreme cases with very small and very large ratios of the relative "strengths" of the two mechanisms, i.e. with very small and very large $|S|$. The structure of rectilinear, sheared turbulence, containing no curvature effects, is well known. This structure was the starting condition for all present curved flows. On the other hand, the asymptotic structure, if such exists, of non-sheared, purely curved flows has not yet been specified with certainty. Such structure might be encountered in flows with very large positive or negative S. In view, however, of the fact that shear stresses produced by shear and by curvature have the same direction, when $S < 0$, and opposite directions, when $S > 0$ (Holloway and Tavoularis, 1992), one may anticipate that it would require a lower magnitude of negative than positive S for a curved shear flow to approach the alleged asymptotic structure of non-sheared, curved turbulence. As explained earlier, the downstream half of the present flow with $S = -0.27$ seems to be very close to this asymptotic structure. Although, qualitatively, the dimensionless Reynolds stresses and the shapes of the spectra are comparable to those in rectilinear shear flows, it is clear that, compared to shear-dominated structure, curvature-dominated structure exhibits a stronger correlation between the streamwise and transverse components at all wavenumbers, with the possible exception of the largest ones. The present study also includes a case with large, although undetermined, positive S ($S > 5$). The structure of this flow was still developing at the last measuring station, but, if one took the risk of extrapolating its coherence (Fig. 11) following the established streamwise trend, one might once more recover the curvature-dominated shape of the case with $S = -0.25$. It then follows that, after an initial period of suppression of turbulence, this flow might also exhibit turbulence growth, provided that all relevant conditions continue to apply.

Rather than the two limiting cases, flows subject to the simultaneous action of shear and curvature, at intermediate values of S, would be of more practical interest. It is conceivable that such flows may approach asymptotic structures, where the dominant eddies would either be of a type intermediate between those in purely sheared and purely curved flows or would consist of two distinct populations, one produced by curvature and the other by shear. If the second possibility were the case, the coexistence of these two types of eddies would be most tenable if these eddies had very different sizes so that their interaction would be weak. Energy production by mean shear and curvature is characterized, respectively, by the time scales $(dU/dn)^{-1}$ and $(U_c/R_c)^{-1}$, the ratio of which is equal to S. Combining the r.m.s. kinetic energy (or approximately, u'), as a velocity scale, with the above time scales, one may estimate, in order-of-magnitude, the size of eddies likely to be produced by the two production mechanisms. Assuming that the same velocity scale continues to characterize both production mechanisms in the curved section, then the curvature-produced eddies and the shear-produced eddies would, typically, have a size ration of $1/S$ and a wavenumber ratio of S. According to this scaling argument, flows with small magnitudes of $|S|$ would contain two types of eddies with substantially different sizes. Furthermore, in flows with small positive S, the two types of eddies would have opposite orientations and curvature-produced eddies would manifest their presence by tending to reverse the sign of the coherence spectrum at small wavenumbers. On the other hand, in flows with small negative S, the two types of eddies would have similar orientations and their individual effects on the coherence would be less distinct.

The present experimental evidence is generally compatible with the above arguments. At the start of curvature, all present flows were shear-dominated and, as a result, their energy containing range would correspond to the spectral region around $k_1 L_{uu} = 1$. The previous arguments lead to the anticipation that the effects of curvature would first become visible in a range around $k_1 Luu = S$. As these effects become more established, they would also start affecting the magnitude of L_{uu}, thus creating further complications. At any rate, in the flow with $S = 0.03$ (Fig. 8), the effects of curvature seem to extend from the lowest measured wavenumber to about $k_1 L_{uu} = 1$ and, while they tend to decrease the coherence, they are not strong enough to cause a sign reversal. It is worth noting that, in this case, eddies corresponding to $k_1 L_{uu} = S = 0.03$ would be larger than the tunnel height and the estimates of $C_{uv}(k_1)$ shown at the lowest wavenumbers ($k_1 L_{uu} = 0.05$) are relatively uncertain. It is clear, however, that the coherence in the neighbourhood of $k_1 L_{uu} = 1$ has effectively resisted the influence of flow curvature, being only slightly different from that in the initial shear flow. One may plausibly conclude that two distinct types of eddies with different coherences and different sizes coexist in this flow. In fact, the difference between these two structures may be greater than what the one-dimensional spectra indicate, because these spectra contain contributions of all smaller-size motions, whose wavenumbers project on the streamwise axis as k_1.

The above conclusion is reinforced by the development of coherence in the flow with $S = 0.10$ (Fig. 9). In this case, $Re\{C_{uv}(k_1)\}$ first increased with increasing wavenumber to reach a near-zero maximum near $k_1 L_{uu} = 0.1$ and then it decreased to a minimum near $k_1 L_{uu} = 1$. This complex behaviour of the coherence seems to persist downstream in spite of the dramatically different structure of the large-scale and small-scale eddies. The maximum of the coherence might be associated with the effects of curvature and the minimum with the effects of shear. The ratio of the respective wavenumbers, where these extremes occured, was approximately equal to S, as predicted by our scaling arguments.

In the flow with $S = 0.65$ (Fig. 10), the effects of curvature on the coherence appeared first around $k_1 L_{uu} = 0.7$ but, unlike the cases with lower S, the coherence spectrum was nearly overturned after some time ($\tau - \tau_0 > 1.1$), so that the coherence at most wavenumbers became positive. Near the last measuring station, the coherence demonstrated a rather weak rate of change, appearing to approach an asymptotic shape. The most interesting aspect of this flow was that, during its development (for example at $\tau - \tau_0 = 0.58$), the coherence of the low wavenumber motions was positive while the coherence of the higher wavenumber motions was negative. While this was not a self-preserving structure, it is clear that turbulent eddies dominating the large-scale structure had an orientation opposite to that of eddies dominating smaller scales.

In the flow with $S > 5$, the effects of curvature first appeared near $k_1 L_{uu} = 0.8$ and then spread to higher and lower wavenumbers until, at a position with $\theta = 0.67$, the entire coherence spectrum turned positive. In this case, the present scaling arguments would predict that curvature would produce eddies at $k_1 L_{uu} > 5$, which is to the right of the initial minimum of the coherence. At these high wavenumbers, however, neither curvature nor shear could have a dominant influence on the eddy structure, because of the strong turbulent distortion associated with the higher wavenumbers. With the exception of the flow with $S = 0.03$, the turbulent kinetic energy in all flows with positive S was decaying, introducing the possibility that some of the structures contributing to the coherence would be unrelated to curvature but they constitute the remnants of some initial, inactive structures, which have been isolated by the decreased net energy production and the persisting dissipation of energy, especially at the smaller eddies. As explained earlier, however, even in these flows, curvature might play an active role in energy production, although it cannot compensate for the energy dissipation, which is strong, as it scales with the initially high turbulence level. Nevertheless, it is possible that the growth of turbulence in these flows may resume farther downstream, where the turbulence level, and thus, the dissipation, would be lower. The absence of net energy production cannot be used as an indication of flow inactivity, in cases where more than one production mechanisms coexist, for reasons explained in the discussion following equation (11). For example, during the development of the flow with $S = 0.65$ (Fig. 10), part of the coherence was positive and part was negative, indicating that large eddies were transporting momentum away from the center of curvature, while small eddies were transporting momentum in the opposite direction. With these two transporting mechanisms acting simultaneously and at proportioned magnitudes, the net momentum transport was zero, so that no turbulence kinetic energy was produced.

Conclusions

Neither shear nor curvature appear to have a strong influence on the fine structure of turbulence, while they both tended to increase the anisotropy of the large-scale structure. In general, curvature modified the coherence of the streamwise and transverse fluctuations, tending to cause momentum to be transported away from the center of curvature. In flows with negative values of S, curvature tended to enhance the relative strength of the large-scale transverse motions and to increase their coherence. In flows with small positive S, curvature diminished the coherence and suppressed the growth of kinetic energy. In flows with large positive S, the curvature effects overcame the tendency for momentum to be transported down the gradient of mean velocity and could cause a net transport away from the center of curvature.

Acknowledgement

Financial support by the Natural Sciences and Engineering Research Council of Canada is gratefully acknowledged.

References

Hinze, J.O. (1975): *Turbulence* (2nd ed.) (McGraw-Hill, New York)

Holloway, A.G.L. and Tavoularis, S. (1992): The effects of curvature on sheared turbulence. J. Fluid Mech. (in press)

Jacquin, L., Leuchter, O., Cambon, C. and Mathieu, J. (1990): Homogeneous turbulence in the presence of rotation. J. Fluid Mech., **220**, 1-52

Mansour, N.N., Shih, T. and Reynolds, W.C. (1991): The effects of rotation on initially anisotropic homogeneous flows. Phys. Fluids A, **3**, 2421-2425

Ramaprian, B.R. and Shivaprasad, B.G. (1978): The structure of turbulent boundary layers along mildly curved surfaces. J. Fluid Mech. **85**, 273-303

Tavoularis, S. and Karnik, U. (1989): Further experiments on the evolution of turbulent stresses and scales in uniformly sheared turbulence. J. Fluid Mech. **204**, 457-478

Townsend, A.A. (1976): *The Structure of Turbulent Shear Flow* (2nd ed.) (Cambridge University Press, Cambridge)

Wyngaard, J.C. (1968): Measurement of small scale turbulence structure with hotwires. J. Sci. Instru., **1**, 1105-1108

Experimental Investigation of the Three-Dimensional Boundary Layer on a Rotating Disk

Howard S. Littell[1] and John K. Eaton[2]

[1] Present address: Shell Development Corporation, Houston, TX, USA
[2] Dept. of Mechanical Engineering, Stanford University, Stanford CA 94305, USA

Abstract

The velocity field above a large spinning disk has been studied using pressure probes and hotwire anemometers. The flowfield consists of a three-dimensional boundary layer due to a crossflow caused by centrifugal forces. The mean flow in both the laminar and turbulent regimes compares well with previous studies of 'infinite' smooth rotating disks, and all six Reynolds stresses and the ten triple products have been measured using established crosswire anemometry techniques. Selected results are compared to previous three-dimensional boundary layer measurements and several important differences are noted.

1. Introduction

Three-dimensional turbulent boundary layers (3DTBL's) are becoming the focus of increased research interest as modelers recognize their practical importance and the key differences between them and their more familiar cousins, the two-dimensional boundary layers. The term three-dimensional boundary layer generally means a wall-attached shear layer in which the flow direction varies with distance from the wall. Thus, regardless of the choice of coordinate system, both velocity components parallel to the wall vary with distance from the wall and both wall-parallel vorticity components are significant.

The turbulence in three-dimensional boundary layers has been studied in a variety of geometries including flow approaching a swept forward-facing step, (Johnston, 1970), flow in ducts simulating infinite swept wings (Elsenaar and Boelsma, 1974, van den Berg *et al.*, 1975, Bradshaw and Pontikos, 1985), flow approaching obstacles (Dechow and Felsch, 1977, Anderson and Eaton, 1989) and axial flow along a rotating cylinder (Driver and Johnston, 1990). In addition, there have been several recent attempts to explore three-dimensional effects on turbulence using direct numerical simulation (Spalart, 1989, Coleman *et al.*, 1990). In cases where an initially two-dimensional boundary layer is turned, thus becoming three-dimensional, it is generally found that the vector formed by the shear stress components parallel to the wall lags behind the strain rate vector. Also, it is frequently found that the turbulent shear stress levels are reduced relative to the turbulent kinetic energy. It is not clear if this latter effect is caused by sudden turning of the flow or if it is a general feature of all boundary layers containing mean flow three dimensionality.

It is difficult to extract general conclusions from the sum of the previous experiments because they are often complicated by extraneous effects such as streamwise pressure gradient and spanwise inhomogeneity. In many 3DTBL experiments all three space coordinates are significant and any analysis of the Reynolds stress transport equations, for instance, requires an extensive data set. The appropriate reference frame for analysis of the data is not apparent because the flow direction varies in all three directions.

We chose to investigate the flow on a rotating disk to simplify the analysis of the resulting data. There is no externally imposed pressure gradient and the flow is axisymmetric so at most only two space coordinates are needed to describe the experiment. In addition, variations in the radial direction are very slow and can be neglected except near transition. The disk flow does develop a substantial skewing across the boundary layer. At its peak the radial velocity component reaches a level of 11% of the local disk speed. An added advantage of using a disk to investigate 3DTBL's is that the Reynolds number can be changed easily by varying the disk speed. There are key differences, though, between a disk boundary layer experiment and previous 3DTBL experiments. First, the skewing is always present rather than developing as a pressure-induced perturbation. Second, turbulence may eventually reach a state of equilibrium with the imposed three-dimensional strain field. Finally, flow rotation adds the complexity of the Coriolis effects to the basic equation set.

The objectives of the present research program are to supply detailed mean velocity data and turbulence measurements up to third-order quantities. We wish to assess if this 3DTBL, which presumably has reached an equilibrium state, is fundamentally different in its turbulence properties than well understood 2DTBLs.

2. Experimental Apparatus

The experimental apparatus consists of a 1 meter diameter aluminum disk mounted on a vertical spindle. The 15 mm thick disk was precision machined and lapped to allow only ± 15 μm vertical deviations when the disk rotates. The disk spins at speeds up to 1500 RPM which corresponds to a tip Reynolds number of 2.5 million. It is driven by a 2 hp DC motor through a belt drive to avoid vibrations. The motor is equipped with solid state speed control allowing smooth variation of the disk speed under manual or computer control. The rotating assembly was carefully balanced by virtue of its precise manufacture, and there is no measurable vibration at any operating speed.

A fixed annular apron approximately 0.3 m wide surrounds the disk to eliminate any interaction between the boundary layers on the top and bottom surfaces. The bottom surface runs in a closed cavity to avoid creating any disturbance in the test cell. A set of 14 spiral-shaped vanes are mounted on the upper surface of the apron to remove the angular momentum of the flow leaving the disk. We observed no significant swirl in the 2.4 m by 2.7 m by 3.0 m isolation cell with the vanes in place.

All experiments are conducted under the remote control of a data acquisition system consisting of an IBM PC-AT with a Metrabyte DAS-20 multifunction card and an SSH-4 simultaneous sample and hold. The computer controls the disk speed and all the measurement systems. Probes are positioned by a stepper motor traverse mounted on a 1″ × 4″ beam located 1 meter above the disk surface. The traverse can rotate the probe around its stem axis with a resolution of 0.9° and translate it normal to the disk with a resolution of 1.6 μm.

Mean flow data are acquired using a three-hole probe in a non-nulling mode as described by Anderson and Eaton (1987). The dynamic and differential pressures are measured using a Setra model 239 (±1.0″H_2O range) or a model 264 (0–10″H_2O range) for high speed dynamic pressure. The transducers are thermally insulated to minimize drift and are calibrated against a micromanometer before each run.

Three different hotwire probes are used. Each probe is mounted in a gooseneck stem allowing yaw of the probe without moving the measurement volume. The single-wire probe uses a Dantec 55P05 boundary layer tip strung with Dantec gold plated 5 μm platinum coated tungsten wire with an active length to diameter ratio of $l/d = 250$. The cross wire probe was custom built. It uses 2.5 μm platinum coated tungsten wire which has been copper plated and subsequently etched for an active length to diameter ratio of $l/d = 200$. The wire spacing is 0.35 mm. The crosswire can be rotated about its axis in 45° increments allowing determination of all six Reynolds stresses. The third probe is a custom dual wire probe which has two wires at ±45° to the stem axis in the same plane, parallel to the disk surface. This probe is used to provide redundant measurements of the flow direction, and was strung with 5 μm platinum coated tungsten wire, copper plated and etched for an active length to diameter ratio of $l/d = 200$.

All hotwires are operated in constant temperature mode by a TSI IFA-100 constant temperature anemometer. The signals are DC shifted and amplified to utilize the full ±5 Volt range on the Metrabyte A/D. Noise is filtered with a Frequency Devices model 901F1 low-pass filter. The test cell temperature is not controlled so it is necessary to calibrate the hotwires for both velocity and temperature variation. This is accomplished by a portable jet apparatus which is placed on the disk and allows precise control of the flow temperature and velocity. A temperature dependent calibration is obtained by varying both the temperature and velocity and fitting King's law in a modification of the temperature correction suggested by Cimbala and Park (1990). In this implementation, the original form of King's law, $Nu = a + b \cdot Re^n$ is used, with fluid properties evaluated using the film temperature: $T_f = \frac{1}{2}(T_a + T_w)$. The calibration is then of the following form,

$$U = \nu(P_{atm}, T_f) \cdot \left[\frac{B \cdot e^2}{k(T_f) \cdot (T_w - T_a)} - E_o^2\right]^{1/n}$$

This calibration is implemented using an empirically obtained wire temperature specific to the type of wire, then B and E_o^2 are found with a least squares fit. The typical RMS deviation of the fit from 20 calibration points varying from 1 to 40 m/s was 0.2% at a single temperature, and < 1.5% for a temperature which varied 10°C.

The mean flow velocity and direction are measured using three different techniques, the three-hole probe corrected for shear and wall proximity, the custom dual wire probe, and finally by yawing the single-wire probe in an automated version of the Cham and Head (1969) technique. For mean flow measurements, 4000 samples are acquired at

100 Hz. The crosswire probe is aligned with the previously measured mean flow direction at each measurement point then 10,000 samples of each wire are acquired in each of the four probe orientations. The measured Reynolds stress and triple product tensors are then rotated into appropriate coordinates. The uncertainty involved in the Reynolds stress data are similar to those reported by Anderson and Eaton (1989) near the disk surface, but farther out in the boundary layer the local turbulence intensities became quite large. Therefore the measurements in the outer part of the boundary layer are subject to question. The analysis of Tutu and Chevray (1975) suggests that the crosswire is subject to very large errors for turbulence intensity greater than 35%. However, their analysis predicts a strong disagreement between single-wire and crosswire measurements for large turbulence intensity. Figure 1 shows a comparison of the longitudinal turbulence intensity as measured by the two probes. The agreement is excellent out to a turbulence intensity of at least 60%, and even beyond this point the differences are of the opposite sign than predicted by the analysis. This may indicate that the flow structure of the disk boundary layer is not of the simple joint Gaussian form used in the analysis, nevertheless the Reynolds stress measurements in the outer part of the boundary layer are certainly subject to higher uncertainty due to the high turbulence intensity.

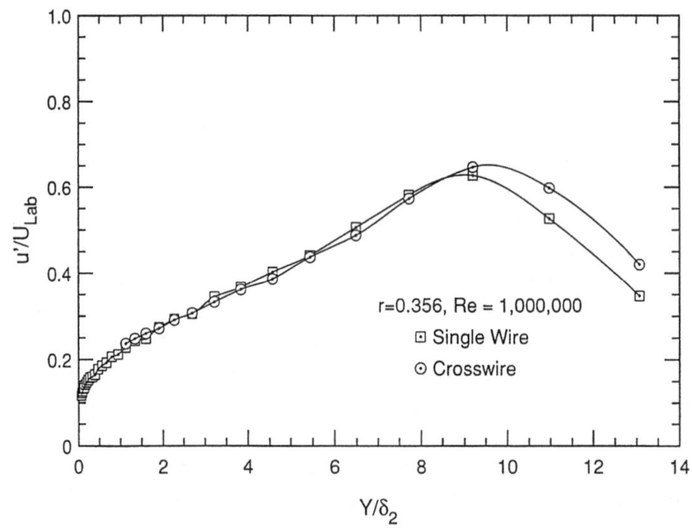

Fig. 1. Comparison of measured turbulence intensity between single boundary layer probe and crosswire.

3. Results and Discussion

The mean velocity and turbulence measurements are presented in a reference frame rotating with the disk so the velocity profiles appear similar to those in a typical wind tunnel experiment. This also gives direct mathematical formulation of the Coriolis effects, and allows the centrifugal forces to be included into the pressure term. The frame of reference is defined with its origin at the radius of interest and is right-handed with $+Y$ directed upwards from the disk surface, $+Z$ pointing at the axis of rotation, and $+X$ directed in the circumferential direction.

One unwanted consequence of the transformation is that small uncertainties in measured flow angle close to the disk lead to very high uncertainties in the flow angle viewed in the rotating reference frame. This is due to the fact that once the local disk speed is subtracted from the measured velocity, the measurement uncertainty is retained as a percentage of the (large) measured value. For this reason the experimentally measured mean values tend to show much more scatter close to the disk than the scatter in the turbulence statistics, which are not subject to this effect.

Both laminar and turbulent cases were examined. The laminar cases are not shown here because they were used primarily for instrumentation qualification. Agreement with an analytic solution was excellent for all three probes especially at the inner radii where the disk appeared to the flow as if it were indeed infinite. The turbulent cases to be presented here are listed in Table 1, along with integral properties of the flow in the X direction. The integral properties were calculated by direct numerical integration of the measured velocity profiles and the skin friction was calculated using the two-dimensional law of the wall with $\kappa = 0.41$. It is apparent from this information that relatively high Reynolds numbers are produced with a small experiment, due to the high speeds involved.

Table 1. Disk Flow Turbulent Cases

Symb.	Radius / Re	U_∞[m/s]	δ_{99}[mm]	δ_2[mm]	Re_{δ_2}	H	G	U_τ[m/s]	ν/U_τ[μm]
	R = 0.235 m								
×	Re = 400,000	26.13	11.60	0.978	1657	1.342	5.522	1.206	12.8
	R = 0.356 m								
⊡	Re = 650,000	28.22	19.22	1.457	2662	1.290	5.055	1.255	12.3
⊙	Re = 940,000	41.01	20.05	1.456	3839	1.269	4.968	1.750	8.9
	R = 0.421 m								
□	Re = 650,000	23.72	21.04	1.712	2645	1.298	5.181	1.051	14.6
○	Re = 1,000,000	36.72	20.61	1.641	3903	1.282	5.191	1.556	9.9
△	Re = 1,300,000	48.14	20.39	1.594	4969	1.270	5.125	1.997	7.7

Figure 2 shows the mean flow profiles in the $+X$ direction plotted using wall scaling. One immediately notices the lack of a wake component, which is likewise absent in the data of Cham and Head (1969). This is not the result of a favorable pressure gradient, since the X direction will not support a mean pressure gradient, but instead may be taken as an indication of structural differences between this 3DTBL and the typical zero pressure gradient 2DTBL.

The primary Reynolds stress $-\overline{u'v'}$ is shown in Figure 3, normalized with the circumferential shear velocity. It appears much the same as a zero pressure gradient 2DTBL, and totally unlike the 3DTBL flows in Table 2 which all exhibit a peak away from the wall. The effect seen in other 3DTBL's can not be ascribed only to an adverse pressure gradient, but is also the result of some type of structural change in the turbulence which will be discussed more fully below.

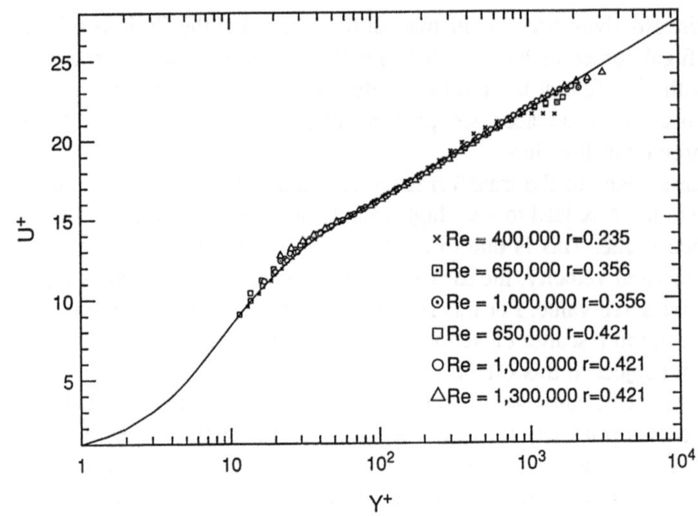

Fig. 2. Mean circumferential velocity in wall units.

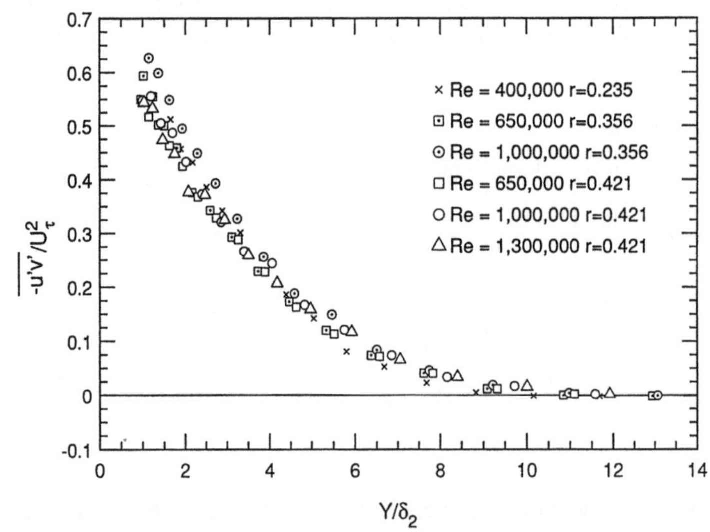

Fig. 3. Primary Reynolds stress $-\overline{u'v'}$, normalized by the shear velocity.

The secondary Reynolds stress $-\overline{v'w'}$ is shown in Figure 4, normalized with the circumferential shear velocity. This quantity is a measure of the effect on the turbulence of the spanwise strain, so it is identically zero in a 2DTBL. It is interesting to note that $-\overline{v'w'}$ has a nearly constant value in the locations the crosswire could acquire data, but must change sign closer to the wall in order to approach the wall shear stress value as expected. The sign change is necessary because the spanwise mean velocity profile has a maximum, and therefore the outer layer sees the opposite shear than the inner

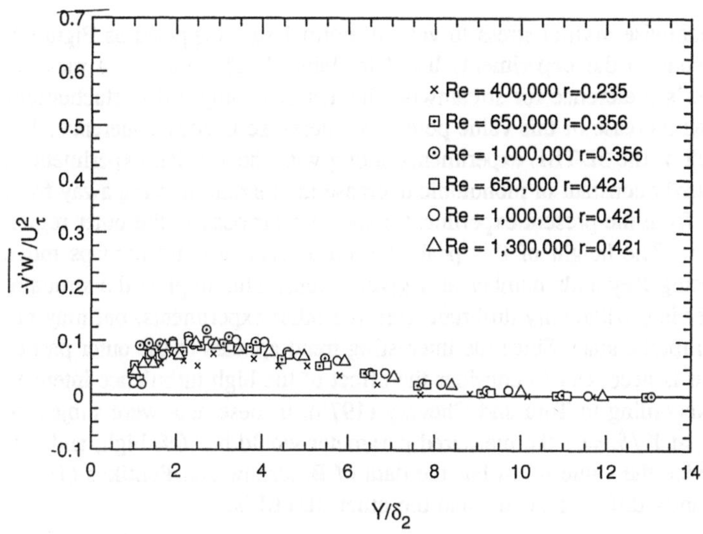

Fig. 4. Secondary Reynolds stress $-\overline{v'w'}$, normalized by the shear velocity.

layer. The sign of the stress is positive in the region plotted because the crossflow is in the -Z direction. These measurements resemble previous 3DTBL experiments plotted in freestream coordinates.

It is always difficult to gain a physical understanding of the pattern of the turbulent fluctuations from single point statistics, and even more so to compare results between experiments. In the process of Reynolds averaging the Navier Stokes equations, quantities (Reynolds stresses) result which are the consequence of average momentum transport by the fluctuations. In attempts to form 'closure' models for these terms, several quantitative relationships between the various turbulent quantities have been found to have a predictable value or behavior in a certain type of flow. These 'structural parameters' allow similarities and differences in fluctuating fields and thus the physics of flowfields to be compared. Selected structural indicators from the present flow will be compared to previous experiments listed in Table 2. It is important to note that three major categories of 3DTBLs are represented: infinite swept wing, obstacle, and shear driven. These combined with Klebanoff's classic 2DTBL should represent the current state of knowledge of the effect of three-dimensionality on an initially 2DTBL from wind tunnel experiments.

Table 2. Reference Experiments

Symb.	Experiment	Flow Type/Designation
A	Anderson/Eaton (1989)	Obstacle, Case I St 4
P	Bradshaw/Pontikos (1985)	Infinite Wing, X=1092
R	Driver/Johnston (1990)	Shear Driven, St 9
E	Elsenaar/Boelsma (1974)	Infinite Wing, X=1095
K	Klebanoff (1955)	2DTBL Re_{δ_2}= 8000

The ratio of streamwise normal stress to vertical normal stress appears as Figure 5, along with comparisons to the experiments listed in Table 2. This ratio is a measure of the boundary layer's preference for streamwise fluctuations compared to fluctuations normal to the wall; an increase of this value points to a decrease in eddy coherence. The behavior seen in each of the 3DTBL experiments, along with the 2DTBL experiment of Klebanoff, show a nearly constant or monotonic decrease in this ratio moving away from the wall. Every case from the present experiment shows a clear peak in the outer region of the boundary layer. The height of this peak is seen to decrease and perhaps move outward with increasing Reynolds number at a given radius. This implies the structure of the boundary layer is substantially different from the other experiments, but may not have reached an asymptotic state. Since the interesting trend appears in the outer part of the boundary layer, it is necessary to consider the effect of the high turbulence intensity on this parameter. According to Tutu and Chevray (1975), if these data were subject to their error estimates, at $Y/\delta_2 = 7$ the measured parameter would be 20% high, and it is seen to be almost twice the value of all but the data of Bradshaw and Pontikos (1985), which show significantly different trends than the other 3DTBL's.

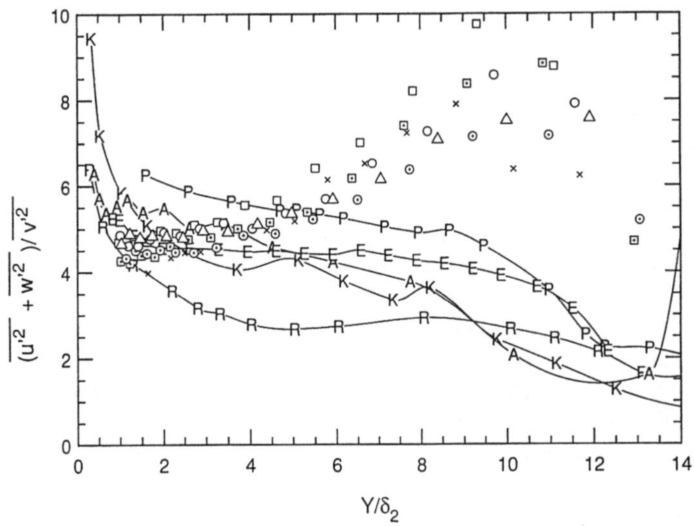

Fig. 5. Ratio of streamwise normal stress to vertical normal stress. (Symbols as in Tables 1 and 2)

Another commonly investigated structural parameter is the ratio of the vector magnitude of the shear stress to twice the turbulent kinetic energy:

$$A_1 = \frac{\sqrt{\overline{u'v'}^2 + \overline{v'w'}^2}}{q^2}$$

This parameter is approximately constant at 0.15 in a 2DTBL even when the layer is strongly distorted by pressure gradient. The wind tunnel experiments plotted in Figure 6 show the typically observed trend that three dimensionality depresses A_1 near the wall, and this depression diffuses outward. As noted above, A_1 is not strongly affected by pressure gradient alone, and the general agreement among 3DTBL experiments implies that three dimensionality affects the flow structure in this qualitatively predictable fashion. The disk flow is vastly different, showing A_1 to decrease almost linearly with distance from the wall for the inner radii, and at least monotonically for the outer radius. The consideration of this effect combined with that in Figure 5 leads one to the conclusion that irrotational (inviscid) motion plays an even bigger role in the outer region of the disk boundary layer than in a wind tunnel layer.

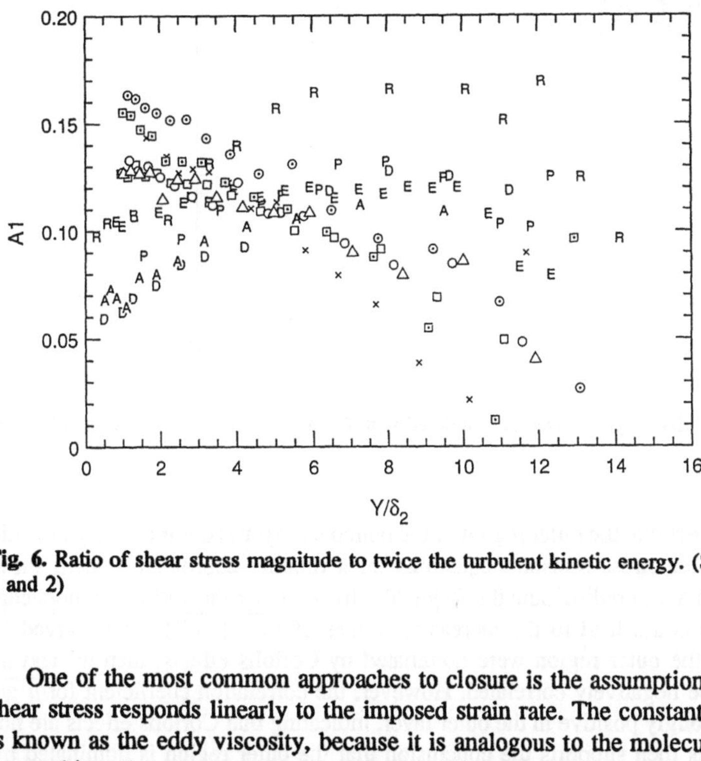

Fig. 6. Ratio of shear stress magnitude to twice the turbulent kinetic energy. (Symbols as in Tables 1 and 2)

One of the most common approaches to closure is the assumption that the Reynolds shear stress responds linearly to the imposed strain rate. The constant of proportionality is known as the eddy viscosity, because it is analogous to the molecular viscosity when inserted into the mean momentum equation. The ratio of the relative effectiveness of the turbulence in transporting streamwise and the spanwise momentum is defined as:

$$N_e = \frac{\tan(\gamma_\tau - \gamma_{fs})}{\tan(\gamma_g - \gamma_{fs})}$$

The angle γ_{fs} is that of the local mean velocity, γ_g is the angle of the velocity gradient vector, and γ_τ is the shear stress vector angle. This may also be thought of as the ratio of local spanwise to local streamwise eddy viscosity. The values obtained for N_e are shown in Figure 7. These data exhibit more scatter than most of the other plots because they contain derivatives of experimental data. As concluded by Anderson and Eaton (1989), low values in N_e (as low as 0.2) are observed in sharply turned experiments such as shear driven or obstacle flows, and a value closer to one in slowly turned flows like the infinite wing. This implies that the disk flow more closely resembles a slowly turned flow or equivalently one which has more time to relax to a new state after the imposition of crossflow. The values near unity indicate that an isotropic eddy viscosity should perform well, and this was indeed the case for a simple crossflow calculation for the disk flow reported by Cham and Head (1969).

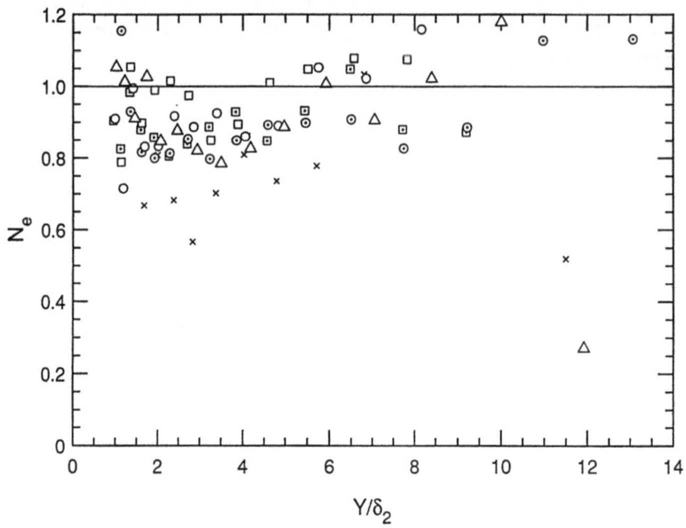

Fig. 7. Ratio of eddy viscosity in coordinates aligned with the local mean velocity. (Symbols as in Tables 1 and 2)

One might suspect that the outer region of the boundary layer is dominated by Coriolis effects. The Reynolds stress transport equations for a rotating coordinate system show that Coriolis effects act to redistribute the Reynolds stress among the various components of the tensor. This could lead to the increasing values of $(\overline{u'^2} + \overline{w'^2})/\overline{v'^2}$ observed in the outer layer. If the outer region were dominated by Coriolis effects, then w' and u' fluctuations must be negatively correlated. However, the correlation coefficient for $\overline{u'w'}$ was found to be strongly positive in the outer layer, indicating that Coriolis effects are not a major factor. This then supports the conclusion that the outer region is dominated by irrotational (inviscid) motions, a conclusion also supported by Cham and Head's (1969) measurement of reduced entrainment.

It remains to address the cause of the reduced level of shear stress in the disk boundary layer. Bradshaw and Pontikos hypothesized that turbulent eddies formed in a 2D boundary layer are tilted out of their preferred orientation by the imposition of three-dimensional skewing. Implicit in their discussion was the assumption that normal boundary layer turbulence would eventually develop if the 3D strain field remained constant for a long enough distance. Anderson and Eaton (1989) theorized that the turbulence is stabilized in the region of peak turbulence production by the presence of crossflow. Eaton (1991) took this further, stating that production is reduced because a fraction of the low speed streaks in the boundary layer are eliminated by the crossflow and because longitudinal vortices are attenuated by the crossflow. Implicit in this argument is the assumption that boundary layer turbulence developed in a three-dimensional strain field is inherently less efficient at extracting energy from the flow field. The present results support the latter hypothesis. The value of A_1 is suppressed below normal 2D levels even though the entire boundary layer development occurs in a nearly constant three-dimensional strain field.

4. Conclusions

The disk boundary layer, while subject to Coriolis effects, appears to be a good configuration for studying the fundamental physics of 3DTBL's. Many of the effects observed in wind tunnel 3DTBL's are also observed here. The A_1 structural parameter is less than in comparable 2D boundary layers and a small lag in the shear stress direction behind the mean velocity gradient angle was noted, behaving not unlike a slowly-turned three-dimensional boundary layer. In the disk flow, A_1 was at a maximum near the wall at close to the 2-D value, but dropped off almost linearly away from the wall. The trend seen for the ratio $(\overline{u'^2} + \overline{w'^2})/\overline{v'^2}$ was also substantially different from 2DTBL's and other 3DTBL's. Both results suggest that the transport of momentum by the Reynolds shear stresses is somehow curtailed for the disk flow compared to other boundary layers. The reduction in A_1 in the outer region of the disk flow boundary layer suggests that the modification of A_1 observed in 3DTBL's is not merely a disequilibrium effect. More research is need to determine the generality of this conclusion. The primary shear stress $-\overline{u'v'}$ normalized by the wall friction velocity was seen to have a value of 0.6 at $y^+ = 110$, which is considerably lower than in a 2DTBL at the same Reynolds number. This indicates a very small, or non-existent, constant Reynolds shear stress region near the wall.

Acknowledgements

This work is supported by the Department of Energy Office of Basic Energy Sciences Grant DE-FG03-86ER13608. The first author would also like to express his appreciation for the support in the form of a scholarship from the Society of Experimental Test Pilots. The numerous discussions with Prof. James P. Johnston and Prof. Peter Bradshaw are especially appreciated.

References

Anderson, S.D. and J.K. Eaton (1989) "An Experimental Investigation of Pressure Driven Three-Dimensional Boundary Layers," *J. Fluid Mech.* **202:** 263–294.

Bradshaw, P. and N.S. Pontikos (1985) "Measurements in the Turbulent Boundary Layer on an 'Infinite' Swept Wing," *J. Fluid Mech.* **159:** 105–130.

Cham, T-S. and M.R. Head (1969) "Turbulent Boundary-Layer Flow on a Rotating Disk," *J. Fluid Mech.* **37**, part 1: 129–147.

Cimbala, J.M. and W.J. Park (1990) "A Direct Hot-Wire Calibration Technique to Account for Ambient Temperature Drift in Incompressible Flow," *Exp. in Fluids* **8**: 299–300.

Coleman, G.N., J.H. Ferziger, and P.R. Spalart (1990) "A Numerical Study of the Turbulent Ekman Layer," *J. Fluid Mech.* **213**: 313–348.

Dechow, R. and K.O. Felsch (1977) "Measurements of the Mean Velocity and of the Reynolds Stress Tensor in a Three-Dimensional Turbulent Boundary Layer Induced by a Cylinder Standing on a Flat Wall," *Proc. 1st Turbulent Shear Flows Symposium*, University Park, Pennsylvania, Apr. 18–20, 1977.

Driver, D.M. and J.P. Johnston (1990) "Experimental Study of a Three-Dimensional Shear-Driven Turbulent Boundary Layer with Streamwise Adverse Pressure Gradient,"*NASA TM* 102211, and *Stanford Univ., Dept. Mech. Eng. Thermosciences Div. Rep.* MD-57.

Eaton, J.K. (1991) "Turbulence Structure and Heat Transfer in Three-Dimensional Boundary Layers," *Proc. 9th Symp. on Energy Engineering Sciences*, Argonne Natl. Lab.

Elsenaar, A. and S.H. Boelsma (1974), "Measurements of the Reynolds Stress Tensor in a Three-Dimensional Turbulent Boundary Layer Under Infinite Swept Wing Conditions," *NLR TR* 74095 U.

Johnston, J.P. (1970) "Measurements in a Three-Dimensional Turbulent Boundary Layer Induced by a Swept, Forward-Facing Step," *J. Fluid Mech.* **42**: 823–844.

Johnston, J.P. (1973) "The Suppression of Shear Layer Turbulence in Rotating Systems," *J. Fluid Mech.* **45**: 229–236.

Klebanoff, P.S. (1954) "Characteristics of Turbulence in a Boundary Layer With Zero Pressure Gradient," *NACA TN* 3178 superseded by *NACA Rep.* 1247.

Spalart, P.R. (1989), "Theoretical and Numerical Study of a Three-Dimensional Turbulent Boundary Layer," *J. Fluid Mech.* **205**: 319–340.

Tutu, N.K. and R. Chevray (1975) "Cross-Wire Anemometry in High Intensity Turbulence," *J. Fluid Mech.* **71**: 785–800.

Reorganization of Coherent Vortices in Shear Layers under the Action of Solid-Body Rotation

O. Métais[1], S. Yanase,[1 2] C. Flores[1], P. Bartello,[1 3] M. Lesieur[1]

[1] Institut de Mécanique de Grenoble, B.P. 53X, 38041 Grenoble-Cedex, FRANCE
[2] Faculty of Engineering, Okayama University, Okayama 700, JAPAN
[3] Recherche en prévision numérique, Service de l'Environnement Atmosphérique, Dorval (Québec), H9P 1J3 CANADA

Abstract

We examine the effects of solid-body rotation, characterized by an angular velocity $\vec{\Omega}$, on a mixing layer and a plane wake (in a plane perpendicular to $\vec{\Omega}$) upon which is superposed a small three-dimensional random perturbation. Using Kelvin's theorem in a frame rotating with $\vec{\Omega}$, and with the aid of arguments based either on the straining of absolute vortex filaments by the ambiant shear, or introducing in a simplified case an exact solution whose results are close to a three-dimensional linear stability analysis, it is proposed that rotation is always stabilizing (with respect to the non-rotating situation) in the cyclonic case. In the anticyclonic case, a slight rotation is destabilizing, with an intensification of the stretching of longitudinal hairpin vortices. This effect is maximum at a local Rossby number of unity according to the nonlinear theory. From the exact solution point of view, this value $R_o = 1$ is the crossover between stability and instability, as already predicted by former studies. At high rotation rates (low Rossby numbers), both cyclonic and anticyclonic vortices are stabilized, in agreement with the Taylor-Proudman theorem. In order to assess these theoretical predictions, we perform three-dimensional numerical simulations. We find a critical Rossby number (based on the maximum initial vorticity of the basic flow) at which maximum three-dimensional anticyclonic destabilization is achieved. In this case, a detailed examination of the structures shows that the flow is highly anisotropic with very elongated streamwise hairpin vortices of small spanwise wavelength. Finally, the same theoretical considerations applied to rotating developed turbulence permit to infer that, for turbulent Rossby numbers of the order of one, cyclonic vortices of axis parallel to $\vec{\Omega}$ are reinforced, while anticyclonic vorticity is disrupted. We present calculations which support this conjecture in the case of initially quasi two-dimensional turbulence. The applications for geophysical or engineering flows are briefly discussed.

1. Introduction

Laboratory experiments on shear flows show stabilizing or destabilizing effects of a constant solid-body rotation $\vec{\Omega}$, according to the cyclonic (with vorticity parallel and of the same sign as the solid-body vorticity $2\vec{\Omega}$) or anticyclonic nature of the eddies considered. Witt and Joubert (1985) found that the wake of a cylinder whose axis is parallel to $\vec{\Omega}$ becomes asymmetric under the effect of rotation. This is confirmed by Chabert d'Hières et al. (1988): they show, at Rossby numbers $U/2\Omega D$ of the order of one (where U is the upstream velocity and D the diameter of the obstacle) that cyclonic vortices are reinforced and anticyclonic destroyed. Instead, at low Rossby number, the wake reorganizes into a very stable Karman Street of alternate vortices. Rotating mixing-layer experiments (Rothe and Johnston 1979; Bidokhti and Tritton 1990) exhibit (at $R_o \approx 1$) the same behaviour according to the cyclonic or anticyclonic nature of the shear layer.

A solid-body rotation of constant angular velocity $\vec{\Omega} = \Omega\vec{z}$ does not have any influence on a two-dimensional flow in the plane (x, y) (see Lesieur 1990). Therefore, the phenomena observed in the laboratory experiments can only be explained by considering the influence of rotation on the growth of three-dimensional perturbations. In this paper, we will develop in Sect. 2 a theory, based on Kelvin's theorem in a rotating frame, explaining, from a nonlinear point of view, the effect of rotation upon the stretching of quasi two-dimensional vortex filaments. We present also an exact solution for a rotating constant shear in a perfect fluid, which reduces to an abridged linearized problem. In Sect. 3, we describe three-dimensional numerical simulations of quasi two-dimensional mixing layers and wakes. Section 4 will deal with quasi two-dimensional isotropic turbulence simulations. In the last section, possible applications of these results to rotating three-dimensional turbulence and to geophysical and engineering situations will be discussed.

2. The Theoretical Model

2.1 Kelvin's Theorem Based Theory

We first try to predict the effects of rotation with the aid of a theoretical model using Kelvin's theorem in the frame rotating with angular velocity $\vec{\Omega}$. The model, first formulated in Lesieur et al. (1991) in the case of the mixing layer, is based upon the concept of absolute vortex filaments strained by the basic velocity. Here, we generalize this model to any rotating shear flow. We consider a two-dimensional basic shear of relative (spanwise) vorticity ω_{2D}, which may vary in space. The local Rossby number is defined as

$$R_o = \frac{|\omega_{2D}|}{|2\Omega|}. \tag{1}$$

The rotation is said to be cyclonic if ω_{2D} and Ω have the same sign and anticyclonic if not. We superpose upon this basic two-dimensional flow a three-dimensional perturbation of low kinetic energy and of vorticity equal to $\vec{\omega}^{(1)}$, such that $|\vec{\omega}^{(1)}| << |\omega_{2D}|$. Without rotation, the corresponding vortex filaments (see Fig. 1) may be stretched longitudinally into a hairpin vortex. This occurs in mixing layers (in the stagnation regions between two concentrations of spanwise vorticity, see Lesieur 1990, for a review), wakes (Lasheras and Meiburg 1991), layers of constant shear (Rogers and Moin 1987), backward-facing steps (see Silveira et al. 1991) and boundary layers (see Herbert 1988, for a review). When rotation is present, Kelvin's theorem applies to the *absolute* vorticity

$$\vec{\omega}_a = (\omega_{2D} + 2\Omega)\vec{z} + \vec{\omega}^{(1)}. \tag{2}$$

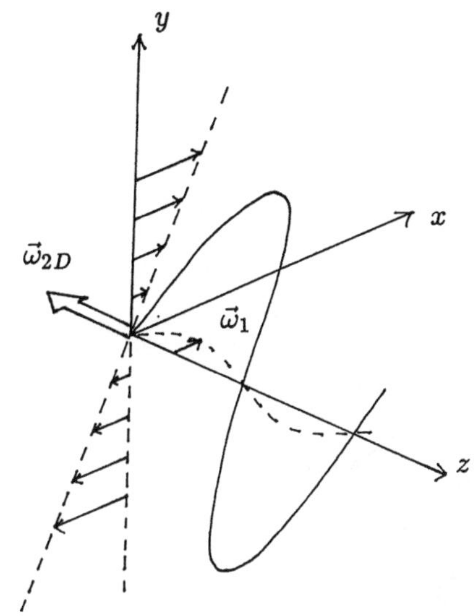

Fig. 1. Stretching of a quasi two-dimensional vortex filament by a local shear; dashed line, initial time; solid line, final hairpin vortex

It is only by a longitudinal straining of the initial *absolute* vortex filaments that longitudinal vorticity (corresponding to three-dimensionalization) may be produced. Several cases have to be envisaged (see Fig. 2, where ω_{2D} is chosen negative)

a) cyclonic rotation ($\Omega < 0$): the absolute vortex filament is closer to the z axis than the corresponding relative vortex filament, implying that cyclonic rotation will have a stabilizing effect with respect to the non-rotating case.

b) anticyclonic rotation ($\Omega > 0$):

• $0 < 2\Omega < -2\omega_{2D}$: the initial absolute vortex filament is more perturbed (three-dimensionally) than its relative counterpart, i.e. the rotation is destabilizing. For $\omega_{2D} + 2\Omega \approx 0$ ($R_o \approx 1$), the initial absolute vortex filaments will be very elongated, since they correspond only to the initial perturbation $\vec{\omega}_1$, which may be associated with three-dimensional residual turbulence. Hence, the basic shear will rapidly stretch this initial vorticity and make it diffuse into the whole flow as it would do with a dye. The resulting stretching of secondary longitudinal vorticity would then react nonlinearly upon the basic velocity. Drastic changes in the flow are thus expected to occur.

• $-2\omega_{2D} \lessapprox 2\Omega$: the anticyclonic rotation becomes stabilizing again.

Notice also that, within this theory, low Rossby numbers correspond to stabilization, whatever the sign of the basic vorticity. This is in agreement with the Taylor-Proudman theorem which states that low-Rossby number flows possess, to lowest order, two-dimensional solutions.

2.2 A Hairpin Stretched in a Constant Shear: an Exact Euler's Solution

Here, we exhibit an exact solution of the Euler equation (in the rotating or non-rotating case), assuming a constant shear (ω_{2D} uniform) in an infinite domain. Hence, no boundary conditions have to be taken care of. This analysis may also be relevant locally, to some

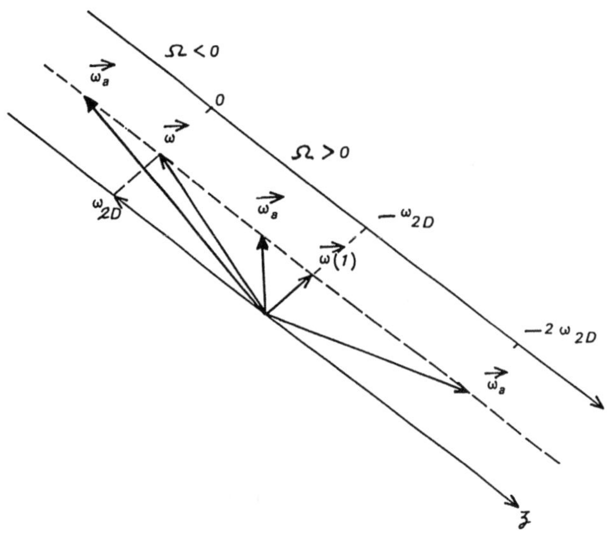

Fig. 2. Initial absolute vortcity for $\Omega = 0$, $\Omega < 0$ (cyclonic rotation), $\Omega > 0$ (anticyclonic rotation); the figure is borrowed from Lesieur 1990, courtesy Kluwer

extent, in other shear-flows such as mixing-layers, wakes or boundary layers. We start with the vorticity equation for a perfect fluid, written as:

$$\frac{\partial}{\partial t}\vec{\omega} + \vec{\nabla} \times [(\vec{\omega} + 2\vec{\Omega}) \times \vec{u}] = \vec{0}. \tag{3}$$

The velocity field is decomposed as:

$$\vec{u} = u_{2D}(y)\vec{x} + \vec{u}^{(1)}(x,y,z,t), \tag{4}$$

where $u_{2D}(y)$ is the basic velocity of vorticity $\omega_{2D} = -du_{2D}/dy$. $\vec{u}^{(1)}(x,y,z,t)$ is a perturbed velocity of projection $\vec{\bar{u}}$ et w_1 respectively on the x,y plane and the $\vec{z}$ axis. Its vorticity $\vec{\omega}^{(1)}$ is projected along $\vec{\bar{\omega}}$ (on the x,y plane) and $\omega^{(1)}$ on $\vec{z}$. Equation (3) writes:

$$\frac{\partial}{\partial t}\vec{\omega}^{(1)} + \vec{\nabla} \times (\vec{\omega}_a^b \times \vec{u}^{(1)}) + \vec{\nabla} \times (\vec{\omega}^{(1)} \times \vec{u}_{2D}) + \vec{\nabla} \times (\vec{\omega}^{(1)} \times \vec{u}^{(1)}) = 0, \tag{5}$$

where

$$\vec{\omega}_a^b = \vec{\omega}_{2D} + 2\vec{\Omega} = (\omega_{2D} + 2\Omega)\vec{z} \tag{6}$$

is the absolute vorticity of the basic flow. Equation (5) is equivalent to:

$$\frac{D_{2D}}{Dt}\vec{\omega}^{(1)} = \vec{\omega}^{(1)}.\vec{\nabla}_H\vec{u}_{2D} + (\omega_{2D} + 2\Omega)\frac{\partial}{\partial z}\vec{u}^{(1)} - \vec{\nabla} \times (\vec{\omega}^{(1)} \times \vec{u}^{(1)}), \tag{7}$$

which, projected in the x,y plane, yields:

$$\frac{D_{2D}}{Dt}\vec{\bar{\omega}} = \vec{\bar{\omega}}.\vec{\nabla}\vec{u}_{2D} - (\omega_{2D} + 2\Omega)\vec{z} \times \vec{\bar{\omega}} + (\omega_{2D} + 2\Omega)\vec{\nabla}_H w_1 - \Pi[\vec{\nabla} \times (\vec{\omega}^{(1)} \times \vec{u}^{(1)})] \tag{8}$$

where Π is the projection operator on the plane perpendicular to $\vec{z}$ and D_{2D}/Dt is the derivative following the basic flow $\partial/\partial t + u_{2D}(y)\partial/\partial x$. Now we look for solutions of the type:

$$\vec{u}^{(1)} = \vec{\tilde{u}} = \Phi(z)[X(t)\vec{x} + Y(t)\vec{y}], \tag{9}$$

where $\Phi(z)$ is an arbitrary function. This perturbed flow (which has no spanwise velocity) may appear somewhat simplistic. It has, however, the interesting property of cancelling the fluctuating nonlinear term $\vec{\nabla} \times (\vec{\omega}^{(1)} \times \vec{u}^{(1)})$. Indeed:

$$\vec{\omega}^{(1)} = \frac{d\Phi}{dz}(-Y\vec{x} + X\vec{y}) \tag{10}$$

$$\vec{\omega}^{(1)} \times \vec{u}^{(1)} = \Phi\frac{d\Phi}{dz}(X^2 + Y^2)\vec{z}. \tag{11}$$

We note also that the helicity of the fluctuating field, $(1/2)\vec{\omega}^{(1)}.\vec{u}^{(1)}$ is zero. On the other hand, the total field posseses a non-zero helicity equal to:

$$\frac{1}{2}\vec{\omega}.\vec{u} = -\frac{1}{2}u_{2D}(y)\frac{d\Phi}{dz}Y. \tag{12}$$

With such a solution, (8) reduces to:

$$\frac{D_{2D}}{Dt}\vec{\tilde{\omega}} = \tilde{\omega}.\vec{\nabla}\vec{u}_{2D} - (\omega_{2D} + 2\Omega)\vec{z} \times \vec{\tilde{\omega}}. \tag{13}$$

This equation may also be obtained from a linear stability analysis (see Lesieur 1990), where the spanwise velocity has been discarded: the first term of the r.h.s corresponds to a passive shearing of the vorticity by the basic shear, and the second term to a rotation around z. A similar model has been proposed by Tritton (1991, private communication).

Let $\tilde{\omega}_1$ and $\tilde{\omega}_2$ be the components of $\vec{\tilde{\omega}}$ on $\vec{x}$ and $\vec{y}$. Equation (13) is equivalent to:

$$\frac{\partial}{\partial t}\tilde{\omega}_1 = 2\Omega\tilde{\omega}_2, \tag{14}$$

$$\frac{\partial}{\partial t}\tilde{\omega}_2 = -(\omega_{2D} + 2\Omega)\tilde{\omega}_1. \tag{15}$$

In the non-rotating case ($\Omega = 0$), $\tilde{\omega}_1$ is conserved and $\tilde{\omega}_2$ increases linearly with time: a hairpin vortex filament will thus be raised to a plane perpendicular to the basic flow. For $\omega_{2D} + 2\Omega = 0$, that is, $R_o = 1$ (anticyclonic case), it is now $\tilde{\omega}_2$ which is constant, while $\tilde{\omega}_1$ grows linearly with time. The result is now a quasi-longitudinal hairpin vortex structure. In the general case, it is obtained:

$$\frac{\partial^2}{\partial t^2}\tilde{\omega}_1 + 2\Omega(\omega_{2D} + 2\Omega)\tilde{\omega}_1 = 0. \tag{16}$$

In the cyclonic case, $\Omega < 0$ and $\Omega(\omega_{2D} + 2\Omega) > 0$. This harmonic oscillator admits periodic solutions for $\tilde{\omega}_1$ and $\tilde{\omega}_2$. The vector of components $\sqrt{-(\omega_{2D} + 2\Omega)}\tilde{\omega}_1$ and $\sqrt{-2\Omega}\tilde{\omega}_2$ rotates counter clockwise about z at a constant speed. As far as the vorticity modulus is concerned, higher stability than in the non-rotating case is achieved. In the anticyclonic case $\Omega > 0$, two situations have to be encountered for: $\omega_{2D} + 2\Omega > 0$ (that is $R_o < 1$), it is now the vector of components $\sqrt{(\omega_{2D} + 2\Omega)}\tilde{\omega}_1$ and $\sqrt{2\Omega}\tilde{\omega}_2$ which rotates about z but with a clockwise rotation. The vorticity modulus grows less than in the case without rotation: the rotation is stabilizing. For $R_o > 1$, on the contrary, $\tilde{\omega}_1$ and $\tilde{\omega}_2$ grow exponentially with time. At large times they become proportional, with:

$$\frac{\tilde{\omega}_2}{\tilde{\omega}_1} = \frac{\sqrt{-2\Omega(\omega_{2D} + 2\Omega)}}{2\Omega}, \tag{17}$$

The maximum stretching is obtained for $R_o = 2$.

Both models show in the cyclonic case a stabilization with respect to no rotation. In the anticyclonic case, they are qualitatively in agreement (destabilization at moderate R_o, restabilization at lower R_o). Since they are both based on certain assumptions, which are not fullfiled for the flows which are going to be computed, we will only ,in the following study, retain the phenomenology.

3. Direct Numerical Simulations of Shear Layers

In order to check the relevance of our theoretical predictions, we have carried out three-dimensional direct-numerical simulations of rotating free shear layers. Two prototypes of flows are considered: the mixing layer and the plane wake. We assume periodicity in the streamwise x direction (temporal hypothesis), and in the spanwise z direction. The initial conditions consist of a low-amplitude random perturbation superposed upon one of two unidirectional basic profiles: a hyperbolic-tangent velocity profile for the mixing layer

$$\overline{u}(y) = U \tanh \frac{2y}{\delta_i} \tag{18}$$

(where δ_i is the initial vorticity thickness) or a gaussian profile for the wake

$$\overline{u}(y) = U_m \exp[-(\log 2)\frac{y^2}{r_m^2}], \tag{19}$$

where r_m is the half deficit velocity width. The initial perturbation also differs for the mixing layer and the wake calculations. In the mixing layer (see Lesieur et al. 1991), there is an initial forcing at the longitudinal fundamental mode k_x, plus a small random 3D isotropic perturbation. In the wake, we superpose two random perturbations: the first one is three-dimensional, of kinetic energy $\epsilon_{3D} U_m^2$, the second one is two-dimensional (z independent) with energy $\epsilon_{2D} U_m^2$, where $\epsilon_{3D} = \epsilon_{2D} = 10^{-5}$. The Rossby number is defined with the aid of the initial relative vorticity:

$$R_o^{(i)} = \frac{2U}{|2\Omega|\delta_i} \tag{20}$$

for the mixing layer. For the wake, the maximum vorticity associated with the gaussian profile is $\approx 0.7U_m/r_m$. Thus, the initial Rossby number is

$$R_o^{(i)} = 0.7\frac{U_m}{|2\Omega| r_m}. \tag{21}$$

The numerical method used is of the pseudo-spectral type. For the mixing layer, a mapping is applied to the y coordinate in order to reject the boundaries at infinity and calculations are performed on a $48 \times 48 \times 24$ grid, containing two fundamental Kelvin-Helmoltz billows. The initial Reynolds number $U\,\delta_i/\nu$ is 150. For the wake, $48 \times 48 \times 48$ modes are used with a Karman Street composed of two pairs of opposite-signed vortices at $Re = U_m r_m/\nu = 180$.

3.1 Mixing Layer

We first consider the mixing layer. The results of these simulations are described in detail in Lesieur et al. (1991). Numerical simulations are performed at $R_o^{(i)} = \infty$, 11, 5, 2, 1, 0.56 and 0.25 for both cyclonic and anticyclonic cases. In Fig. 3, we show the time development of $< w'^2 >$, the variance of the spanwise velocity (z-component), which indicates the degree of three-dimensionality. The brackets denote the average over the entire computational domain. Figure 3 a) shows the cyclonic cases and 3 b) the anticyclonic cases. Figure 3 a) demonstrates that cyclonic rotation is increasingly stabilizing as the initial Rossby number decreases from infinity to 0.5. Afterwards the

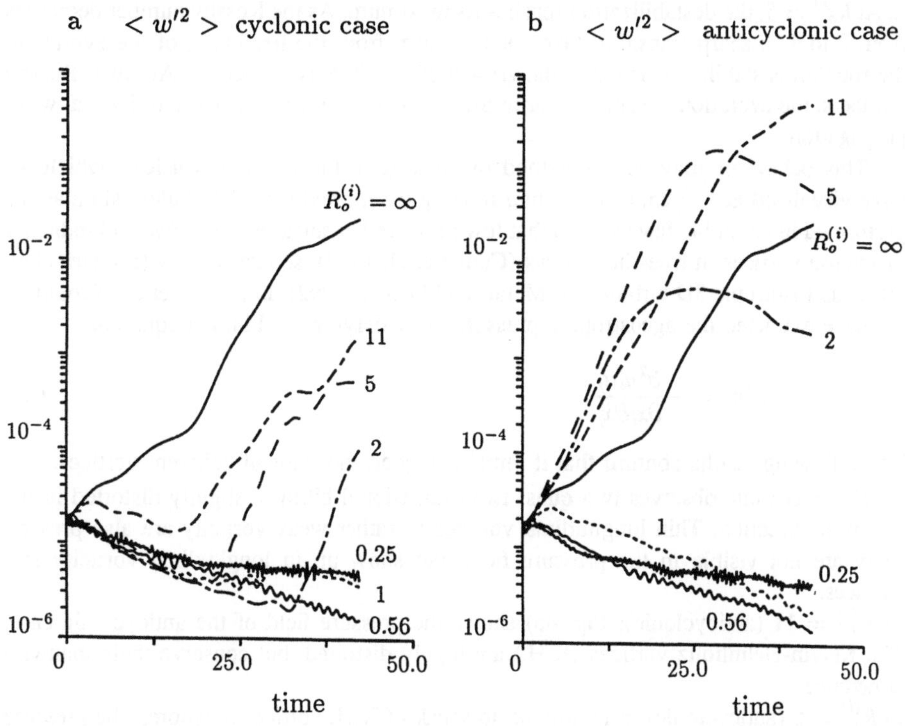

Fig. 3. Mixing layer; time evolution of the spanwise component of the velocity fluctuation ($< w'^2 >$), $R_o^{(i)} = \infty$, 11, 5, 2, 1, 0.56, 0.25, a) Cyclonic case; b) Anticyclonic case (borrowed from Lesieur et al. 1991, courtesy Phys. Fluids)

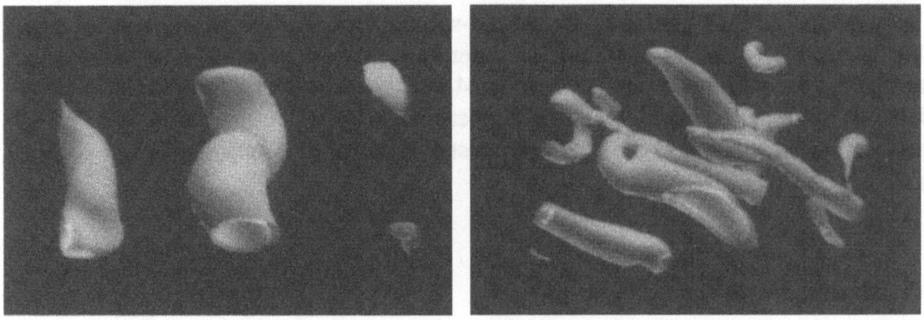

Fig. 4. a) Mixing layer, constant pressure surfaces (anticyclonic case, $t = 25.4$) : $R_o^{(i)} = 11$
b) Mixing layer, constant pressure surfaces (anticyclonic case, $t = 25.4$) : $R_o^{(i)} = 2.5$

energy exhibits oscillations which are the signature of inertial waves. The presence of the waves leads to a reduction of the dissipation rate, which explains why the $R_o^{(i)} = 0.25$ curve decreases more gently than the $R_o^{(i)} = 0.56$ one.

Figure 3 b) shows an extremely different behaviour in the anticyclonic case. The initial growth rate of $< w'^2 >$ is higher than that of the non-rotating case for $R_o^{(i)} = 11, 5$ and 2. At $R_o^{(i)} = 5$, the destabilization reaches its maximum. As the Rossby number decreases from 2 to 1, a sharp transition takes place, since, from the beginning of the evolution, the rotation is stabilizing as far as the growth of $< w'^2 >$ is concerned. At lower Rossby numbers, the evolution is very similar to the cyclonic case, with evidence of inertial wave propagation.

This behaviour may be understood by looking at the coherent vortices, which we have visualized at $t = 25.4$. In the non-rotating case, previous 3D turbulent simulations performed in Grenoble have shown that low pressure iso-surfaces are a powerful indicator of intense vortices in free-shear flows (Comte et al. 1992), separated flow (Silveira et al. 1991) and isotropic 3D turbulence (Métais and Lesieur 1992). In the presence of rotation, we have extracted the ageostrophic pressure, p_a satisfying the Poisson equation

$$\nabla^2 p_a = -\frac{\partial^2 u_i u_j}{\partial x_i \partial x_j}. \tag{22}$$

The following results confirm that it remains a good indicator of coherent vortices.
a) $R_o^{(i)} = \infty$: one observes two quasi two-dimensional billows, slightly distorted in the spanwise direction. Thin longitudinal vortices of rather weak vorticity are also present. They are not visible in the pressure field, but show up in longitudinal vorticity iso-surfaces.
b) $R_o^{(i)} = 11$ (anticyclonic): Fig. 4a) shows the pressure field of the anticyclonic case. The Kelvin-Helmholtz vortices (K.H.) are highly distorted, but conserve their spanwise coherence.
c) $R_o^{(i)} = 5$ (anticyclonic): it is difficult to speak of K.H. vortices anymore. The pressure contours now show (see Fig. 4b) very anisotropic structures of small spanwise wavelength which are extremely elongated in the streamwise direction.

At low Rossby number, in both cyclonic and anticyclonic cases, a very strong two-dimensionalization is observed.

3.2 Plane Wake

Cyclonic and anticyclonic vortices are now simultaneously present in the computational domain. Figure 5a) shows the $t = 60$ vorticity contours for the cyclonic spanwise vorticity (light grey) and for the anticyclonic spanwise vorticity (white), at $R_o^{(i)} = 2.5$, which statistics will show as corresponding to maximum anticyclonic three-dimensionalization. Both cyclonic eddies exhibit spanwise oscillations due to the strong interaction with the three-dimensional anticyclonic eddies. In Fig. 5b) are displayed low-p_a isosurfaces at $t = 30$. The anticyclonic eddies are in the foreground and the cyclonic ones in the background. As in the mixing layer, very elongated streamwise structures are visible on the anticyclonic side. We have checked that they correspond to concentration of vortex lines of near zero absolute spanwise vorticity which are stretched by the ambiant shear. At low Rossby number, both sides are two-dimensionalized, as shown in Fig. 6.

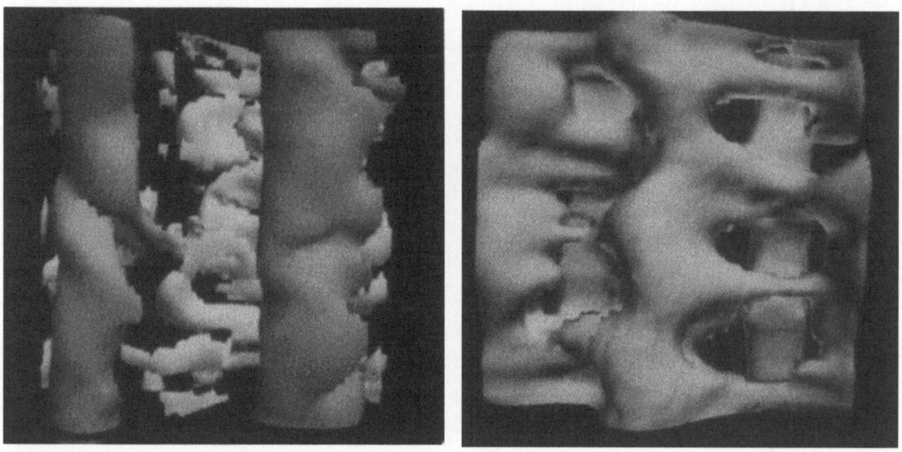

Fig. 5. a) Plane wake: vorticity contours for the cyclonic spanwise vorticity (light gray) and for the anticyclonic spanwise vorticity (white) for $R_o^{(i)} = 2.5$ at $t = 60$ b) Plane wake: constant pressure surfaces for $R_o^{(i)} = 2.5$ at $t = 30$

Figure 7 is the analogue of Fig. 3 in the case of the wake. $< w'^2 >$ designates a spatial average of the spanwise velocity variance on the points of the computational domain at which the vorticity is cyclonic for Fig. 7a) and anticyclonic for Fig. 7b). The Rossby numbers considered here are $R_o^{(i)} = \infty$, 7, 2.5, 1.3 and 0.7. On the anticyclonic side, the time evolution of $< w'^2 >$ exhibit features similar to those observed for the anticyclonic mixing layer. For $R_o^{(i)} = 7$and 2.5, the perturbation exponentially grows in an initial stage, then saturates and eventually decreases. The Rossby number based upon the instantaneous mean velocity profile decreases as time evolves: the numerical simulations give a value for R_o close to unity at the time of saturation of the perturbation. The maximum anticyclonic three-dimensionalization occurs at $R_o^{(i)} = 2.5$.

On the cyclonic side (Fig. 7a), the rotation is initially stabilizing for all simulated values of the Rossby number. Later on, for $R_o^{(i)} = 7$ and 2.5, the growth of $< w'^2 >$ is stronger than the one observed without rotation: this comes from the contribution of

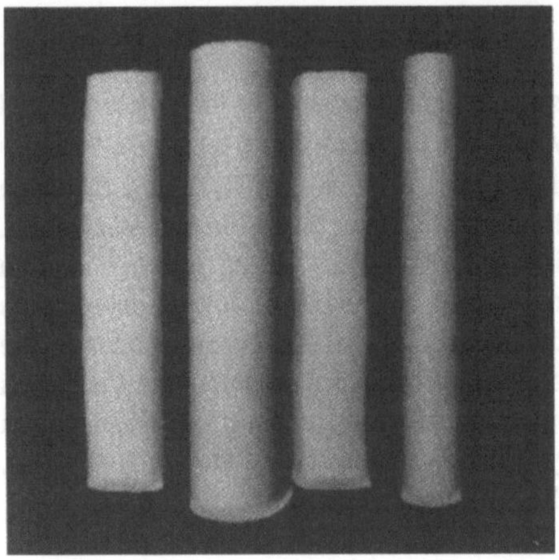

Fig. 6. Plane wake: constant pressure surfaces for for $R_o^{(i)} = 0.5$ at $t = 30$

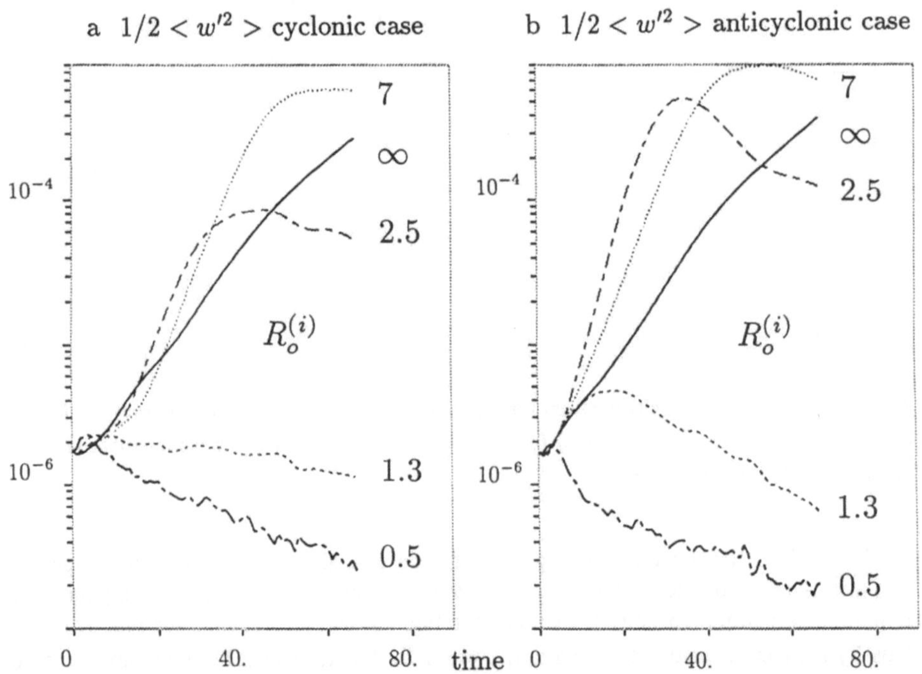

Fig. 7. Same as Fig. 3 for the plane wake ($R_o^{(i)} = \infty, 7, 2.5, 1.3$ and 0.5). $< w'^2 >$ designates a spatial average of the spanwise velocity variance on the points of the computational domain at which the vorticity is cyclonic for Fig. 7 a) and anticyclonic for Fig. 7 b)

cyclonic regions which are generated by the strong three-dimensionalization on the side of the wake where the spanwise vorticity was originally anticyclonic. Furthermore, the cyclonic eddies of the Karman-Street become more three-dimensional than in the non-

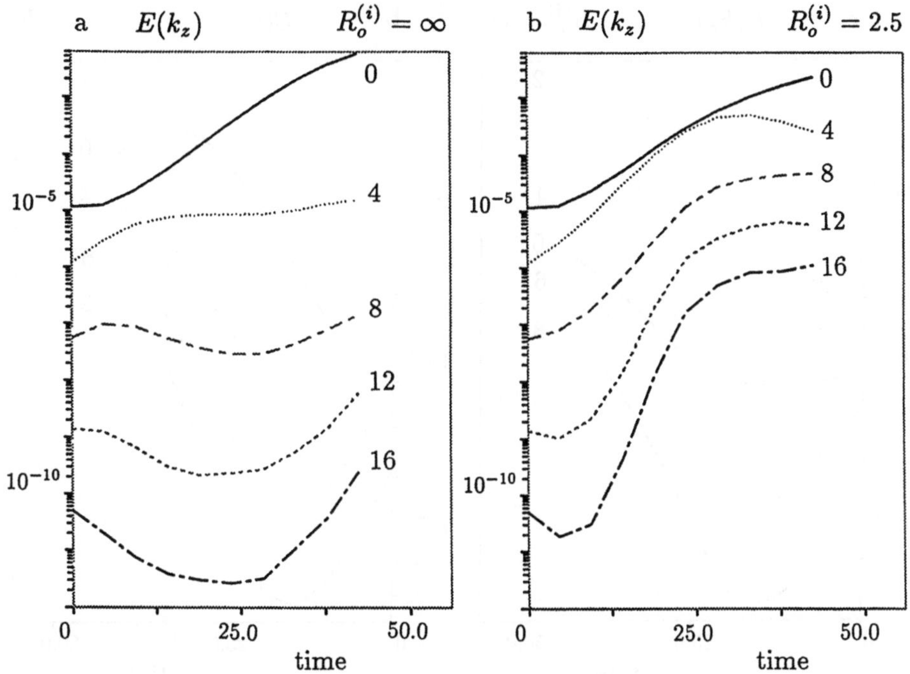

Fig. 8. Time evolution of the total kinetic energy ($1/2 < u'^2+v'^2+w'^2 >$) at spanwise wavenumbers $k_z = 0, 4, 8, 12$ and 16 integrated over all wavenumbers k_x and k_y. Figure 8 a) non-rotating case; Figure 8 b) rotating case ($R_o^{(i)} = 2.5$)

rotating case because of their interaction with the highly three-dimensional anticyclonic ones.

In Fig. 8, we present the time evolution of the total kinetic energy ($1/2 < u'^2 + v'^2 + w'^2 >$) at the spanwise wavenumbers $k_z = 0, 4, 8, 12$ and 16 integrated over all wavenumbers k_x and k_y. Figure 8 a) corresponds to the non-rotating case and Fig. 8 b) to the most destabilized case ($R_o^{(i)} = 2.5$). The growth of the two-dimensional energy (corresponding to $k_z = 0$) is slightly reduced by the rotation. On the other hand, in the rotating case, the energy growth is more and more pronounced as k_z gets larger. This indicates a strong three-dimensionalization of the wake (due to the anticyclonic side) with generation of small scales in the spanwise direction.

Figure 9 is the analogue of Fig. 8, but the total kinetic energy is now integrated over k_y and k_z wavenumbers and we consider the time evolution of the energy at the longitudinal wavenumbers $k_x = 0, 2, 4, 6$ and 8. The wavenumber $k_x = 2$ is the fundamental mode corresponding to the 2 pairs of alternate-signed primary vortices of the Karman street. The energy corresponding to this mode grows strongly in the non-rotating case. Its growth is slightly reduced in the presence of rotation, where the most striking feature is the rapid growth of the energy at $k_x = 0$. The destabilization of the anticyclonic eddies gives rise to structures of small spanwise wavelength but which are very elongated in the streamwise direction. This is confirmed by the linear stability analysis performed by Yanase et al. (1990).

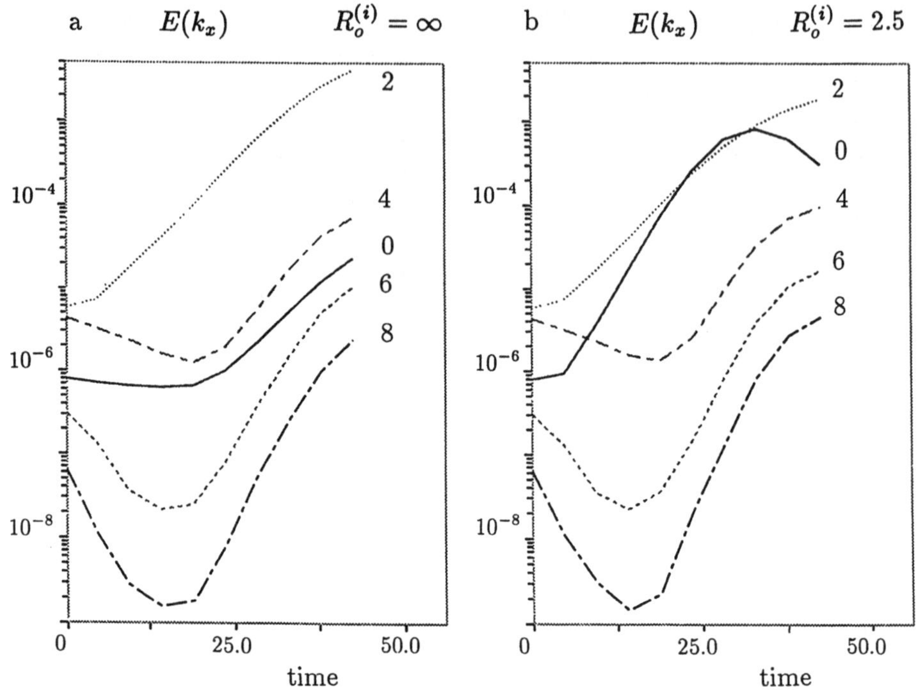

Fig. 9. Time evolution of the total kinetic energy at streamwise wavenumbers $k_x = 0, 2, 4, 6$ and 8 integrated over all wavenumbers k_y and k_z. Figure 9 a) non-rotating case; Figure 9 b) rotating case ($R_o^{(i)} = 2.5$)

4. Quasi Two-Dimensional Turbulence

The above theoretical model does not hold strictly for homogeneous rotating turbulence. However, we will apply the phenomenology to vortex filaments of vorticity $\vec{\omega}$ parallel to the rotation axis, assuming they are stretched by some local ambiant shear. Here, pseudo-spectral triply-periodic numerical simulations (64^3) illustrating the effects of rotation on quasi 2D vortices in homogeneous flows are performed.

The initial vortices are obtained from a preliminary simulation of decaying two-dimensional turbulence, where we recall that 2D dynamics (with velocity perpendicular to $\vec{\Omega}$) are independent of rotation. This simulation was integrated over approximately 30 initial large-eddy turnover times. The resulting fields characterize the, by now familiar, signature of 2D turbulence in that the vorticity has wrapped up into intense coherent vortices, surrounded by quiescent regions. In this context the principal mechanism for nonlinear transfer is the quasi-passive straining of small-scale vorticity by the large-scale shear (Lesieur and Herring 1985). In this sense, even though the prescribed mean-flow shears of Sect. 3 are no longer imposed, each vortex can be considered to be embedded in the shear induced by its neighbours, implying that the reasoning of Sect. 2 remains applicable.

The quasi 2D vortices were perturbed with an isotropic three-dimensional broad-band random velocity field with spectrum $E(k) = Ak \exp -k^2/64$, where A was chosen such that the perturbation had 10^{-3} of the energy and 10^{-2} of the enstrophy of the 2D vortices. The simulations were then continued at various Ω for 30 τ, where $\tau = 2\pi/[\omega_{2D}]_{\rm rms}$ is the initial large-scale turnover time. Figure 10 displays the vorticity fields at the end of the simulations with $R_o^{(i)} = \infty$, 1 and 0.1, where we define the initial Rossby number as

$$R_o^{(i)} = \frac{[\omega_{2D}]_{\rm rms}}{|2\Omega|}. \tag{23}$$

It can be seen that without rotation, a number of narrow longitudinal vortices are formed at the stagnation points between 2D eddies. They increase in intensity and eventually attain vorticities rivalling those of the initial quasi 2D structures, which three-dimensionalize beginning at the small scales. When $R_o^{(i)} = 1$, the anticyclonic 2D vortices three-dimensionalize more rapidly than in the non-rotating case, whereas the cyclones remain quasi two-dimensional throughout the simulation. The remnants of the anticyclonic vortices appear as isotropic 3D structures aligned in vertical columns. This is a result of the fact that the large scales remain quasi 2D in accord with the Taylor-Proudman theorem, since at these scales the Rossby number is small. At $R_o = 0.1$ the energy of the three-dimensional part of the flow actually *decreases* from the initial value, indicating stability of both cyclones and anticyclones (as shown on Fig. 10c).

5. Discussion

The numerical simulations qualitatively confirm the predictions of our theoretical models: the cyclonic eddies are stabilized, while the anticyclonic eddies are three-dimensionally destabilized at a critical Rossby number. The striking feature of the calculations is that the explosive anticyclonic three-dimensionalization leads to very anisotropic vorticity structures. The vorticity originally oriented in the span rotates by an angle of 90° and is ultimately directed in the streamwise direction. The results obtained here present some strong similarities with what is observed in the boundary layer over a curved plate: when the plate is concave, the angular rotation induced by the curvature and the vorticity associated with the boundary-layer velocity profile are of opposite direction. The consequent centrifugal instability (Görtler instability) results in the formation of intense streamwise vortices. For a convex plate, the boundary layer is stabilized.

In a separate study to be published elsewhere Bartello et al. (1992) have performed calculations of rotating homogeneous turbulence from isotropic 3D initial conditions characterizing fully-developed non-rotating turbulence. They demonstrate that rotation acts to organize cyclonic regions (i.e. with $\vec{\omega} \cdot \vec{\Omega} > 0$) into quasi 2D coherent structures, while anticyclonic regions show no such tendency when $R_o^{(i)} = [\vec{\omega} \cdot \vec{\Omega}]_{\rm rms}/2\Omega^2 \approx 1$. The situation bears a resemblance to the rotating turbulence experiments of Hopfinger et al. (1982), where 3D structures generated by an oscillating grid at the bottom of their tank two-dimensionalized towards the top, where the Rossby number was smaller. In this experiment, the quasi-two-dimensional vortices were all cyclones. It is possible, following our simulations, that these cyclones were generated at a distance from the grid

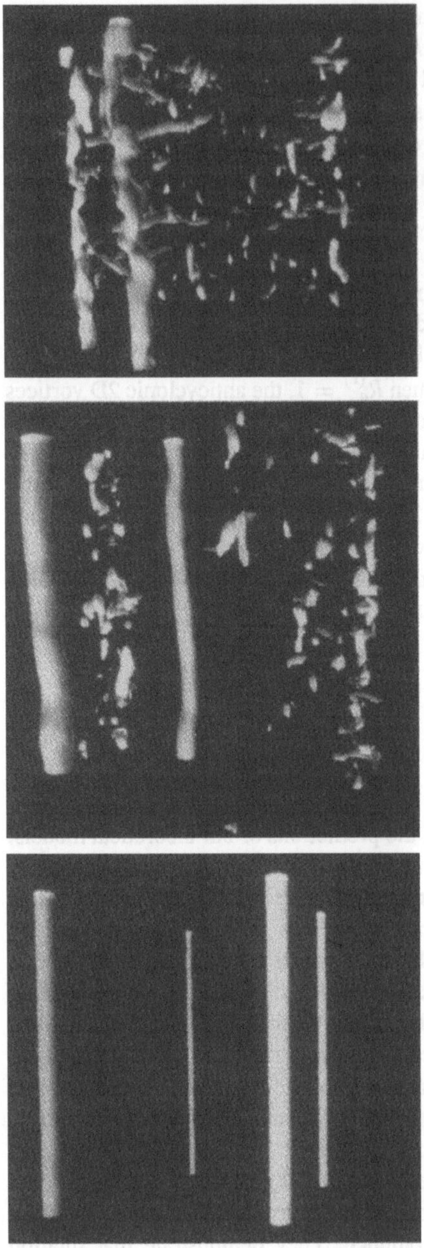

Fig. 10. Vorticity field from simulations of quasi two-dimensional turbulence at 30 initial turnover times, showing surfaces of constant $\omega_z = \omega_o > 0$ (dark grey), $\omega_z = -\omega_o < 0$ (light grey) and $(\omega_x^2 + \omega_y^2)^{1/2} = \omega_o$ (intermediate grey), where $\omega_o = 2\omega_{rms}$ at a) $R_o = \infty$, b) $R_o = 1$ and c) $R_o = 0.1$

such that the local Rossby number should be ≈ 1: the upper vertical coherence, in zones with lower R_o, would then be a Taylor-Proudman effect due to the underneath $R_o = 1$ flow. The emergence of symmetry-breaking organized structure from chaotic initial fields is clearly of primordial importance in understanding rotating turbulence. It is felt that the arguments outlined in Sect. 2, emphasizing the stretching of absolute vorticity filaments by the ambient flow, can be enlightening in this respect.

There are numerous situations in atmospheric or oceanic flows to which these results might apply: in the atmosphere, clouds patterns in the lee of islands often show asymmetric wakes, with cyclonic vortices reinforced (Etling 1991). The corresponding Rossby numbers are the order of unity. Another example concerns the generation of cyclonic storms behind the Tibesti in Africa (G.Sommeria 1991, private communication). In the ocean, wakes or capes are shear layers which behave like the above simulated mixing-layers, and may be very stable or unstable according to the sign of their vorticity. Coastal currents may also behave like rotating boundary layers, and should be investigated along the theoretical and numerical lines pointed out in the present study.

Another very important field of applications concerns turbo-machinery: the high rotation rates imply, in general, low Rossby numbers: this is very encouraging for the development of direct numerical simulations in this domain, since the flows are expected to be quasi-two-dimensional. Notice finally that some important applications of the $R_o = 1$ case should be found, for enhancing or diminishing turbulent diffusion.

Acknowledgements

The authors are grateful to W.T. Ashurst, J.J. Riley, D.J. Tritton for many enlightening discussions.

References

Bartello, P., Métais, O., Lesieur, M. (1992) " Coherent Structures in Rotating Three dimensional Turbulence", in preparation

Bidokhti, A.A., Tritton, D.J. (1990) "The Structure of a Turbulent Free Shear Layer in a Rotating Fluid", submitted to the J. Fluid Mech.

Chabert d'Hières, G., Davies, P.A., Didelle, H. (1988) "Laboratory Studies of Pseudo-Periodic Forcing Due to Vortex Shedding from an Isolated Solid Obstacle in an Homogeneous Rotating Fluid", in 20th Int. Liège Colloquium on Ocean Hydrodynamics, 2-6 May, Elsevier Publisher.

Comte, P., Lesieur, M., Lamballais, E. (1992) "Large and Small-Scale Stirring of Vorticity and a Passive Scalar in a 3D Temporal Mixing Layer", submitted to Phys. Fluids A

Etling, D. (1991) "Mesoscale Vortex Shedding From Large Islands: a Comparison to Laboratory Experiments in Rotating Stratified Flows", Meteorology and Atmospheric Physics, in press

Herbert, T. (1988) "Secondary Instability of Boundary Layers", Ann. Rev. Fluid Mech., Vol. 20, pp. 487–526

Hopfinger, E.J.,Browand, F.K., Gagne, Y. (1982) "Turbulence and Waves in a Rotating Tank", J. Fluid Mech., Vol. 125, pp. 505-534

Lasheras, J.C., Meiburg, E. (1991) "On the Three-Dimensional Dynamics of Coherent Vortical Structures Forming in Free, Shear Flows", in Turbulence and Coherent Structures, ed. by O. Métais and M. Lesieur, Kluwer Academic Publishers, pp. 91–111

Lesieur, M., 1990, Turbulence in Fluids, Kluwer Academic Publishers, Dordrecht, 1987, revised edition 1990

Lesieur, M., Herring, J.R. (1985) "Diffusion of a Passive Scalar in Two-Dimensional Turbulence", J. Fluid Mech., Vol. 161, pp. 77-95

Lesieur, M., Yanase, S. , Métais, O. (1991) "Stabilizing and Destabilizing Effects of a Solid-Body Rotation upon Quasi-Two-Dimensional Shear Layers", Phys. Fluids A, Vol., pp. 403–407

Métais, O., Lesieur, M. (1992) " Spectral Large-Eddy Simulation of Isotropic and Stably-Stratified Turbulence", J. Fluid Mech, in press

Rogers, M.M., Moin, P. (1987) "The Structure of the Vorticity Field in Homogeneous Turbulent Flows", Phys. Fluids, Vol. 30, pp. 2662–2671

Rothe, P.H., Johnston, J.P. (1979) "Free Shear Layer Behavior in Rotating Systems", J. Fluids Eng., Vol. 101, pp. 171–120

Silveira Neto, A., Grand, D., Métais, O., Lesieur, M. (1991) "Large-Eddy Simulation of the Turbulent Flow in the Downstream Region of a Backward-Facing Step", Phys. Rev. Lett., Vol. 66, pp. 2320–2323

Witt, H.T., Joubert, P.N. (1985) "Effect of Rotation on a Turbulent Wake", in Proc. 5th Symposium on Turbulent Shear Flows, Cornell, pp. 21.25–21.30

Yanase, S., Riley, J.J., Lesieur, M., Métais, O. (1990) "The Effects of Rotation on Transition in a Two-Dimensional Shear Layer", Bulletin of the American Physical Society, Vol. 35, No 10, p. 2277

Index of Contributors

MIX
Papier aus verantwortungsvollen Quellen
Paper from responsible sources
FSC® C105338

If you have any concerns about our products,
you can contact us on
ProductSafety@springernature.com

In case Publisher is established outside the EU,
the EU authorized representative is:
Springer Nature Customer Service Center GmbH
Europaplatz 3, 69115 Heidelberg, Germany

Printed by Libri Plureos GmbH
in Hamburg, Germany